Springer Series in
MATERIALS SCIENCE 112

Springer Series in

MATERIALS SCIENCE

Editors: R. Hull R. M. Osgood, Jr. J. Parisi H. Warlimont

The Springer Series in Materials Science covers the complete spectrum of materials physics, including fundamental principles, physical properties, materials theory and design. Recognizing the increasing importance of materials science in future device technologies, the book titles in this series reflect the state-of-the-art in understanding and controlling the structure and properties of all important classes of materials.

99 **Self-Organized Morphology in Nanostructured Materials**
Editors: K. Al-Shamery and J. Parisi

100 **Self Healing Materials**
An Alternative Approach to 20 Centuries of Materials Science
Editor: S. van der Zwaag

101 **New Organic Nanostructures for Next Generation Devices**
Editors: K. Al-Shamery, H.-G. Rubahn, and H.Sitter

102 **Photonic Crystal Fibers**
Properties and Applications
By F. Poli, A. Cucinotta, and S. Selleri

103 **Polarons in Advanced Materials**
Editor: A.S. Alexandrov

104 **Transparent Conductive Zinc Oxide**
Basics and Applications in Thin Film Solar Cells
Editors: K. Ellmer, A. Klein, and B. Rech

105 **Dilute III-V Nitride Semiconductors and Material Systems**
Physics and Technology
Editor: A. Erol

106 **Into The Nano Era**
Moore's Law Beyond Planar Silicon CMOS
Editor: H.R. Huff

107 **Organic Semiconductors in Sensor Applications**
Editors: D.A. Bernards, R.M. Ownes, and G.G. Malliaras

108 **Evolution of Thin-Film Morphology**
Modeling and Simulations
By M. Pelliccione and T.-M. Lu

109 **Reactive Sputter Deposition**
Editors: D. Depla and S. Mahieu

110 **The Physics of Organic Superconductors and Conductors**
Editor: A. Lebed

111 **Molecular Catalysts for Energy Conversion**
Editors: T. Okada and M. Kaneko

112 **Atomistic and Continuum Modeling of Nanocrystalline Materials**
Deformation Mechanisms and Scale Transition
By M. Cherkaoui and L. Capolungo

113 **Crystallography and the World of Symmetry**
By S.K. Chatterjee

114 **Piezoelectricity**
Evolution and Future of a Technology
Editors: W. Heywang, K. Lubitz, and W.Wersing

115 **Lithium Niobate**
Defects, Photorefraction and Ferroelectric Switching
By T. Volk and M.Wöhlecke

116 **Einstein Relation in Compound Semiconductors and Their Nanostructures**
By K.P. Ghatak, S. Bhattacharya, and D. De

117 **From Bulk to Nano**
The Many Sides of Magnetism
By C.G. Stefanita

118 **Extended Defects in Germanium**
Fundamental and Technological Aspects
By C. Claeys and E. Simoen

Volumes 50–98 are listed at the end of the book.

(Continued after index)

Mohammed Cherkaoui · Laurent Capolungo

Atomistic and Continuum Modeling of Nanocrystalline Materials

Deformation Mechanisms and Scale Transition

Springer

Mohammed Cherkaoui
Georgia Institute of Technology
School of Mechanical Engineering
Atlanta, GA 30332
mcherkaoui@me.gatech.edu

Laurent Capolungo
Los Alamos National Laboratory
1675A 16th Street
Los Alamos, NM 87544
laurentc@lanl.gov

Series Editors

ISBN 978-1-4419-4286-9 e-ISBN 978-0-387-46771-9
ISSN 0933-033X
DOI 10.1007/978-0-387-46771-9

Printed on acid-free paper

springer.com

Preface

This book was motivated by the extensive amount of literature dedicated to nanocrystalline (NC) materials published over the last two decades. The authors have been greatly interested in this new emerging field and wished to provide a comprehensive state-of-the-art text on the matter. Therefore, this oeuvre is suited for graduate students and research scientists in mechanical engineering and materials science. All chapters are written such that they can be read independently or consecutively.

Since their discovery in the early 1980s, NC materials have been the subject of great attention, for they revealed unexpected fundamental phenomena, such as the breakdown of the Hall-Petch law, and suggested the possibility of reaching the ever-so-challenging large-ductility/high-yield stress compromise. Although the problem of describing the behavior of NC materials is still challenging, numerous fundamental, computational, and technological advances have been accomplished since then. Most of these are presented in this book. By raising the difficulties and remaining problems to solve, the book highlights new directions for research to develop rigorous and complete multi-scale methods for NC materials.

The introduction of this book chronologically summarizes the different advances in the field. Chapter 1 is dedicated to the presentation of the most commonly employed processing methods. Chapter 2 presents the microstructures of NC materials as well as their elastic and plastic responses. Additionally, Chapter 6 introduces a discussion of several plastic deformation mechanisms of interest. In all other chapters, modeling techniques and advanced fundamental concepts particularly relevant to NC materials are presented. For the former, continuum micromechanics, molecular dynamics, the quasi-continuum method, and nonconventional finite elements are discussed. For the latter, grain boundary models and interface modeling are discussed in dedicated chapters. Given the vast diversity of subjects encompassed in this book, references are provided for readers interested in more specialized discussion of particular subjects. Applications of each concept and method to the case of NC materials are presented in each chapter. The last two chapters of this book are dedicated to more advanced material and aim at showing original methods allowing multi-scale material's modeling.

The authors wish to thank the editor and the formidable group of – unfortunately anonymous – reviewers for their support, rigorous comments, and insightful discussions.

Atlanta, GA, USA — Mohammed Cherkaoui
Los Alamos, NM, USA — Laurent Capolungo

Contents

Introduction

Major technological breakthroughs engendering significant impact on modern society have occurred during this past century. These novelties have emerged in areas as diverse as transportation, telecommunications, construction, etc. Recall that only 20 years ago, the Internet, global positioning, electric-powered cars, and so forth were either pure theory or reserved to a then much-envied small pool of the population. In the early 20th century, automotive and aerospace engineering were the stuff of popular and scientific fantasy and interest because they literally created a revolution, contributing to the "flattening of the world." The last part of the past century has seen the same sort of interest being directed towards device minimization, in its general sense. An unquestionable example is that of cellular phones and computers, whose dimensions and weight have been substantially optimized since their introduction on the market. Recently, a summit was reached with the creation of micro-electromechanical systems (MEMS). Devices such as resonators, actuators, accelerometers, and gyroscopes can already be fabricated with micrometer dimensions. These are already used in industry. The "trend" to minimize devices and structures and the subsequent successes has lead to new fields of science all encompassed in the generic term *nanotechnologies*. In a general way, one could define nanotechnologies as all devices and materials with either dimensions or characteristic dimensions in the range of several nanometers up to several hundred nanometers.

The reader is certainly aware of what a nanometer represents in terms of units. However, it is important to assess the physical "smallness" of the nanometer. For example, a single particle of smoke still has dimensions more than a thousand times larger than a nanometer. A nanometer is approximately equal to three interatomic distances in a copper crystal. Keeping the above remark in mind, one can easily suspect nanomaterials and nanotechnologies to reveal novel and never-before-observed phenomena.

Interestingly, the "infinitesimal" has been a perpetual subject of fascination, intensive reflection, and often sthe ource of advances in all fields of science. In mathematics, the not-so-simple yet crucial, idea of integration results from the conceptualization of the infinitesimally small. Indeed, supposing a function f from the real line to the real line, the integration of this function is based on the consideration that the real line is an infinite sequence of real values and the distance between two consequent values is infinitesimal. Similarly, the concept of atom, the etymology of which is from the Greek word *atomos* "non-cut," attributed to Leucippus of Miletus and Democritus of Abdera, is dated from 500 B.C. and is clearly still subject to ongoing studies. Nowadays, owing to the increase in computing resources and to the amelioration of experimental apparatus such as the transmission electron microscope, the observation and numerical modeling of atoms and groups of atoms with complex arrangements are commonly performed in most research laboratories. Even nanotechnologies that may seem recent and whose early development is often assumed to date from the late 1990s can actually be traced back to the middle of the 20th century. Indeed, in 1959, Richard Feynman discussed in detail in a talk entitled, "There Is Plenty of Room at the Bottom," the possibility of encrypting the totality of the *Encyclopedia Britannica* on the head of a pin. During World War II, nanoparticles smaller than ~5 nm could already be synthesized in Japan.

Although unremarkable to the "untrained eye," simultaneously to the minimization of devices, materials have also been the subject of massive investigations aiming at refining their microstructure. The idea being that most phenomena are dependent on characteristic dimensions (e.g., time, length). Indeed, let us consider the following experiments: (1) a person walks slowly into the ocean and (2) the same person falls at high speed from a wakeboard into the ocean. The perception of the reaction of the water on the body of the subject will clearly be different due to the change in characteristic dimensions: time. Similar reasoning can be applied to the reaction, or more precisely to the behavior of materials which can largely differ depending on the characteristic dimensions. One of the most notable effects observed in polycrystalline materials (i.e., materials composed of agglomerates of crystals) is that predicted by the Hall-Petch law describing the increase in yield strength proportional to the inverse of the square root of the grain size. With the above size-dependent yield strength, decreasing the characteristic dimensions (e.g., crystal size) of a copper sample from 100 microns down to 1 micron would lead to an increase in the yield strength on the order of 250%. This example brings to light the importance of size effects in materials which are unquestionably an efficient way to improve the response of materials. The second route of improvement typically results from the addition of different substances in an initially pure material. This is the case of dopants in semiconductors. The remarkable size effect mentioned in the above has driven the scientific community to further refine the microstructure of materials down to nanometric dimensions. These materials are referred to as nanostructured (NS) materials.

Since the early 1990s, a broad range of NS materials – exhibiting outstanding mechanical, electrical, and magnetic properties – have been synthesized. For example, ZnO nanorods and nanobelts, typically obtained via solid-vapor thermal sublimation, exhibit high piezoelectric coefficient, on the order of 15–25 pm/V, which suggest promising applications in sensors and actuators.

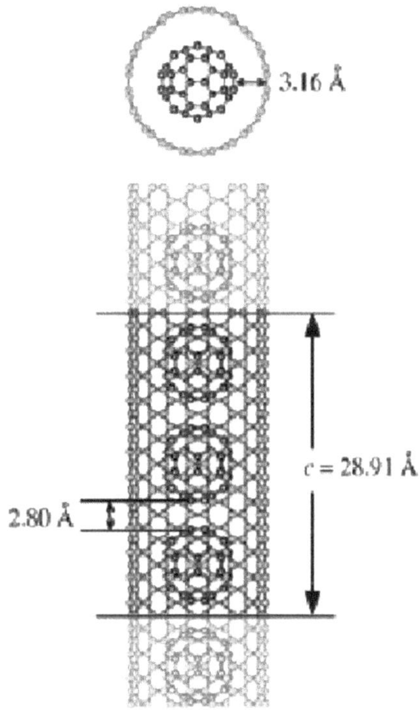

Fig. 1 Multiwalled carbon nanotube

Similarly, multiwalled carbon nanotubes (see Fig. 1), whose tensile properties are measured by attaching them to tips of AFM cantilever probes, exhibit tensile strength ranging from 11 to 63 GPa [1]. Hence, multiwalled carbon nanotubes are outstanding candidates for reinforcement in composite materials.

The appeal of NS materials is not limited to the potential applications that may result from the adequate use of their superior properties but is also driven by the novel fundamental phenomena occurring solely in these materials. The most renowned example is the breakdown of the Hall Petch law which will be discussed in more details throughout this book.

These novel phenomena, underlying the occurrence of unknown deformation mechanisms, have suggested a particular interest in the scientific community. This is especially the case of nanocrystalline materials, to be introduced in the following section, for which numerous technical papers debating on their structure, mechanical response and deformation mechanisms were published since their creation in the late nineteen eighties.

Let us first clearly define the type of NS material this book is dedicated to, and present a short history of the advances in the field in order to help the reader better comprehend and judge of the many remaining challenges to be faced in the area.

What Are Nanocrystalline Materials?

Owing to the large variety of fabrication processes, which will be discussed in detail in the following chapter, a vast diversity of NS materials can be synthesized. Indeed, NS materials present an opportunity to mix substances which were so far not miscible. As an example, Ag-Fe alloys, which are typically immiscible substances in the solid state, can be fabricated via inert gas condensation using two evaporators [2] (this technique will be discussed in the following chapter).

A classification of nanocrystalline materials (see Fig. 2), based on their chemical composition and crystallite geometry, was proposed in Gleiter's pioneering work [3]. NS materials can be divided in three families: (1) layer shaped, (2) rod shaped, and (3) equiaxed crystallite. For each family the composition of the crystallites can vary. All crystallites can have same structure, or a different composition. Also, the composition of the crystallites can be different from that of the boundaries, or more generally of that of the interphase (the phase between crystallites). Finally, the crystallites can be dispersed in a matrix of different composition.

Different fabrication processes are used to fabricate different families and categories of nanostructured materials. For example, nanocrystalline Ni Co/CoO functionally graded layers with mean grain size ranging from ~10 to

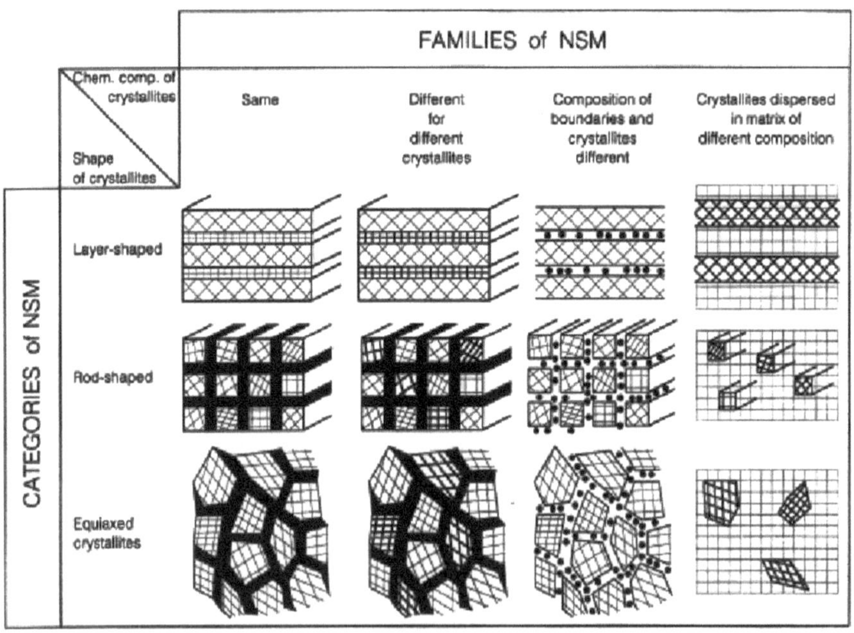

Fig. 2 Classification of nanostructured materials as proposed by Gleiter [3]

$\sim$40 nm are processed via electrodeposition followed by cyclic oxidation and quenching [4], while nanocrystalline Ni can be processed solely via electrodeposition (among others).

This book focuses on equiaxed nanostructured materials with crystallites having similar constitution. Depending on the size of the crystallites (also referred to as grain cores), a particular nomenclature, generally accepted by the community, is used. Hence, throughout this book, nanostructured materials with equiaxed crystallites and mean grain size larger than $\sim$100 nm and smaller than 1 micron will be referred to as ultrafine grain materials, while nanostructured materials with equiaxed crystallites and mean grain size smaller than $\sim$100 nm will be referred to as nanocrystalline materials.

Although the microstructure of nanocrystalline (NC) materials is to be presented in detail in a later chapter, let us briefly comment on the particular features of NC materials. Three constituents compose NC materials: (1) grain cores also referred to as crystallites, (2) grain boundaries, and (3) triple junctions also referred to as triple lines. Grain cores exhibit a crystalline structure (e.g., face center cubic, hexagonal compact, body center cubic). Grain boundaries correspond to regions of junction between two grains. It has a structure that depends on the orientations of the adjacent grains and on the shape of the grains. Therefore, grain boundaries can exhibit either an organized structure, yet different from that of the crystallites, or a much less ordered structure. This is dependent on several factors. One of the most influential factors is the fabrication process. Also, most defects (e.g., impurities, pores, vacancies) are localized within the grain boundaries and triple junctions. The latter are regions where more than two grains meet. Interestingly, they typically do not exhibit particular atomic order. Grain boundaries and triple junctions constitute an interphase and have a more or less constant thickness on the order of $\sim$1 nm. This means that a decrease in the grain size leads to an increase in the volume fraction of interphase. In the case of coarse grain polycrystalline materials, with grain size larger than 1 micron, the volume fraction of interphase is typically less than 1% while in the case of NC materials, the volume fraction of interphase can be as high as 40–50% (depending on the grain size). This is one of the most striking features of NC materials.

A Brief History

In order to build appreciation for the critical modeling and experimental issues and points of interest concerning NC materials, it is appropriate here to present a brief history of NC materials which obviously does not have the vocation to be exhaustive.

Nanocrystalline materials were first fabricated in 1984 in pioneering work of Gleiter and Birringer, who first produced samples with grain sizes ranging from 1 to 10 nm and immediately discussed the extremely high ratio of volume fraction of interface to grain core [5, 6]. Let us note that successful synthesis of nanoparticles could already be achieved in the late 1940s (for further details the reader is encouraged to read the review by Uyeda). The microstructure of these novel materials was also the subject of interest because neither long-range nor short-range structural order in the interphase was revealed by X-ray diffraction and Mössbauer microscopy.

In 1987, the first diffusivity measures at relatively low temperature (~360 K) on 8 nm grain size NC materials produced by vapor condensation reported a self-diffusion coefficient 3 orders of magnitude larger than that of grain boundary self-diffusion [7, 8]. Similarly, studies on the diffusivity of silver in NC copper with 8 nm grain size revealed diffusivity coefficients 2–4 orders of magnitude higher than measured in a copper bicrystal. Hence, the existence of a novel solid state structure in the interphase was suggested [9]. Moreover, the mixture of apparently nonmiscible elements was already discussed.

These first results were quickly followed by an extensive series of experiments (e.g., positron annihilation, X-ray diffraction) revealing what was referred to as an "open structure" for grain boundaries and characterized by the presence of voids and vacancies within the interphase region [10, 11]. Let us note here that these experiments were performed on nanocrystalline metals with grain size smaller than 10 nm.

In 1989, hardness measurements on NC Cu and Pd produced by inert gas condensation (to be presented in a later chapter) reveal a deviation from the Hall-Petch law. Precisely, these experiments revealed that below a critical grain size NC metals exhibit a negative Hall-Petch slope. This means that, contrary to the prediction given by the Hall-Petch law (i.e., a decrease in the grain size leads to an increase in the yield strength proportional to the inverse of the square root of the grain size), the yield strength can decrease with decreasing grain size providing the crystallites are smaller than a critical value. This "breakdown" of the Hall-Petch law was suggested to result from rapid diffusion throughout grain boundaries, similar to the process predicted by Coble but activated at room temperature.

The experimental results mentioned in the above are of primary importance because NC materials appeared, then, to be capable of reaching an excellent strength/ductility compromise. This would emerge from the high-yield strength obtained prior to the breakdown of the Hall-Petch law and from exceptional diffusion coefficients at room temperature (suggesting the possibility of superplastic deformation). Consequently, NC materials were soon considered by many as a technological niche.

Simultaneously, the novel properties of nanocrystalline materials brought to light numerous fundamental questions. Among others, limited data available in the early 1990s were not sufficient to establish, on the basis of rigorous statistical analysis, the certainty of the occurrence of the breakdown of the Hall-Petch law, or the abnormal diffusivity coefficients reported. Similarly, considering the high interphase-to-grain-core volume fraction ratio, one may wonder what is the role of grain boundaries and triple junctions to the viscoplastic deformation of NC materials? Does the interphase region actively participate in the deformation? What is the structure of grain boundaries in nanocrystalline materials? Typically, in coarse-grained metals, dislocation activity (nucleation, storage, annihilation) drives the plastic deformation. Is it the case in nanocrystalline materials? Precisely, how is dislocation activity affected by grain size? What is the relationship to superplastic deformation?

Since the early 1990s, the scientific community has focused on simultaneously improving the fabrication processes and models (both computational and theoretical) in order to elucidate the long list of challenging questions listed in the above (among others). As will be shown throughout this book, considerable progress was achieved since the appearance of NC materials. For example, molecular dynamics simulations (both two-dimensional columnar and fully three-dimensional) and quasi-continuum studies, to be discussed in detail in upcoming chapters, revealed some of the details of NC deformation (e.g., grain boundary dislocation emission, grain boundary sliding). NC materials are particularly well suited for numerical simulations via molecular dynamics. Indeed, performing a back-of-the-envelope calculation, a cubic 20 nm sized copper grain contains approximately 220,000 atoms, which is well below the maximum number of atoms that one would simulate with molecular statics (at zero Kelvin) or molecular dynamics. From a purely theoretical standpoint, numerous phenomenological models were developed to investigate the effect of particular mechanisms (e.g., grain boundary sliding, vacancy diffusion, grain boundary dislocation emission). Also, particular attention was paid to the theoretical description of grain boundaries from structural unit models for example.

Finally, the fabrication processes have been systematically improved over the past decade in order to produce defect-free samples (e.g., low porosity, low contamination, etc.). As a result, the mechanical response of NC materials has clearly improved over the 20 years or so since the synthesis of the first sample. Indeed, early traction tests on NC Cu samples in the quasi-static regime exhibited limited ductility (tensile strain $< 5\%$) while the latest experiments on cold-rolled cryomilled NC Cu exhibit more than 40% ductility.

Modeling Tools

One of the particularities of NC materials is that their characteristic lengths and time scale stand at the crossroads of that of several modeling techniques (micromechanics, molecular statics, molecular dynamics, and nonconventional finite elements). Consequently, detailed understanding of size effects and novel phenomena occurring in nanocrystalline materials can be reached solely via the use of complimentary approaches relying on detailed observations, fundamental models at the atomic and mesoscopic scale (the scale of the grain), and complex computer-based models.

Figure 3 presents the range of application of the most commonly used modeling techniques as a function of characteristic length (vertical axis) and time scale (horizontal axis). First, computational models based on molecular statics (at 0 K) and dynamics are typically used to predict the displacements, position, and energies of a given number of atoms, ranging from a few to several hundred thousand, subjected to externally applied boundary conditions (e.g., temperature, displacement, pressure). These simulations rely on the description

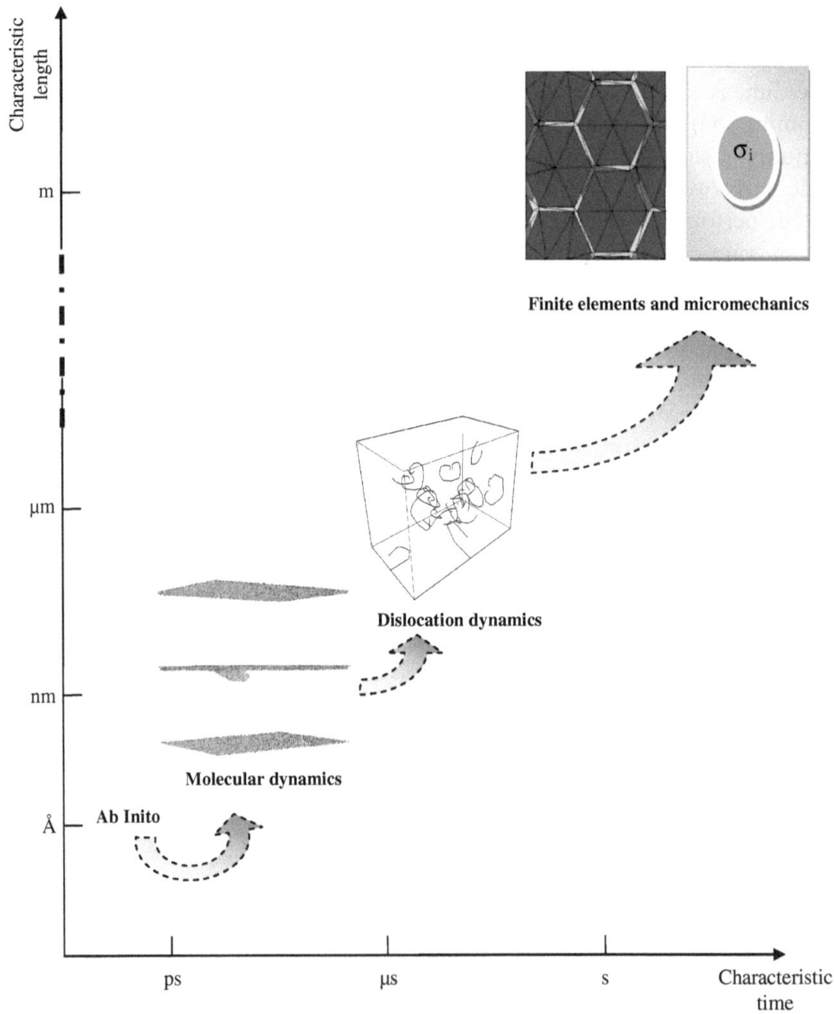

Fig. 3 Schematic of the range of applications of the most commonly used computational and theoretical modeling techniques

of the interatomic potential from which the attractive or repulsive forces can be calculated. The interatomic potentials are typically based on ab initio calculation. It is fitted to a relatively large number of parameters (e.g., interatomic distance, stacking fault energies, etc.). Owing to the large number of operations to be performed simultaneously, the characteristic lengths and time of molecular simulations are limited. For example, simulations are rarely performed in real time larger than $\sim$200 ps. This is due to the limitation on the calculation time-steps which must remain smaller than the period of vibration of atoms (on the order of the femtosecond). Hence, molecular dynamics simulations aiming at studying viscoplastic deformation mechanisms are limited to extremely high strain rates or applied stresses on the order of several GPa. Alternatively, molecular static simulations present the clear advantage of not being limited to small computation steps. However, the simulations are limited to zero Kelvin. Nonetheless, molecular simulations are crucial for they provide valuable information on the motion of atoms which cannot be trivially observed via transmission electron microscopy.

At the microscopic scale, dislocation dynamics simulations can provide useful information as to the intricacies of the dislocation interactions in nanocrystalline materials. Dislocation dynamics are based on the equations of motion of dislocation lines which are typically modeled as a concatenation of smaller dislocation segments. The nodes, or junction between the segments, are the points of interest where the equations of motions are applied. Considerable progress was made in the field such that, nowadays, dislocation dynamics can be applied to complex problems (e.g., cracks). However, to date, dislocation dynamics models are limited to low dislocation densities and representative volume elements on the order of a couple micrometers cubed. One of the major remaining limitations of discrete dislocation dynamics is that of the treatment of interfaces, which has yet to be addressed. Clearly, this limits the application of such methods to study NC materials. Similarly, models based on phase field theory (e.g., constrained energy minimization of a variational formulation) can successfully predict the details of dislocation interactions. While these models present the advantage of being less computationally intense than dislocation dynamics simulations, published work in the literature is often limited to single slip.

At much larger time and length scales, micromechanics and finite elements analyses can predict macroscopic properties and responses of NC materials from a set of parameters extracted from both experiments and models based on the techniques mentioned earlier. In the case of finite elements, precise predictions of stress and strain fields can be obtained. However, the description of the statistical distribution of grain and grain boundary misorientations is often prevented due to computational times. On the other hand, micromechanical models (e.g., mixture rules, Taylor's model, Mori-Tanaka, self-consistent schemes, generalized self-consistent schemes) inherently account for the statistical microstructural features of the material. However, a rigorous description of the grain geometry is typically not obtained with these models. Recently, micromechanical models were solved via Fast Fourier Transform (FFT) coupled with Voronoi tessellation. This has allowed us to overcome the limitations mentioned above – at the expense of calculation time.

The transfer of information between the different time and length scales, corresponding to the range of applications of each modeling technique, is the keystone to successful modeling of NC materials. One of the major difficulties is bridging information from the scale of atomistic simulations to the micron scale, where large quantities of defects interact. This challenge is often referred to as the micron gap (see Fig. 3). In the last decade, several techniques, which will be presented in this book, have been proposed to perform scale transitions between the different time and length scales.

This book aims at summarizing some of the most important advances in the field in terms of modeling, both theoretical and computational, and fabrication process prospective. The objective here is clearly not to make an exhaustive list of all published work to date but to present and discuss the foundations, limitations, and possible evolutions of existing techniques.

References

1. Yu, M., O. Lourie, M. Dyer, K. Moloni, T.F. Kelly, and R.S. Ruoff, Science **287**, (2000)
2. Gleiter, H., Journal of Applied Crystallography **24**, (1991)
3. Gleiter, H., Acta Materialia **48**, (2000)
4. Wang, L., J. Zhang, Z. Zeng, Y. Lin, L. Hu, and Q. Xue, Nanotechnology **17**, (2006)
5. Gleiter, H. and P. Marquardt, Zeitschrift fur Metallkunde **75**, (1984)
6. Birringer, R., H. Gleiter, H.P. Klein, and P. Marquardt, Physics Letters A **102A**, (1984)
7. Horvath, J., R. Birringer, and H. Gleiter, Solid State Communications **62**, (1987)
8. Birringer, R., H. Hahn, H. Hofler, J. Karch, and H. Gleiter, Diffusion and Defect Data – Solid State Data, Part A (Defect and Diffusion Forum) **A59**, (1988)
9. Schumacher, S., R. Birringer, R. Strauss, and H. Gleiter, Acta Metallurgica **37**, (1989)
10. Zhu, X., R. Birringer, U. Herr, and H. Gleiter, Physical Review B (Condensed Matter) **35**, (1987)
11. Jorra, E., et al., Philosophical Magazine B (Physics of Condensed Matter, Electronic, Optical and Magnetic Properties) **60**, (1989)

Chapter 1
Fabrication Processes

As a preliminary note, let us acknowledge that the initial microstructure of a nanocrystalline (NC) sample – which defines its mechanical and thermal responses – is dependent on its processing route. Therefore, models with adequate predicting capabilities must originate from a clear description of the material's microstructure. Since different processing routes may lead, for example, to materials with different amounts of defects, it is capital to acquire a fairly good knowledge on the relationship between fabrication process and resulting microstructure. In doing so, the analysis of model predictions can be adequately discussed with respect to experimental observations. For this purpose this chapter is entirely dedicated to fabrication processes.

Let us also acknowledge here that NC materials cannot yet be produced in quantities sufficient for large-scale industrial applications, and samples available for experiments are produced in a relatively limited number of laboratories. It is thus a complex exercise to describe the various fabrication processes, for the resulting microstructures are dependent on the set of fabrication parameters used in each laboratory. Nonetheless, owing to the increasing documentation available, useful information relating the "trends" in the microstructural features with respect to the synthesis route can be obtained. Those will be presented in this chapter.

Fabrication processes can be broadly classified into two different categories as shown in Fig. 1.1: (1) single-step processes and (2) two-step processes.

Single-step processes allow the direct synthesis of NC materials. Electrodeposition, typically used in the thin coating industry, severe plastic deformation (except for ball milling), and crystallization of an amorphous metallic glass are one-step fabrication processes. There are several one-step severe plastic deformation-based processes; the two most widely used techniques are (1) high-pressure torsion (HPT), and (2) equal channel angular pressing (ECAP). These two routes are based on the grain refinement of an initially coarse sample via the application of large strains. Those approaches are typically referred to as "top-down" processes.

All other synthesis processes (e.g., physical vapor deposition, ball milling, etc.) involve, first, the synthesis of nanoparticles and, second, the compaction/consolidation of the nanoparticle powder typically under high pressure.

M. Cherkaoui, L. Capolungo, *Atomistic and Continuum Modeling of Nanocrystalline Materials*, Springer Series in Materials Science 112,
DOI 10.1007/978-0-387-46771-9_1,

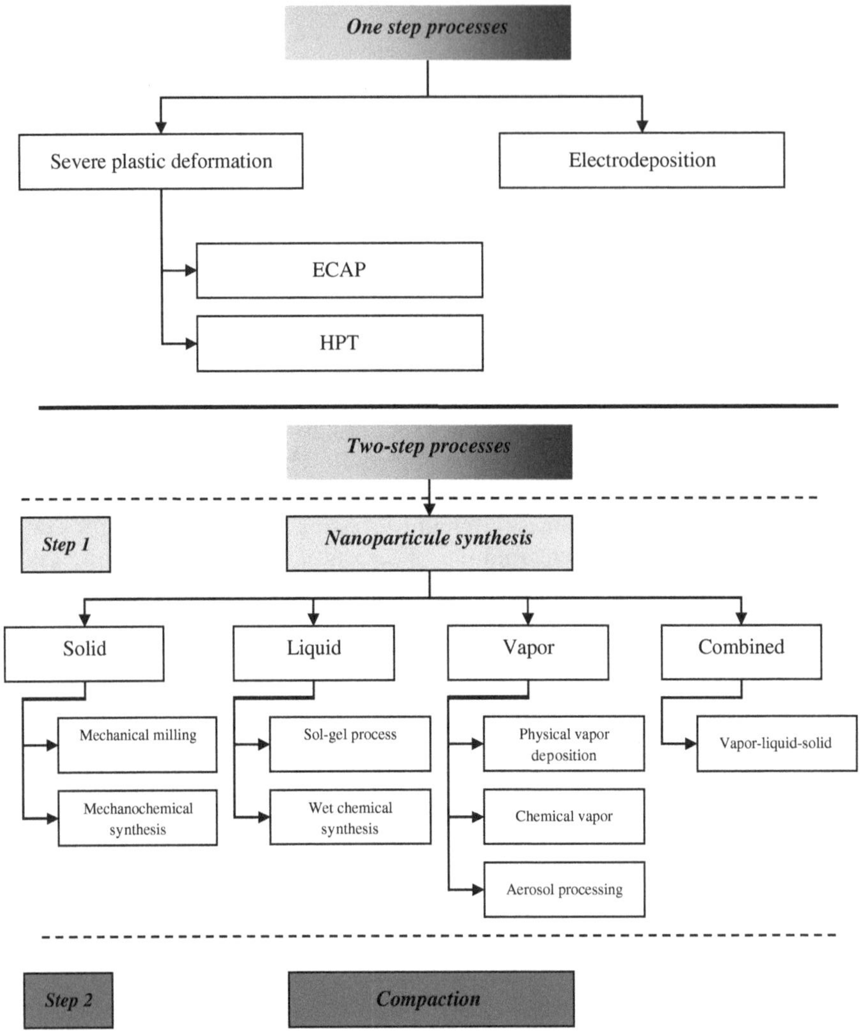

Fig. 1.1 Fabrication processes for nanocrystalline materials

Nanoparticle synthesis can be subdivided into three steps: (1) nucleation, (2) coalescence, and (3) growth. Four routes can be used to fabricate nanoparticles; vapor, liquid, solid, and combined vapor liquid solid. The compaction step avers to be delicate since nanoparticles exhibit a peculiar thermal stability, and particle contamination remains a critical issue. In particular, rapid grain growth can occur during the compaction step. The consolidation step has remained one of the major challenges over the past decade. The synthesis of fully dense samples with high purity and desired grain size is complex. Great progress, to be presented later in the text, has been made over the past decade.

The objective of this chapter is obviously not to make an exhaustive description of all fabrication processes available. Solely the most widely used processes will be presented, that is: HPT, ECAP, electrodeposition, crystallization from an amorphous glass, mechanical alloying (also referred to as mechanical attrition), and physical vapor deposition.

1.1 One-Step Processes

Let us first focus on processes allowing the fabrication of nanocrystalline samples without the use of a compaction/consolidation step. Although these processes might appear at first as more appealing due to their a priori simplicity, they do also present some limitations to be discussed here.

1.1.1 Severe Plastic Deformation

Severe plastic deformation corresponds to the application of large deformations (much larger than unity) to a coarse-grain bulk sample. It engenders considerable microstructural refinement. Hence, it is what we can refer to as a "top-down approach," as opposed to a "bottom-up approach," where the nanostructure is built from the assembly of atoms. Contrary to cold rolling, the sample thickness and height remain constant during severe plastic deformation in order to prevent materials' relaxation.

Typically, these approaches are more time efficient than other fabrication processes and present the major advantage of leading to fully dense samples of relatively large size (several centimeters in all directions) and almost perfect purity. However, the smallest grain size achievable with severe plastic deformation is typically on the order of ~80–100 nm while other techniques such as inert gas condensation can lead to samples with much smaller grain size, on the order of ~5–20 nm.

All processes involving severe plastic deformation –ECAP, HPT, cyclic extrusion-compression cylinder covered compression, and so forth – are based on the same core idea, which is to introduce a large number of dislocations into the as-received sample via the application of large strains into an initially coarse grain sample. Dislocations will rearrange and form high-angle grain boundaries thus leading to finer grain size. The resulting microstructures will differ depending on the fabrication process.

1.1.1.1 ECAP

Equal channel angular pressing (ECAP), also referred to as equal channel angular extrusion, simply consists of extruding a square or circular bar into a die with two connected channels with relative orientation angle denoted by ϕ

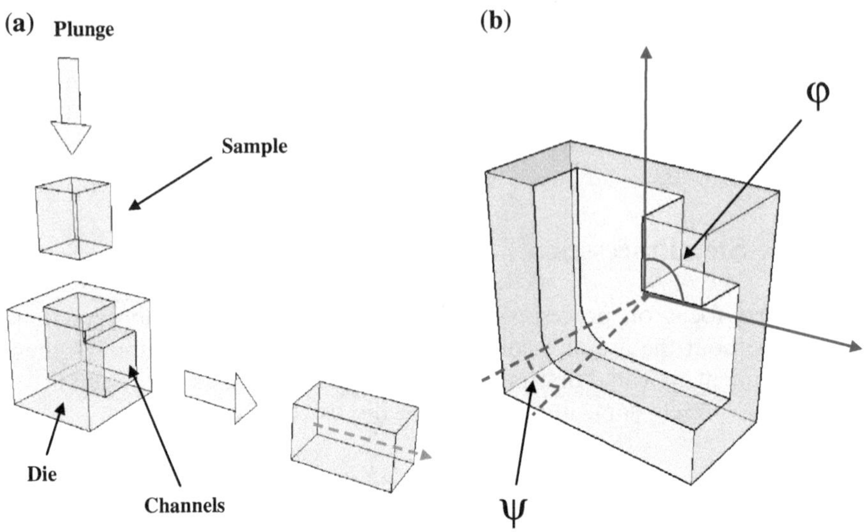

Fig. 1.2 (**a**) Schematic of the ECAP process for a rod, (**b**) cut of the die showing the channels' geometry

and with outer arc of curvature, where the two sections of the channels intersect, denoted by ψ (see Fig. 1.2) [1].

The sample introduced in the channels has dimensions larger than bulk nanocrystalline samples obtained by two-step methods. Indeed, an extruded rectangular sample generally contains more than a 1000 micron-sized grains on its sides. The extrusion process engenders extremely large shear strains (larger than unity) within the sample. In order to produce a sample with a microstructure as homogeneous as possible, ideally one would like to introduce a homogeneous state of strain within the sample. Considering the geometry of the channels, it is quite obvious that a simple extrusion step may not lead to homogeneous strains within the samples. However, the combination of multiple extrusion steps (or passes) with rotation of the sample between the passes leads to a more homogeneous state of strain.

The net strain, denoted ε_N imposed on the bar depends on the angle between the channels and the angle of intersection of the curvatures of the channel. The latter is also referred to as the curve angle. Several models were developed to evaluate the equivalent strain in the sample as a function of the two geometrical parameters and the number of passes, denoted N. Among the most popular propositions, Iwahashi et al. [2] predict the following evolution of the equivalent strain with respect to the above-mentioned variables:

$$\varepsilon_N = \frac{N}{\sqrt{3}}\left[2\cot\left(\frac{\phi}{2}+\frac{\psi}{2}\right)+\psi\operatorname{cosec}\left(\frac{\phi}{2}+\frac{\psi}{2}\right)\right] \tag{1.1}$$

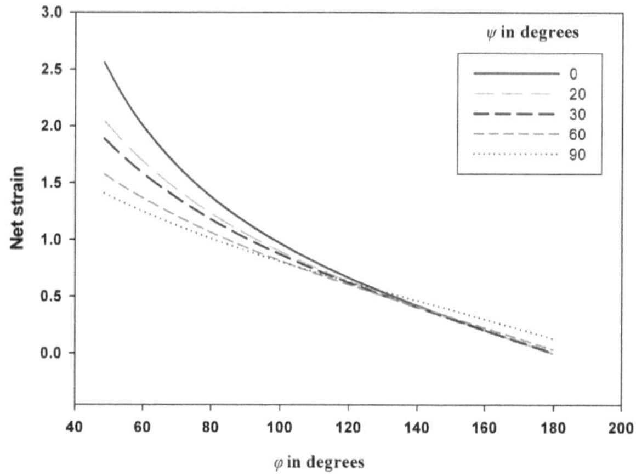

Fig. 1.3 Evolution of the equivalent strain after one pass as a function of ϕ and ψ [2]

A plot of Equation (1.1) is presented in Fig. 1.3. The die angle ϕ has the largest influence on the equivalent strain achieved after each pass. Indeed, at a 0 curve angle, a change in the die angle from 180 to 50 degrees leads to a five-fold increase in the net strain imposed on the sample. Obviously, one would ideally select the smallest die angle ϕ in order to obtain the largest strain within the sample. However, in practice, angles larger than 90 degrees, yet relatively close to that value, are used for the two following reasons: (1) in the case of relatively hard materials it is delicate to use dies with angle smaller or equal to 90 degrees without introducing cracks within the die, and (2) experiments revealed that a 90 degree die angle is more favorable in producing a well-defined equiaxed microstructure. Although the inner and outer arcs of curvature where the two sections of the channel intersect are less critical in order to achieve large deformations, these angles have some influence on the homogeneity of the plastic deformation. Typically, it is recommended to use an inner angle of 0 degrees and an outer angle of ~20 degrees (as shown in Fig. 1.2b).

As mentioned in the above, samples are extruded several times in order to further refine their microstructure and to improve the homogeneity of the state of strain (and thus of the microstructure). Four different routes, corresponding to the rotation of the bar between two consecutive passes, can be employed: (A) the sample is not rotated between passes, (B_A) the sample is alternately alternatively rotated by a ±90 degree angle about its longitudinal axis (denoted by the greenarrow in Fig. 1.2b), (B_C) the sample is rotated by a 90 degree rotation angle between passes and the rotation direction is kept constant, and (C) the sample is rotated by 180 degrees between passes. Sample extraction after a pass can be tedious. Hence, novel dies, such as the rotary dies, have recently been introduced to minimize the number of extractions. Also, several samples can be concatenated within the channels in order to decrease the number of extractions to be performed. Conceptually though, the samples are subjected to the same constraints whether or not a rotary die is used.

Depending on the selected route, different shears will be introduced on different "slip systems" (not to be confused with actual slip systems from conventional crystallography). Routes B_C and C are referred to as redundant routes, for after every even number of passes the shear strain is restored on a slip system. With this remark, it is natural to expect a dependence on the microstructure evolution with the processing route. Experiments have shown that route B_C is most efficient in producing equiaxed microstructures [3].

Microstructure

Transmission electron microscopy (TEM) associated with hardness measurement revealed interesting information related to the microstructure evolution during multi-pass processing. First, the initial state of the material does not influence the resulting microstructure of the sample since after two passes the effect of annealing does not affect hardness measurements. Second, grain refinement occurs mainly during the first two passes. Choosing route B_C it was observed via TEM that a grain refinement from 30 microns to $\sim$200 nm can be achieved over the course of the first two passes while subsequent passes tend to homogenize the grain size [4]. In terms of grain shape, routes A and C lead to elongated grains while route Bc leads to more equiaxed grains.

The mechanisms of grain refinement are not yet well known. It was suggested in several studies that dislocations which do not initially present any regular organization will rearrange to create dislocation walls (which can be pictured here as planes of high density of dislocations) forming elongated cells. The newly formed dislocations will later be blocked on the subgrain walls which will break up and reorient to form high-angle grain boundaries and lead to microstructural refinement. The previously mentioned hypothesis is also supported by experimental measures of the grain boundary misorientation angles during multi-pass ECAP. Figure 1.4 presents plots of the misorientation angle of ECAP processed Cu after zero, two, four, and eight passes. One can observe that the initial microstructure is composed mostly of high-angle grain boundaries and of grain boundaries with angles larger than 30 degrees and the amount of low-angle grain boundaries is limited. However, one can observe that after two passes, the sample has a larger low-angle grain boundaries content. This does indeed confirm the hypothesis mentioned in the above, suggesting the formation of cells delimited by low-angle grain boundaries. Increasing the number of passes to four and then eight results in an increase in the fraction of high-angle grain boundaries. This does indeed suggest that the walls of the cells have split and rearranged into high-angle grain boundaries.

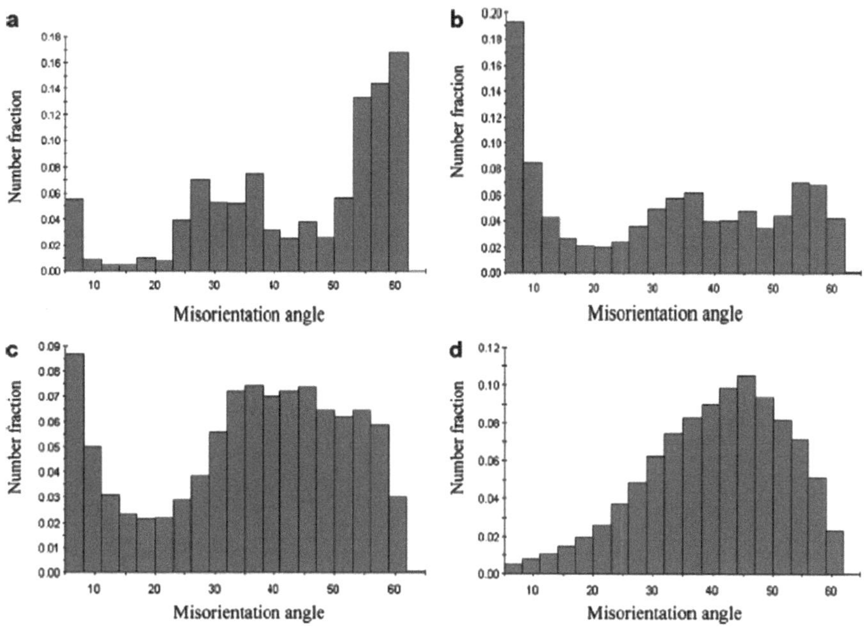

Fig. 1.4 Evolution of the grain boundary misorientation angle in ECAP-processed Cu samples: (**a**) initial configuration, (**b**) after two passes, (**c**) after four passes, and (**d**) after eight passes. Extracted from [4]

1.1.1.2 High-Pressure Torsion

The second most popular one-step severe plastic deformation process consists of the simultaneous application of high pressures and torsion (HPT) to an initially coarse grain sample. Similarly to ECAP, the finest grain size that can be reached is on the order of ~180–100 nm. Thus, this method is limited to the fabrication of ultra fine grain materials. Nonetheless, it has the great advantage of being a fairly simple process leading to slightly larger samples than that obtained via electrodeposition and other methods involving a consolidation step. The disc-shaped samples are typically smaller than that processed via ECAP and have diameter in the range of ~2 cm and thickness on the order of ~0.2–10 mm [5].

The apparatus, schematically shown in Fig. 1.5, is fairly simple and consists of a die with a cylindrical hole which will receive the disc-shaped sample. The sample is pressed by a plunger under high pressures, on the order of several GPa. Simultaneously, large strains are imposed by the rotation of the plunger [6]. Let us note that some apparatuses allow the sample to relax on its side [7]. Typically, large twists, on the order of ~5 turns, are applied to sample to obtain the desired microstructure.

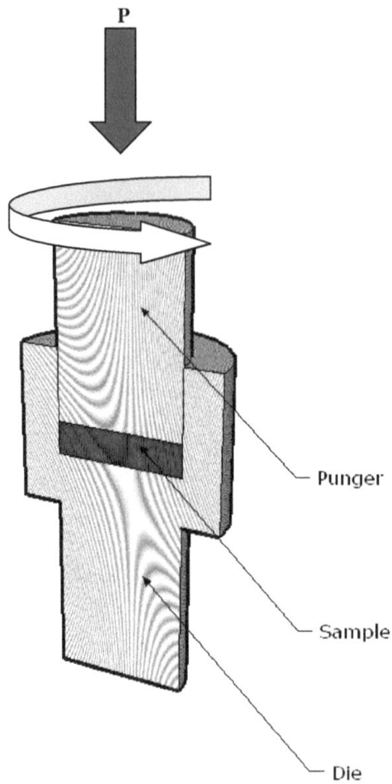

Fig. 1.5 Schematic of the cut of an HPT apparatus

Due to its geometry, HPT leads to highly inhomogeneous strains within the sample. Typically, the maximum shear strain, denoted γ_{max} is estimated with:

$$\gamma_{max} = \frac{2\pi rn}{t} \tag{1.2}$$

where t denotes the sample thickness, r is the radius, and n is the number of turns. From the above equation it can be readily concluded that extremely large strains that can reach up to 700 are applied during the process. Hence, substantial microstructural changes are to be expected from such large strains.

Microstructure

Grain refinement during HPT occurs in a similar fashion as in ECAP. From an experimental standpoint, TEM observations exhibit SAD (selected area diffraction) patterns evolving, with increasing number of turns, from a nonuniform elongated spot-like figure to a more uniform and clearly defined SAD pattern. Hence, the evolution in microstructure can be described as follows. First, subgrains joined by low-angle grain boundaries are formed. With increasing strain, these subgrains split and form high-angle grain boundaries. The resulting grain boundaries present zigzags and facets. Also, the final microstructure does not reveal the presence of twins in Copper samples, which is expected since the process is a top-down approach and since Cu presents medium stacking fault energy. Although dislocations are reported to be hard to find in the samples, some regions present high defect densities, presumably due to dislocation debris.

Samples produced by HPT present a well-defined texture, representative of the preferential grain orientations. Figure 1.6 presents several X-ray diffraction (XRD) measurements of a Cu sample subjected to a 5 GPa pressure and to 0, 1/2, 1, 3, and 5 turns. Typically, a Cu sample with randomly oriented grains will exhibit a (111) to (200) peak high ratio of 2.17. After ½ turn the height peak ratio decreases, which is a consequence of the very large pressure applied to the sample. However, when the number of turns is increased, it can clearly be observed that the height peak ratio is clearly increasing. This reveals a notable change in the texture of the sample. Finally after 5 turns the peak ratio reaches a maximum much larger than 2.17. This reveals that the grain orientation can clearly not be considered random.

1.1.2 Electrodeposition

Electrodeposition is a technique typically used in the thin coating industry which simply consists of introducing both an anode and a cathode in an electrolytic bath containing ions to be deposited on a substrate. The deposition

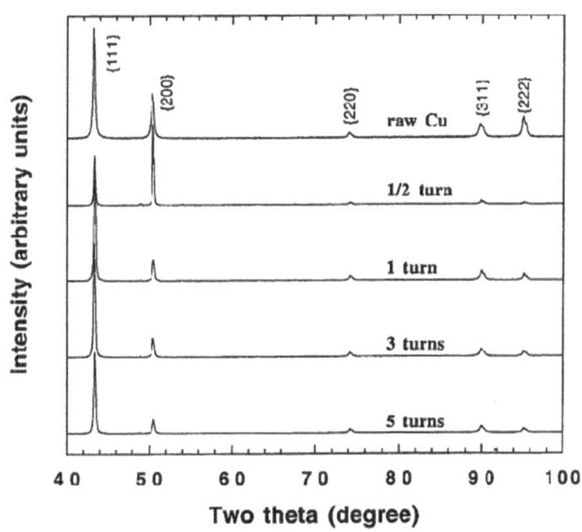

Fig. 1.6 XRD diffraction patterns of Cu sample submitted to different turns [7]

process results from the oxidization of the anode. Owing to its simplicity, this technique is one of the most used NC fabrication processes. Moreover, the deposition rates are relatively high and the process allows the synthesis of NC materials with grain sizes smaller than ~20 nm.

The smallest grain size achievable is dependent on the bath composition, on the current intensity, and on the pH. The objective is obviously to facilitate grain creation rather than grain growth. For example, an increase in the pH typically results in a reduced grain size. This was shown in work by Ebrahimi et al. [8] on NC Ni. Grain size is also influenced by the substrate. For example, Ni deposited on cold-laminated Cu exhibits larger grain size than Ni deposited on heat-treated Cu.

This process has the advantage of allowing fairly good control of the grain size distribution, which typically exhibits low variance. However, the resulting microstructure frequently exhibits a well-pronounced texture. For example, in experiments by Cheung et al. [9] on electrodeposited Ni, a strong (100) texture was reported. Although, let us note that as shown in work by Ebrahimi et al. [8], the texture becomes less pronounced when the grain size is decreased. One of the major limitations related to the use of electrodeposition stands in the limited purity of samples. Indeed, the electrolytic bath tends to introduce impurities within the sample.

1.1.3 Crystallization from an Amorphous Glass

Metallic composite materials reinforced with crystalline nanoparticles can be processed via the devitrification of a bulk metallic glass (BMG). Let us note that this fabrication process does not lead to pure nanocrystalline samples but to nanostructured alloys. Nonetheless, the resulting material exhibits interesting properties.

Metallic glasses are typically produced with a rapid solidification process such as melt spinning in which the cooling rate is on the order of $\sim 10^4 - 10^7 \text{Ks}^{-1}$[10]. In doing so, crystallization is prevented during the formation of the metallic glass. The amorphous nature of the material can be verified via TEM and XRD observations. BMG typically exhibit high fracture toughness, relatively large hardness, and a large elastic domain. For example, $Zr_{41.2}Ti_{13.8}Cu_{12.5}Ni_{10}Be_{22.5}$ fabricated via the melt spinning technique was reported to exhibit a 600 Hv Hardness (~1.9 GPa yield stress) and a fracture stress on the order of 770 MPa [11].

Interestingly, most BMG exhibit a wide supercooled liquid region – corresponding to the thermal stability region of the material bounded in between the glass temperature and the crystallization temperature – on the order of 50–60 K. As discussed by Louzguine-Luzgin and Inoue [12], the glassy→liquid transition is still matter of debate in that the structures in the glassy and liquid region do not exhibit major differences. Hence, the glassy-liquid transition may or not be perceived as a first-order transformation

[12]. The wide supercooled liquid region offers an opportunity to fabricate nanostructured composite materials via primary crystallization, which can be achieved in two ways: (1) thermal treatment and (2) mechanical crystallization.

In the case of crystallization resulting from mechanical constraints, it was observed that nanocrystals develop in shear bands. Furthermore, several studies have suggested that the nucleation of nanocrystals within shear bands results from the enhanced free volume localized within the shear bands. The particle density, distribution, and size result from the control over grain growth and nucleation, which are the two fundamental phenomena ruling the crystallization process. As discussed in review by Perepezko [13], these phenomena rule, more generally, the glass formation such that BMG fabricated under nucleation control will exhibit a distinct glass transition temperature – exhibited by an endothermic signal – and crystallization temperature which is characterized by an exothermic signal; such difference is not observed in the case glass formation by growth control [10, 13]. The endothermic and exothermic signals can be clearly observed in Fig. 1.7 corresponding to the measure – performed via differential scanning calorimetry (hereafter DSC)- of heat flow of $(Cu_{0.5}Zr_{0.425}Ti_{0.075})_{99}Sn1$ under a $0.67Ks^{-1}$ heating rate.

In terms of effective properties, the nucleation and growth of nanocrystals within an initially amorphous BMG was shown to improve the ductility of the metallic glass. Maximum deformation of ~8% was reached in $(Cu_{0.5}Zr_{0.425}Ti_{0.075})_{99}Sn1$ samples and the presence of crystallites at 3% plastic strain was clearly observed via high-resolution transmission electron microscopy (HRTEM). The effect of crystallites in the initial structure on the response of nanostructure alloys prepared by devitrification can be summarized as follows. A relatively small volume fraction of

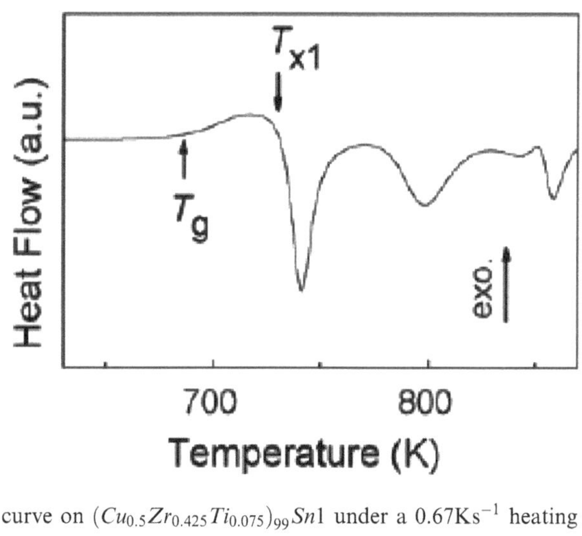

Fig. 1.7 DSC curve on $(Cu_{0.5}Zr_{0.425}Ti_{0.075})_{99}Sn1$ under a $0.67Ks^{-1}$ heating rate. Extracted from [14]

dispersed crystallites (~20%) typically leads to a slight increase in the fracture toughness and in the hardness of the material. However, larger volume fractions of crystallites lead to a large increase in the hardness and a decrease in the fracture toughness [11].

1.2 Two-Step Processes

Synthesis techniques mentioned in the above present the clear advantage of being fairly simple. However, these processes have limitations such as the minimum grain size achievable (SPD) and presence of impurities (electrodeposition). The second approach, discussed here below, consists of first producing and then assembling a large number of nanoparticles. To that end, several methods can be used. The most frequently used ones are presented here.

1.2.1 Nanoparticle Synthesis

Several techniques, such as physical vapor deposition, chemical vapor condensation, mechanical alloying (attrition), and sol-gel can be used to produce metallic nanograins and/or ceramic nanoparticles (see Fig. 1.1). In this section, only the following methods used to fabricate metallic nanograins are treated: mechanical alloying and inert gas condensation. Some aspects of nanoceramics processing will also be briefly discussed. In order to fabricate bulk NC samples, the synthesized nanograins must be consolidated via different techniques presented later in this chapter.

Depending on the synthesis process, nanograins can be joined directly into a micron-sized particle or within a nanoparticle. In the latter, a nanoparticle is typically composed of a couple of grains while in the former a micron-sized particle is composed of several nanograins. Except for mechanical alloying, all processes described here lead to the synthesis of nanoparticles. Since the powder to be compacted differs in the case of mechanical attrition synthesis and other techniques, the consolidated sample shall have a microstructure and hence properties that also depend on the synthesis method.

1.2.1.1 Mechanical Alloying

Mechanical alloying (MA) is a “top-down” process belonging to the family of severe plastic deformation techniques. It allows the refinement of a coarse grain powder, with grain diameter on the order of ~50 μm, via the cyclic fracture and welding of powder particles. Several types of mills such as standard mills, ball mills, or shaker mills can be used. Ball mills are usually preferred to other types of mills. MA is extremely appealing due to its simplicity and ease of use. A vast diversity of nanocrystalline alloys (e.g., Fe-Co, Fe-Pb, Al-Mg, etc.), including

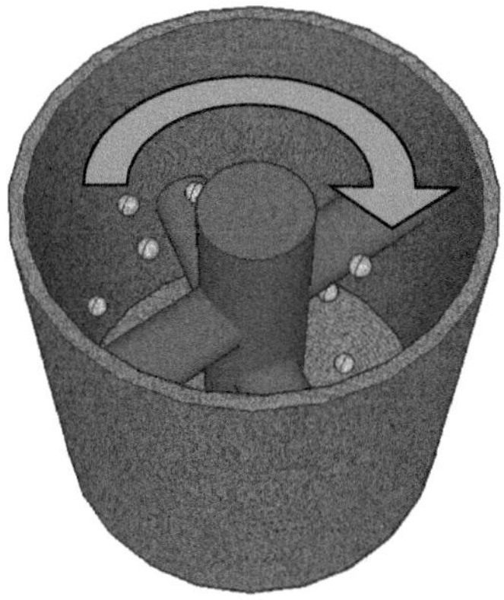

Fig. 1.8 Schematic of a ball mill

immiscible systems such as Pb-Al [15] and even ceramics, can be fabricated by mechanical alloying [16] which opens a vast range of opportunities for NC materials processed by MA.

Let us describe the procedure of ball milling. A schematic of a ball mill is presented in Fig. 1.8. The coarse grain powder is introduced into a sealed vial containing milling balls, which can be made of various different materials (e.g., ceramics or steel). Steel milling balls are typically preferred to ceramic balls.

The rotating rod is activated at relatively high frequency in order to input a substantial amount of energy to the balls. Large strains are imposed on the powder particles at every entrapment of a particle between two balls (see Fig. 1.9). This leads to a refinement in the grain size. The entrapment of powder

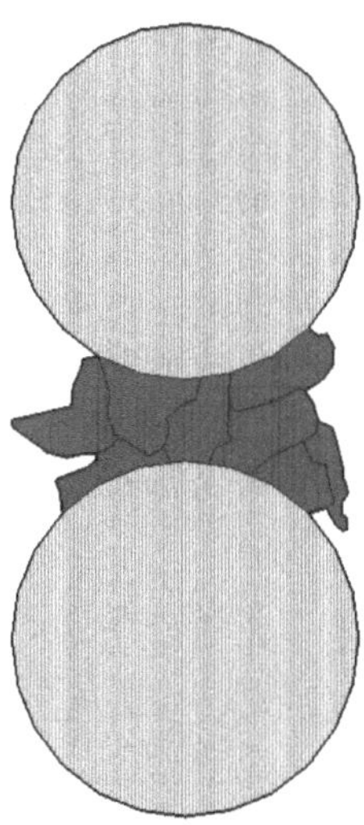

Fig. 1.9 Schematic of the powder entrapment process

particles between milling balls creates severe plastic deformation of the particles, which typically exhibit a flattened shape after several hours of milling. Details of the refinement process are presented in the following section.

Typically, the temperature rise during the milling process does not exceed 200 degrees. In order to avoid contamination of the particle powders with oxygen, nitrogen, and humidity, the process is commonly performed in an inert gas atmosphere such as Ar or He. Also, ductile materials tend to coalesce by welding. This is typically avoided by adding small quantities of a process control agent such as methanol, stearic acid, and paraffin compounds.

The energy input into the particles, that contain an increasing number of grains with decreasing grain size, will lead to the continuous fracture and welding of particles with one another. Typically, in the case of cryomilling (milling in a liquid nitrogen environment), the first hours of milling are dominated by the welding of the particles, which will consequently tend to grow, while further milling is dominated by the fracture of particles, which size will decrease yet remain on the order of several microns. Recall that, contrary to the particle size, the grain size is continuously decreasing during the process.

Grain Refinement Mechanism

Let us now discuss the grain refinement mechanism. Following detailed X-ray analysis, it was found that grain refinement occurs in three distinct stages (see Fig. 1.10) [17]. Recall that the initial powder size ranges on the order of several microns. Hence, in the case of metals, dislocation activity is still expected to prevail during plastic deformation. Initially, plastic deformation is localized in shear bands and results in strain at the atomic level in the order of 1–3% for metals and compounds, respectively. These shear bands result from cross-slip of dislocations which dominate the plastic deformation. In the second stage, the dislocation arrangements will recombine to create low-angle grain boundaries within the shear bands. Hence, the initial grain will become subdivided in subgrains with dimensions in the nanometer range. After further milling, the areas composed of small subgrains will extend throughout all grains. Also, during the deformation, the formation of multiple twins was observed by TEM in Cu powder and shear bands can be generated at the tip of the twin

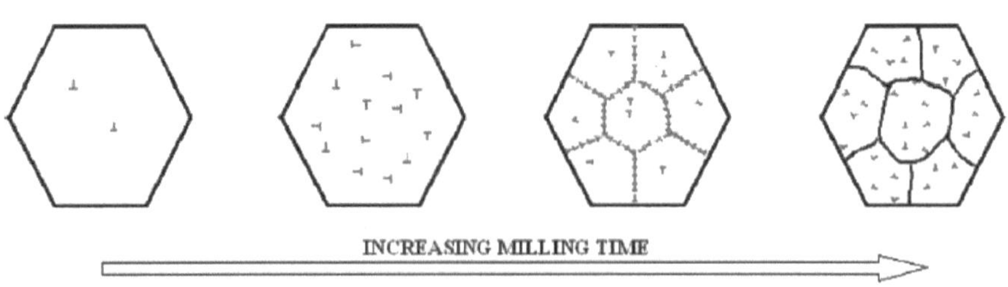

Fig. 1.10 Schematic of the grain refinement process during mechanical attrition [20]

boundaries [18]. It was conjectured that the formation of twins results from the applied shear stresses that can become larger than the critical shear stress for twin formation. Finally, in the last stage, large-angle grain boundaries are created via the reorientation of low-angle grain boundaries. The size of the nanograins is limited by the stress imposed by the ball mill on the grain.

Let us note that dislocations can be observed within the newly formed nanograins [18]. The resulting grain boundary structures are generally ordered, curved, and present excess strain [19]. However, disordered grain boundary regions can also be present within the nanograin clusters.

As mentioned above, the grain refinement rate can be severely decreased when the milling process is performed in a liquid nitrogen environment [21]. This process is referred to as cryogenic milling/cryomilling. For extensive review on cryomilling the reader is referred to reference [20]. Milling in a liquid hydrogen environment is usually performed at temperatures on the order of ~70 K. And a grain size refinement from ~50 μm to ~20 nm can be achieved in ~15 h depending on the fabrication process parameters and on the material. Also, as shown experimentally, cryomilling leads to much lower sample contamination emerging, for example, from the wear of the steel milling balls [22]. However, let us note that in some cases the contamination of the powder can improve the thermal stability of the condensed material. In the case of cryomilled particles, the compaction step must be preceded by nitrogen evaporation.

The final grain size depends on the ball to powder ratio (typically on the order of 10–1), the milling time, the type of powder (e.g., ceramic, metal), the size of the milling balls (on the order of a quarter-inch diameter), and the milling frequency (several hundreds of revolution per minute). As shown in Fig. 1.11,

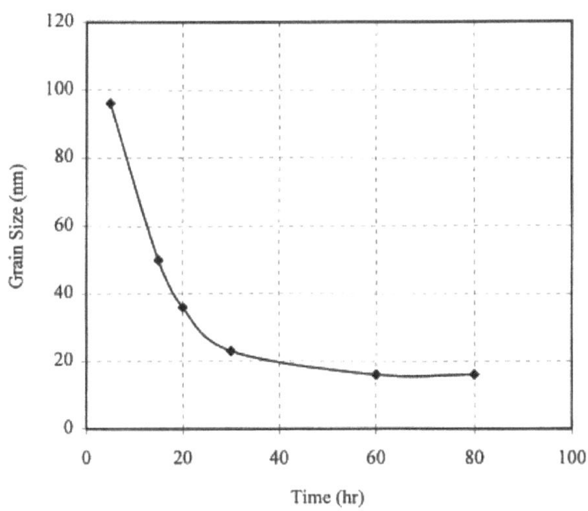

Fig. 1.11 Experimental grain size versus milling time measurements for iron powders, extracted from [23]

presenting the grain size to milling time dependence of NC iron powders, the grain size typically decreases sharply during the first ~20 h of milling. After significant milling time (>~20 h), the decrease in grain size with increasing milling time becomes less pronounced until a plateau is reached. The finest achievable grain size is referred to as the steady state grain size.

The effect of temperature was studied on pure Fe, prepared by low-energy ball milling. It was shown that the steady state grain size exhibits only a weak dependence on the milling temperature [24]. Hence, an increase in the milling temperature leads to a small increase in the steady state grain size.

Also, let us recall here that due to the presence of debris from the milling process and to the addition of control agents, the purity of the samples can be affected [22]. The ball-to-powder ratio affects the average grain size, as shown in Fig. 1.12. As expected, at a given milling time an increase in the ball-to-powder ratio will increase the chances of collision between two balls and a given particle (composed of several grains), which in turns leads to a smaller average grain size.

In order to emphasize the importance of the details of the processing route let us allow ourselves to a little digression. First, let us recall that the primary appeal of NC materials is that the fabrication of the first few samples of these novel materials could potentially reach the usually antonymic compromise of high strength and high ductility. However, as will be discussed in the following chapter, much work is still necessary in order to reach this goal. In most early studies on NC materials, the fabricated samples exhibited much higher yield stress than their coarse grain counterparts but unfortunately also exhibited low ductility with less than ~3% elongation. However, this compromise may be reached by combining cryomilling to room temperature milling [25, 26]. Indeed,

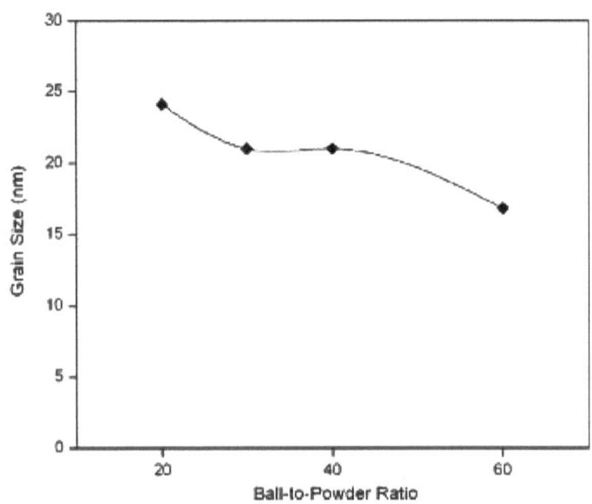

Fig. 1.12 Grain size versus ball-to-powder ratio in a cryomilled Ti alloy. Extracted from [22]

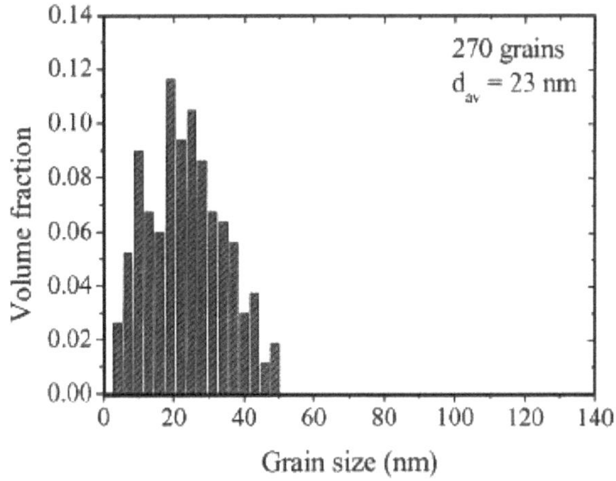

Fig. 1.13 Grain size distribution obtained after mixed cryomilling and room temperature milling of Cu on a count of 270 grains [26]

after 10 h of the mixed milling procedure the nanoparticle powder obtained exhibited a mean grain size of 23 nm with a fairly narrow grain size distribution (as shown in Fig. 1.13). The fabricated samples exhibited no artifacts due to poor density or contamination. Alternatively, one recent groundbreaking study has revealed NC samples – prepared by ball milling and compaction in an Ar environment – capable of reaching up to 50% deformation [27].

1.2.1.2 Physical Vapor Deposition

Physical vapor deposition (PVD) has shown to be a very efficient nanoparticle synthesis process. In this section, only inert gas condensation (IGC) will be presented for it is the most frequently used PVD method. IGC was one of the first techniques with electrodeposition and mechanical alloying used to fabricate nanocrystalline materials [28, 29]. The production of nanograins via IGC is more complex than in other methods presented above. For ease of comprehension, a schematic of one of the many possible existing IGC devices is presented in Fig. 1.14. The metallic gas, evaporated from two sources, condenses in contact with cold inert gas atoms leading to the creation of atom clusters which are transported by convection onto a cold finger refrigerated with liquid nitrogen. The nanoparticles are then collected from the cold finger, using a scraper, prior to being compacted in the low- and high-pressure compaction units.

As extensively described in review by Gleiter [31], the synthesis of nanograins via IGC, which is a "bottom-up" approach, can be divided into three steps: (1) evaporation of the metal source, (2) condensation of the vaporized metal, and (3) growth and collection of nanoparticle clusters. These steps are presented here.

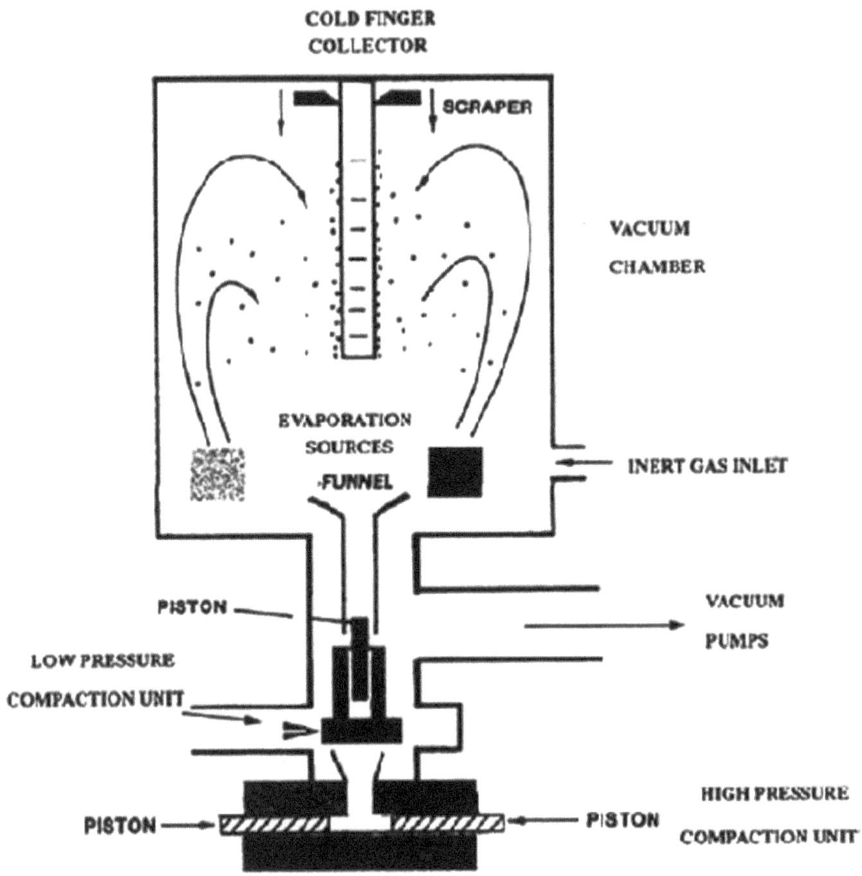

Fig. 1.14 Schematic of an IGC device, extracted from [30]

Evaporation of the Metal Source

Metal evaporation has been subject to study since the 1940s [32]. Conceptually, it is based on the following idea: at a given pressure, metal evaporation can be performed simply by increasing the temperature of the sample. This operation is commonly performed in a high vacuum chamber backfilled with an inert gas (typically Ar, Xe, or He). Several techniques can be used to evaporate the metal source. Among others, resistive heating, ion sputtering, plasma/laser heating, radio-frequency heating, and ion beam heating are the most commonly used processes. Although the integrated devices in IGC units may slightly differ from those presented here below, let us shortly describe some of the existing metal evaporation devices.

One of the simplest metal evaporation apparatuses is a resistive heater coil in which the source metal is placed (see Fig. 1.15) [33]. The resistive heater is not

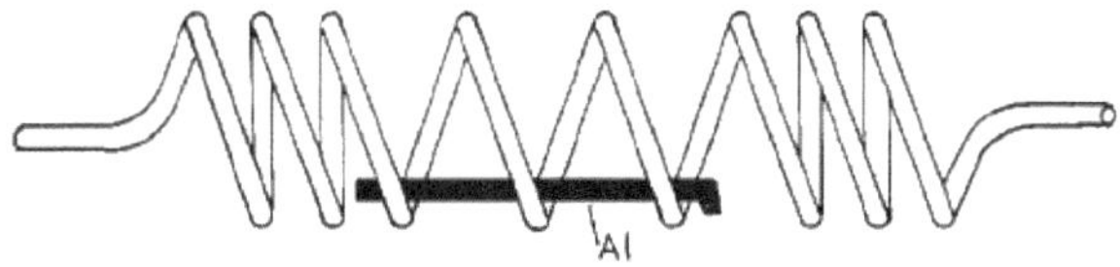

Fig. 1.15 Resistive heating evaporation source, image extracted from [33]

necessarily a coil. For example, it can be shaped as a "w" boat. Note here that, as shown in Fig. 1.15, the pitch is narrower at the extremities of the coil in order to avoid erratic flow of the melted metal (in this case Al).

The coil can be used only for a limited number of evaporation cycles, on the order of 5–10. The amount of powder produced by one evaporation cycle is usually on the order of 1 mg. The particle size can be controlled to some extent by the temperature of the metal source and the pressure of the inert gas.

Metal can also be vaporized from a crucible heated by a graphite element [34]. This method was introduced by Grandvquist and Buhram in 1976, who synthesized Fe, Al, Cr, Co, Zn, Ga, Mg, and Sn particles. The particle plant designed by Grandvquist and Buhram is presented in Fig. 1.16. The apparatus is composed of a glass cylinder fitted to water tubes which will cool a Cu plate on which the nanoparticles will be collected. A crucible containing the metal sample is placed within the tube near a graphite element heated by an optical pyrometer. The vaporization process is performed under Ar atmosphere at 0.4–05 Torr (~50 Pa.). The resulting grain size distribution is well described with a log normal distribution.

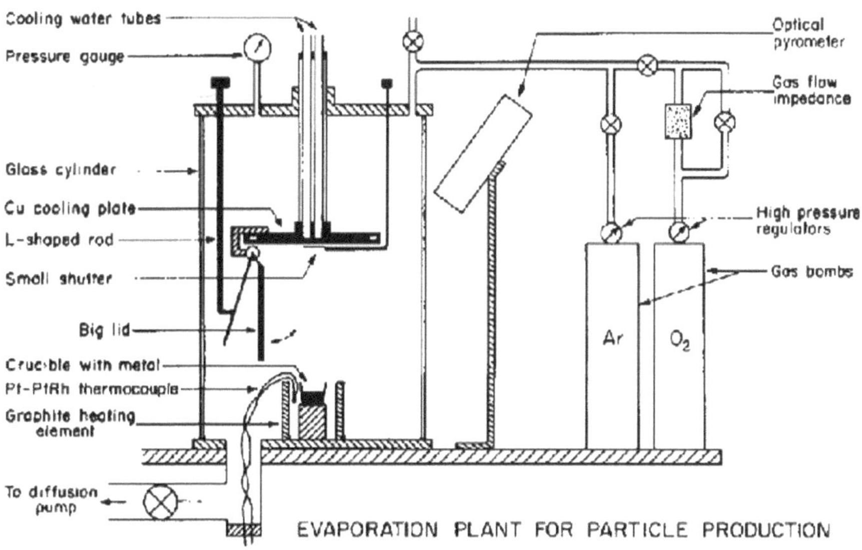

Fig. 1.16 Schematic of the apparatus used for evaporation from a crucible [34]

Electron beam evaporation typically consists of heating a melt confined in a water-cooled container via an energy input from an electron beam guided with a magnetic field [35]. The melt liquid circulates due to temperature gradients and surface tensions. The diameter of the melted spot is dependent on both the distance from the beam outlet and the beam power [36]. Typically, the electron beams delivers a beam power ranging from 20 to 400 MW under a current ranging from ~50 to 400 mA. The resulting production rate is on the order of a few grams per hour. Note here that this is considerably lower than the production rate obtained from mechanical attrition.

Condensation of the Vaporized Metal

Following the vaporization of the metal, condensation will occur by collision of the vaporized metal with the inert gas. The condensed particles will form a "smoke" (supersaturated vapor) because condensation is localized near the metallic source. Generally, a given particle of smoke contains a single crystal. The specific features of the smoke (e.g., shape) depend on the inert gas pressure, the gas density, and the evaporation temperature. For example, when a metal is evaporated at very low pressure ranging from a few torr to ~100 torr one can observe a candle-shaped smoke (see Fig. 1.17).

In most cases, the metal smoke can be divided into four zones: (1) the inner zone, (2) the intermediate zone, (3) the outer zone, and (4) the vapor zone. However, let us note that in the case of vaporization from a crucible only a

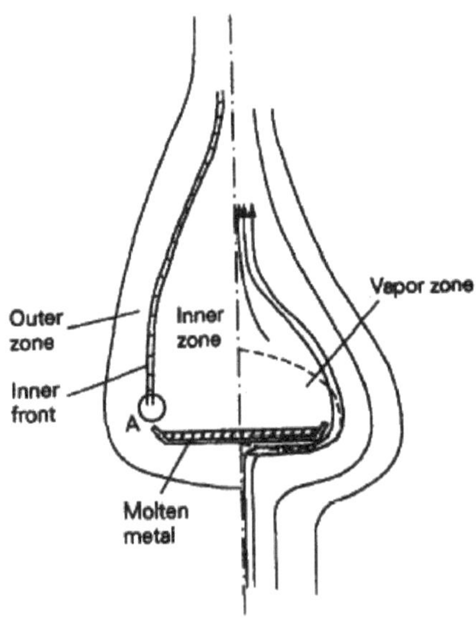

Fig. 1.17 Schematic of a typical smoke [32]

single region can be observed. Experimentally it was shown that particles in the inner zone are smaller than that of the intermediate zone (located between the inner front and the outer zone). This is presumably due to the fact that the particle growth mechanism is assisted by the diffusion from the vapor zone below the inner zone. The outer zone is formed by the vapor formed below the inner zone and then convected upward. Let us recall here that this discussion does not apply to evaporation methods from a crucible, where no vapor can goes downward.

Growth and Collection of Nanoparticle Clusters

The last step in the physical vapor deposition process corresponds to the growth and collection of the nanoparticle powder. There are two particle growth mechanisms: (1) absorption of vapor atoms within the vapor zone which is the zone where the supersaturated vapor exists and (2) coalescence of particles. The second mechanism is known to occur when the particles are small.

Typically, the collection of the powder is performed by scraping the powder from its fixation surface. As shown in Fig. 1.18, particle collection can affect the microstructure of the sample. Indeed, one can observe spiral like shape possibly engendered by the scrapper used during the collection of particles produced by condensation on liquid nitrogen cooled cold finger. Although not shown here, as shown by HRTEM observations, each spiral contains nanograins with poor particle bonding [37].

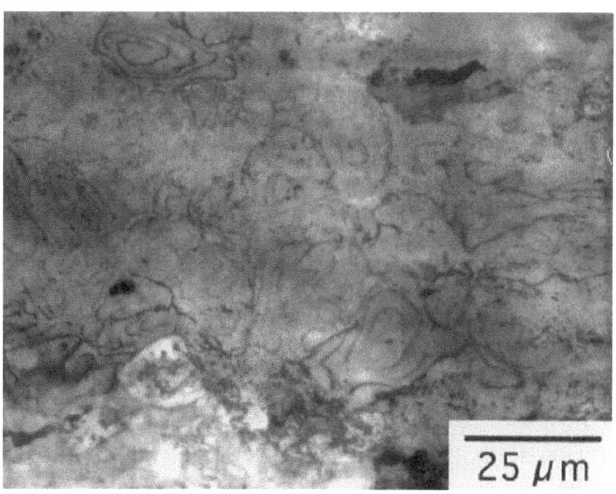

Fig. 1.18 Spiral morphology revealed by chemical etching of compacted nanocrystalline Cu [37]

1.2.2 Powder Consolidation

Following the synthesis of nanoparticle powder, which can take the form of nanograin agglomerates in the case of ball milling processed particles, a consolidation step is necessary to ensure of the bonding of particles. Its objective is to produce a compact structure with the desired density and optimal particle bonding. In the case of materials designed for structural application, for example, a high density, close to the theoretical density, is desired. The theoretical density is the density of the perfect lattice. However, considering the fact that nanocrystalline materials are largely composed of grain boundaries (depending on the grain size), which have a different structure than that of the perfect lattice, the use of the theoretical density to assess of the quality of the consolidation may appear inappropriate.

The consolidation process is clearly dependent on the external temperature and pressure. Several strategies can be used to consolidate the nanoparticle powder. This includes warm compaction, cold compaction (cold sintering), sintering (typically preceded by a compaction step), hot isostatic pressing (HIP), and so forth. For extensive descriptions of the several existing consolidation processes, the reader is referred to the review by Gutmanas [38].

The external conditions applied to the powder (e.g., pressure, temperature) will lead to elastoviscoplastic deformation of the powder via dislocation glide (in the case of metals), diffusion processes (vacancy diffusion, dislocation climb), or grain boundary sliding. The activity of each mechanism depends on the material considered, its microstructure, and the above-mentioned parameters.

This step is extremely delicate for nanoparticles, which, owing to the large number of atoms located on their surface, are sensible to contamination. For example, during cryomilling of Ti alloy an increase in nitrogen and oxygen content and Fe emerging from the wear of the balls was measured experimentally [22]. Powder contamination can severely affect the structure of the resulting nanograin agglomerate, which will clearly affect the response of the material. This was shown in studies of NiAl alloys contaminated with Fe and Cr [39]. Also, depending on the compaction method, additional difficulties may arise from the particular thermal stability of nanograins. Indeed, annealing experiments on nanocrystalline Ni showed that rapid grain growth can occur at temperatures as low as $\sim$350 K, which is approximately 20% of the melting temperature of pure Ni [40]. Hence, it is relatively delicate to retain the nanofeatures of the initial powder after consolidation.

Much progress was made since the early 1990s in the consolidation process. Among others the major challenges were to produce defect free (e.g., impurities, cracks) and fully dense samples with grain size remaining within the nanorange. Indeed, due to the low thermal stability of nanograins, the application of even moderate temperature fields to the powder samples often lead to abnormal grain growth. For example, in the mid-1990s hot isostatic pressing of Fe powder with 150 nm grain size lead to 1000 nm fully dense grained bulk "nanocrystalline" samples [41]. However, let us note that the "nano" feature was lost at the detriment of the density. By the late 1990s, nanocrystalline Fe produced by hot isostatic pressing with 9 nm grain size and 94.5% of the theoretical density could already be produced. Let us note here that, as mentioned above, a nanocrystalline sample with 9 nm grain size cannot reach a 100% theoretical density due to the intercrystalline regions (grain boundaries and triples junctions) which volume fraction is no longer negligible and which also have a lower density [42].

The grain size distribution depends on both the nanograin synthesis process and the consolidation technique used. The effect of grain size distribution, although not regarded as of primary importance in early studies on NC materials, can affect the response of the sample. Indeed, considering a log-normal grain size distribution with a given mean grain size and varying variance, it was predicted via a Taylor type of model that, depending on the variance, the ultimate stress can drop by several hundred MPa [43]. Usually the grain size distribution is measured by XRD (e.g., Fourier transform of the diffraction peaks, Monte Carlo, etc.) and/or TEM [44]. However, the evaluation of the grain size distribution remains a complicated exercise and a very limited set of data, too limited to draw conclusions, were reported for in situ consolidated nanograins powders. Typically, samples exhibit log normal distribution with a relatively small variance. For example, a log normal distribution with mean grain size 5.3 nm and variance 1.9 nm was reported for cold compacted nanocrystalline Pd [45]. Let us now describe the most commonly used consolidation techniques.

1.2.2.1 Cold Compaction

Cold compaction has proved to be an efficient way to proceed to the consolidation step. It consists of applying a high pressure, on the order of ~1 GPa, at low temperatures to the powder which was previously loaded into a die. The obtained compact is typically referred to as a green compact with associated green properties (e.g., density) The resulting consolidated nanocrystalline material depends on several compaction parameters such as the initial powder density, the imposed pressure, and the die shape.

Several deformation mechanisms are involved in the densification of the powder. First, the particles will slide relatively to one another. Considering the fact that the particles have a size on the order of ~10–80 nm, the number of interparticle contacts will be clearly higher in the case of nanoparticle powder compared to that of coarse grain powders. Hence, the motion of particles via sliding is more difficult in the case of nanoparticle powder to friction. Also, as shown in finite element–based simulations, each particle will deform elastoplastically under the applied pressure and an increase in the applied pressure engenders a higher density [46].

1.2.2.2 Sintering

Sintering consists of exposing the nanoparticle powders to a relatively high temperature, remaining below the melting point, under no pressure. Typically, the compound to be sintered is exposed to the relatively high temperature for a duration varying from several minutes to several hours. Traditionally, the sintering process is preceded by a compaction step at low applied pressures and temperatures, in the range of 50 MPa to ~1 GPa, in order to obtain a green compound with adequate green density.

Typically, this compaction process is not optimal in the case of nanocrystalline metals. Indeed, as mentioned in the above, metal nanoparticles exhibit extremely low thermal stability and grain growth would occur during sintering. This would lead to the loss of the nano-features of the material. Let us note, however, that bulk nanocrystalline Cu and Fe with ~70 nm grain size were obtained via cold isostatic pressing followed sintering under particular conditions. The obtained densities vary from 60 to 90% of the theoretical density [47]. A comparative experiment on consolidation of Fe nanograins in nanosized particles and in micron-sized particles clearly showed that sintering at high temperatures does not lead to similar densities as hipping or cold compaction [48]. Typically, much lower densities are obtained in the case of sintering at high temperatures. For example, nanocrystalline Al was produced by compaction of aluminum powders with 53 nm grain size by two techniques: (1) cold compaction and (2) sintering at various temperatures ranging from 200 to 635 degrees Celsius for a short time of 40 min (in order to preserve the grain size) [49]. While the density of sintered Al was typically 96%, with a peak at 98%, the density of the cold-compacted nanocrystalline Al was 99%, which is to date the highest density obtained for nanocrystalline metals. Also, the densest sintered specimen exhibits much lower maximum elongation, about 4%, while the cold compacted sample exhibits a 7.7% maximum elongation. In the case of ceramics, such as erbium, for example, excellent final densities in the range of 97% of the theoretical density can be observed [50].

1.2.2.3 Hot Isostatic Pressing

The consolidation of nanoparticle powders can also be achieved via hot isostatic pressing (HIP), which consists of applying high pressures, on the order of several GPa, to a powder that is simultaneously submitted to a relatively high temperature, yet remaining well below the melting point. This processing method allows the exposition of the sample to lower temperatures than of sintering in order not to activate grain growth. HIP presents some interesting peculiarities enabling the consolidation of samples with remarkably high densities and small grain size. For example, this method has shown to be successful in fabricating porosity-free FeAl alloys with 98% density. The FeAl powder produced by mechanical alloying was subjected to a pressure of 7.7 GPa and temperature on the order of 100 degrees Celsius for 180 s, and the obtained bulk sample had a 23 nm grain size. Typically, at such high temperature, one would not expect to conserve nanosized grains [51]. However, grain growth is highly suspected to occur via diffusion and the higher diffusivity of nanocrystalline materials would clearly be causing the abnormal grain growth phenomenon. Since diffusivity decreases with pressure, the application of high pressure will penalize the grain growth phenomenon.

1.3 Summary

Nanocrystalline and ultrafine grained materials can be processed either by one-step processes or by two-step processes. In the latter case, samples are produced via consolidation of nanocrystalline powder.

One-step processes (e.g., ECAP, HPT, electrodeposition) are typically less time consuming than two-step processes. ECAP and HPT yield large samples with both high density and high purity. However, these processes do not allow fabrication of nanocrystalline sample. The minimum grain size achievable lies in the neighborhood of ~20. Contrary to HPT and ECAP, electrodeposition can yield samples with very fine average grain size ($d < \sim 10$ nm). However, the sample thickness is typically limited to a few hundred microns. On the one hand, the sample ductility is typically limited by its purity, which can be compromised by the electrolytic bath. On the other hand, the ductility of electrodeposited sample can be improved by controlling the grain size distribution. In general, a wider grain size distribution leads to an improved ductility.

In two-step processes, nanocrystalline powder can be synthesized by various methods. The most commonly used methods are mechanical alloying, which is a severe plastic deformation mechanism, and physical vapor deposition. The purity of nanocrystalline powder can usually be controlled by processing in an inert gas environment or in a liquid nitrogen environment. The second step consists of compacting the powder, typically via HIP or cold compaction, to obtain a bulk sample with dimensions typically in the order the centimeter. The compaction step is critical for it is desirable to keep the nanofeature of the powder. As opposed to the sample density, which remains critical, the grain size distribution can generally be controlled during the compaction step.

References

1. Langdon, T.G. and R.Z. Valiev, Progress in Materials Science **51**, (2006)
2. Iwahashi, Y., J. Wang, Z. Horita, M. Nemoto, and T.G. Langdon, Scripta Materialia **35**, (1996)
3. Xu, S., G. Zhao, Y. Luan, and Y. Guan, Journal of Materials Processing Technology **176**, (2006)
4. Mishra, A., B.K. Kad, F. Gregori, and M.A. Meyers, Acta Materialia **55**, (2007)
5. Lowe, T.C. and R.Z. Valiev, JOM **52**, (2000)

6. Furukawa, M., Z. Horita, M. Nemoto, and T.G. Langdon, Materials Science and Engineering A **324**, (2002)
7. Jiang, H., Y.T. Zhu, D.P. Butt, I.V. Alexandrov, and T.C. Lowe, Materials Science and Engineering A **290**, (2000)
8. Ebrahimi, F., G.R. Bourne, M.S. Kelly, and T.E. Matthews, Mechanical properties of nanocrystalline nickel produced by electrodeposition. Nanostructured Materials, **11(3)**, 343–350, (1999)
9. Cheung, C., F. Djuanda, U. Erb, and G. Palumbo, Electrodeposition of nanocrystalline Ni-Fe alloys. Nanostructured Materials, **5(5)**, 513–52, (1995)
10. Wu, R.I., G. Wilde, and J.H. Perepezko. *Glass formation and primary nanocrystallization in Al-base metallic glasses*. Cincinnati, OH, USA: Elsevier, (2001)
11. Gravier, S., L. Charleux, A. Mussi, J.J. Blandin, P. Donnadieu, and M. Verdier, Journal of Alloys and Compounds **434–435**, (2007)
12. Louzguine-Luzgin, D.V. and A. Inoue, Journal of Alloys and Compounds **434–435**, (2007)
13. Perepezko, J.H., Progress in Materials Science **49**, (2004)
14. Zhang, T. and H. Men, Journal of Alloys and Compounds **434–435**, (2007)
15. Zhu, M., X.Z. Che, Z.X. Li, J.K.L. Lai, and M. Qi, Journal of Materials Science **33**, (1998)
16. Jiang, J.Z., R. Lin, S. Morup, K. Nielsen, F.W. Poulsen, F.J. Berry, and R. Clasen, Physical Review B (Condensed Matter) **55**, (1997)
17. Fecht, H.J., Nanostructured Materials **1**, (1992)
18. Huang, J.Y., Y.K. Wu, and H.Q. Ye, Acta Materialia **44**, (1996)
19. Huang, J.Y., X.Z. Liao, and Y.T. Zhu, Philosophical magazine **83**, (2003)
20. Witkin, D.B. and E.J. Lavernia, Progress in Materials Science **51**, (2006)
21. Lee, J., F. Zhou, K.H. Chung, N.J. Kim, and E.J. Lavernia, Metallurgical and Materials Transactions A (Physical Metallurgy and Materials Science) **32A**, (2001)
22. Zuniga, A., S. Fusheng, P. Rojas, and E.J. Lavernia, Materials Science and Engineering A (Structural Materials: Properties, Microstructure and Processing) **430**, (2006)
23. Khan, A.S., Z. Haoyue, and L. Takacs, International Journal of Plasticity **16**, (2000)
24. Tian, H.H. and M. Atzmon, Acta Materialia **47**, (1999)
25. Cheng, S., et al., Acta Materialia **53**, (2005)
26. Youssef, K.M., R.O. Scattergood, K.L. Murty, and C.C. Koch, Applied Physics Letters **85**, (2004)
27. Khan, A.S., B. Farrokh, and L. Takacs, Materials Science and Engineering: A **489**, (2008)
28. Fougere, G.E., J.R. Weertman, and R.W. Siegel. *On the hardening and softening of nanocrystalline materials*. Cancun, Mexico, (1993)
29. Nieman, G.W., J.R. Weertman, and R.W. Siegel. *Mechanical behavior of nanocrystalline Cu, Pd and Ag samples*. New Orleans, LA, USA: TMS – Miner. Metals & Mater. Soc., (1991)
30. Meyers, M.A., A. Mishra, and D.J. Benson, Progress in Materials Science **51**, (2006)
31. Gleiter, H., Progress in Materials Science **33**, (1989)
32. Uyeda, R., Progress in Materials Science **35**, (1991)
33. Singh, A., Journal of Physics E (Scientific Instruments) **10**, (1977)
34. Granqvist, C.G. and R.A. Buhrman, Journal of Applied Physics **47**, (1976)
35. Westerberg, K.W., M.A. McClelland, and B.A. Finlayson, International Journal for Numerical Methods in Fluids **26**, (1998)
36. Bardakhanov, S.P., A.I. Korchagin, N.K. Kuksanov, A.V. Lavrukhin, R.A. Salimov, S. N. Fadeev, and V.V. Cherepkov, Materials Science and Engineering: **B 132**, (2006)
37. Agnew, S.R., B.R. Elliott, C.J. Youngdahl, K.J. Hemker, and J.R. Weertman, Materials Science and Engineering A: Structural Materials: Properties, Microstructure and Processing **285**, (2000)
38. Gutmanas, E.Y., Progress in Materials Science **34**, (1990)

39. Murty, B.S., J. Joardar, and S.K. Pabi, Nanostructured Materials **7**, (1996)
40. Klement, U., U. Erb, A.M. El-Sherik, and K.T. Aust, Materials Science and Engineering A (Structural Materials: Properties, Microstructure and Processing) **A203**, (1995)
41. Alymov, M.I. and O.N. Leontieva. *Synthesis of nanoscale Ni and Fe powders and properties of their compacts*. Stuttgart, Germany, (1995)
42. Rawers, J., F. Biancaniello, R. Jiggetts, R. Fields, and M. Williams, Scripta Materialia **40**, (1999)
43. Zhu, B., R.J. Asaro, P. Krysl, K. Zhang, and J.R. Weertman, Acta Materialia **54**, (2006)
44. Ortiz, A.L., F. Sanchez-Bajo, and F.L. Cumbrera, Acta Materialia **54**, (2006)
45. Reinmann, K. and R. Wurschum, Journal of Applied Physics **81**, (1997)
46. Hyoung Seop, K. *Densification modelling for nanocrystalline metallic powders*. Taipei, Taiwan: Elsevier, (2003)
47. Dominguez, O., Y. Champion, and J. Bigot. *Mechanical behavior of bulk nanocrystalline Cu and Fe materials obtained by isostatic pressing and sintering*. Chicago, IL, USA: Metal Powder Industries Federation, Princeton, NJ, USA, (1997)
48. Livne, Z., A. Munitz, J.C. Rawers, and R.J. Fields, Nanostructured Materials **10**, (1998)
49. Sun, X.K., H.T. Cong, M. Sun, and M.C. Yang, Metallurgical and Materials Transactions A (Physical Metallurgy and Materials Science) **31A**, (2000)
50. Lequitte, M. and D. Autissier. *Synthesis and sintering of nanocrystalline erbium oxide*. Stuttgart, Germany, (1995)
51. Krasnowski, M. and T. Kulik, Intermetallics **15**, (2007)

Chapter 2
Structure, Mechanical Properties, and Applications of Nanocrystalline Materials

2.1 Structure

Nanocrystalline (NC) materials are composed of grain cores with well-defined atomic arrangement (e.g., face center cubic, body center cubic) joined by an interphase region composed of grain boundaries and higher-order junctions (e.g., triple junctions, quadruple junctions). Early experiments on nanocrystalline materials have shown that the interphase region and particularly grain boundaries exhibit an almost grain size–independent thickness [1]. Hence, as the grain size is decreased, the volume fraction of the interphase region increases. Supposing a tetracaidecahedral grain shape, corresponding to a realistic grain shape, the following expressions of the volume fraction of interphase (e.g., grain boundaries and triple junctions), grain boundaries, and triple junctions can be derived [2].

$$f_{in} = 1 - \left[\frac{(d-w)}{w}\right]^3, f_{gb} = \frac{3w(d-w)^2}{d^3}, f_{tj} = f_{in} - f_{gb} \tag{2.1}$$

where the subscripts *in, gb,* and *tj* refer to the interphase, the grain boundaries, and the triple junctions, respectively.

Note here that early X-ray measurements estimated the volume fraction of interphase to about ~30% with a mean grain size equal to 10 nm [3]. This measure lies well within predictions presented in Fig. 2.1. It can be observed that the volume fraction of interphase increases sharply when the grain size is in the nanocrystalline range (e.g., grain diameters smaller than ~100 nm). Also, notice that the volume fraction becomes non-negligible when the grain size is smaller than ~10 nm. Hence, it is easy to comprehend the importance of the interphase region in NC materials for the material properties are directly dependent on the microstructure of the sample, which depends itself on the fabrication process.

M. Cherkaoui, L. Capolungo, *Atomistic and Continuum Modeling of Nanocrystalline Materials*, Springer Series in Materials Science 112,
DOI 10.1007/978-0-387-46771-9_2,

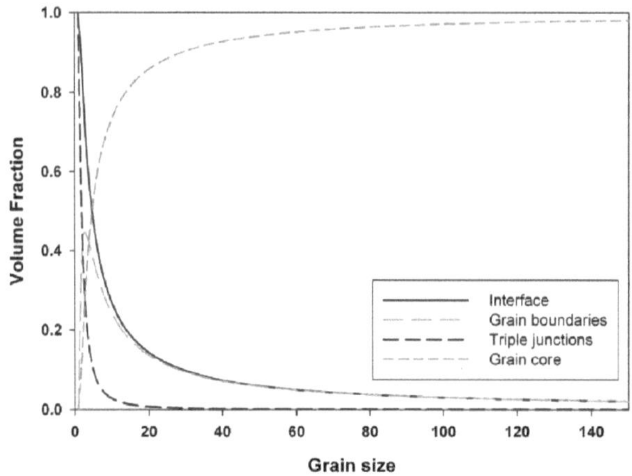

Fig. 2.1 Evolution of volume fractions of interface, grain boundaries, triple junctions, and grain cores with the grain size in nm

2.1.1 Crystallites

Independent of the fabrication process and grain size, grain cores exhibit a crystalline structure (e.g., face center cubic, body center cubic, hexagonal close packed [hcp]) up to the grain boundary. Interestingly, the lattice parameter of NC materials was reported to be size dependent. Precisely, X-ray diffraction measurements on Cu samples processed by equal channel angular pressing (ECAP) revealed that the lattice parameter within the grain cores is decreased by 0.04% [4]. It was suggested that the compressive stress imposed by nonequilibrium grain boundaries is the source of this reduced lattice parameter. The same conclusion was reached on samples fabricated by several different processes. Let us note that the lattice strain is typically more pronounced in the vicinity of grain boundaries and triple junctions.

2.1.1.1 Dislocations

Dislocation density measurements have been subject to controversial debate with reported values of dislocation density varying from $10^{15}\,m^{-2}$ to zero. Figure 2.2 presents high-resolution transmission electron microscopy (HRTEM)image of electrodeposited Ni with average grain size of ~30 nm prior to deformation [5]. The bright and dark field images (Fig. 2.2a, b) exhibit a crystalline structure devoid of dislocations and impurities indicating a low initial dislocation density within the grain cores. As shown in Fig. 2.2.c, the occasional presence of dislocation loops can be observed as well as the presence of twins. The same conclusion was also reached in the case of 20 nm grained nanocrystalline Pd processed by inert gas condensation followed by

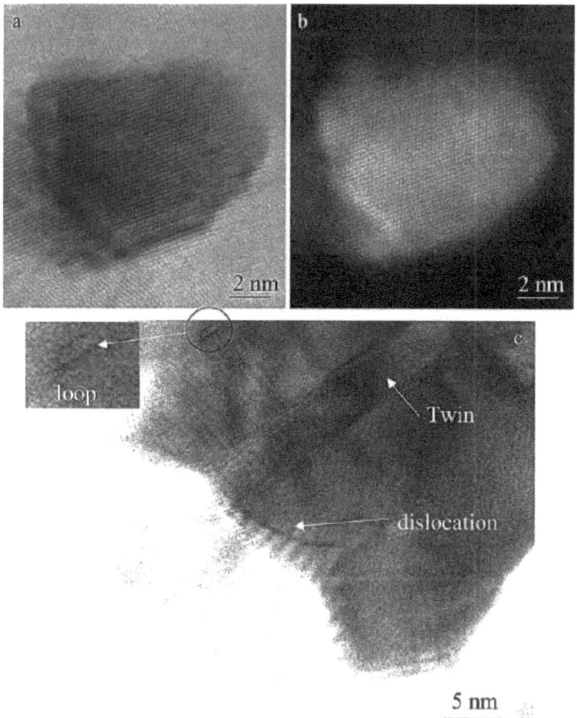

Fig. 2.2 HRTEM image of a grain core [5]

compaction [6]. However, in the case of ECAP-processed NC Cu, with grain size 150 nm, the initial dislocation density was reported in the order of $\sim 2.10^{15}\,m^{-2}$ and was about 20 times larger than that of the reference sample used the X-ray diffraction analysis [4]. A high initial dislocation density on the order of $1.10^{15}\,m^{-2}$ was also reported for nanocrystalline Ni processed by high-pressure torsion (HPT) [7]. However, let us note here that in the case of materials processed by severe plastic deformation processes, such as ECAP and HPT, grain refinement results from the large strains imposed to a coarser-grained sample. Thus, the high dislocation density measured experimentally is to be expected. Finally, let us recall that the minimum grain size achieved by severe plastic deformation is rarely smaller than $\sim$100 nm, which falls into what is referred to as the ultrafine range, where dislocation activity is similar to that of coarser-grain materials. Finally, a dislocation density on the order of $\sim 5.10^{15}\,m^{-2}$ was reported for 15 nm grain inert gas condensation processed nanocrystalline copper [8]. Also, the same authors report average dislocation spacing close to the grain size. This signifies that a given grain will initially contain zero to 1 dislocation loop. Hence, in general, within a given grain core, the dislocation density is severely reduced compared to that of coarse-grain materials.

Consequently, dislocation activity, which is typically governed by dislocation storage and dislocation annihilation in coarse grain materials, is expected to decrease within grain cores in the case of nanocrystalline materials.Dislocation storage is an athermal process, corresponding to the pinning of a dislocation on a sessile obstacle (e.g., defect, stored dislocation, grain boundary), and leading to a decrease in the mean free path of mobile dislocations. Typically, strain hardening models such as the first model from Kocks and Mecking and subsequent evolutions account for the effect of grain boundaries and the effect of stored dislocations [9–12]. The effect of stored dislocations on the mobility of dislocations is accounted for via the principle of material scaling, introduced by Kuhlman Wilsdorf. Essentially, it introduces proportionality relations between the dislocations' mean free path and the dislocation density. However, so far, it has not been shown experimentally that the principle of similitude remains valid in the case of nanocrystalline materials. Dynamic recovery, which is a thermally activated mechanism, typically written with an Arrhenius type of law, is treated in phenomenological manner.

2.1.1.2 Twins

As mentioned in earlier sections, the fabrication process has great effect on the resulting microstructure. Hence, depending on the fabrication process, two NC samples with equal mean grain size can exhibit different microstructures (e.g., grain size distribution, grain boundary misorientations, impurities, pores, etc.). One of the most remarkable examples is the presence of mechanical twins in NC materials. Recall here that a twin corresponds to a mirror symmetry lattice reorientation with respect to a twinning plane. Indeed, even in face-centered cubic (FCC) metals, such as Cu and Al, which present enough slip systems (12) for dislocation glide to occur – as opposed to metals in the hcp system, in which, due to the crystal's low symmetry, twinning is a common feature of plastic deformation in coarse grain polycrystals and single crystals – twin boundaries can still be observed.

Let us note here that the presence of twins within the grain cores is directly dependent on the fabrication process. Indeed, ECAP and HPT processed nanocrystalline materials rarely exhibit the presence of twins while materials processed via inert gas condensation (IGC), electrodeposition, and mechanical alloying typically lead to the presence of twins. In Fig. 2.3 one can observe nanocrystalline Cu grain core containing a "giant step," the step is delimited by the arrowheads on the HRTEM image [13]. The stepped region is highly incoherent.

2.1.1.3 Stacking Faults

Although no quantitative data are available as to the number of stacking faults, that is the break of the sequence of close-packed planes, transmission electron microscopy (TEM)experiments and X-ray diffraction (XRD)followed by

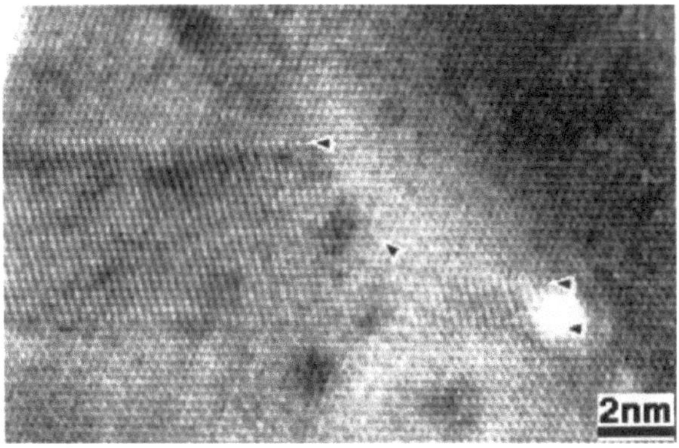

Fig. 2.3 Cu cryomilled grain core containing a stepped twin [13]

calculation of the α_{warren} probability of faults have revealed valuable information on the matter [14–17]. Calculation of the probability of faults on nanocrystalline Cu and Pd samples with grain size ranging from 5 to 25 nm and from 3 to 18 nm, respectively, have revealed that in the initial structure exhibits an almost null stacking fault probability. However, this does not signify that stacking faults are not present in the initial structure. Indeed, stacking faults can be observed in TEM experiments [15] but their initial presence is rather scarce. The fault probability was shown to increase with plastic deformation. This is clearly shown in the rolling experiment on IGC-synthesized nanocrystalline Pd. Indeed, in Fig. 2.4, presenting the evolution of the stacking fault parameter with strain for an ultrafine grain Pd sample and nanocrystalline Pd sample with grain size ~33 nm, one can clearly see that the stacking fault parameter increases sharply with deformation until it reaches a plateau. The value of the stacking fault parameter is consistently higher in the case of the NC samples. Although this measure is purely qualitative, it reveals an interesting phenomenon. That is, the activity of dislocations is driven by the motion or interaction of Shockley partial dislocations (which necessarily result in the presence of stacking faults). Moreover, it was also suggested that twinning deformation mode may be caused by the stacking faults led by Shockley partial dislocations.

2.1.2 Grain Boundaries

The microstructure of grain boundaries has been subject to a long-lasting debate. Recall that the first studies by Gleiter and co-workers on small-grained nanocrystalline materials, with grain size in the neighborhood of 10 nm, exhibited an *open structure*of grain boundaries which were consequently referred to as anomalous with respect to that of coarse grained materials.

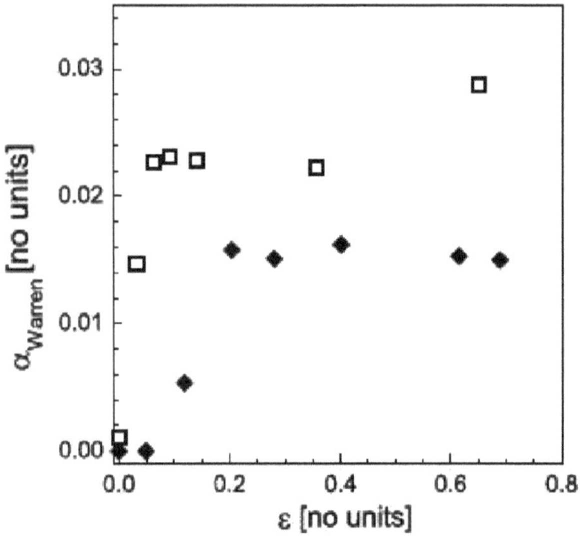

Fig. 2.4 Evolution of the stacking fault parameter with strain for UFG PD (*in bold*) and nanocrystalline IGC Pd

Although this will be described in detail in Chapter 5, let us briefly discuss here the modeling of grain boundaries in coarse-grained polycrystalline materials. Grain boundaries can be regarded as particular groups of geometrically necessary dislocations. Indeed, dislocations can generally be put into two categories: (1) statistically stored dislocations, and (2) geometrically necessary dislocations. Statistical dislocations are present as a consequence of hardening, which results in the decrease of the mean free path of dislocations. Some other dislocations referred to as geometrically necessary must be present within the material in order to accommodate for local lattice curvature changes. Grain boundaries are regions of high change in lattice curvature. Hence, they can be regarded as regions of high density of geometrically necessary dislocations.

First, the grain boundary thickness or width is known not to exhibit major size effects and can be regarded as constant and equal to approximately 3–4 perfect lattice spacing (~0.6–1 nm). Also, grain boundaries are regions of lower atomic density. This leads to the presence of strain fields within the grain cores induced by those within the grain boundaries. A simple model based on the scattering cross-section measurements and neglecting porosity effects leads to an estimate of density for grain boundaries of 60–70% of that of the perfect lattice.

Regarding the detailed microstructure of grain boundaries, two schools are opposed. The first one suggests an open structure of grain boundaries where no atomic order is present while the second school of thought regards grain boundaries as a much more defined phase which in most cases can be described by structural unit models (see Chapter 5). Let us consider the limit case where the grain size takes the theoretical value zero; in that particular configuration one cannot expect any particular atomic ordering of the "interphase." Now taking the other extreme where a sample would be constructed of simply two grains delimited by a single grain boundary, one would expect a much more organized grain boundary microstructure. In the case of nanocrystalline materials with grain size larger than ~10 nm, so that the triple junction volume fraction does not come into account, one would then expect to find well-defined grain boundary regions, pertaining to the second school of thought, and other interphase regions exhibiting less order. As shown in TEM observations on nanocrystalline Pd with ~10 nm grain size processed by a physical vapor deposition technique, the grain boundary microstructure is not homogeneous within the material. In Fig. 2.5, presenting a HRTEM image of a NC Pd sample processed by physical vapor deposition, some regions, such as region A-B, present a well-ordered grain boundary, while region D-E presents no particular order and region B-C exhibits a grain boundary with changing character. Let us note here that the sample presented has a small grain size, even in the nanocrystalline regime, hence one could probably suppose that an increase in the grain size may lead to more order in the grain boundary region.

As the various fabrication processes differ vastly and due to the limited data on the grain boundary structure, which is inherent to the difficulty in preparing samples for observations, it is rather difficult to discuss grain boundary microstructure in its general sense. However, outstanding observations performed by Huang and co-workers revealed that, in the case of materials processed by severe plastic deformation, both one-step and two-step processes (e.g., HPT, ECAP, and ball milling), grain boundaries are usually high-energy and exhibit strains and steps or curves [13].

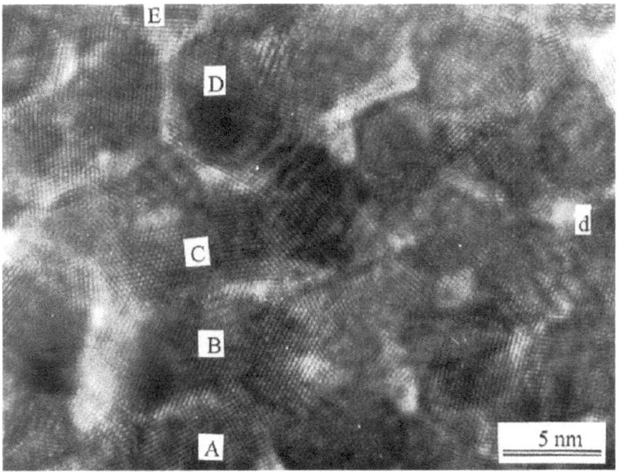

Fig. 2.5 HRTEM image of a nanocrystalline Pd sample. Extracted from [18]

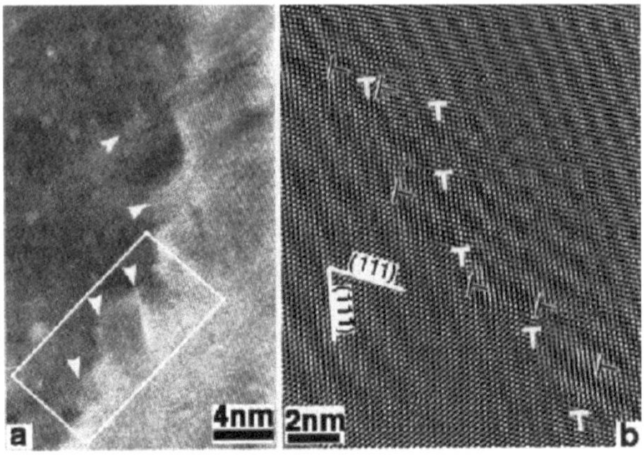

Fig. 2.6 Small angle grain boundary with steps and stacking faults (**a**) and zoom on the selected region revealing the presence of extrinsic stacking faults (**b**) [13]

In Fig. 2.6(a) one can observe a small-angle asymmetric grain boundary with a 2 degree misorientation angle. One can easily observe the presence of strips which are representative of thin twin or stacking faults. Now, looking at Fig. 2.6(b), corresponding to a zoom on the selected window of Fig. 2.6(a), one can observe the presence of slightly disassociated dislocations which are responsible for the presence of the stacking fault or thin twins within the adjacent grain cores. It is thus clear that the grain boundary structure has a great influence on the microstructure of the sample and this influence is not limited to that on the interphase. Finally, let us note that small-angle grain boundaries are known to be dislocation sources operating in a manner similar to that of a traditional Frank and Read source.

As mentioned in Chapter 1, most grain boundaries are large-angle grain boundaries. Similar to the case of small-angle grain boundaries, large-angle grain boundaries typically present facets or steps that correspond to extra-atomic layers. This can be observed in Fig. 2.7 presenting a large-angle grain

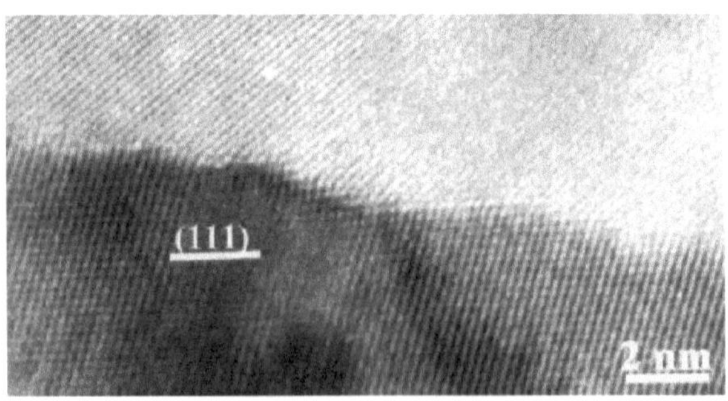

Fig. 2.7 HRTEM image of a high-angle stepped grain boundary in cryomilled Cu [13]

boundary observed in cryomilled Cu. The observed steps are 4–5 atomic layers thick and are lying on the (111) plane. These steps can also be regarded as large ledges. Let us note that in the early 1960s, J.C.M. Li in his pioneering theoretical work, suggested that grain boundary ledges can act as dislocation donors. Hence, upon emitting a dislocation, a grain boundary ledge corresponding to a single layer of extra atoms would be annihilated. As will be shown later, the role of these steps may not be limited to that of dislocation donors.

Most of the defects in nanocrystalline materials are localized within the grain boundaries, which is especially the case of small pores and large flaws that can be as long as one micron. In the case of IGC-processed samples, during the outgassing step followed by warm compaction, it was clearly shown that gas can remain trapped within the pores at pressures high enough to stabilize the pore.

2.1.3 Triple Junctions

Triple junctions are regions where more than two grains meet. Considering the fact that the atomic positions in a grain boundary are directly dependent on the relative five degrees of freedom of the two grains composing the grain boundary resulting in a particular spatial organization of the atoms, it is expected that the position of atoms localized within a triple junction will clearly depend on the relative orientation of the neighboring grains. TEM observations revealed that no regular organization of the atoms can be observed in a triple junction. This can be clearly observed in region denoted *d* in Fig. 2.5. Also, as in the case of grain boundaries, triple junctions are regions of concentrated defects such as pores, flaws, and impurities.

2.2 Mechanical Properties

Nanocrystalline materials exhibit fascinating properties which are intimately linked to their particular microstructure characterized by a large volume fraction of grain boundaries. One of the most acknowledged and studied peculiarities of nanocrystalline materials is the extremely high yield strength that can be reached with small grain size. Indeed, a typical NC sample will exhibit yield strength up to 7 times larger than its coarse grain counterpart with the same composition.

Let us recall that when decreasing the crystallite size to the nanorange, one hopes to reach a great if not an optimal compromise between strength and ductility. This has not yet been reached, but giant steps were taken in that direction over the past decade. More than the grain size/yield strength dependence, nanocrystalline materials exhibit other size-dependent properties. Some are expected, such as the size-dependent elastic constants and others needing detailed modeling. This is the case of the strain rate sensitivity discussed below.

Also, as nanosized particles exhibit poor thermal stability and since grain boundaries in nanocrystalline materials are typically high-energy grain boundaries, a particular size effect in the thermal response of nanocrystalline materials is expected. This particular subject still requires a great deal of investigation to understand the underlying phenomenon.

A word of caution is necessary when analyzing experimental data on nanocrystalline materials. First, as will be presented below, most available data exhibit large discrepancies. This has unfortunately lead to a great deal of debate among modelers. Hence, prior to analyzing data on the mechanical or thermal response of a sample, it is crucial to cautiously analyze the fabrication process and resulting microstructure. Indeed, as shown in Chapter 1, the sample microstructure is a direct consequence on the fabrication process which so far is particular for each, mostly academic, laboratory. Second, the measurement of several properties of NC materials is rather complicated. Let us cite two stringent examples.

Typically, the yield stress of a sample is measured by tensile test and subsequent application of the 0.2% offset rule. However, in the case of NC materials, the samples are typically of reduced dimensions and it is not always possible to perform a tensile test on the samples. Hence, nanohardness measurements are often performed and the yield stress is simply deduced by dividing the hardness by 3. This is a commonly acceptable approximation in the case of coarse grain materials. However, it has been reported that, in the case of NC materials, hardness measurements consistently lead to higher values of the yield strength than obtained by tensile tests. Moreover, hardness measurements are very inhomogeneous within the material. Also, as can be observed in Fig. 2.8, the effect of artifacts such as porosity is far from being negligible. Indeed, one can see that powder compacts with densities lower than 99.5% exhibit hardness on the order of 30% lower than samples with higher density.

Second, let us take the example of the estimation of the grain size. The two most frequently used methods are (1) observation via TEM experiments and

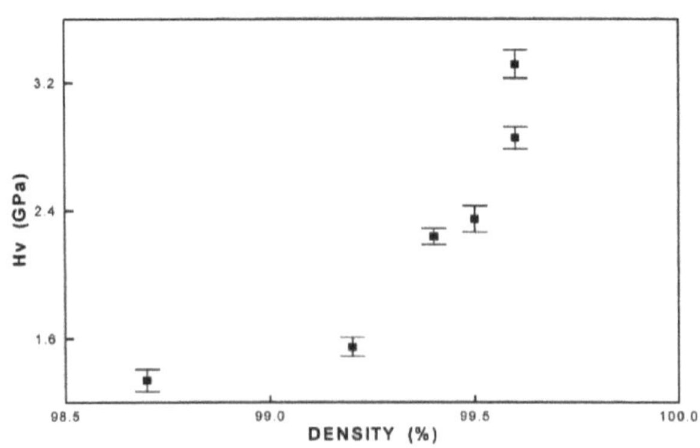

Fig. 2.8 Hardness versus density of the powder compact. Extracted from [19]

(2) XRD measurements combined with the use of the Scherrer formula. The first method consists of preparing a thin sample for observation in a transmission electron microscope. Although the sample preparation is rather delicate, ion milling is an effective method of preparation. Then, the grain size is measured on a given number of grains. Note that the number of grains observed must be sufficient for the estimated grain size to be representative of the actual mean grain size of the sample. Also, different regions of the sample must be selected because the grain size may be highly inhomogeneous within the material. Finally, the grain shape, which is certainly not ideally spherical, adds to the difficulty of mean grain size estimation. The second method consists of preparing a sample for XRD analysis and using the well-known Scherrer formula given by [20, 21]:

$$d = \frac{K\lambda}{B\cos\theta} \tag{2.2}$$

Here, K is the Scherrer constant, λ the X-ray wavelength, *B*is the integral breadth of a reflection located at 2θ. Grain size measurement from XRD and TEM observations rarely leads to the same predictions. Let us note, however, that the two measures remain in the same ballpark. However, for modeling purposes precise values are often required. Keeping in mind this word of caution, let us now present the mechanical and thermal response/properties of nanocrystalline materials.

2.2.1 *Elastic Properties*

The elastic response of a material is directly correlated to the interatomic bonds within the sample and on atomic structure/ordering. Since the volume fraction of interphase (e.g., triple junctions and grain boundaries) can increase up to 10-fold in the case of nanocrystalline materials compared to that of coarse-grain materials, and since grain boundaries exhibit a structure different from the perfect crystal lattice, it is natural to expect a size effect in the elastic response of nanocrystalline materials. Also, due to the fact that the grain boundary density is smaller than that of a perfect crystal, revealing a more *open structure*, one expects a decrease in the elastic constants of nanocrystalline materials. This can be observed in Fig. 2.9, presenting experimental measures from several different teams, of the Young's modulus of pure Cu sample as a function of grain size. Indeed, one can notice that for grain size smaller than 40 nm, corresponding to a volume fraction of interphase larger than ~10%, a decrease in Young's modulus ranging from ~6 to ~30% is exhibited by NC materials. However, let us note that some of the lower values are likely to be biased by poor consolidation.

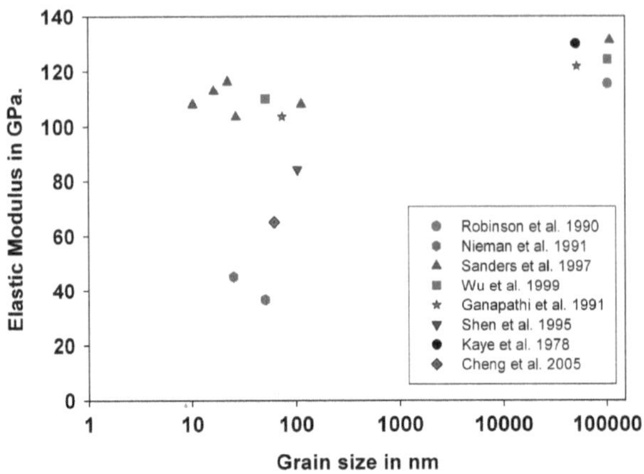

Fig. 2.9 Experimental measurements of Young's modulus as a function of grain size

2.2.1.1 Yield Stress

Coarse grain polycrystalline metals are known to exhibit a size-dependent yield stress obeying the Hall-Petch law [22, 23]. It predicts an increase in the yield stress proportional to the inverse of the square root of the grain size and is given by:

$$\sigma_y = \sigma_0 + \frac{K_{HP}}{\sqrt{d}} \tag{2.3}$$

Here, σ_0 is the friction stress, σ_y is the yield strength of the material, K_{HP} is the Hall-Petch slope, and d is the grain size. Modeling of the Hall-Petch law has been subject to intensive studies over the past decades. All models are based on the dislocation-dislocation interaction. First, models based on the pile-up of dislocations localized at the grain boundaries were developed [23]. However, body-centered cubic materials, in which dislocation pile-ups do not occur, are known to respect the Hall-Petch law. Second, J.C.M. Li proposed a model accounting for the Hall-Petch law based on the emission of dislocations by grain boundary ledges [24]. In Li's model, a dislocation emitted from a grain boundary ledge, corresponding to a step or extra atomic layer localized at the grain boundary, will interact with a dislocation forest in the vicinity of the grain boundary. The dislocation density within the forest is then related to the grain boundary misfit angle, which is itself dependent on the grain boundary ledge density. Murr and Venkatesh dedicated substantial time and effort in showing a dependence of the yield strength on the grain boundary ledge density as predicted in Li's theoretical work [25–28]. Although the ledge density affects the yield stress of the material, it was also shown that with the fabrication processes used then, the ledge density decreased with grain size. Hence, Li's theory was shown to need further refinement. Finally, models based on the strain gradient engendered by the presence of geometrically necessary dislocations were also successful in modeling the Hall-Petch law [29, 30].

The appeal of the Hall-Petch law is evident. Let us consider the case of pure copper, which typically exhibits a Hall-Petch slope of 0.11 $MPa \cdot \sqrt{m}$. Starting from a 1 μ grain size material with 180 Mpa yield stress, and decrease the grain size to. say. 50 nm, according to the Hall-Petch law, the yield stress of the fine-grained copper sample will be 561 MPa. In other words, the yield strength is multiplied by a factor of 3.

Recall that when the grain size is decreased to the nanorange, experiments on nanocrystalline samples produced by various fabrication processes have revealed that below a critical grain size, the yield stress deviates from the Hall-Petch law. Precisely, the critical grain size is $d_c \approx 25\,nm$ and below d_c the Hall-Petch slope can either decrease or even become negative.

A limited number of data are available to precisely describe the breakdown of the Hall-Petch law and, as mentioned in the beginning of this section, most available data are inconsistent due to (1) the different type of measurements methods (e.g., tensile tests, compressive tests, hardness measurements) and (2) the presence of artifacts within the samples. Indeed, due to poor particle bonding the yield stress and maximum elongation of nanocrystalline samples differs largely in compression tests and in tensile tests. Figure 2.6 presents the experimental measurements of the yield stress with the inverse of the square root of the grain size. Although the data presented in Fig. 2.10 exhibit noticeable scatter, one can clearly observe a deviation from the Hall-Petch law (represented by the dashed line). Let us note that to be consistent a measure of the size effect in the yield stress shall be performed with a single fabrication process allowing variation of the sole grain size parameter.

The breakdown of the Hall-Petch law has been subject to vigorous debate. This is easily understandable by looking at Fig. 2.10. Indeed, since most nanocrystalline samples present artifacts it is rather delicate to impede the observed breakdown of the Hall-Petch law as an intrinsic characteristic of nanocrystalline materials or as resulting from the previously mentioned defects. Moreover, thanks to a better control on the processing routes, the quality of samples has tremendously improved over the past decade and the critical grain size has continuously decreased. However, with consistent and meticulous modeling, a general agreement as to the fact/artifact breakdown of the Hall-Petch law was reached.

Currently, the general consensus on the evolution of yield stress with grain size is the following (see Fig. 2.11). In the case of polycrystalline materials with grain size ranging from several microns down to ~100 nm, the Hall-Petch law is respected. When the grain size ranges from ~100 nm down to ~25 nm a decrease in the Hall-Petch slope is expected. However, the slope is expected to remain positive. Finally, a negative Hall-Petch slope is expected when the grain size is smaller than a critical grain size that is believed to be in the neighborhood of ~10 nm. Hence, this suggests that experiments showing a breakdown of the

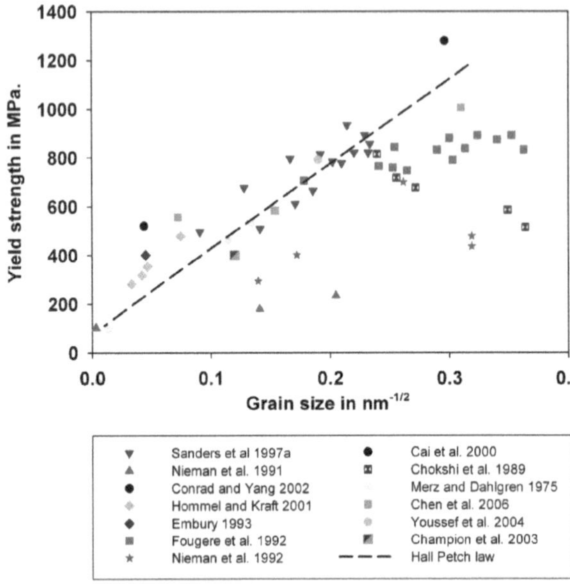

Fig. 2.10 Experimental data presenting yield stress as the function of the inverse of the square root of the grain size

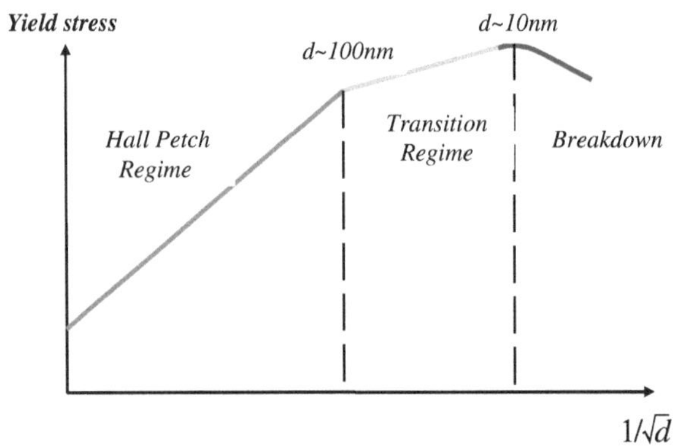

Fig. 2.11 Plot of the expected grain size dependence of yield stress for ideal samples

Hall-Petch law occurring at a critical grain size in the order of ~25 nm may be hindered by artifacts such as poor particle bonding or contamination.

2.2.2 *Inelastic Response*

2.2.2.1 Ductility

Due to poor sample quality, the first samples exhibited limited ductility with maximum elongation rarely exhibiting 2–3%, and the few samples exhibiting larger maximum deformation did not reach the expected yield strength. Hence, the capability of nanocrystalline materials to exhibit a ductile behavior was severely questioned [31]. However, with the progress in fabrication processes and particularly in consolidation of nanocrystalline powders, samples with narrow grain size distributions and bimodal distribution exhibited relatively large ductility and extremely high yield strength [31–38]. This is shown in Fig. 2.12 presenting the yield strength of Cu samples as a function of maximum elongation from various sources (date, fabrication process, and grain sizes are presented in the legend). One can easily judge the tremendous progress made over the past decade. While first samples exhibited 2–3% ductility, the most recent samples are now capable of deforming up to 50% with much higher yield strength than coarse grain materials. The latter were fabricated by ball milling in an inert gas environment and graphite plates were placed in the compression dies to ensure no sample contamination.

As shown in Fig. 2.12, the early NC samples typically exhibited limited ductility. Indeed, most samples typically exhibit a maximum elongation smaller than 5% deformation. This has been one of the most limiting factors preventing industrial applications of NC materials as structural materials. The limited ductility of these NC samples is rather abnormal in the case of coarse-grained materials; a grain refinement typically results in an enhanced ductility of the materials. Indeed, a microcrack has more chance of being stopped by a barrier – such as a grain boundary – in more refined samples. The presence of defects in the as processed samples naturally impacts the ductility of NC materials. For example, one would expect electrodeposited samples containing residues such as S and O atoms to exhibit a borderline brittle behavior. This can be seen in NC Ni samples produced by electrodeposition, which exhibit close to no plastic response prior to failure (see the TEM image presented in Fig. 2.13) [39].

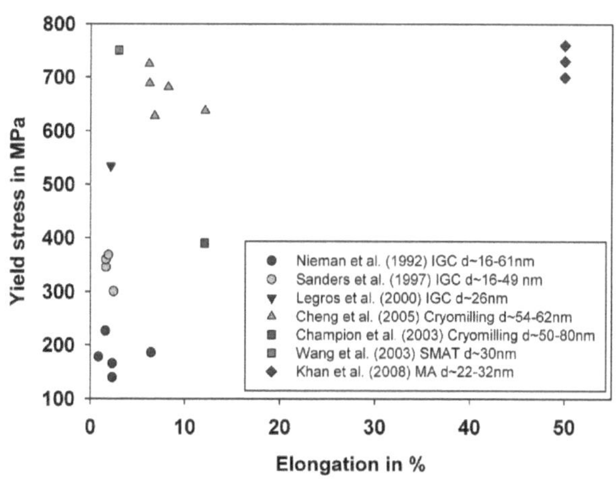

Fig. 2.12 Experimental data presenting a yield strength vs. elongation plot

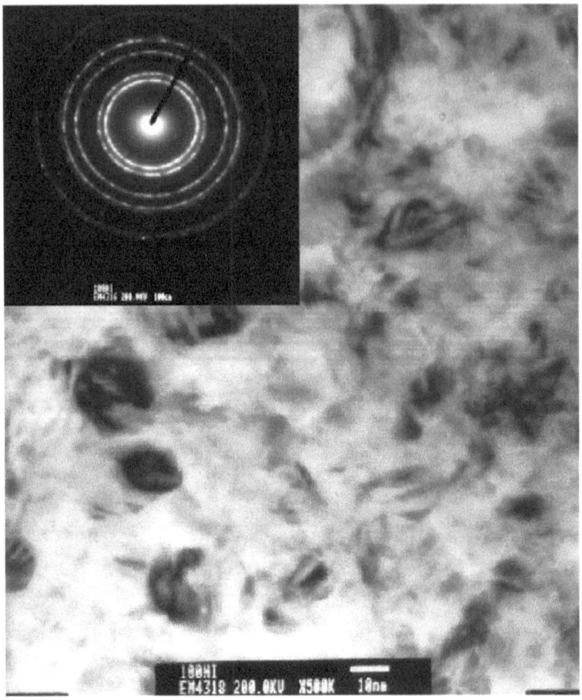

Fig. 2.13 TEM image of a NC electrodeposited Ni samples deformed in tension

Clearly, the superior ductility of the samples of Khan et al. results from the high purity resulting from meticulous sample preparation.

2.2.2.2 Flow Stress

Active plastic deformation mechanisms in NC materials are expected to differ from that of coarse-grain materials. This is due to the fact that the dislocation density, activity, and grain boundary volume fraction largely differ in these two classes of materials. Moreover, mechanisms that are not expected to be active at room temperature and in the quasi-static range are suggested to participate to the deformation of NC materials. This is the case of grain boundary sliding and deformation twinning, for example.

NC materials exhibit particular inelastic response that is often qualified as quasi or almost elastic perfect plastic. This is shown in Fig. 2.14, presenting a true stress vs. true strain curve of a NC Cu sample with 50 nm grain size and of coarse-grain Cu sample. It can clearly be seen that while the coarse grain sample exhibits significant strain hardening – engendered by dislocation activity – the NC sample exhibits a near-perfect elastic plastic response. The plastic response can be decomposed into three regions: (1) work hardening domain with

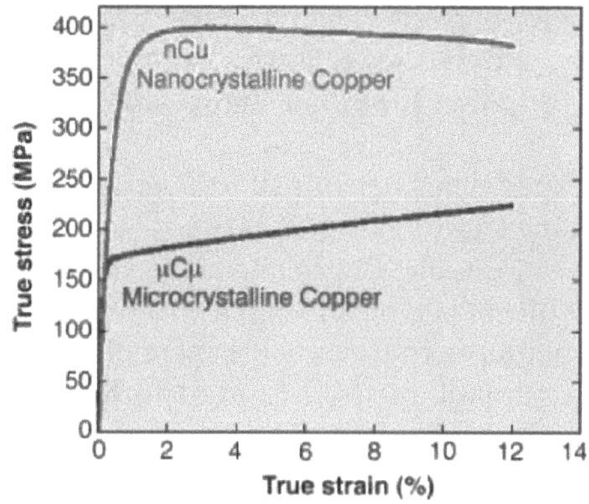

Fig. 2.14 Experimental true stress true strain curve of nanocrystalline Cu with 50 nm grain size and coarse grain Cu [34]

decreasing strain exponent towards zero, (2) plastic yielding domain at constant flow stress, and (3) plastic yielded with linear softening. This reinforces the idea that the active plastic deformation mechanisms in NC materials differ from those in coarse-grain polycrystalline materials.

2.2.2.3 Strain Rate Sensitivity

In the thermal activation regime, the behavior of metallic materials is often phenomenologically can be described with use of a power law (e.g., $\dot{\varepsilon} = \dot{\varepsilon}_0(\sigma/\sigma_{critt})^{1/m}$, the inverse of this law is also used), which is an approximation of exponential laws, accounting for the thermally activated nature of the deformation mechanisms. A well-known example is that of the description of the effect of dislocation glide [11]. The exponent *m*, used in power laws, is referred to as the strain rate sensitivity and typically considered constant during deformation in continuum models. In fact, the strain rate sensitivity parameter varies slightly during deformation (due to the change in activation volume and flow stress).

Let us recall that the strain rate sensitivity is typically used to determine active plastic deformation mechanism. For example, $m = 1$ typically corresponds to the activity of Coble creep, that is the steady state vacancy diffusion along the grain core/grain boundary interface. Similarly, $m = 0.5$ suggests the activation of grain boundary sliding. Hence, a change in hardening coefficient is an element suggesting a change in the nature of the dominant plastic deformation process. It is usually given by:

$$m = \frac{\sqrt{3}kT}{v\sigma} \tag{2.4}$$

Here, k, T, ν, and σ refer to the Boltzmann constant, the absolute temperature, the activation volume, and the uniaxial tensile stress. Note here that depending on the expression of the power law, some authors define *m* as the inverse of the present definition.

Strain rate jump experiments performed on several NC samples have clearly shown an increase in the strain rate sensitivity compared to that of their coarse grain counterparts. For example, Cheng and co-workers report a value of 0.027 for 62 nm grain Cu while *m* is typically equal to 0.006 in coarse-grain Cu [31]. Numerous experiments have confirmed the increase in strain rate sensitivity with decreasing grain size [40]. Figure 2.15 presents literature data showing the evolution of the strain rate sensitivity as a function of grain size [7, 31, 35, 40, 41]. An obvious increase in the strain rate sensitivity parameter with a decrease in grain size can be observed. It has been suggested in a relatively large number of models that the grain size dependence of the strain rate sensitivity parameter results from a decrease in the activation volume [31, 41].

2.2.2.4 Thermal Stability

Nanocrystalline materials exhibit abnormal thermal stability characterized by rapid grain growth at temperatures above a critical value (which is obviously dependent on the material considered). This issue avers to be critical for – as discussed in previous chapter dedicated to fabrication processes – the synthesis of NC materials may require temperature treatment. For example, this would be the case of a sample fabricated with a two-step process. Therefore, it is relatively difficult to retain the nanofeatures of the material during its fabrication. Moreover, the abnormal temperature stability of NC materials also impedes their use in the industry. Indeed, as the grain size of the material increases, its response will change – and more than likely deteriorate for the particular application considered.

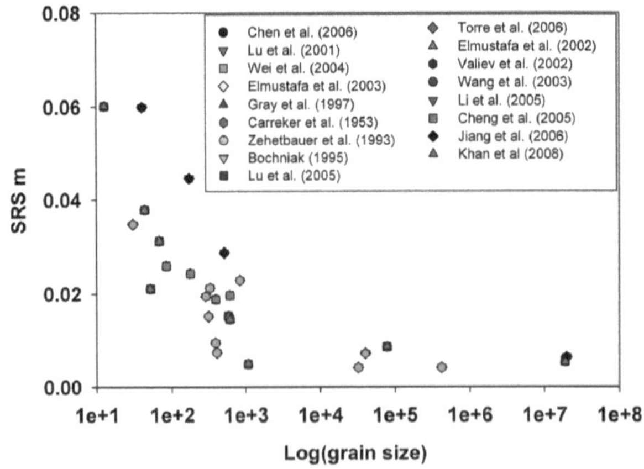

Fig. 2.15 Strain rate sensitivity parameter as a function of grain size. Extracted from [41]

Grain growth in conventional polycrystals can be either homogeneous – in the sense that the grain size distribution remains rather uniform – or not. In the former case, the evolution of grain growth with annealing time (at constant temperature) is given empirically by a parabolic law of the following form [42]:

$$d^{1/n} - d_0^{1/n} = kt \qquad (2.5)$$

Here, d and d_0 denote the instantaneous grain size and the initial grain size, respectively. $t$denotes the time and k is the temperature-dependent grain growth constant. The rate of growth exponent is typically equal to 2. However, some deviations have been observed. Also, grain growth is typically initiated at $0.5\,T/T_m(T_m$ denotes the melting temperature). Typically, the grain growth constant is related to the grain boundary mobility. For example, this is the case in Hillerts' model based on the idea – generally accepted – that the grain boundary mobility is proportional to the pressure difference resulting from its curvature [42]. As discussed in work by Lu, one would expect the thermal instability of a polycrystals – characterized by the smallest temperature at which grain growth sets off – to decrease as the grain size decreases. However, this is not necessarily the case for nanocrystalline materials which typically exhibit an higher than expected critical temperature. For example, 20 nm NC aluminum prepared by mechanical attrition exhibit a stable grain size until $0.72\,T/T_m$ [43]. Several explanations have been proposed to explain such a phenomenon. For example, the grain boundary mobility may be decreased in NC materials due to solid impurities causing drag. Generally, the following abnormal thermal effects are found to occur in NC materials:

- The starting temperature, the peak temperature and the activation energy increase with decreasing grain size.
- Discontinuous grain growth occurs at a critical temperature. At this critical temperature, the rate of grain growth increases drastically. This can be seen in annealing experiments by Song et al. [44]. Figure 2.16a presents the

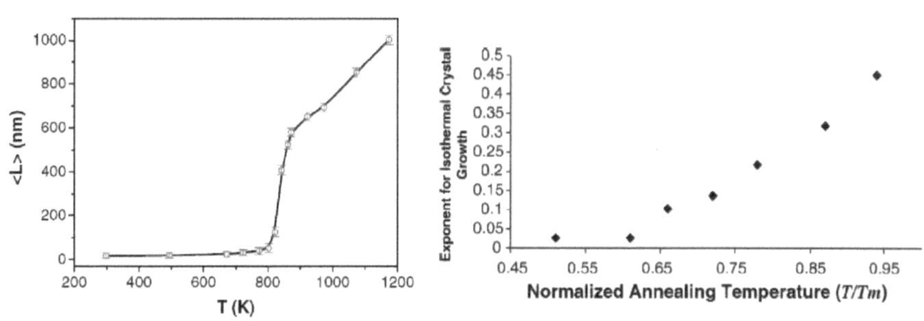

Fig. 2.16 (**a**) Evolution of mean grain size as a function of annealing temperature (pure nanocrystalline Co), extracted from [44]; (**b**) best fit growth exponent as a function of annealing temperature [43]

evolution of the mean grain size of a pure NC Co sample subjected to 1 h annealing as a function of annealing temperature.

- The grain growth exponent – chosen for each annealing temperature to obtain a best fit of the average grain size vs. annealing time curve – increases with the normalized annealing temperature to reach a value close to the typical ½ value for conventional metals [43]. This can be observed in Fig. 2.12.b presenting the evolution of the growth exponent as function of annealing temperature for NC Al samples.
- During an annealing experiment at a given constant temperature, the evolution of the average grain size as a function of time is characterized by a change in the grain growth exponent. Precisely, the growth rate decreases monotonically with time.

Several models were developed to rationalize the four points mentioned in the above. Some of the most acknowledged models are that of Fecht [45] and Wagner [46]. Both models establish thermal properties of grain boundaries based on the idea (which is yet to be shown experimentally) that grain boundaries present an excess volume compared to a perfect crystal. Recently, Song et al. [44] introduced a model combining the two approaches used by Fecht and Wagner and proposed a convincing explanation of the abnormal thermal effects in NC materials. For the sake of comprehension, the aforementioned model will be described in what follows. First, if V denotes the grain boundary atomic volume and V_0 denotes the atomic volume of a perfect crystal, the excess volume of grain boundaries can be expressed as follows:

$$\Delta V = \frac{V}{V_0} - 1 \tag{2.6}$$

This excess volume is thought to decrease with an increase in the grain size. Therefore, as the grain size is decreased the volume fraction of grain boundaries increases – this was seen previously – as well as the excess volume of grain boundaries. Assuming the thermal features of grain boundaries to be similar to that of a dilated crystal, a universal equation of state and the quasi-harmonic Debye approximation are combined to predict the evolution of the excess enthalpy, excess entropy, and excess free energy as a function of the excess volume. The quantity of interest here is the excess free energy which is predicted to evolve as shown in Fig. 2.17.

In agreement with experiments (see Fig. 2.16), it is predicted that there is a critical excess volume ΔV_c – and consequently a critical grain size – at which the discontinuous grain growth occurs. When the excess volume is larger than the critical excess volume (e.g., the grain size is smaller than a critical value), the excess free energy is smaller than the maximum value and the material is in a more stable state than at smaller excess volumes (e.g., larger grain size). The converse reasoning is also true. When the excess volume is equal to the critical

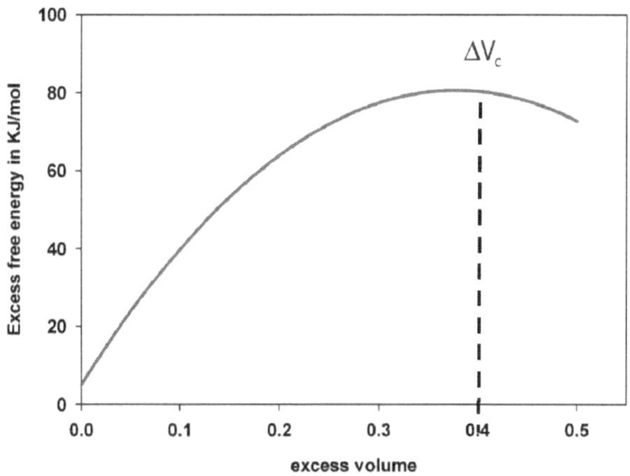

Fig. 2.17 Schematic of the evolution of the excess free energy of grain boundaries with excess free volume

value, the system is not thermally stable and thermal activation alone could destabilize the system. Therefore, one expects to observe a critical temperature at which the rate of grain growth changes abruptly.

An effective stabilization method consists of adding impurities or dopants to a pure mixture. For example, nanocrystalline Al was prepared by mechanical attrition in both a nylon and a stainless steel media. Mechanical attrition in the nylon media is clearly expected to lead to impurities within the sample. The onset of grain growth occurred at 0.72 T/T_m and 0.83 T/T_m in the stainless steel and nylon media, respectively. Indeed, the addition of dopants is expected to decrease excess free energy of grain boundaries. This was already predicted in Gibbs pioneering work where the evolution of the grain boundary energy, γ, evolves with the dopant coverage (that is the amount of dopant in the grain boundary), Γ, and its chemical potential, μ, as follows [47]:

$$d\gamma = -\Gamma \, d\mu$$

Recent molecular simulations – using the isothermal-isobaric (NPT) ensemble (see Chapter 4) – on high-angle bicrystal interfaces have shown the effect of the amount of dopant and its radius on the grain boundary energy. Such effects are shown in Fig. 2.18 [48]. It can be seen that a decrease in the dopant radius leads to a decrease in the grain boundary energy. Similarly, an increase in the dopant coverage leads to a decrease in the grain boundary excess energy. Interestingly, Fig. 2.18 suggests that there is a critical dopant coverage – such that the excess free energy of grain boundaries is null – (function of the dopant radius) which would stabilize grain boundaries.

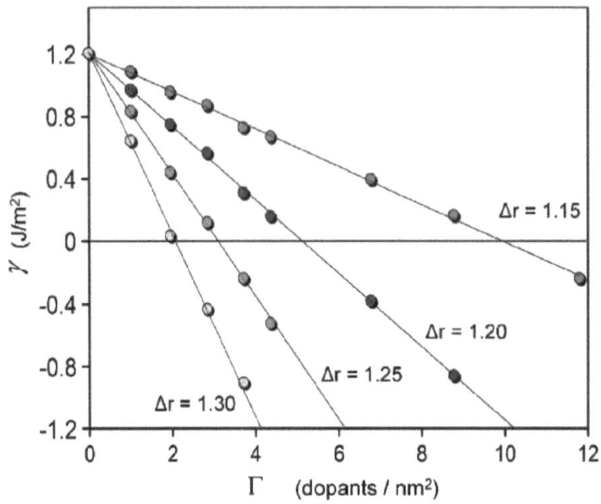

Fig. 2.18 Grain boundary energy as a function of dopant segregation for several dopant radii. Extracted from [48]

2.3 Summary

Nanocrystalline materials exhibit a particular microstructure characterized by a large volume fraction of grain boundaries and triple junctions. Nanosized grain cores retain a crystalline structure presenting lattice strains. Triple junctions present a structure devoid of regular organization while the structure of grain boundaries can exhibit changing character. Grain boundaries typically present an excess volume. Most fabrication processes lead to high large-angle grain-boundary contents.

NC materials exhibit several peculiarities. First, the evolution of yield stress with grain size does not respect the Hall-Petch law. Below a critical grain size d $\sim$20 nm the yield stress decreases with decreasing grain sizes.

Second, the quasistatic response of NC materials largely differs from that of coarse-grain materials. Indeed, the strain rate sensitivity of NC materials is higher than that of coarse grain polycrystalline materials. Also, while coarse grain materials exhibit strain hardening, NC materials exhibit a pseudo-elastic perfect plastic response.

Third, the ductility of NC materials was shown to be severely affected by the materials' purity. However, ductility can be improved by tailoring the grain size distribution. High-purity, bimodal grain size distributions, and wide distributions lead to larger elongation to failure.

Finally, the thermal response of NC materials is characterized by a regime of rapid grain growth at a critical temperature. The latter depend on the material processed. This can be prevented by adding dopants to the sample during fabrication.

References

1. Champion, Y. and M.J. Hytch, The European Journal of Applied Physics **4**, (1998)
2. Palumbo, G., S.J. Thorpe, and K.T. Aust, Scripta Metallurigica et Materialia **24**, (1990)
3. Birringer, R., Materials Science and Engineering A **117**, (1989)
4. Zhang, K., I.V. Alexandrov, and K. Lu. *The X-ray diffraction study on a nanocrystalline Cu processed by equal-channel angular pressing*. Kona, HI, USA: Elsevier, (1997)
5. Kumar, K.S., S. Suresh, M.F. Chislom, J.A. Horton, and P. Wang, Acta Materialia **51**, (2003)
6. Straub, W.M., T. Gessman, W. Sigle, F. Phillipp, A. Seeger, and H.E. Schaefer, Nanostructured Materials **6**, (1995)
7. Torre, F.D., P. Spatig, R. Schaublin, and M. Victoria, Acta Materialia **53**, (2005)
8. Ungar, T., S. Ott, P.G. Sanders, A. Borbely, and J.R. Weertman, Acta Materialia **46**, (1998)
9. Estrin, Y. and H. Mecking, Acta Metallurgica **32**, (1984)
10. Kocks, U.F., Transactions of the ASME (1976)
11. Kocks, U.F. and H. Mecking, Progress in Materials Science **48**, (2003)
12. Mecking, H. and U.F. Kocks, Acta Metallurgica **29**, (1981)
13. Huang, J.Y., X.Z. Liao, and Y.T. Zhu, Philosophical Magazine **83**, (2003)
14. Sanders, P.G., A.B. Witney, J.R. Weertman, R.Z. Valiev, and R.W. Siegel, Journal of Engineering and Applied Science **A204**, (1995)
15. Mingwei, C., M. En, K.J. Hemker, S. Hongwei, W. Yinmin, and C. Xuemei, Science **300**, (2003)
16. Markmann, J., et al., Scripta Materialia **49**, (2003)
17. Liao, X.Z., F. Zhou, E.J. Lavernia, D.W. He, and Y.T. Zhu, Applied Physics Letters **83**, (2003)
18. Ranganathan, S., R. Divakar, and V.S. Raghunathan, Scripta Materialia **27**, (2000)
19. Sun, X., R. Reglero, X. Sun, and M.J. Yacaman, Materials Chemistry and Physics **63**, (2000)
20. Patterson, A.L., Physical Review **56**, (1939)
21. Scherrer, P., Gottinger Nachrichten **2**, (1918)
22. Hall, E.O., Proceedings of the Physical Society of London **B64**, (1951)
23. Petch, N.J., Journal of Iron Steel Institute **174**, (1953)
24. Li, J.C.M., Transactions of the Metallurgical Society of AIME **227**, (1963)
25. Murr, L.E., Materials Science and Engineering **51**, (1981)
26. Murr, L.E. and E. Venkatesh, Metallography **11**, (1978)
27. Venkatesh, E.S. and L.E. Murr, Scripta Metallurgica **10**, (1976)
28. Venkatesh, E.S. and L.E. Murr, Materials Science and Engineering **33**, (1978)
29. Ashby, M.F., Philosophical Magazine **21**, (1970)
30. Cheong, K.S. and E.P. Busso, Discrete dislocation density modelling of single phase FCC polycrystal aggregates. Acta Materialia, **52(19)**, 5665–5675, (2004)
31. Cheng, S., et al., Acta Materialia **53**, (2005)
32. Yinmin, W., C. Mingwei, Z. Fenghua, and M. En, Nature **419**, (2002)
33. Youssef, K.M., R.O. Scattergood, K.L. Murty, and C.C. Koch, Applied Physics Letters **85**, (2004)
34. Champion, Y., C. Langlois, S. Guerin-Mailly, P. Langlois, J.L. Bonnentien, and M.J. Hytch, Science **300**, (2003)
35. Khan, A.S., B. Farrokh, and L. Takacs, Materials Science and Engineering: A **489**, (2008)
36. Legros, M., B.R. Elliott, M.N. Rittner, J.R. Weertman, and K.J. Hemker, Philosophical Magazine A: Physics of Condensed Matter, Structure, Defects and Mechanical Properties **80**, (2000)
37. Nieman, G.W., J.R. Weertman, and R.W. Siegel. *Mechanical behaviour of nanocrystalline Cu, Pd and Ag samples*. New Orleans, LA, USA: TMS – Miner. Metals & Amp; Mater. Soc., (1991)

38. Sanders, P.G., J.A. Eastman, and J.R. Weertman, Acta Materialia **45**, (1997)
39. Yim, T., S. Yoon, and H. Kim, Materials Science & Engineering. A, Structural materials **449–451**, (2007)
40. Chen, J., L. Lu, and K. Lu, Scripta Materialia **54**, (2006)
41. Asaro, R.J. and S. Suresh, Acta Materialia **53**, (2005)
42. Hillert, M., Acta Metallurgica **13**, (1964)
43. De Castro, C.L. and B.S. Mitchell, Materials Science and Engineering A **396**, (2005)
44. Song, X., J. Zhang, L. Li, K. Yang, and G. Liu, Acta Materialia **54**, (2006)
45. Fecht, H.J., Physical Review Letters **65**, (1990)
46. Wagner, M., Physical Review B (Condensed Matter) **45**, (1992)
47. Gibbs, J.W., *The collected works*. Green and Co, New York, (1928)
48. Millet, P.C., R.P. Selvam, and A. Saxena, Acta Materialia **55**, (2007)

Chapter 3
Bridging the Scales from the Atomistic to the Continuum

3.1 Introduction

Although some understanding seems to be emerging on the influence of grain size on the strength of nanocrystalline (NC) materials, it is not presently possible to accurately model or predict their deformation, fracture, and fatigue behavior as well as the relative tradeoffs of these responses with changes in microstructure. Even empirical models predicting deformation behavior do not exist due to lack of reliable data. Also, atomistic modeling has been of limited utility in understanding behavior over a wide range of grain sizes ranging from a few nanometers ($\sim$5 nm) to hundreds of nanometers due to inherent limitations on computation time step, leading to unrealistic applied stresses or strain rates, and scale of calculations. Moreover, the sole modeling of the microstructures is hindered by the need to characterize defect densities and understand their impact on strength and ductility. For example, nanocrystalline materials processed by ball milling of powders or extensive shear deformation (e.g., equal channel angular extrusion [ECAE]) can have high defect densities, such as voids, and considerable lattice curvature. Accordingly, NC materials are often highly metastable and are subject to coarsening. Recently, processing techniques such as electrodeposition have advanced to the point to allow the production of fully dense, homogeneous, and low defect material that can be used to measure properties reliably and reduce uncertainty in modeling associated with initial defect densities [53].

Identification of the fundamental phenomena that result in the "abnormal" mechanical behavior of NC materials is a challenging problem that requires the use of multiple approaches (e.g., molecular dynamics and micromechanics). The abnormal behavior in NC materials is characterized by a breakdown of the Hall-Petch relation [30, 57], i.e., the yield stress decreases for decreasing grain size below a critical grain diameter. Also, recent experiments [79] revealed that, in the case of face-centered cubic (FCC) NC materials, a decrease in the grain size engenders an increase in the strain rate sensitivity. Recent work by Asaro and Suresh [2] successfully modeled the size effect in the strain rate sensitivity, or alternatively in the activation volume, by considering the effect of dislocation

M. Cherkaoui, L. Capolungo, *Atomistic and Continuum Modeling of Nanocrystalline Materials*, Springer Series in Materials Science 112,
DOI 10.1007/978-0-387-46771-9_3, © Springer Science+Business Media, LLC 2009

nucleation from stress concentrations at grain boundaries. Although experimental observations and molecular dynamics (MD) simulations suggest the activity of local mechanisms (e.g., Coble creep, twinning, grain boundary dislocation emission, grain boundary sliding), it is rarely possible to directly relate their individual contributions to the macroscopic response of the material. This is primarily due to the fact that the scale and boundary conditions involved in molecular simulations are several orders of magnitude different from that in real experiment or of typical polycrystalline domains of interest. In addition, prior to predicting the global effect of local phenomena, a scale transition from the atomic scale to the mesoscopic scale must first be performed, followed by a second scale transition from the mesoscopic scale to the macroscopic scale. Micromechanical schemes have been used in previous models and have proven to be an effective way to perform the second scale transition [10, 12, 33]. However, the scale transition from the atomistic scale to the mesoscopic scale is a more critical and complex issue. The present chapter will raise the difficulties in performing systematic scale transitions between different scales, especially from atomistic to mesoscopic. The chapter will also highlight succinctly the promising methodologies that may be able to succeed at this challenging issue of bridging the scales. A few of these methodologies are developed and discussed in detail in later chapters of the book.

3.2 Viscoplastic Behavior of NC Materials

The viscoplastic behavior of NC materials has been subject to numerous investigations, most of which are focused on the role of interfaces (grain boundaries and triple junctions) and aimed at identifying the mechanisms responsible for the breakdown of the Hall-Petch relation. Within this context, the viscoplastic behavior of NC materials relies on a generic idea in which grain boundaries serve as softening structural elements providing the effective action of the deformation mechanisms in NC materials. Therefore any modeling attempts toward the viscoplastic behavior of NC materials face the problem of identification of the softening deformation mechanisms inherent in grain boundaries as well as the description of their competition with conventional lattice dislocation motion.

The nature of the softening mechanism active in grain boundaries is still subject to debate [8, 9, 41, 42, 87]. Konstantinidis and Aifantis [41] assumed that the grain boundary phase is prone to dislocation glide activities where triple junctions act as obstacles and have the properties of disclination dipoles. Tensile creep of nanograined pure Cu with an average grain of 30 nm prepared by electrodeposition technique has been investigated at low temperatures by Cai et al. [9]. The obtained creep curves include both primary and steady state stages. The steady state creep rate was found to be proportional to the effective stress. The activation energy for the creep was measured to be 0.72 eV, which is close to that of grain boundary diffusion in NC Cu. The experimental creep rates are of the same order of magnitude as those calculated from the equation for Coble creep. The existence of threshold stress implies that the grain boundaries of the nanograined Cu samples do not act as perfect sources and sinks of atoms (or vacancies). Hence, the rate of grain boundary diffusion is limited by the emission and absorption of atoms (or vacancies). The results obtained suggest that the low temperature creep of nanograined pure Cu in this study can be attributed to the interface controlled diffusional creep of Coble creep type. The creep of cold-rolled NC pure copper has been investigated in the temperature range of 20–50°C and different stresses by Cai et al. [8]. The average grain size of rolled samples was 30 nm. The author concluded that the creep behavior is attributed to grain boundary sliding accommodated by grain boundary diffusion. Coble-type creep behavior operating at room temperature was also revealed by the experimental studies of Yin et al. [87] performed on porosity-free NC nickel with 30 nm grains produced by an electrodeposition processing. Kumar et al. [42] studied the mechanisms of deformation and damage evolution in electrodeposited, fully dense, NC Ni with an average grain size of ~30 nm. Their experimental studies consist of (i) tensile tests performed in situ in the transmission electron microscope and (ii) microscopic observations made at high resolution following ex situ deformation induced by compression, rolling, and nanoindentation. The obtained results revealed that deformation is instigated by the emission of dislocations at grain boundaries whereupon voids and/or wedge cracks form along grain boundaries and triple junctions as a consequence of transgranular slip and unaccommodated grain boundary sliding. The growth of voids at separate grain boundaries results in partial relaxation of constraint, and continued deformation causes the monocrystalline ligaments separating these voids to undergo significant plastic flow that culminates in chisel-point failure.

Overall, for NC materials with grain sizes ranging from ~100 nm down to ~15 nm, theoretical models, molecular simulations, and experiments suggest three possible mechanisms governing their viscoplastic responses. The reader should refer to Chapters 5 and 6 for more details.

First, the softening behavior of NC materials may be attributed to the contribution of creep phenomena, such as Coble creep [14], accounting for the steady state vacancy diffusion along grain boundaries [36, 37, 38, 64]. This hypothesis is motivated by several experimental observations and models which revealed that creep mechanisms could operate at room temperature in the quasistatic regime [8, 9, 87]. However, more recent work has suggested that the observation of creep phenomena could be due to the presence of flaws in the initial structure of the samples, leading to non-fully dense specimens [45].

Second, both MD simulations [80] and experimental studies [35] have shown that solid motion of grains (e.g., grain boundary sliding or grain rotation) is one of the primary plastic deformation mechanisms in NC materials. For example, MD simulations on shear of bicrystal interfaces [80] showed that grain boundary sliding could be appropriately characterized as a stick-slip mechanism. Moreover, grain boundary sliding could operate simultaneously with interface dislocation emission

[42, 79]. Discussion in the literature has focused on the possible accommodation of these mechanisms by vacancy diffusion [42, 73]. However, the most recent studies tend to show that the grain boundary sliding and grain rotation mechanisms are not accommodated by vacancy diffusion. For example, ex situ TEM observations of electrodeposited nickel [42] clearly show the creation of cracks localized at grain boundaries. Recently, an interface separation criterion was introduced to predict the observed low ductility of NC materials with small grain sizes ($<\sim$50 nm) [81]. The authors indicated that a detailed description of the dislocation emission mechanism could improve their model predictions.

Third, molecular dynamics simulations on 2D columnar structures [86], 3D nanocrystalline samples [17], and planar bicrystal interfaces [65, 69, 70] suggest that interfacial dislocation emission can play a prominent role in NC material deformation [75, 86]. The grain boundary dislocation emission mechanism was first suggested by Li in order to describe the Hall-Petch relation [44]. In this model, dislocations are emitted by grain boundary ledges which act as simple dislocation donors in the sense that a ledge can emit a limited number of dislocations equal to the number of extra atomic planes associated with the height of the ledge. Once the dislocation source is exhausted, the ledge is annihilated and the interface becomes defect free. Recent work has indicated that planar interfaces (without ledges or steps) can also emit dislocations, as exhibited by models based on energy considerations [29] and atomistic simulations on bicrystal interfaces [65, 69, 70]. Moreover, MD simulations of 2D columnar and 3D nanocrystalline geometries lead to similar conclusions regarding the role of the interface on dislocation emission [75, 86]. The latter have also shown that grain boundary dislocation emission is a thermally activated mechanism, although there are differences in the definition of the criterion for emission of the trailing partial dislocation. A mesoscopic model accounting for the effect of thermally activated grain boundary dislocation emission and absorption has recently been developed and shows that the breakdown of the Hall-Petch relation could be a consequence of the absorption of dislocations emitted by grain boundaries [11]. The model also raises the question of the identification of the primary interface dislocation emission sources (e.g., perfect planar boundary, ledge).

Clearly, atomistics are most useful to characterize the structure of grain boundaries and unit processes of dislocation emission, ledge formation, absorption, and transmission. The large length and time scales of polycrystalline responses preclude application of atomistics and necessitate a strategy for bridging scales based on continuum models. However, conventional continuum crystal plasticity, whose basic concepts are discussed in Chapter 7, is inadequate for this purpose for a number of reasons, most notably in its inability to distinguish the effects of grain boundary character on interfacial sliding and dislocation nucleation/absorption processes. Grain boundaries are treated as geometric boundaries for purposes of compatibility in conventional theory. Moreover, although continuum micromechanics approaches have been developed that incorporate grain boundary surface effects that play a role in the inverse Hall-Petch behavior in nanocrystalline metals, there are problems with conventional models such as inability to factor in dislocation sources in nucleation-dominated regimes, and inability to predict appropriate concentrations of stress at grain boundary ledges and triple junctions.

Moving toward an appropriate theory of cooperative response of nanocrystalline materials requires a combination of three modeling elements: molecular statics/dynamics, continuum crystal plasticity theory, and self-consistent micromechanics. Such a theory should be able to model kinetics of dislocation nucleation and motion properly, as well as coarsening and shear banding phenomena. The latter is a challenge that requires the notion of cooperative slip localization to be introduced over many grains.

Therefore, developing a framework that can link scales of atomic level grain boundary structure with emission of dislocations, grain boundary-dislocation interactions, and grain boundary sliding processes, informing the structure of a self-consistent modeling methodology of anisotropic elastic-plastic crystals that can handle both bulk dislocation activity and grain boundary sliding induced by atomic shuffling/rearrangement or grain boundary dislocation motion, is still a challenging problem to overcome. The resulting theory should be founded on consideration of the surface area to volume ratio in polycrystals, along with accurate accounting for surface energies and activation energy estimates for various nucleation sources, which affect the change to grain boundary-mediated deformation processes at grain sizes below several hundred nanometers. Also, the effect of grain size distribution has to be considered [88]. Figure 3.1 shows the

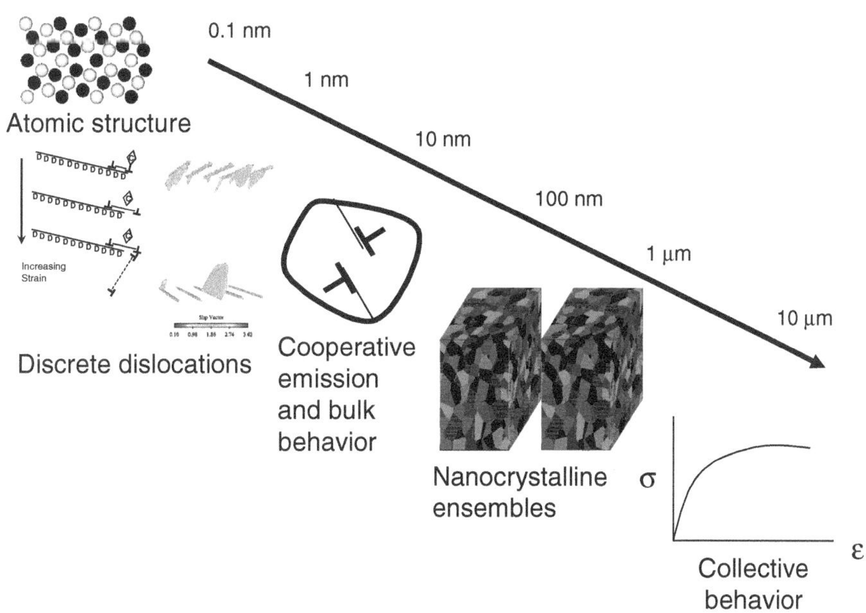

Fig. 3.1 Multiple length scales to be considered in mechanism-based self-consistent multi-scale modeling of nanocrystalline materials

scales involved in the multi-scale modeling of such kind of frameworks, ranging from the interatomic scale that characterizes grain boundary structure, region of excess energy and ledges, or triple junctions to individual grains that limit transit of dislocations to large sets of nanocrystalline grains, producing collective strength, work hardening, and ductility properties of interest.

3.3 Bridging the Scales from the Atomistic to the Continuum in NC: Challenging Problems

The link between atomic level and grain boundary structures in NC materials can be considered under the so-called field of mesomechanics, which focuses on the behavior of defects rather than that of atoms. Mesomechanics approaches are needed to complement atomistic methods and to provide information about defect interaction and the kinetics of plastic deformation. Such fundamental information can then be transferred to the continuum level to underpin the formulation of flow and evolutionary behavior of continuum-based constitutive equations. This type of multi-scale material design capability will require a few challenges to be overcome. One of the most powerful mesomechanics methods is dislocation dynamics, where considerable progress has been made during the past two decades owing to a variety of conceptual and computational developments. It has moved from a curious proposal to a full and powerful computational method. In its present stage of development, dislocation dynamics have already addressed complex problems, and quantitative predictions have been validated experimentally. Progress in three-dimensional dislocation dynamics has contributed to a better understanding of the physical origins of plastic flow and has provided tools capable of quantitatively describing experimental observations at the nanoscale and microscale, such as the properties of thin films, nanolayered structures, microelectronic components, and micromechanical elements [27].

New and efficient computational techniques for processing and visualizing the enormous amount of data generated in mesomechanical and continuum multi-scale simulations must be developed. Then, the issue of computational efficiency must be addressed so that truly large-scale simulations on thousands of processors can be effectively performed.

It should be noticed that the behavior of NC materials can be undertaken within the framework of nonlocal formulations that originally been developed to predict size effects in conventional polycrystalline materials (e.g. [2, 15, 16, 21, 22, 26, 66, 67]. These approaches will require improved and more robust numerical schemes to deal with a more physical description of dislocation interaction with themselves and with grain boundaries or other obstacles in NC materials.

The issues discussed above, in addition to the ever-increasingly powerful and sophisticated computer hardware and software available, are driving the development of multi-scale modeling approaches in NC materials. It is expected that, within the next decade, new concepts, theories, and computational tools will be developed to make truly seamless multi-scale modeling a reality. In this chapter, we briefly outline the status of research in each component that enters in building a multi-scale modeling tool to describe the viscoplastic behavior of NC materials. Two major components will be addressed in the present chapter and individually discussed in the coming chapters:

First, the chapter will discuss methodologies that rely on the ability of atomistic studies in computing structures and interfacial energies for boundaries to provide a link between the atomistic level and defects that govern the deformation mechanisms of NC materials. This will be successively discussed in the following sections:

- Computing structure and interfacial energies for boundaries
- Kinetics of dislocation nucleation and motion
- Mesoscopic simulations of nanocrystals

Second, the chapter will discuss the possible ways in incorporating the mesocopic information generated by the above studies in classical continuum micromechanics frameworks to account for grain boundary structures. This will be highlighted in the following sections:

- Thermodynamic construct for activation energy of dislocation nucleation and competition of bulk and interface dislocation structures
- Kinetics of grain boundary-bulk interactions, emission, and absorption of dislocations
- Incorporation of GB network into micromechanics scheme

3.3.1 Mesoscopic Studies

3.3.1.1 Computing Structure and Interfacial Energies of Boundaries

Computing structure and interfacial energies of boundaries is a required preliminary step to model kinetics of dislocation nucleation and motion properly In view of the focus on building multi-scale models for NC materials, avoiding for this purpose, complexities associated with impurities, substitutional atoms, or second phases, simple FCC pure metals such as Cu and Al are mainly taken as model materials to perform the atomistic studies. For both materials, embedded atom potentials (EAM) have been developed previously [47] that are appropriate for modeling dislocation nucleation and dissociation into Shockley partial dislocations associated with stacking faults. Accordingly, an algorithmic platform can be established that can serve as a useful basis for later extension to more complex alloy systems. The EAM describes the nondirectional character of bonding in Cu quite well, and hence provides more realistic consideration of grain boundary and dislocation core structures. Hence, consideration of Cu facilitates thorough and rigorous characterization of multi-scale model from the atomistic scale up. Two critical properties that must be well characterized by the interatomic potential to model dislocation nucleation and defect structures are the intrinsic and unstable stacking fault energies. For example,

Rittner and Seidman [62] showed that predicted interface structure can vary depending on the magnitude of the intrinsic stacking fault energy. Mishin et al. [47] report an intrinsic stacking fault energy of 44.4 mJ/m^2 and an unstable stacking fault energy of 158 mJ/m^2 for Cu, both of which compare favorably with experimental evidence and quantum calculations presented in their work.

Computing structure and interfacial energies of boundaries and modeling kinetics of dislocation nucleation was recently the original work of Spearot et al. [69]. Their contribution relies on a methodology that builds grain boundaries in Cu and Al bicrystals through a two-step process: (1) nonlinear conjugate gradient energy minimization using a range of initial starting positions and (2) equilibrating (annealing) to a finite temperature using Nosé-Hoover dynamics. The grain boundary energy is calculated over a defined region around the bicrystal interface after the energy minimization procedure. Figure 3.2 (a) and (b) show the grain boundary energy at 0 K as a function of misorientation angle for interfaces created by symmetric rotations around the [001] and [110] tilt axes, respectively. Grain boundary structures predicted from energy minimization calculations for several low-order coincident site lattice (CSL) interfaces in copper are shown in Fig. 3.3. Atoms shaded white are in the

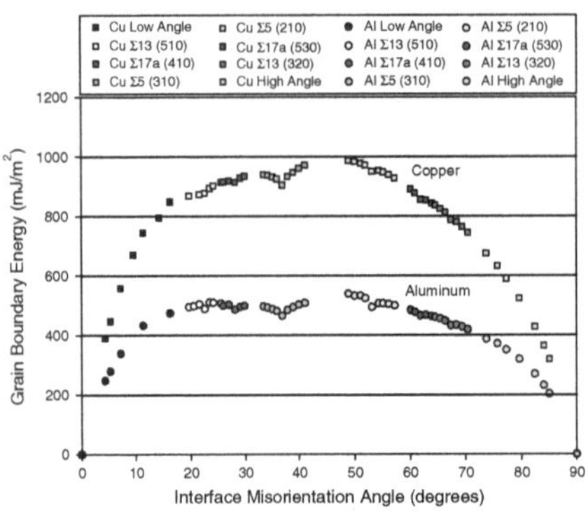

Fig. 3.2 Interface energy as a function of misorientation angle for symmetric tilt (**a**) [001] and (**b**) [110] copper and aluminum grain boundaries [69]

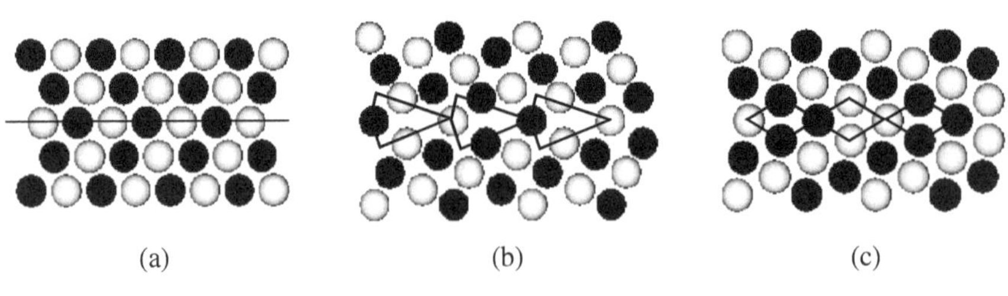

(a) (b) (c)

Fig. 3.3 Grain boundary interface structures for low-order CSL interfaces in copper: (**a**) Σ3 {111}/[110], (**b**) Σ5 {210}/[001] and (**c**) Σ11 {113}/[110] [69]

[001] plane, while atoms shaded black are in the [002] plane. The structural units for each grain boundary are outlined for clarity. The calculated structures for the Σ3 {111}/[110] and Σ11 {113}/[110] interface structures are mirror symmetric across the interface plane, while the Σ5 {210}/[001] structure shows a slight asymmetric character. The structural unit for the Σ5 {210}/[001] interface in Fig. 3.2 (b) is commonly defined as B (cf. [5]).

Figure 3.3(a) and (b) show detailed views of the planar and stepped Σ5 {210} interface structures. The viewing direction is along the [001] crystallographic axis (Z-direction) and atom positions are projected into the X-Y plane for clarity. Snapshots of the atomic configurations at the interface are taken after the isobaric-isothermal equilibration procedure at 0 bar and 10 K. The structure of each interface can be readily identified by shading atoms according to their respective {001} atomic plane, as indicated in the legend of Fig. 3.3. The planar 53.1° interface in Fig. 3.3(a) is composed entirely of B structural units, in agreement with previous atomistic simulations that employ embedded-atom method interatomic potentials [5]. It is noted that two configurations are commonly observed for this structural unit, the other being termed the B structural unit [71]. The B structural unit is identical to that shown in Fig. 3.3(a); however, an additional atom is located in the center of the "arrowhead" shaped feature. Supplementary energy minimization calculations are performed to verify that the copper Σ5 {210} boundary composed entirely of B structural units is accurate. Energy minimization calculations report an interfacial energy of 950 mJ/m^2 for the boundary composed entirely of B structural units, which is lower than all other potential configurations for this particular misorientation. Thus, the interface configuration shown in Fig. 3.3(a) is appropriate.

Figure 3.4 shows the grain boundary structure for a copper 41.1° [001] interface. The interface is comprised of structural units from both Σ5 {210}/[001] and Σ5 {310}/[001] interfaces. The Σ5 {310}/[001] structural unit is commonly defined as C, thus the 41.1° grain boundary interface has a |CCB.CCB| structure [71].

(a)

(b)

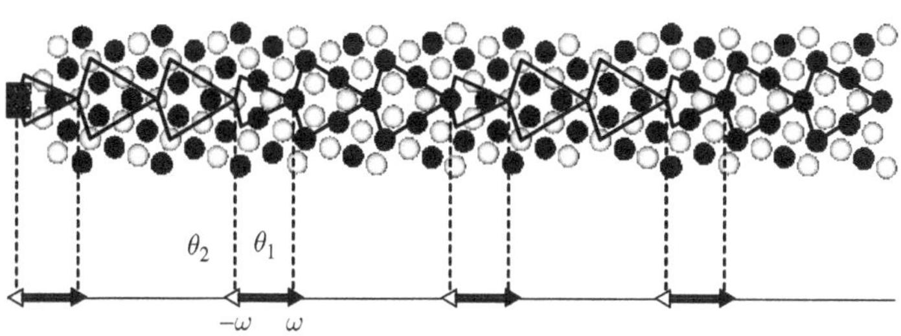

Fig. 3.4 (**a**) Grain boundary structure for a 41.1° [001] interface in copper. The interface is comprised of structural units from Σ5 (210) and Σ5 (310) boundaries. (**b**) Disclination/dislocation representation of interface [69]

3.3.1.2 Kinetics of Dislocation Nucleation and Motion

Molecular dynamics simulations are also adopted to (1) observe dislocation nucleation from the planar and stepped bicrystal interfaces, (2) compute the stress required for dislocation nucleation, and (3) estimate the change in interfacial energy associated with the nucleation of the first partial dislocation during the deformation process. The aim is to have MD simulations provide an appropriate set of values for use in the proposed continuum model for nanocrystalline deformation. Dislocation nucleation from ledges or steps along the interface plane is considered a primary cause of the initiation of plastic deformation in the model of Spearot et al. [69].

Figure 3.5 shows emission of dislocations computed from MD within a periodic unit cell for a Σ11 symmetric tilt boundary in Cu. Clearly we are interested in stresses and activation energies necessary for dislocation nucleation/emission from both planar and stepped boundaries. To compute the stress required for dislocation nucleation, both the planar and stepped interface models are subjected to a sequence of steps of increasing applied uniaxial

Fig. 3.5 Snapshots of dislocation emission during uniaxial tension of Cu for the Σ11 (113) 50.5° grain boundary model for a depth of 32.52 lattice units [69]

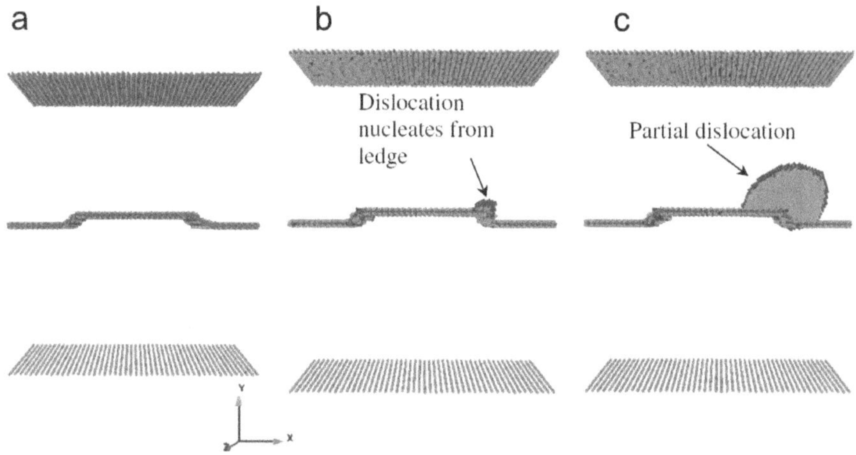

Fig. 3.6 Nucleation of a partial dislocation loop during uniaxial tension of the planar Σ5 (210) 53.1° interface model at 10 K. Atoms are colored by the centrosymmetry parameter [13]

tension perpendicular to the interface plane. A similar procedure has been used in the literature to determine the stress required for dislocation nucleation in nanocrystalline samples (cf. [86]). For example, Fig. 3.6 shows the essentially athermal (10 K) nucleation of a dislocation at a ledge in a Σ5 {210} 53.1° boundary symmetric tilt boundary [001]. Clearly, MD simulations are capable of capturing the first partial dislocation as it is nucleated from the interface. This partial dislocation is nucleated on one of the primary {111}/<112> slip systems, in agreement with that predicted using a Schmid factor analysis of the lattice orientation (cf. [32]). The core of the nucleated partial dislocation (which is shown in blue) has both edge and screw character, while the leading partial dislocation is connected back to the interface by an intrinsic stacking fault (shown in green). Nucleation of the trailing partial dislocation from the interface is not observed during the simulation time.

This is characteristic of MD simulations of dislocation nucleation in copper and has been discussed at length by Van Swygenhoven and colleagues [17, 25]. To determine the magnitude of the resolved shear stress that acts on the slip plane in the direction of the partial dislocation nucleation, the unixial state of stress is resolved onto the activated {111} plane in the <112> slip direction. This stress is calculated as 2.58 GPa. If additional tensile deformation is applied to the interface model, it is noted that dislocation nucleation will occur at other sites along the interface plane. In addition, the nucleated dislocation shown in Fig. 3.3 (c) will propagate through the periodic boundary.

Images of partial dislocation nucleation from the *stepped* interface with 53.1° misorientation are shown in Fig. 3.6. MD simulations reveal that dislocation nucleation originates from the interface ledge and occurs on one of the primary {111}<112> slip systems. The leading partial dislocation, which has both edge and screw components, is connected back to the interface via an intrinsic stacking fault. Even though the dislocation is nucleated at the interface step, the dislocation moves along the activated slip plane, eventually incorporating regions of the interface away from the ledge (as shown in Fig. 3.6 (bottom right)). To determine the magnitude of the stress that acts on the slip plane in the direction of the partial dislocation nucleation, the unixial state of stress is resolved onto the activated {111} plane in the <112> slip direction. The stress required for partial dislocation nucleation was calculated as 2.45 GPa.

3.3.1.3 Mesoscopic Simulations of Nanocrystals

There are two objectives to be met by atomistic modeling of ensembles of nanocrystals under mesoscopic simulations. The first objective is to build representative polycrystalline structures by energy minimization to determine the distribution of grain boundary character. Since deformation of nanocrystals tends towards control of interfaces, it is necessary to understand whether there is an expectation in the change of grain boundary character with mean grain size in polycrystals. We may speculate that the fraction of special boundaries will increase as average grain size decreases because the system energy becomes increasingly dependent upon minimization of the boundary energy. For example, certain CSL boundaries have been shown to have a substantially lower energy than those boundaries with non-CSL orientations [18, 63, 82]. Commensurate with a higher fraction of special boundaries would be a more faceted nature of boundaries. This will provide direct input into a continuum model in terms of the statistical distribution of dislocation sources, since each grain boundary source and mediation effect will have different activation energy barrier strength. Moreover, the activation volume depends on grain size or feature spacing (cf. [2]), and can be estimated with atomistics. The second objective is to validate the continuum micromechanics model over a relatively limited range of nanocrystalline grain size lying in the range of the transition from bulk to boundary-mediated deformation.

Mesoscopic simulations of nanocrystals can be carried out by MD methods that rely on building Voronoi tesselated 3D grain structures with appropriate grain distribution functions. A conjugate gradient energy minimization procedure is then required, followed by finite temperature equilibration. Within tessellations of microstructure and assignment of misorientation distribution, a misorientation-dependent interfacial energy penalty function may be introduced to build the initial structure prior to energy minimization, with the goal of enhancing existing algorithms that consider only facet size and no differential energies among facets in the tesselation. The use of a columnar nanocrystalline structures can be adopted for better visualization and interpretation of the mechanisms that contribute to grain growth, diffusion, and deformation processes at high temperatures with respect to the [110] CSL boundaries. After obtaining the minimum energy configuration for the nanocrystalline grains, further simulations are necessary to highlight the effect of application of and reaction to mechanical deformation on the atomic structure. This portion is of interest in answering more questions concerning the effect of length scales in nanomaterials: (1) How does dislocation nucleation occur as a function of grain size? (2) What are the specific deformation mechanisms and how do these compare with the findings of Yamakov and co-workers with respect to their columnar nanocrystalline structures? (3) How do these compare with the recent findings of Van Swygenhoven and co-workers (cf. [24, 25]) with respect to their 3D Voronoi nanocrystalline structures? (4) How do the coincident site lattice boundaries affect the nucleation of full and partial dislocations from grain boundaries as a function of grain size?

3.3.2 Continuum Micromechanics Modeling

Significant advances in multiscale modeling are essential to understand and model larger scale cooperative deformation phenomena among grains, including strengthening mechanisms and localization of shear deformation. Methods that combine molecular and continuum calculations still a challenging problem to model relevant deformation phenomena across length scales ranging from tens of nanometers to hundreds of nanometers. However, it must be emphasized that this must be done in the context of a rigorous continuum defect field theory capable of accepting quantitative information from atomistic calculations and high resolution experiments. New, specialized modeling tools must be developed since existing bulk plasticity models, including conventional crystal plasticity, are of limited use in modeling the behavior of sets of nanocrystalline grains (say, 10–100 grains) since they are too phenomenological in character to accept detailed information regarding grain boundary structure. Moreover, the use of dislocation dynamics to bridge the atomistic and continuum descriptions has its own fundamental limitations of time and length scales, not to mention the difficulty of incorporating the complex variety of dislocation nucleation mechanisms and interactions with grain boundaries that characterize nanocrystalline materials. The present chapter will discuss how molecular statics and dynamics calculations performed in the mesoscopic studies can support development of continuum models for dislocation nucleation, motion and interaction of statistical character which can then serve in the context of a micromechanics scheme as a viable alternative to explicit simulations in NC. Recent contributions that rely on the concept of combining atomistics and continuum micromechanics are developed in Chapters 8 and 9.

3.3.2.1 Thermodynamic Construct for Activation Energy of Nucleation and Competition of Bulk and Interface Dislocation Structures

As mentioned in the above, plastic deformation in NC materials results from the competitive activity of grain boundary sliding [35] and grain boundary dislocation emission [75, 76]. Recent experimental studies on physical-vapor deposited NC materials also suggest the possible accommodation of grain boundary sliding by the penetration of a dislocation, nucleated and emitted from a grain boundary, into the grain boundary opposite to the dislocation source [46].

Let us recall here that NC materials with small grain sizes on the order of ~30 nm, in which grain boundary dislocation emission is expected to be active, have been experimentally reported to be initially dislocation free (except for the dislocation structural units, or structural dislocation units, constructing the grain boundaries) [89]. The dislocation emission process is fairly complex [69, 74, 86]:

- First, a leading partial dislocation is nucleated and propagates by growth within the grain cores on favorable slip systems. As shown in work by Warner et al. [80], based on a quasi-continuum study coupling both finite elements and molecular statics, prior to the emission of the leading partial dislocation, the grain boundary can sustain significant atomic shuffling. This is the case of grain boundaries containing E structural units.
- Second, the emitted dislocation will propagate into the grain core. In the case of NC materials produced by physical vapor deposition, electrodeposition, and ball milling followed by compaction, which are to date the three only fabrication processes enabling the fabrication of fine-grained NC materials, the grain cores are defect free. Let us note that depending on the fabrication process, twins can observed within the initial structure of grain cores [42]. However, these twins can be treated as mobile grain boundaries and their presence shall consequently lead to lower mean free paths of mobile dislocations. Let us acknowledge recent molecular simulations of the interaction of screw dislocation with twin boundaries which revealed that a screw dislocation can either be absorbed in a twin boundary or cut through the twin boundary [34]. Moreover, a criterion function of the faults difference was introduced to predict the interaction of twin boundaries and dislocations. This study can be considered as a first approach in order to understand the details of the dislocation/grain boundary collision process. In all cases an emitted dislocation will propagate until it reaches either a grain boundary or a twin boundary. Since post mortem observation of NC Ni samples produced by electrodeposition have revealed solely the occasional presence of dislocation within the grain cores, the emitted dislocation must penetrate into the grain boundary [42].
- Following the penetration of the leading partial dislocation the grain boundary dislocation source can nucleate a trailing partial dislocation which will annihilate the stacking fault upon propagating within the grain core. However, in most cases and even in high-stacking fault energy materials such as Al, molecular simulations dot not predict the emission of the trailing partial dislocation [83]. Experimentally, an increase in the number of stacking faults has been measured during plastic deformation [46]. However, this increase is not pronounced enough to confirm predictions from molecular simulations. Hence, to date there is no accepted theory or model enable to rigorously define a criterion for the emission of the trailing partial dislocation. The

molecular simulations on bicrystal interface will clearly help bringing new element to the debate as discussed in the previous section of mesoscopic studies.

Simultaneously, several other issues related to the grain boundary dislocation emission process deserve special attention:

1. What are the most prominent grain boundary dislocation sources (e.g., perfect planar grain boundaries, grain boundary ledges, triple junctions)?
2. Does grain boundary sliding affect the emission of dislocations?
3. What is the macroscopic effect of dislocation emission from grain boundaries?

Clearly, the dislocation emission mechanism is much localized and a continuum model of the mechanism must take into account the local nature of the phenomenon (e.g., dependence on grain boundary misorientation angle). Since molecular simulations are the only tool able to provide the required details on the dislocation emission process, it is capital to develop a methodology capable of receiving information from molecular simulations.

MD simulations on two-dimensional columnar structures [85, 86], fully three-dimensional structures [24, 25, 85, 86] and bicrystal interfaces [68, 69, 70] have revealed the thermally activated nature of the dislocation emission process. Consequently the dislocation emission mechanism can be described at the continuum level with well accepted theories based on statistical mechanics.

Locally, the emission of a dislocation by a grain boundary source, which could either be a typical disclination unit or a grain boundary ledge [50, 51, 77], shall have two effects: (1) from the conservation of the Burger vector, it should lead to a net strain (significant or not) on the structure of the grain boundary and (2) it will create a dislocation flux from the grain boundary to the grain core. Therefore, appropriate tools are required for modeling the effect of dislocation emission on the strain within the grain boundary as well as for kinetics of boundary-bulk interactions, emission and absorption.

From statistical mechanics [4], the effect of a given process is typically written as the product of an activation rate term, accounting for the probability of success of the process and for the frequency at which the phenomenon occurs, and of second term describing the average effect of the phenomenon. Hence, in a general case the strain rate engendered by the activity of a thermally activated mechanism, noted $\dot{\gamma}$, can be written as follows [13]:

$$\dot{\gamma} = \gamma_0 \nu P \tag{3.1}$$

Here, γ_0 is the average strain engendered by the event considered, ν is the frequency of attempt, and P denotes the probability of successful emission. Adopting the thermodynamic description proposed in early work by Gibbs, the probability of success given by a Boltzmann distribution and noted P, is described in a phenomenological manner as follows [13]:

$$P = \exp\left(-\frac{\Delta G_0}{kT}\left(1 - \left(\frac{\sigma}{\sigma_c}\right)^p\right)^q\right) \tag{3.2}$$

Here, ΔG_0, σ_c, p, and q represent the free enthalpy of activation, the critical emission stress, and two parameters describing the shape of the grain boundary dislocation emission resistance diagram. Physically, the free enthalpy of activation represents the energy that must be brought to the system at a given temperature for an event, in our case a dislocation emission, to be successful. The event is said to be successful if a dislocation initially in a stable configuration reaches an unstable configuration with positive driving force.

As discussed here below, the statistical description of thermally activated mechanism appears to be well suited for receiving information directly from molecular simulations. This provides an opportunity to perform the scale transition from the atomistic scale to the scale at which continuum micromechanics can be adopted.

In recent molecular dynamics simulations on perfect planar (210)Σ5 bicrystal interface and on a bicrystal interface with same misorientation but containing a ledge, it was shown that the difference in the excess energy of the grain boundary at the initial undeformed state and at the state in which the emitted dislocation has reached an unstable configuration with positive driving force can provide a good estimate of the free enthalpy of activation [13].

The details of the calculation of the excess energy are given here below. Figure 3.7 presents a schematic of the bicrystal constructed in molecular dynamic simulation and a schematic of the energy profile. The excess energy is given by [48, 55, 84]:

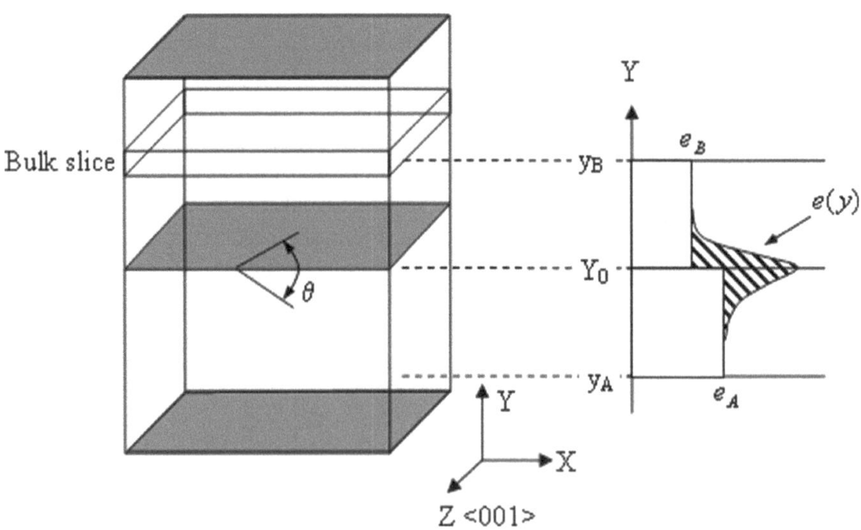

Fig. 3.7 Schematic illustration of the calculation of interface "excess" energy

$$E^{\text{int}} = \sum_{i=1}^{N_A} [e_i - e_A] + \sum_{i=1}^{N_B} [e_i - e_B] \tag{3.3}$$

Here, N_A and N_B are the number of atoms in regions A and B, respectively. The bulk energies, e_A and e_B, are determined by averaging the individual atomic energies of a "slice" of atoms positioned sufficiently far away from the interface such that the presence of the boundary is not detected (beyond y_A or y_B in Fig. 3.7). Also, as mentioned in the previous section, it was shown that the critical emission stress at zero Kelvin, denoted σ_c can be calculated from simple tensile simulation on the NPT ensemble.

The evaluation of the free enthalpy of activation and of the critical emission stress at zero Kelvin enable the estimation of the probability of successful emission presented in Fig. 3.8. The parameters p = 1 and q = 1.5 are chosen so that the dislocation emission resistance diagram has a rather abrupt profile. It is shown that for this particular geometry, the dislocation emission process is activated at very high values of the local stress in the grain boundaries, ranging from ~2450 MPa, in the case of a stepped interface, to ~2580 MPa, in the case of a perfect planar interface.

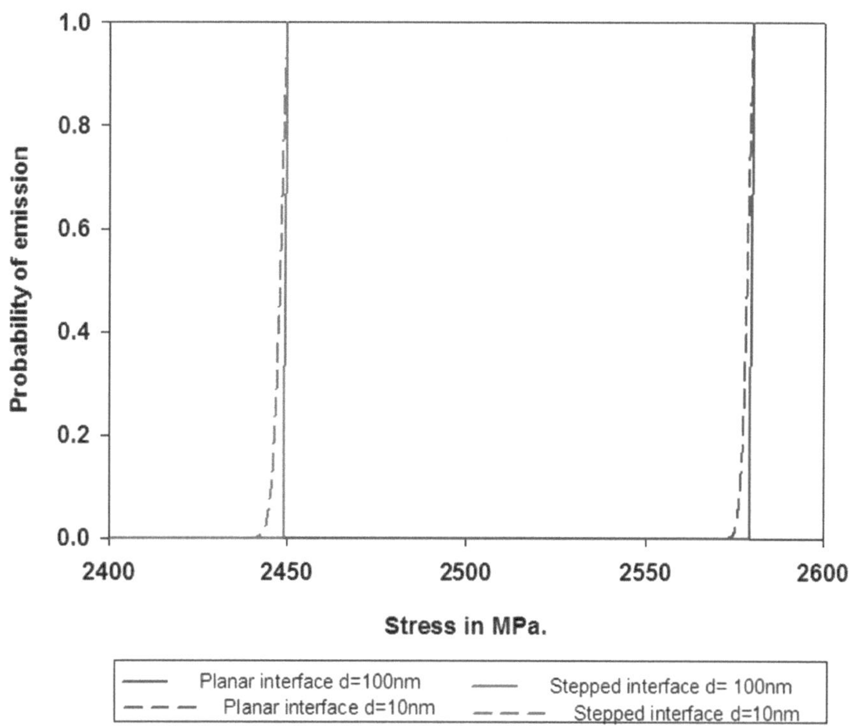

Fig. 3.8 Predicted probability of successful dislocation emission with respect to the Von Mises stress

Hence, these simulations reveal that grain boundary ledges are more prone to emit dislocation than perfect planar grain boundaries. However, let us note here that these simulations are limited to the case of a single misorientation angle and consequently only 1 out of the 5 degrees of freedom of the grain boundary is not null. Obviously, these simulations shall be extended to a wider range of grain boundary misorientations in order to draw conclusions. Also, as mentioned in the above, up to date the parameters p and q have not yet been calculated from molecular simulations.

The frequency of attempt of dislocation emission could be calculated from molecular statics simulations. However, as mentioned in discussion by Van Swygenhoven et al. [75], molecular simulations often predict the emission of a single leading partial dislocation within the grain cores of nanocrystalline materials, leaving behind a stacking fault in the material. Also, an increase in the total stacking faults of NC materials was measured experimentally [46]. However, if as predicted by molecular simulations, dislocation activity is incomplete, the amount of stacking faults measured shall be much higher. Let us recall here that the few experimental data available revealed that NC material with small grain sizes in the order of $\sim$30 nm (which are either produced by ball milling, electrodeposition, or physical vapor deposition) have an initial microstructure which is virtually dislocation free. Moreover, no conclusive experimental data have shown that grain boundary sliding, accommodated or not by diffusion mechanisms, is active in the size range. Hence, the above discussion suggests that molecular simulation cannot yet quantitatively capture the complete activity of dislocation emission.

Hence, it is proposed to develop a continuum model in order to approximate as reasonably as possible the frequency of attempt of dislocation emission. As discussed by Ashby [4], the emission frequency is bound by two extreme values ν_0, representing the dislocation bound frequency in the case of discrete obstacles, and ω_A, representing the atomic frequency. Several models were already developed to approximate the frequency of activation in the case of discrete obstacles [23, 28]. For example, Granato et al. [28] predicts a frequency in the order of $10^{11}/s$.

It is proposed here to evaluate the average strain engendered by a single dislocation emission event from continuum based reasoning. Typically grain boundaries are described with dislocation of disclination structural unit models. Let us recall here that disclinations, first introduced by Volterra [78], are linear rotational defects (see Fig. 3.9), the strength of which is given by Frank's vector, denoted in Fig. 3.9 [63].

Similarly to dislocation, which can either have a twist or an edge character, a disclination can either have a twist or a wedge character (see Fig. 3.9). Tilt boundaries are composed of a series of wedge disclinations.

The emission of a dislocation will lead to a change in the strength of the disclinations localized in the vicinity of the source which engenders a net plastic strain. Plastic deformation in the grain boundary would accordingly occur via local rotation of the two adjacent grains composing the grain boundary.

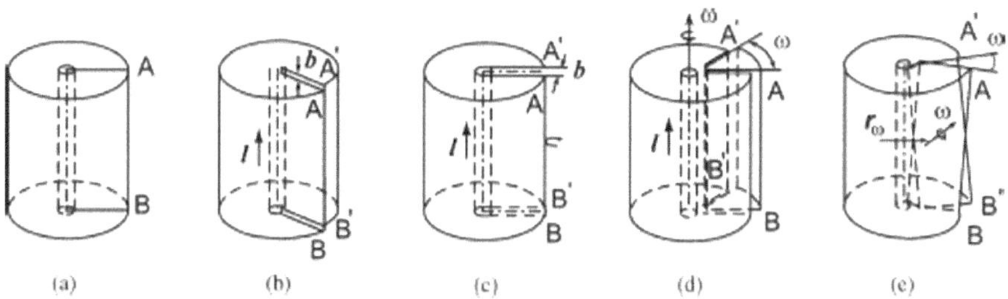

Fig. 3.9 (**a**) Perfect cylindrical volume element, (**b**) twist dislocation, (**c**) edge dislocation, (**d**) wedge disclination, and (**e**) twist disclination [63]

Theoretical work based on the disclination dipole construction of grain boundaries has already been developed and was able to discuss qualitatively the activity of grain boundary dislocation emission [29]. However, these first studies need further extensions to estimate the net strain engendered by an emission process. Also, molecular simulations on bicrystal interfaces have revealed that upon emitting a dislocation a perfect planar interface can generate a ledge [69].

An alternative approach consists of considering grain boundaries as regions of high concentrations of geometrically necessary dislocations. Indeed, grain boundaries are regions in the material presenting curvatures in the crystalline network. As described first by Nye [56] and later in Ashby's work [3], these curvatures can directly be related to the presence of dislocations, referred to as geometrically necessary. Hence, the net strain within the grain boundary engendered by the emission of a single dislocation could also be evaluated by investigating the effect of a decrease in the GND density on the curvature of the crystalline network. However, let us note that this approach would be more suited for the description of low angle grain boundaries in which dislocation cores can be identified.

3.3.2.2 Kinetics of Boundary-Bulk Interactions, Emission, and Absorption

As mentioned above, the dislocation emission process leads to a dislocation flux from the grain boundary region into the grain core. Also the converse, which corresponds to the penetration of a dislocation present within the grain core into the grain boundary, is strongly expected to occur. Moreover, it is of primary importance to characterize at the continuum level the effect of the presence of stacking faults on the emission and propagation of the trailing partial dislocation. Let us recall that these stacking faults are induced by the propagation of the leading partial dislocation within grain cores.

Fortunately, the initial dislocation density within the grain cores of NC materials is extremely low. Hence, dislocation networks interactions do not appear as being of primary importance. Consequently, typical strain hardening theories [39, 40, 52] based on the simultaneous activity of athermal dislocation storage, engendering a decrease in the mean free path of dislocations, and on the thermally activated dislocation annihilation mechanism are not suited in the case of NC materials.

The kinetics of deformation cannot be appropriately described without rigorous models describing the coupling of dislocation emission from the grain boundaries, dislocation penetration within the boundaries, dislocation glide within the grain cores, and grain boundary sliding. Three key aspects shall be considered:

- Dislocation stability within grain cores
- The effect of stacking faults on the emission of the trailing partial dislocation
- The effect of dislocation penetration on the deformation of grain boundaries

Supposing an initial microstructure with grain cores devoid of dislocations, which is particularly the case in NC materials produced via electrodeposition, once a dislocation is nucleated and propagates within the grain cores it can either be absorbed within the grain boundary opposite to the dislocation source or be stored within the grains. The latter is less likely to happen. Establishing a stability criterion for an emitted dislocation will directly let us evaluate the probability of dislocation absorption. Clearly, from the conservation of Burger's vector, the penetration of an emitted dislocation will lead to plastic deformation within the grain boundary. Simultaneously, the propagation of the leading partial dislocation leaves a stacking fault within the grain core which could have two effects: (1) increasing the resistance to dislocation glide within the grain cores and (2) impeding or facilitating the nucleation and emission of the trailing partial dislocation.

Qin et al. [58, 59] proposed a model for the stability of dislocation within grain cores. The proposed reasoning is fairly simple and based on the stress fields localized in the grain boundary area, and engendered by the local lattice expansion present at grain boundaries. The lattice expansion was measured experimentally on samples produced with various processes [72]. At equilibrium the stresses applied by the grain boundaries are equal to Peierls stresses. It is then shown (see Fig. 3.10) that a decrease in the grain size leads to a decrease in the surface area in

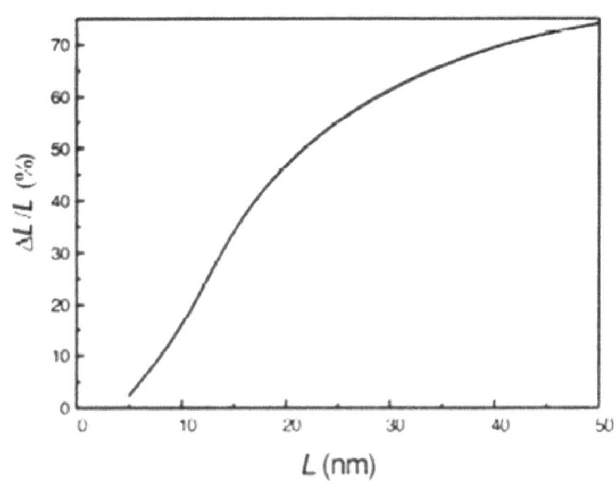

Fig. 3.10 Ratio of the length of stability of a dislocation over the grain size, denoted L, with respect to the grain size [60]

which the dislocation can be stable [60]. Note here that in the work of Qin et al., the elastic modulus of the grain boundaries is dependent on the excess volume within the grain boundaries, and decreases with the grain size [58, 59].

Recently, Asaro and Suresh [2] developed a model to predict the transition from typical dislocation glide dominated plasticity to grain boundary dislocation emission plasticity occurring in NC materials. By minimizing the energy of an extended dislocation, accounting for the energies of the two partial dislocations, of their interaction and of the stacking fault, the authors derive the equilibrium distance between the two partial dislocations and the yield stress within the grain cores. While this study is the first of the kind to establish a criterion for the emission of full dislocation from grain boundaries, it does not account for the thermally activated nature of the grain boundary dislocation emission process.

The previously described model, coupled with the MD simulations, shall facilitate the modeling of emission criterion for the trailing partial dislocation. Alternatively, Van Swygenhoven et al. [75] have shown via MD simulations that the ratio of the stable stacking fault energy over the unstable stacking fault energy has an influence on the emission of the trailing partial dislocation.

Finally, following molecular dynamics simulations focusing on the dislocation penetration process, a model will be developed at the continuum level to quantify the net strain resulting from dislocation penetration events. The model will be based on the disclination structural unit description of grain boundaries. Indeed, from the conservation of Burger's vector, the dislocation penetration mechanism, will directly lead to an increase in the strength of the wedge disclination. However, a priori and without MD simulations, it is impossible to assess of the details of the penetration process.

3.3.2.3 Incorporation of Grain Boundary Network into Self-Consistent Scheme

From the characterization of the grain boundary dislocation emission mechanism, of the stability of dislocation within grain cores (driving the penetration of an emitted dislocations) and of the effect of stacking faults on the emission of trailing dislocations, constitutive laws describing the behavior of both grain cores and grain boundaries can be established. In order to develop a model capable of predicting the behavior of NC materials and able to receive information on the microstructure, three issues must be addressed:

1. How to perform the scale transition from the mesoscopic scale to the macroscopic scale?
2. How to introduce the grain boundary geometry within the continuum model?
3. How to account for the effect of grain boundary sliding?

Finite elements and micromechanics are the two possible ways to perform the scale transition. Although finite elements can reveal higher level of details than traditional micromechanics (e.g., nonhomogeneous stresses and strain fields within the grain cores and grain boundaries), its use is rather costly in terms of computation time. Moreover, it is fairly complex, if not impossible, to recreate an exact microstructure with the same statistical distribution of the grain boundaries as observed experimentally. Hence, recent micromechanical models, which ineluctably account for the statistical description of the microstructure, that have proven to be effective in the case of modeling of NC materials [10, 11, 33] can be of great interest. As discussed in the following section, the selected micromechanical scheme can be extended to account for the peculiarity of the geometry of grains and grain boundaries.

The micromechanical approach is based on a composite description of the material which is typically represented as a two phase material composed of (1) an inclusion phase representing grain cores and (2) a matrix phase representing grain boundaries and triple junctions. Also three-phase models have recently been used to predict the quasi-static purely viscoplastic response of NC materials [6], where a coated inclusion is embedded in an effective homogeneous material, and the coating represents both grain cores and triple junctions while the inclusion represents grain cores. Three-phase models are well suited to describe materials in which diffusion mechanisms, such as Coble creep, and sliding of phases, such as grain boundary sliding, are activated.

The extension of Kroner's method to the case of inhomogeneous elastic-viscoplastic materials was used in past studies to predict the effect of the activity of Coble creep on the breakdown of the Hall-Petch law [10]. In this approach, the elastic response of the material is decomposed as the sum of the contribution of a spatially independent term and a fluctuation term. Similarly, the same decomposition is performed for the viscoplastic response. This scheme has the benefit of being fairly simple in its implementation but does lead to stiffer responses than the secant elastic-viscoplastic scheme used by Berbenni et al. [7], which accounts for the spatial and time coupling of the solution fields.

The micromechanical scheme developed by Berbenni et al. [7] was also used to estimate the effect of the combined effect of grain boundary dislocation emission and penetration [11, 13]. The following method is used to homogenize the behavior of the NC material.

First, The elastic moduli are decomposed into a uniform part and a fluctuating part. In order to ensure the compatibility and equilibrium in the representative volume element (RVE), Kunin's projection operators [43] are used to transform the fields on the space of possible solutions. The self-consistent approximation is applied to the projected equations. In self-consistent schemes, the properties of the homogeneous equivalent medium are obtained by imposing that the spatial average of the nonlocal contributions is equal to zero. At this stage, the system cannot be solved because the viscoplastic strain field is still to be determined. The problem is solved by translating the local viscoplastic strain rate about a non-necessarily uniform but compatible strain rate which is chosen to be the self-consistent solution for a polycrystalline material displaying a purely viscoplastic behavior.

Second, the global behavior is obtained by performing the homogenization step, which consists of averaging the local fields over the volume and setting the averaged fields equal to the macroscopic fields. However, let us note that this scheme does not account for possible strain or stress jumps at the interface that occur during sliding of grains. Also, the previously mentioned models are based on Eshelby's [19] solution to the inclusion problem in which inclusions are supposed ellipsoidal which leads to homogeneous stress and strain states.

As discussed in Chapter 7, micromechanical schemes are based on Eshelby's solution to the inclusion problem, which is obtained via the use of Green's functions [19, 20, 49], and inclusions are assumed, for simplicity, to be ellipsoidal. This assumption leads to a homogeneous solution of the inclusion problem which induces the homogeneity of the localization tensors. Hence with traditional micromechanical approaches the predicted stress and strain fields in all phases are homogeneous. Typically a higher level of refinement is not required to obtain acceptable predictions of the global behavior of the material. However, previous work as shown that (1) dislocation emission necessitates high values of stresses which cannot be predicted with Eshelbian schemes [11] and (2) triple junctions are regions of high stress concentrations [6]. New solutions to the inclusion problems are necessary to consider other grain shapes and to account for the effect of grain boundary ledges.

References

1. Aifantis, E.C., The physics of plastic deformation. International Journal of Plasticity **3**, 211–247, (1987)
2. Asaro, R.J. and S. Suresh, Mechanistic models for the activation volume and rate sensitivity in metals with nanocrystalline grains and nano-scale twins. Acta Materialia **53**(12), 3369–3382, (2005)
3. Ashby, M.F., The deformation of plastically non homogeneous materials. Philosophical Magazine **21**, 399–424, (1970)
4. Ashby, M.F., A first report on deformation-mechanism maps. Acta Metallurgica (pre 1990), **20**, 887, (1972)
5. Bachurin, D.V., R.T. Murzaev, et al., Atomistic computer and disclination simula-tion of [001] tilt boundaries in nickel and copper. Fizika Metallov i Metallovedenie **96**(6), 11–17, (2003)
6. Benkassem, S., L. Capolungo, M. Cherkaoui, Mechanical properties and multi-scale modeling of nanocrystalline materials. Acta Materialia **55**(10), 3563–3572, (2007)
7. Berbenni, S., Favier, V. et al., Micromechanical modeling of the elastic-viscoplastic behavior of polycrystalline steels having different microstructures. Materials Science and Engineering A, **372**(1–2), 128–136, (2004)
8. Cai, B., Q.P. Kong, et al., Creep behav-ior of cold-rolled nanocrystalline pure cop-per. Scripta Materialia **45**(12), 1407–1413, (2001)
9. Cai, B., Q.P. Kong, et al., Low tempera-ture creep of nanocrystalline pure copper. Materials Science and Engineering A **286**(1), 188–192, (2000)
10. Capolungo, L., M. Cherkaoui, et al., A self consistent model for the inelastic de-formation of nanocrystalline materials. Journal of engineering materials and tech-nology **127**, 400–407, (2005)
11. Capolungo, L., M. Cherkaoui, et al., On the elastic-viscoplastic behavior of nanocrystalline materials. International Journal of Plasticity **23**(4), 561–591, (2007)

12. Capolungo, L., C. Jochum, et al., Ho-mogenization method for strength and ine-lastic behavior of nanocrystalline materials. International Journal of Plasticity **21,** 67–82, (2005)
13. Capolungo, L., D.E. Spearot, et al., Dis-location nucleation from bicrystal interfaces and grain boundary ledges: Relationship to nanocrystalline deformation. Journal of the Mechanics and Physics of Solids, **55**(11), November, 2007, 2300–2327, (2007)
14. Coble, R.L., A Model for Boundary Diffusion Controlled Creep in Polycrystal-line Materials. Journal of Applied Physics **34**(6): 1679–1682, (1963)
15. Dai, H. and D.M. Parks, Geometrically-necessary dislocation density and scale-dependent crystal plasticity. In A. S. Khan (ed.), *Proceedings of Plasticity '97: The Fifth International Sym-posium on Plasticity and its Current Applications*, 17–18, Juneau, Alaska. Neat Press, (1997)
16. Dai, H., *Geometrically-necessary dislocation density in continuum plasticity theory, FEM implementation and applications*. Ph.D. thesis, Massachusetts Institute of Technology, Department of Mechanical En-gineering, (1997)
17. Derlet, P.M. and H. Van Swygenhoven, Length scale effects in the simulation of deformation properties of nanocrystalline met-als. Scripta Materialia **47**(11), 719–724, (2002)
18. Douglas, E.S., I.J. Karl, and D.L. McDowell, Nucleation of dislocations from [0 0 1] bicrystal interfaces in aluminum. Acta Materialia **53**(13), 3579–3589, (2005)
19. Eshelby, J.D., The determination of an ellispoidal inclusion and related problems. Proceedings of the Royal Society of London **A241,** 376–396, (1957)
20. Eshelby, J.D. *Elastic inclusions and inhomogeneities*. North Hol-land, (1961)
21. Fleck, N.A., G.M. Muller, M.F. Ashby, and J.W. Hutchinson, Strain gradient plastic-ity: theory and experiment. Acta metallur-gica et Materialia **42,** 475–487, (1994)
22. Fleck, N.A. and J.W. Hutchinson, Strain gradient plasticity. Advances in Ap-plied Mechanics, **33**, 295–361, (1997)
23. Friedel, J., Physics of Strength and Plasticity, MIT Press, Boston, (1969)
24. Froseth, A., H. Van Swygenhoven, et al., The influence of twins on the mechani-cal properties of nc-Al. Acta Materialia **52,** 2259–2268, (2004)
25. Froseth, A.G., P.M. Derlet, et al., Dis-locations emitted from nanocrystalline grain boundaries: Nucleation and splitting dis-tance. Acta Materialia **52**(20), 5863–5870, (2004)
26. Gao, H., Y. Huang, W.D. Nix, and J.W. Hutchinson, Mechanism-based strain gradi-ent plasticity–I. Theory. Journal of the Mechanics and Physics of Solids **47**, 1239–1263, (1999)
27. Ghoniem, N.M., E. Busso, et al., Mul-tiscale modelling of nanomechanics and micromechanics: an overview. Philosophical Magazine **83**, 3475–3528, (2003)
28. Granato, A.V., K. Lucke, et al., Journal of Applied Physics **35,** 2732, (1964)
29. Gutkin, M.Y., I.A. Ovid'Ko, et al., Transformation of grain boundaries due to disclination motion and emission of dislocations pairs. Materials Science and Engineering **A339,** 73–80, (2003)
30. Hall, E.O., The deformation and aging of mild steel. Proceedings of the Physical Society of London **B64**, 747, (1951)
31. Hoover, W.G. (Ed.), Proceedings of the the international school of physics - Enrico Fermi- Molecular Dynamics Simulations of Statistical Mechanical Systems. North Holland, Amsterdam, (1985)
32. Hosford, W.F., *The Mechanics of Crystals and Textured Polycrys-tals*. New York, Oxford University Press, (1993)
33. Jiang, B. and G.J. Weng, A generalized self consistent polycrystal model for the yield strength of nanocrystalline materials. Journal of the Mechanics and Physics of Solids, **52,** 1125–1149, (2004)
34. Jin, Z.H., P. Gumbsch, et al., The inter-action mechanism of screw dislocations with coherent twin boundaries in different face-centred cubic metals. Scripta Materialia **54**(6), 1163–1168, (2006)

35. Ke, M., S.A. Hackney, et al., Observa-tions and measurement of grain rotation and plastic strain in nanostructured metal thin films. Nanostructured Materials **5**, 689–697, (1995)
36. Kim, B.-N., K. Hiraga, et al., Viscous grain-boundary sliding and grain rotation accommodated by grain-boundary diffusion. Acta Materialia **53**(6), 1791–1798, (2005)
37. Kim, H.S. and Y. Estrin, Phase mixture modeling of the strain rate dependent mechanical behavior of nanostructured materi-als. Acta Materialia **53**, 765–772, (2005)
38. Kim, H.S., Y. Estrin, et al., Plastic de-formation behaviour of fine grained materi-als. Acta Materialia **48**, 493–504, (2000)
39. Kocks, U.F., Laws for work hardening and low temperature creep. Journal of Engineering Materials and Technology Transactions of ASME **98**(1), 76–85, (1976)
40. Kocks, U.F. and H. Mecking, Physics and phenomenology of strain hardening. Progress in Materials Science **48**, 171–273, (2003)
41. Konstantinidis, D.A. and E.C. Aifantis, On the "anomalous" hard-ness of nanocrystalline materials. Nanos-tructured Materials **10**, 1111–1118, (1998)
42. Kumar, K.S., S. Suresh, et al., Defor-mation of electrodeposited nanocrystalline nickel. Acta Materialia **51**, 387–405, (2003)
43. Kunin, I.A., *Elastic media with microstructure II: Three dimensional mod-els*. Springer-Verlag, Berlin, (1983)
44. Li, J.C.M., Petch relation and grain boundary sources. Transactions of the Met-allurgical Society of AIME **227**, 239, (1963)
45. Li, Y.J., W. Blum, et al., Does nanocrystalline Cu deform by Coble creep near room temperature? Materials Science and Engineering A **387–389**, 585–589, (2004)
46. Markmann, J., P. Bunzel, et al., Micro-structure evolution during rolling of inert-gas condensed palladium. Scripta Materialia **49**(7), 637–644, (2003)
47. Mishin, Y., D. Farkas, et al., Intera-tomic potentials for monoatomic metals from experimental data and ab initio calcula-tions. Physical Review B **59**, 3393–3407, (1999)
48. Muller, P. and A. Saul, Elastic effects on surface physics. Surface Science Reports **54**(5–8), 157, (2004)
49. Mura, T., *Micromechanics of defects in solids*. Kluwer Academic Publisher, Dordrecht, (1993)
50. Murr, L.E., Strain induced dislocation emission from grain boundaries in stainless steel. Materials Science and Engineering **51**, 71–79, (1981)
51. Murr, L.E. and E. Venkatesh, Contrast phenomena and identification of grain boundary ledges. Metallography **11**, 61–79, (1978)
52. Nes, E., Modelling of work hardening and stress saturation in FCC metals. Pro-gress in Materials Science **41**, 129–193, (1997)
53. Nieh, T.G. and J.G. Wang, Hall Petch relationship in nanocrystalline Ni and Be-B alloys. Intermetallics **13**, 377–385, (2005)
54. Nose, S., A molecular dynamics method for simulations in the canonical ensemble. Molecular physics **52**, 255–268, (1984)
55. Nozieres, P. and D.E. Wolf, Interfacial properties of elastically strained materials. I. Thermodynamics of a planar interface. Zeitschrift fur Physik B (Condensed Matter) **70**(3), 399, (1988)
56. Nye, J.F., Some geometric relations in dislocated crystals. Acta Metallurgica **1**: 153–162, (1953)
57. Petch, N.J., The cleavage strength of polycrystals. Journal of Iron Steel Institute, **174**, 25–28, (1953)
58. Qin, W., Z. Chen, et al., Dislocation pileups in nanocrystalline materials. Journal of Alloys and Compounds **289**, 285–288, (1999)
59. Qin, W., Z.H. Chen, et al., Crystal lat-tice expansion of nanocrystalline materials. Journal of Alloys and Compounds **292**, 230–232, (1999)
60. Qin, W., Y.W. Du, et al. Dislocation stability and configuration in the crystallites of nanocrystalline materials. Journal of Al-loys and Compounds **337**, 168–171, (2002)

61. Qu, J., The effect of slightly weakened interfaces on the overall elastic properties of composite materials. Mechanics of Materials **14**(4), 269–281, (1993)
62. Rittner, J.D. and D.N. Seidman, <110> symmetric tilt grain-boundary structures in FCC metals with low stacking-fault ener-gies. Physical Review B: Condensed Matter **54**(10), 6999, (1996)
63. Romanov, A.E., Mechanics and physics of disclinations in solids. European Journal of Mechanics – A/Solids **22**(5), 727–741, (2003)
64. Sanders, P.G., M. Rittner, et al., Creep of nanocrystalline Cu, Pd, and Al-Zr. Nanostructured Materials **9**(1–8), 433–440, (1997)
65. Sansoz, F. and J.F. Molinari, Mechani-cal behavior of Sigma tilt grain boundaries in nanoscale Cu and Al: a quasicontinuum study. Acta Materialia **53**, 1931–1944, (2005)
66. Shi, M.X., Y. Huang, and K.C. Hwang, Plastic flow localization in mechanism-based strain gradient plasticity. International Journal of Mechanical Sciences **42**, 2115–2131, (2000)
67. Shu, J.Y. and N.A. Fleck, Strain gradi-ent crystal plasticity: size-dependent defor-mation of bicrystals. Journal of the Mechan-ics and Physics of Solids **47**, 297–324, (1999)
68. Spearot, D.E., Atomistic Calculations of Nanoscale Interface Behavior in FCC metals. Woodruff School of Mechanical En-gineering. Georgia Institute of Technology, Atlanta, 276, (2005)
69. Spearot, D.E., K.I. Jacob, et al., Nuclea-tion of dislocations from [001] bicrystal interfaces in aluminum. Acta Materialia **53**, 3579–3589, (2005)
70. Spearot, D.E., K.I. Jacob, et al., Dislo-cation nucleation from bicrystal interfaces with dissociated structure. International Journal of Plasticity **23**(1), 143–160, (2007)
71. Sutton, A.P. and V. Vitek, On the struc-ture of tilt grain boundaries in cubic metals. I. Symmetrical tilt boundaries. Philosophical Transactions of the Royal Society of London A **309**(1506), 1–36, (1983)
72. Van Petegem, S., F. Dalla Torre, et al., Free volume in nanostructured Ni. Scripta Materialia **48**, 17–22, (2003)
73. Van Swygenhoven, H. and A. Caro, Plastic behavior of nanophase Ni: a molecu-lar dynamics computer simulation. Applied Physics Letters **71**(12), 1652, (1997)
74. Van Swygenhoven, H., A. Caro, et al., Grain boundary structure and its influence on plastic deformation of polycrystalline FCC metals at the nanoscale: a molecular dynamics study. Scripta Materialia **44**, 1513–1516, (2001)
75. Van Swygenhoven, H., P.M. Derlet, et al., Stacking fault energies and slip in nanocrystalline metals. Nature Materials **3**, 399–403, (2004)
76. Van Swygenhoven, H., M. Spaczer, et al., Microscopic descritpion of plasticity in computer generated metallic nanophase samples: a comparison betwwen Cu and Ni. Acta Metallurgica **47**, 3117–3126, (1999)
77. Venkatesh, E.S. and L.E. Murr, The in-fluence of grain boundary ledge density on the flow stress in Nickel. Materials Science and Engineering **33**, 69–80, (1978)
78. Volterra, V., Ann. Ecole Normale Supérieure de Paris **24**, 401, (1907)
79. Wang, Y.M., A.V. Hamza, et al., Acti-vation volume and density of mobile dislo-cations in plastically deforming nanocrystal-line Ni. Applied Physics Letters **86**(24), 241917, (2005)
80. Warner, D.H., F. Sansoz, et al., Atom-istic based continuum investigation of plas-tic deformation in nanocrystalline copper. International Journal of Plasticity **22**(4), 754, (2006)
81. Wei, Y.J. and L. Anand, Grain-boundary sliding and separation in polycrys-talline metals: application to nanocrystalline fcc metals. Journal of the Mechanics and Physics of Solids **52**(11), 2587, (2004)
82. Wolf, D., Structure-energy correlation for grain boundaries in F.C.C. metals-III. Symmetrical tilt boundaries. Acta Metallurgica et Materialia **338**(5), 781–790, (1990)
83. Wolf, D., V. Yamakov, et al., Deforma-tion of nanocrystalline materials by molecu-lar dynamics simulation: relationship to ex-periments? Acta Materialia **53**, 1–40, (2005)

84. Wolf, D.E. and P. Nozieres, Interfacial properties of elastically strained materials. II. Mechanical and melting equilibrium of a curved interface. Zeitschrift fur Physik B (Condensed Matter) **70**(4), 507, (1988)
85. Yamakov, V., D. Wolf, et al., Deforma-tion twinning in nanocrystalline Al by mo-lecular dynamics simulation. Acta Materi-alia **50**, 5005–5020, (2002)
86. Yamakov, V., D. Wolf, et al., Length-scale effects in the nucleation of extended disloca-tions in nanocrystalline Al by mo-lecular-dynamics simulation. Acta Materi-alia **49**(14), 2713–2722, (2001)
87. Yin, W.M., S.H. Whang, et al., Creep behavior of nanocrystalline nickel at 290 and 373 K. Materials Science and Engineer-ing A **301**(1), 18–22, (2001)
88. Zhu, B., R.J. Asaro, et al., Effects of grain size distribution on the mechanical re-sponse of nanocrystalline metals: Part II. Acta Materialia **54**(12), 3307–3320, (2006)
89. Zhu, Y.T. and T.G. Langdon, Influence of grain size on deformation mechanisms: An extension to nanocrystalline materials. Materials Science and Engineering: A **409**(1–2), 234–242, (2005)

Chapter 4
Predictive Capabilities and Limitations of Molecular Simulations

Atomistic simulations – in which the position, velocity, and energy (among others) of each atom within a group of atoms subjected to various types of external constraints (e.g., displacement, temperature, stress) can be predicted – are particularly suited to study the response of nanocrystalline (NC) materials. Indeed, the size of numerically generated microstructures, typically varying from $\sim 10^5$ up to $\sim 3.10^6$ atoms, is sufficient to study both local processes, such as the emission of a dislocation from bicrystals, and larger scale processes, such as grain growth via grain boundary coalescence. The amazing predictive capabilities provided by atomistic simulations are unfortunately limited (1) by their computational cost and (2) by the description of the interaction between atoms via use of an energy potential function.

While the use of molecular dynamics (MD) and statics codes may appear quite complex, the fundamental idea is quite simple and consists of simultaneously solving the equations of motions for a group of atoms. As will be shown, the equations of motion are usually augmented to account for boundary conditions while ensuring that the system could reach all possible acceptable states. The relation between atomistic simulations and statistical mechanics will be discussed in greater detail. The force applied by all atoms surrounding a given atom – within a given range – is given by an inter-atomic potential. Ideally, the latter should capture the electronic environment around each atom. Clearly, such task is one of the greatest challenges associated with atomistic simulations. Typically, the best performing potentials can successfully reproduce many intrinsic material properties: elastic constants, stacking fault energy, etc. Interatomic potential will be presented in the first section of this chapter. The equations of motions as well as solution algorithms for each type of statistical ensemble considered will be the subject of the following section. Finally, the importance of boundary conditions will be discussed prior to showing some particularly interesting studies.

Numerous books and manuals have been dedicated to MD such that the objective of this chapter is not to present an extensive review of all existing molecular dynamics methods, atomistic potentials, and distributed codes – such as LAMMPS (Large-scale Atomic/Molecular Massively Parallel Simulator),

M. Cherkaoui, L. Capolungo, *Atomistic and Continuum Modeling of Nanocrystalline Materials*, Springer Series in Materials Science 112,
DOI 10.1007/978-0-387-46771-9_4, © Springer Science+Business Media, LLC 2009

NAMD (NAnoscale Molecular Dynamics), WARP, etc. – but to present the fundamentals of atomistic simulations in order to allow the reader to have an overall understanding on the subject and on some of its most important subtleties. For complete reviews on the matter, the reader is referred to books dedicated to both MD simulations [1, 2], numerical methods [3], and statistical mechanics [4–6]. Most of the concepts introduced in this chapter can be illustrated with simple simulations. Since the use of molecular codes requires in-depth knowledge of all parameters setting the simulation conditions, the reader is referred to nanohub.org, which allows performance of relatively simple simulations in a user friendly fashion.

4.1 Equations of Motion

In the simplest fashion, atomistic simulations solve the equations of motion of a group of atoms subjected to different types of constraints. As mentioned in the above, depending on the system considered – in the sense of thermodynamics – the equations of motion need to be augmented to ensure that the results are statistically representative. Such level of detail is not necessary in the present section and we will consider the simplest case where the system of equation to be solved is as follows:

$$\boldsymbol{F}_i = m_i \ddot{\boldsymbol{r}}_i \qquad \forall i \in \Omega \tag{4.1}$$

Here, $\boldsymbol{F}_i$, m_i, $\boldsymbol{r}_i$ denote the force applied on atom "i", the mass of the atom and the position of the atom. Time derivatives are denoted with the dot symbol, vectors are denoted with bold symbols, and subscripts "'i" will refer to atom "'i". In other words (4.1), which must be solved for each atom belonging to system Ω, represents a set of three equations.

Additional constraints (e.g., temperature bath, etc.) are often imposed on the physical system studied. This is the case, for example, of simulations performed in the canonical and isobaric–isothermal ensemble (which are presented later in this chapter). In these cases, the augmented equations of motion are derived from use of both of the Lagrangian and Hamiltonian reformulations of classical mechanics. For excellent review on the matter, the reader is referred to [7]. Let us derive the Hamiltonian expression of the equations of motions from Lagrange's result in the simple case were no additional constraint is imposed on the physical system. First, the system's Lagrangian is denoted $L(\boldsymbol{r},\boldsymbol{p})$ with $\boldsymbol{r} = \{\boldsymbol{r}_1, \boldsymbol{r}_2 ..., \boldsymbol{r}_N\}$ and $\boldsymbol{p} = \{\boldsymbol{p}_1, \boldsymbol{p}_2 ..., \boldsymbol{p}_N\}$, N denote the number of atoms composing the system. $\boldsymbol{p}_i = m_i \dot{\boldsymbol{r}}_i$ denotes the momentum of atom i. The Lagrangian of the physical system is written as the difference between its kinetic energy K and its potential energy V:

$$L(\boldsymbol{r},\boldsymbol{p}) = K(\boldsymbol{p}) - V(\boldsymbol{r}) \tag{4.2}$$

In the present case, the kinetic and potential energy of the system are given by given:

$$K(\boldsymbol{p}) = \sum_{i=1}^{N} \frac{\boldsymbol{p}_i \cdot \boldsymbol{p}_i}{2m_i} \quad \text{and} \quad V(\boldsymbol{r}) = \sum_{i=1}^{N} U(\boldsymbol{r}_i) \tag{4.3}$$

Here, $U(\boldsymbol{r}_i)$ denotes the potential energy of atom i, as given by the interatomic potential to which the following section is dedicated. With the above definitions, let us use the principle of virtual work to derive Lagrange's equation. For the sake of generality let us use generalized coordinates $q_1, q_2, \ldots q_m$ which correspond here to each of the m independent variables the system's Lagrangian depends on, or in other words has $\boldsymbol{r}_1 = r_1(q_1, q_2, \ldots q_m)$. The principle of virtual work states that the virtual work associated with virtual displacements imposed on a system in equilibrium is null. Denoting the virtual displacements vector and virtual work $\delta\tilde{\boldsymbol{r}}_i$ and $\delta\tilde{W}$, respectively, one has:

$$\delta\tilde{W} = 0 = \sum_{i=1}^{N} (\boldsymbol{F}_i - m_i\ddot{\boldsymbol{r}}_i) \cdot \delta\tilde{\boldsymbol{r}}_i \tag{4.4}$$

The elementary virtual displacements can be written as:

$$\delta\tilde{\boldsymbol{r}}_i = \sum_{j-1}^{m} \frac{\partial\tilde{r}_i}{\partial q_j} \delta q_j \tag{4.5}$$

Introducing (4.5) into (4.4) one has:

$$\delta\tilde{W} = \sum_{j=1}^{m} \sum_{i=1}^{N} \boldsymbol{F}_i \frac{\partial\tilde{r}_i}{\partial q_j} \delta\tilde{q}_j - \sum_{j=1}^{m} \sum_{i=1}^{N} m_i\ddot{\boldsymbol{r}}_i \frac{\partial\tilde{r}_i}{\partial q_j} \delta\tilde{q}_j \tag{4.6}$$

Let us now relate the second term on the right hand side of equation (4.6) to the system's kinetic energy. With (4.3) the partial derivative of the system's kinetic energy of with respect to $\dot{q}_j$ is given by:

$$\frac{\partial K}{\partial \dot{q}_j} = \sum_{i=1}^{N} m_i\dot{\boldsymbol{r}}_i \cdot \frac{\partial\tilde{\dot{r}}_i}{\partial \dot{q}_j} \tag{4.7}$$

Taking the time derivative of (4.7) and using the chain rule, one obtains after some algebra:

$$\frac{d}{dt}\left(\frac{\partial K}{\partial \dot{q}_j}\right) = \sum_{i=1}^{N} \left(m_i\ddot{\boldsymbol{r}}_i \cdot \frac{\partial\tilde{r}_i}{\partial \dot{q}_j} + m_i\dot{\boldsymbol{r}}_i \frac{\partial\tilde{\dot{r}}_i}{\partial q_j} \right) \tag{4.8}$$

Acknowledging the following identity; $\partial K/\partial q_j = \sum_{i=1}^{N} \left(m_i \dot{r}_i (\partial \tilde{\dot{r}}_i / \partial q_j)\right)$ and with (4.8) one obtains the following relation:

$$\frac{d}{dt}\left(\frac{\partial K}{\partial \dot{q}_j}\right) - \frac{\partial K}{\partial q_j} = \sum_{i=1}^{N} m_i \ddot{r}_i \frac{\partial r_i}{\partial q_j} \tag{4.9}$$

Introducing (4.9) into (4.6) and supposing $\boldsymbol{F}_i$ to be conservative – such that $\boldsymbol{F}_i = -\nabla U(\boldsymbol{r}_i)$ – one obtains:

$$\delta \tilde{W} = -\sum_{j=1}^{m} \left(\frac{\partial V}{\partial q_j} - \frac{d}{dt}\left(\frac{\partial K}{\partial \dot{q}_j}\right) - \frac{\partial K}{\partial q_j}\right) \delta \tilde{q}_j \tag{4.10}$$

Finally, since the potential energy does not depend on position, Lagrange's equations is obtained by setting the term in parenthesis of (4.10) to zero – this is the only solution which respects (4.10) for all kinematically admissible $\delta \tilde{q}_j$-:

$$\frac{d}{dt}\left(\frac{\partial \boldsymbol{L}}{\partial \dot{q}_j}\right) = \frac{\partial \boldsymbol{L}}{\partial q_j} \qquad \forall j \in [1, m] \tag{4.11}$$

Equation (4.11) is used to derive the equations of motion of the physical system of interest. For example, applying (4.11) to the case of each particle position, one retrieves (4.1). Note that the scope of application of Lagrange's equations goes far beyond the case of atomistic simulations. Let us now introduce the system's Hamiltonian which is written as the sum of the kinetic and potential energy:

$$H(\boldsymbol{r}, \boldsymbol{p}) = K(\boldsymbol{p}) + V(\boldsymbol{r}) \tag{4.12}$$

The Hamiltonian and Lagrangian differential can be written as follows:

$$\left.\begin{aligned} dH &= \sum_{i=1}^{N} \left(\tfrac{\partial H}{\partial r_i} dr_i + \tfrac{\partial H}{\partial p_i} dp_i\right) + \tfrac{\partial H}{\partial t} \\ dL &= \sum_{i=1}^{N} \left(\tfrac{\partial L}{\partial r_i} dr_i + \tfrac{\partial L}{\partial \dot{r}_i} d\dot{r}_i\right) + \tfrac{\partial L}{\partial t} \end{aligned}\right\} \tag{4.13}$$

Using Lagrange's equations (4.11) and acknowledging $p_i = \partial L / \partial \dot{r}_i$ one can rewrite the differential of the Lagrangian as follows:

$$d\left(\sum_{i=1}^{N} p_i \dot{r}_i - L\right) = \sum_{i=1}^{N} (-\dot{p}_i dr_i + \dot{r}_i dp_i) - \frac{\partial L}{\partial t} \tag{4.14}$$

Finally, by identification with the Hamiltonian's differential, the equations of motion can be rewritten as follows:

$$\left.\begin{aligned}\dot{r}_i &= \frac{\partial H}{\partial p_i} \\ \dot{p}_i &= -\frac{\partial H}{\partial r_i}\end{aligned}\right\} \tag{4.15}$$

4.2 Interatomic Potentials

Of interest here is the description of the force perceived by each atom within a physical system. Such a problem is far from being trivial since each atom will interact with all other atoms within the system. Moreover, the interaction between atoms, given by an interatomic potential, is uniquely defined by the local time-dependent electron density [8, 9]. Therefore, a rigorous solution of the problem defined by Equation (4.1) would require solving Schrödinger's equations in addition to the equations of motions. Such rigorous computations are referred to as ab initio simulations. Several codes, such as SIESTA, have been developed to solve such complex problems. As one would expect, ab initio simulations are extremely time consuming, which limits their use to relatively small ensembles – on the order of thousands of atoms. Instead, in molecular dynamics simulations, the interaction between atoms is described with a potential function which approximates the exact interactions between atoms.

Let us first simplify the problem and assume that the energy of a given atom can be decomposed such that the interaction between pairs of atoms can be regarded independently. If $U(\boldsymbol{r}_{ij})$ denotes the contribution of atom i potential's energy due to its interaction with atom j – where $\boldsymbol{r}_{ij} = \boldsymbol{r}_i - \boldsymbol{r}_j$ denotes the distance between atom i and atom j – the force between the two atoms is simply given by:

$$\boldsymbol{f}_{ij} = -\nabla U(\boldsymbol{r}_{ij}) \tag{4.16}$$

The force applied on atom i shall then account for all possible pairs of atoms that can be formed with atom i. Therefore, if N denotes the total number of atoms in the system, one has:

$$\boldsymbol{F}_i = \sum_{j=1, j \neq i}^{N} \boldsymbol{f}_{ij} \tag{4.17}$$

With the paired atoms approximation and the set of Equations (4.1) (4.2), (4.3), (4.4), (4.5), (4.6), (4.6), (4.7), (4.8), (4.9), (4.10), (4.11), (4.12), (4.13), (4.14), (4.15), (4.16) and (4.17) the problem becomes that of defining a function $U(\boldsymbol{r}_{ij})$, physically representing the potential energy between atom i and atom j, which satisfactorily approximates the interaction between pair of atoms without requiring to consider quantum effects. Additionally, the interatomic potential must yield acceptable predictions of the materials intrinsic properties, such as the elastic constants, vacancy formation energies, etc. The most often used potential arising from this pair-interaction approximation, namely the Lennard-Jones potential, will be presented in next section.

Then, two refined methods, the embedded atom method (EAM) and the approach of Finnis-Sinclair, which are more suited to represent inter-atomic interaction within metals, will be presented. In these models, the effect of each atoms *environment* will be considered additionally to pair interactions.

4.2.1 Lennard Jones Potential

One of the simplest potential is the Lennard-Jones (LJ) potential. It is well suited for inert gases at low densities and is typically the first potential used when interatomic forces are unknown. Importantly, as will be shown later, it is not a suited potential for metals and for charged particles. In particular, the LJ potential leads to the following relations between the materials elastic constants (in Voigt notation): $C_{11} = C_{22}$ and $C_{44} = 0$. To circumvent this limitation, a volume-dependent energy term is typically added to the expression of the LJ potential. The most refined potential – such as the EAM potential [10, 11] and Finnis-Sinclair [12] potentials – share some common ground with the LJ potential. Therefore, while this potential is not typically used to model the response of metals, it is an ideal starting point to discuss the atomic interaction modeling. The LJ potential is expressed as follows:

$$U(r_{ij}) = 4\varepsilon\left[\left(\frac{\sigma}{r_{ij}}\right)^{12} - \left(\frac{\sigma}{r_{ij}}\right)^{6}\right], \qquad r_{ij} < r_c \qquad (4.18)$$

In (4.18) ε defines the strength of the bonds and σ defines a length scale, as shown in the above when the strength of the bond is null when the distance between the two atoms is larger than the cut off distance r_c. Therefore, in calculating the force exerted on atom i, given by Equation (4.17), only the effects of atoms at distances smaller than the cut-off distance should be considered. Typically, the critical distance r_c is chosen such that the attractive tail of the potential is neglected (e.g., $U(r_c) = 0$) leading to the following choice $r_c = 2^{1/6}\sigma$. Figure 4.1, presents the evolution of the bond's strength as a function of distance in the simpler case of water molecules ($\varepsilon = 0.6501 KJ/mol$, $\sigma = 0.31 nm$). As shown, the interaction between two atoms, as given by the LJ potential, is composed of a repulsive part at small distances. The repulsion between atoms tends to infinity as the interatomic distance tends to zero, which ensures that atom collision is prevented. The interatomic potential also accounts for the attraction between atoms when their spacing is larger than an equilibrium distance. This attraction term, corresponding to the second term on the right-hand side of Equation (4.18), corresponds to the Van Der Waals force.

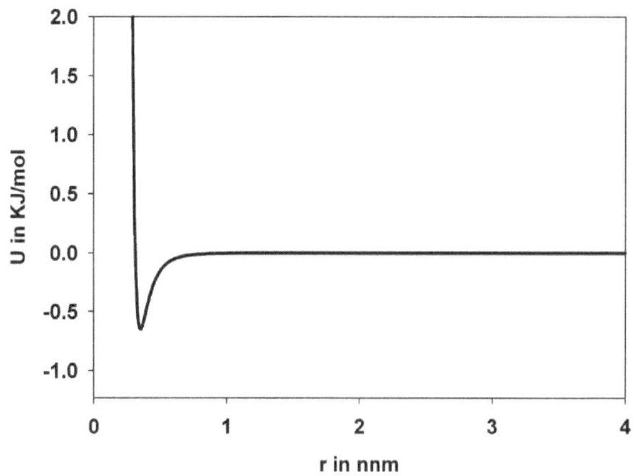

Fig. 4.1 Evolution of potential energy as a function of distance

4.2.2 Embedded Atom Method

As mentioned in the above, simple pair-potentials present limitations (e.g., predictions of the elastic properties of metals, etc.) which arise mainly from the fact that the local atomic charge is not accounted for. The potential to be described in this section – referred to as EAM potential – overcomes this limitation. In essence, it is based on the notion of quasi-atom and on density functional theory (DFT). In what follows, the important notion behind the EAM potential will be presented. For details on the matter, the reader is referred to the original work of Daw and Baskes [10].

Consider an initially pure metal in which an impurity is introduced. The potential of the host metal is uniquely defined by its electron density and the potential of the impurity is dependent on its position and charge. Therefore, the potential of the host metal with impurity is expected to depend on both previously mentioned contributions. Let E denote the energy of the host with impurity, one can write:

$$E = f_{Z,r}[\rho_h(r)] \tag{4.19}$$

Here f is a functional to be defined, Z and r define the type of impurity and its position. The host electron density, which depends on position, is denoted with $\rho_h(r)$. Equation (4.19) is referred to as the Stott-Zarembra corollary. Daw and Baskes extended the Stott-Zarembra corollary by supposing that each atom within a pure metal can be considered as an impurity. With this assumption, the energy of atom i within the system (e.g., a pure metal) is written as the sum of the contribution of an embedding function, accounting for the effect of the electron density, and of pair-wise interactions:

$$U_i = F_i^{emb}\left[\rho_{h,i}(\boldsymbol{r}_i)\right] + \frac{1}{2}\sum_{i \neq j} \phi_{ij}\left(\boldsymbol{r}_{ij}\right) \tag{4.20}$$

Here, F_i defines an embedding function, which relation to f in Equation (4.19) is not explicit, and ϕ_{ij} defines a pair interaction function – similar to that introduced in previous section describing the LJ potential. So far, functions F_i,$\rho_{h,i}(r)$ and $\phi_{ij}(r_{ij})$ are unknown. The embedding function and pair interaction functions are not uniquely defined.

Several expressions for the above mentioned functional have been proposed in the literature [13]. The EAM potential was recently extended to improve its predictive power in the case of metals of non-full electron shells [14]. For the sake of comprehension only the original proposition by Daw and Baskes will be introduced here since subsequent updates rely on it. First, $\rho_{h,i}(r)$ can be approximated as a linear superposition of spherically averaged electron densities:

$$\rho_{h,i}(\boldsymbol{r}) = \sum_{j \neq i} \rho_j^a\left(\boldsymbol{r}_{ij}\right) \tag{4.21}$$

Here ρ_j^a denotes the contribution of atom j to the density of atom i. ρ_j^a is obtained by further approximation. That is, the following average is taken:

$$\rho_j^a(\boldsymbol{r}) = N_s \rho_s^a(\boldsymbol{r}) + N_d \rho_d^a(\boldsymbol{r}) \tag{4.22}$$

Here N_s and N_d denote the approximate number of outer electrons in the s and d shells. These can be obtained by fit of the heat of solution of hydrogen within the metal, for example. The electron densities on the s and d shells $\rho_s^a(\boldsymbol{r})$ and $\rho_d^a(\boldsymbol{r})$ are then given by DFT calculations.

The definition of the potential energy of atom i, given by Equation (4.20) is equivalent to defining the pair-wise interaction function and the embedding function. The former, in its original form, was simply given by:

$$\phi_{ij}\left(\boldsymbol{r}_{ij}\right) = \frac{\mathbf{Z}_i\left(\boldsymbol{r}_{ij}\right)\mathbf{Z}_j\left(\boldsymbol{r}_{ij}\right)}{\left|\boldsymbol{r}_{ij}\right|} \tag{4.23}$$

Here, $Z(r)$ defines the effective charge of a given atom. Finally, the problem becomes that of finding two functions: the embedding function, F_i^{emb}, and a function giving, $Z(r)$, an appropriate evolution of the effective charge of an atom. These two functions are typically obtained via empirical fit of intrinsic material properties which are uniquely defined by the potential. For example, the embedding function can be fitted with a third order spline function. The following approximation was originally made: $Z(r) = Z_0(r)(1 + \beta r^{\nu})e^{-\alpha r}$.

As discussed previously, both the embedding function and the charge function are obtained via fit of intrinsic material properties. Therefore, let us relate the interatomic potential to some experimentally measurable intrinsic material properties. For the sake of simplicity let us restrict ourselves to the case of the lattice constants and elastic constants. Let the superscript ' denote spatial derivatives. The lattice constant is simply obtained by the equilibrium condition: $dU_i/dr = 0$. Applying the equilibrium condition to (4.20):

$$A_{ij} + \left(F^{emb}\left(\rho_{h,i}\right)\right)' V_{ij} \tag{4.24}$$

with

$$A_{ij} = \frac{1}{2}\sum_m \phi'_m \frac{r_i^m r_j^m}{r^m} \text{ and } V_{ij} = \frac{1}{2}\sum_m \rho'_m \frac{r_i^m r_j^m}{r^m} \tag{4.25}$$

Here r_i^m denotes the *ith* component of the position vector of atom m. In the case of the elastic constants, the algebra is more involved (the complete derivation is left as an exercise and can be based from the elastic strain energy). After some algebra, one obtains:

$$C_{ijkl} = \left(B_{ijkl} + F^{emb\,\prime}(\bar{\rho})W_{ijkl} + F^{emb\,\prime\prime}(\bar{\rho})V_{ij}V_{kl}\right)/\Omega_0 \tag{4.26}$$

where

$$B_{ijkl} = \frac{1}{2}\sum_m \frac{\left(\phi''_m - \phi'_m\right)}{r^m} \frac{r_i^m r_j^m r_k^m r_l^m}{\left(r^m\right)^2} \tag{4.27}$$

$$W_{ijkl} = \frac{1}{2}\sum_m \frac{\left(\rho''_m - \rho'_m\right)}{r^m} \frac{r_i^m r_j^m r_k^m r_l^m}{\left(r^m\right)^2} \tag{4.28}$$

Rewriting (4.26) in the Voigt notation, one can see that if only pair interactions were considered – that is all contributions F and its derivative are set to 0 – then one obtains $C_{11} = C_{22}$ and $C_{44} = 0$. Using the lattice constant, elastic constants, vacancy formation energy, sublimation energy, etc, the constants required to obtain an empirical expression of $Z(r)$ and F^{emb} can be obtained.

4.2.3 Finnis-Sinclair Potential

Alternatively to the EAM potential and more recent EAM-based potentials, Finnis and Sinclair [12] proposed an empirical N-body type potential which, while similar, in essence, to the EAM potential was developed. The energy per atom at a given position is written as the sum of an N-body term and a pair potential term:

$$U_i = U_N + U_P \tag{4.29}$$

The N body contribution is given by:

$$U_N = -Af(\rho) \tag{4.30}$$

with

$$\rho = \sum_{i \neq 0} \Phi(\boldsymbol{r}_i) \tag{4.31}$$

Here, ρ is the local electronic charge density and function f is chosen such that it mimics the results of tight binding theory; therefore: $f(\rho) = \sqrt{\rho}$. Φ, which is an unknown function, similarly to the embedding function in the case of the EAM potential, can be interpreted as the sum of squares of overlap integrals. Core-core repulsion interaction in (4.29) is given by:

$$U_P = \frac{1}{2} \sum_{i \neq 0} V(r_i) \tag{4.32}$$

Using similar reasoning as presented previously, the interatomic potential can be related to macroscopic properties: elastic moduli, equilibrium volume, cohesive energy. For example, parabolic and fourth-order polynomials were chosen to fit the cohesive potential and the pair potential functions in the case of body-centered cubic (BCC)transition metals.

4.3 Relation to Statistical Mechanics

In Section 4.1, dedicated to the different formulations of the equations of motion (e.g., classical, Lagrangian and Hamiltonian), the system of equations to be solved (4.1) simply consisted of the dynamic equilibrium without further constraint. In other words, the physical system was assumed not to interact with its environment.

Limiting ourselves to thermodynamically isolated systems – although in some cases it may be desired to isolate the system – can clearly hinder the predictive capabilities of a numerical simulation. For example, in the case of metals – and for reasons mentioned in the introduction to this chapter – more particularly of NC materials, most of the information of interest concerns the processes activated during their plastic response (e.g., grain boundary dislocation emission, grain boundary migration, etc.). Numerically, the time step used in a molecular dynamics simulation must necessarily be in the order of the femtosecond such that, among others, atom collision is prevented. Recall that the period of vibration of atoms is in the order of the Debye frequency. Therefore, simulating the plastic response of NC materials during a tensile test in the quasistatic regime would use several years of computational time. Typically, one simulates rarely more than a hundred picoseconds of real time. To overcome such limitations, one can either impose larger strain rates, on the order of 10^7 to $10^9/s$, or stresses in the order of several GPa to reach the plastic regime in a reasonable computational time. Such high strain rates remain far from those imposed during experiments – even in the case of shock loading. Therefore, the occurrence of a simulated mechanism, for example, may not be relevant at a much lower strain rate. Suppose a metallic bar was subjected to a tensile load at $10^7/s$ strain rate. Considering a system composed solely of the solid bar (e.g., its environment is disregarded), during loading under these conditions one would necessarily expect large temperature and pressure fluctuations which are very likely to activate mechanisms (e.g., diffusive mechanisms, for example) irrelevant to the targeted study. During an actual test, the solid bar would also be subject to an externally imposed temperature and pressure arising from its environment. Therefore, if one were to consider a new system, consisting of the solid bar in its environment, temperature or pressure fluctuations (or ideally both) would clearly be diminished such that extraneous artifacts – additionally to the high strain rates or stress – would not have to be considered.

The relation between statistical mechanics and molecular dynamic simulations arises from the following considerations. Consider a physical system, interacting or not with its environment. The overall state of the system is uniquely defined by its thermodynamic state variables. For a given set of such independent variables – assume constant volume, energy, and number of atoms, for example – there are multiple atomic configurations leading to the same state. Each of these acceptable configurations is referred to as a microstate. Depending on the variables held fixed, the statistical distribution of microstates will be different. In order to account for external temperature baths or pressure, the equations of motion of each atom composing the physical system must be augmented while leading to the same statistical distribution given by statistical mechanics.

In what follows, several ensembles corresponding to collection of microstates will be introduced. Their relationship to thermodynamic quantities will be presented. The latter, is of great importance since it is often, if not always, necessary to relate pressure, or temperature from an ensemble of matter. Moreover, these ensembles and respective distributions will serve as the steppingstone to augment the equations of motion. In what follows the three most typical ensembles will be presented: (1) the microcanonical ensemble NVE, (2) the canonical ensemble NVT, and (3) the isobaric-isothermal ensemble NPT.

4.3.1 Introduction to Statistical Mechanics

Consider a system composed of N atoms in a given macrostate – defined in terms of typical thermodynamic quantities such as N, P, T, S, V, etc. – There are numerous configurations at the atomistic scale, referred to as microstate, leading to the same macrostate [15]. From the point of view of quantum mechanics, the microstate of the system – defined by the knowledge of each atoms position and momentum – cannot be known in a deterministic manner due to the uncertainty principle:

$$\Delta r \Delta p \sim h \tag{4.33}$$

Here h is the Planck constant. Since a deterministic description of the system would violate Heisenberg's principle, a statistical approach may be used instead. The ensemble of possible realization of a system – which is a 2 N dimensional space – is referred to as the phase space. Each point in the phase space –a phase point – corresponds to a given microstate. Under a given macroscopic condition, such as a fixed entropy S for example, the subspace of possible phase points (e.g. microstates), is a hyperspace of the phase space. The probabilistic approach can be introduced with knowledge of the probability that for a given imposed macrostate, the system is in a given admissible phase point. An ensemble can be defined as a large group of systems, all in the same macrostate, but in different microstates. Let us now proceed by introducing the following two postulates:

Postulate 1: Consider a thermodynamic quantity Q. The long-time average of Q is equal to the its ensemble average if the number of members in the ensembles tends to infinity.

Postulate 2: All microstates with same energy are equi-probable.

The first postulate, corresponds to the ergodicity hypothesis and states that over an infinite amount of time a system will probe all possible microstates respecting the imposed constraints. The second postulate infers that the probability of occurrence of a given phase point is dependent on energy. For a given energy E with and necessary allowance δE arising from the uncertainty principle, the total number of possible microstates of a conservative system is given by:

$$W(N, V, E) = \frac{1}{N! h^{3N}} \int_V d^{3N}\boldsymbol{r} \int_{-\infty}^{+\infty} d^{3N}\boldsymbol{p}\,\delta(H(\boldsymbol{r}, \boldsymbol{p}) - E) \tag{4.34}$$

In Equation (4.34), Planck's constant and the factorial term arise from the uncertainty principle and from the fact that identical particles cannot be distinguished, respectively. In some cases it may be desirable to reason with state densities $\Omega(N, V, E)$ defined as follows:

$$\Omega(N, V, E) = \frac{1}{N! h^{3,N}} \int_V d^{3N}\boldsymbol{r} \int_{-\infty}^{+\infty} d^{3N}\boldsymbol{p} \tag{4.35}$$

At any given time t, an ensemble can be described with a probability density function (p.d.f), which will depend on the ensemble considered. Let $\rho(x^N, t)$

denote the p.d.f. and $x^N = (r^N, p^N)$ denote a phase point. The probability $P(R,t)$ of a finding phase points in a region R of the phase space is thus given by:

$$P(R,t) = \int_{x^N \in R} \rho(r^N, p^N, t) d^{3N}x \tag{4.36}$$

From the definition of the state density function, one of the most fundamental equationS of statistical mechanics, referred to as the Liouville equation, can be derived. The subspace of admissible phase points is defined by a bounding surface S_{bound}. The Liouville equation is based on the "incompressibility" of the state density function. In other words, the rate of increase of state points in the volume defining admissible phase points is equal to the net amount of state points exiting the surface:

$$\oint_{S_{bound}} n \bullet \dot{x}^N \rho(x^N, t) dS = -\frac{\partial}{\partial t} \int_V \rho(x^N, t) d^{3N}x \tag{4.37}$$

Using the divergence theorem one obtains:

$$\dot{x}^N \bullet \nabla_{x^N} \rho(x^N, t) + \frac{\partial \rho(x^N, t)}{\partial t} = 0 \tag{4.38}$$

Recalling expression (4.12) derivatives of the Hamiltonian can be introduced in (4.38) such that Liouville equation, relating the systems Hamiltonian to the state distribution function, can be established:

$$\frac{\partial \rho(x^N, t)}{\partial t} + \sum_{i=1}^{N} \left(\frac{\partial H}{\partial p_i} \frac{\partial}{\partial q_i} - \frac{\partial H}{\partial q_i} \frac{\partial}{\partial p_i} \right) \rho(x^N, t) = 0 \tag{4.39}$$

4.3.2 The Microcanonical Ensemble (NVE)

The first ensemble to be considered, referred to as microcanonical ensemble, corresponds to the case were the physical system of interest consist of N atoms occupying a constant volume V and were the overall system's energy, denoted E, is constant. The ensemble refers to a collection of systems with same thermodynamic state – in the present case with same N, V, and E – but each system is different at the molecular level. Let us now relate the NVE ensemble to thermodynamic properties. First, consider a rigid isolated volume, V containing N atoms and with overall energy E (see Fig. 4.2).

The probability, P, of being in a given state is thus: $P = 1/W(N,V,E)$. Consider now the same system as in Fig. 4.2(a) and divide the system in two

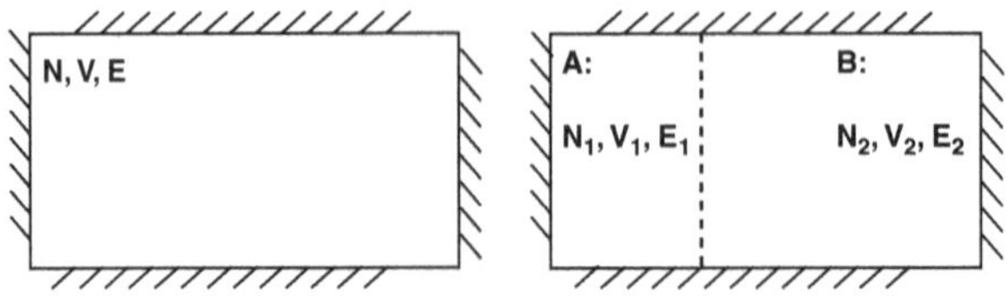

Fig. 4.2 (**a**) Isolated rigid system, (**b**) Isolated rigid system divided in two subsystems

subsystems A and B (Fig. 4.2(b)). The total number of microstates such that system A has an energy E_1 and system B its complementary is then:

$$W(N_1, V_1, E_1) \bullet W(N - N_1, V - V_1, E - E_1) \tag{4.40}$$

Following Boltzmann hypothesis stating that "the entropy of a system is related to the probability of its being in a quantum state" one can introduce the entropy of the system S. The entropy is an extensive property (e.g. $S_{AB} = S_A + S_B$) which can be defined as follows:

$$S = k_B \ln(W(N, V, E)) \tag{4.41}$$

We can verify that with this definition, the entropy respects $S_{AB} = S_A + S_B$. Indeed with equations (4.40) and (4.41) one has:

$$\begin{aligned} S_{AB} = k_B \ln(W(N, V, E)) = k_B(\ln(W(N_1, V_1, E_1)) \\ + \ln(W(N - N_1, V - V_1, E - E_1))) = S_A + S_B \end{aligned} \tag{4.42}$$

From the definition of the entropy given by (4.41) all other thermodynamic quantities of interest can be derived:

$$\begin{aligned} \text{Temperature} : \frac{\partial S}{\partial E} &= \frac{1}{T} \\ \text{Pressure} : \frac{\partial S}{\partial V} &= -\frac{P}{T} \\ \text{Chemical potential} : \frac{\partial S}{\partial N} &= \frac{\mu}{T} \end{aligned} \tag{4.43}$$

From (4.43) and the definition of entropy, calculation on the NVE ensemble can be related to thermodynamic properties of interest. Practically, in the case of the NVE ensemble, unless the entropy can be written as an explicit function of N, V or E – such as in the trivial case of noninteracting particles – (4.43) is of relatively limited use to relate simulation observable quantities to thermodynamic properties. Numerically, the calculation of pressure is typically performed via use of the virial stress – to be introduced in upcoming section – which will be introduced in a later section. However, for other ensembles considered such as the canonical ensemble, explicit relations between "simulation observable" quantities and thermodynamic quantities can be obtained.

4.3.3 The Canonical Ensemble (NVT)

The second ensemble of interest in this chapter is the canonical ensemble where the number of atoms considered, the volume and the temperature are constant. Similarly to the microcanonical ensemble, it is typically used to impose velocity constraints (e.g., strain rates) to the systems while maintaining the temperature constant. From the standpoint of thermodynamics this corresponds to setting the system of interest in a temperature bath as shown in Fig. 4.3.

Recalling postulate 4.2, the probability that the system is in a microstate – defined by all the atomic positions and momenta r_i and p_i – is given by the following ratio:

$$P = \frac{\Omega_{bath}(E - H(r_i, p_i))}{\sum_{microstates} \Omega_{bath}(E - E_{sys})} \tag{4.44}$$

When the number of microstates becomes large, expression (4.44) should be written in terms of integrals. Since the system's energy is small compared to the total energy, the natural log of $\Omega_{bath}(E - E_{sys})$ can be expanded around E, thus leading to a p.d.f. Taking the exponential of the resulting expansion and with (4.43) leads to the Maxwell Boltzmann distribution:

$$\rho = \frac{e^{-\frac{H(\mathbf{r},\mathbf{p})}{kT}}}{\int_{\text{microstates}} e^{\frac{H(\mathbf{r},\mathbf{p})}{kT}}} \tag{4.45}$$

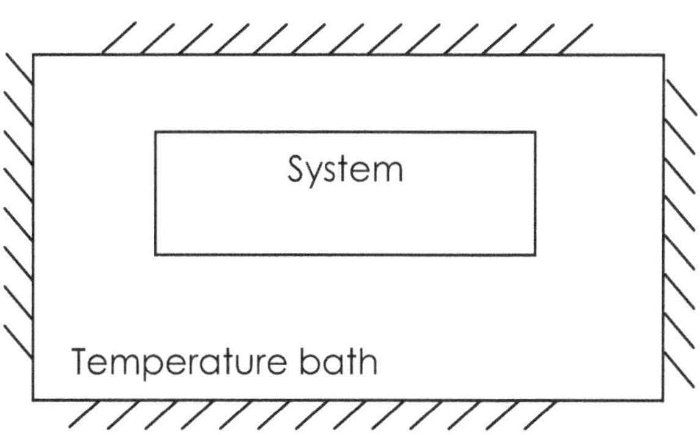

Fig. 4.3 System in a temperature bath

Let us introduce the partition function:

$$Z(N,V,T) = \int_{\text{microstates}} e^{\frac{H(r,p)}{kT}} \tag{4.46}$$

Z is related to the Helmholtz free energy as follows:

$$F = E - TS = k_B T \ln Z(N,V,T) \tag{4.47}$$

With (4.47) it can be seen that as opposed to the NVE ensemble where the total system's energy is conserved, in the canonical ensemble, it is the Helmholtz free energy which is conserved. From the Maxwell Boltzmann distribution, (1.45), a particularly interesting property, referred to as the equi-partition of energy, can be deduced. Let us re-write the Hamiltonian, given by Equation (4.12), as the sum of a squared term and of a remnant term such that:

$$H(r_i, p_i) = H'(r_i, p_i) + \lambda p_1^2 \tag{4.48}$$

Here λ is a multiplying factor and p_1^2 is the momentum of a given atom in a given direction. Let us now consider the ensemble average of the quantity λp_1^2. Recall that the average of a random variable x which distribution is given by a probability distribution function $f(x)$ is given by: $\langle x \rangle = \int_{-\infty}^{\infty} x f(x) dx$. Extending the previously mentioned property to a 3 N dimension, the ensemble average of λp_1^2 is thus given by:

$$\langle \lambda p_1^2 \rangle = \frac{\int d^{3N}r d^{3N}p \lambda p_1^2 e^{-\frac{H(r,p)}{k_B T}}}{\int d^{3N}r d^{3N}p e^{-\frac{H(r_i,p_i)}{k_B T}}} = \frac{k_B T}{2} \tag{4.49}$$

As shown by Equation (4.49), in the canonical ensemble, any squared term appearing in the Hamiltonian, such as the atoms kinetic energy, will contribute equally to the system's temperature. Using the above relation and considering each atoms contribution, temperature can be related to the systems kinetic energy, K^e, as follows:

$$K^e = \frac{3N}{2} k_B T \tag{4.50}$$

Practically, the overall linear or angular momenta (or both) may have to be set to zero to avoid rigid motion of the system. In that case, 3 or 6 degrees of freedom shall be removed from (4.50). This is typically used to obtain a measure of temperature (regardless of the ensemble considered). Importantly it can be seen that the overall systems temperature could be controlled via rescaling the kinetic energy. This will be discussed in more detail later.

4.3.4 *The Isobaric Isothermal Ensemble (NPT)*

The last ensemble of interest, referred to as isobaric isothermal ensemble, corresponds to a system in both a temperature and pressure bath. This ensemble is thus more suited to impose pressure control, rather than displacement control over the system. Using similar reasoning as in the case of the NVT ensemble, a partition function can be defined as follows:

$$Z_p(N, P, T) = \int_V \int_{microstates} e^{\frac{E-PV}{k_B T}} \tag{4.51}$$

Such that the system's states obey the following probability distribution functions:

$$\rho = \frac{e^{-\frac{H(r_i, p_i) - pV}{k_B T}}}{Z_P(N, P, T)} \tag{4.52}$$

Due to the similarities in the probability distribution functions of the NVT and NPT ensembles, the system temperature is related to the kinetic energy by (4.50). The case of pressure will be discussed in following section. Using similar reasoning as in the above, it can be shown that the isobaric isothermal ensemble conserves the Gibbs free enthalpy.

4.4 Molecular Dynamics Methods

In previous section, several statistical ensembles have been introduced. It was shown that the relevant statistical ensemble depends on the environment surrounding the physical system studied. In this section, the objective is to introduce some of the most frequently used modeling methods, which all consist of augmenting the Hamiltonian – or equivalently the Lagrangian –, such that (1) the effect of the environment on the physical system can be accounted for, and (2) the resulting system of equations obeys the state density distribution of the ensemble it is supposed to represent.

4.4.1 *Nosé Hoover Molecular Dynamics Method*

In this section the mathematical description of the canonical ensemble (e.g., NVT) will be presented [16–19]. The idea here is to modify the expression of the equations of motion as expressed in (4.12) – using either the Lagrangian or Hamiltonian formulation – such that the system's temperature remains constant during a simulation. With the relation between the system's kinetic energy and temperature (e.g., Equation (4.50)) it can be easily seen that the constant temperature condition could be respected by rescaling each particle's velocity. Unfortunately, this simple approach does not reproduce the statistical canonical ensemble. In other words, by performing a simple temperature rescaling, some admissible microstates will never be reached during a simulation. To overcome this limitation, Nosé introduced a canonical ensemble molecular dynamics method. This idea is to introduce an extraneous degree of freedom, s, to the system of equation. It can be regarded as an external system. Similar notations as in previous section are used where the particle position is denoted r_i and denoting its velocity v_i. The external system interacts with the studied system as follows:

$$\boldsymbol{v}_i = s\dot{\boldsymbol{r}}_i \tag{4.53}$$

With s the Lagrangian, L, of the new extended system is written such as to incorporate – similarly to the nonaugmented equations of motions – both the potential energy and kinetic energy contributions:

$$L = \sum_i \frac{m_i}{2} s^2 \dot{r}_{i^2} - V(r) + \frac{Q}{2}\dot{s}^2 - (1+f)kT_{eq}\ln(s) \tag{4.54}$$

Here, the potential and kinetic energy contributions arising from the external degree of freedom are $\frac{Q}{2}\dot{s}^2$ and $(1+f)kT_{eq}\ln(s)$, respectively. T_{eq} is the thermostat's temperature. Note that the interaction between the physical system and the external system is capture in the first term on the right hand side of Equation (4.54). f represents the number of degrees of freedom of the physical system. Its actual value depends on the problem studied. Using the Lagrange's equation (4.11), the system's equations of motion – to be numerically integrated – can be derived. Recalling equation (4.2), one obtains both for the particles and for the new degree of freedom:

$$\ddot{\boldsymbol{r}}_i = -\frac{1}{m_i s^2}\frac{\partial U}{\partial r_i} - \frac{2\dot{s}}{s}\dot{\boldsymbol{r}}_i \tag{4.55}$$

and

$$Q\ddot{s} = \sum_i m_i s \dot{\boldsymbol{r}}_i^2 - \frac{(f+1)kT_{eq}}{s} \tag{4.56}$$

An appropriate choice of Q is critical. If Q is too small, the coupling between the external system and the physical system will be weak. Alternatively, if Q is too large complete sampling of the phase space may not be permitted.

Let us now see, as shown by Nosé [18], that the newly formed system of equations accurately represents the canonical distribution. First, using expressions (4.34) and (4.45) and recalling the expression of the partition function for the NVT ensemble, it can be seen that the new partition function of the extended system can be written as follows:

$$\begin{aligned} Z' &= \tfrac{1}{N!h^{3N}} \int_V d^{3N}r \int_{-\infty}^{+\infty} d^{3N}p \int_{-\infty}^{+\infty} ds \int_{-\infty}^{+\infty} dp^s \delta\Bigg(\sum_i \tfrac{p_i^2}{2m_i s^2} + U(r) \\ &+ \tfrac{p_s^2}{2Q} + (1+f)kT_{eq} \ln(s) - E\Bigg) \end{aligned} \tag{4.57}$$

As shown in the above equation, the system's Hamiltonian accounts for the contribution of the external system. p_s denotes the momentum associated with the external degree of freedom s. The particle momentum can be rewritten as follows: $p_i' = p_i/s$ Using the previous expression of the momentum and the following relation $\delta(g(s)) = \delta(s - s_0)/g'(s)$, where$s_0$is the zero of $g(s)$, one can rewrite (4.57) as follows:

$$Z' = \frac{1}{(f+1)kT_{eq}} \frac{1}{N!} \int dp^s \int d^{3N}p \int_V d^{3N}r \exp\left[-\left(H(p', r) + \frac{p^{s2}}{2Q} - E\right)/kT_{eq}\right] \tag{4.58}$$

Performing the integration with respect to p^s one obtains the following relation between the partition function of the extended system and the partition function of the NVT ensemble Z:

$$Z' = \frac{1}{(1+f)} \left(\frac{2\pi Q}{kT_{eq}}\right)^{1/2} \exp(E/kT_{eq})Z \tag{4.59}$$

Performing an ensemble average on any static quantity with the partition function (4.59) will lead to the average in the canonical ensemble.

While Nosé's approach can describe the canonical ensemble, it is computationally intensive due to variable rescaling resulting from the introduction of the external variable s. In the spirit of Nosé, Hoover proposed the following molecular dynamics method for the canonical ensemble:

$$\dot{r}_i = \frac{p_i}{m} \tag{4.60}$$

$$\dot{p}_i = F_i - \varsigma p_i \tag{4.61}$$

$$\dot{\varsigma} = \nu_T^2 \left(\frac{T}{T_{ext}} - 1\right) \tag{4.62}$$

ν_T is a numerical parameter which choice is based on arguments similar to that of the choice of Q. With Equations (4.60), (4.61), and (4.62) it can be seen that the additional difficulty of Nosé's method arising from the fact that the variables have to be rescaled, is removed while the system will still obey the canonical distribution. This can be shown via similar reasoning as presented previously (e.g., Equations (57), (4.58), and (59)). This set of equation is the most frequently used to simulate the canonical ensemble.

4.4.2 Melchionna Molecular Dynamics Method

In a previous section, a system of equations allowing control of the physical system's temperature while respecting the canonical ensemble was introduced. Obviously, in the case of the isobaric-isothermal ensemble, additional difficulty arises from the fact that it is now desired to control the pressure (or stress) imposed on the physical system. Anderson and Nosé, and then Hoover, first proposed extensions of the canonical molecular dynamic method described by (4.60), (4.61), and (4.62). Hoover's equations of motion in the case of pressure constraints are easier to implement than that of Anderson and Nose. As discussed by Melchionna et al., the previously mentioned approaches do not satisfy the isobaric-isothermal ensemble. They proposed the following approach based on Hoover's constraint method [20]:

$$\dot{r}_i = \frac{p_i}{m} + \eta(r_i - R_0) \tag{4.63}$$

$$\dot{p}_i = F_i - (\varsigma + \eta)p_i \tag{4.64}$$

$$\dot{\varsigma} = \nu_T^2\left(\frac{T}{T_0} - 1\right) \tag{4.65}$$

$$\dot{\eta} = \frac{v_p}{NkT_{ext}} V(P - P_{ext}) \tag{4.66}$$

The derivation of (4.63), (4.64), (4.65), and (4.66) is also based on Lagrange's equation. Here v_p is a numerical parameter, P_{ext} is the external pressure, and $R_0 = \sum_i m_i r_i / \sum_j m_j$ is the center of mass. It can be proved – via similar reasoning as presented in previous section – that the system of equation (4.63), (4.64), (4.65), and (4.66) respects the isobaric-isothermal ensemble. Additionally, let us note that with this approach, the relative simplicity of Hoover's approach is conserved. As in the case of the Nosé-Hoover approach, the choice of ν_T and v_p should be made such as to reduce oscillations in temperature and pressure in the neighborhood of the desired values. Due to those oscillations, it is critical when simulating a physical system in the isobaric isothermal as well as in the canonical ensemble to test that the selected values of ν_T and v_p do not introduce oscillations causing artifacts. This can be seen in Fig. 4.4, presenting the evolution of pressure as function of time for different values of v_p The system studied corresponds to a cube of copper containing 4000 atoms. As v_p increases, the oscillation frequency is increased. For all three values of v_p, deviations from the desired pressure can be observed. To overcome such limitation, additional

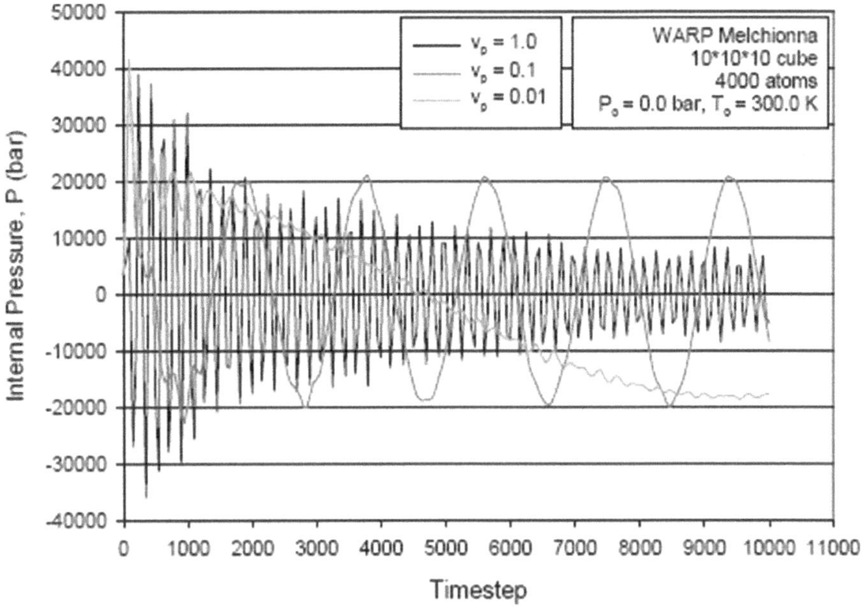

Fig. 4.4 Evolution of pressure as a function time during an isobaric-isothermal equilibration using Melchionna et al.'s approach [21]

damping coefficients can be added to the system. These options are already available in most codes such as LAMMPS.

4.5 Measurable Properties and Boundary Conditions

In most cases, it is desirable to access thermodynamic quantities, or other variables, allowing to easily assess the system state. The knowledge of temperature and pressure are clearly required when using the NPT ensemble. Recall that temperature can be measured through the overall kinetic energy with Equation (4.50). As will be shown, pressure is typically measured via the virial stress. Additionally, it may be necessary to observe materials defects – such as stacking faults or dislocations. For such purpose, a relatively simply measure, particularly suited for highly symmetric systems (such as the FCC) and referred to as centro-symmetry parameter will be introduced.

4.5.1 Pressure: Virial Stress

Pressure and stress are usually "measured" via use of the virial stress [22]. Let Ω_i denote the volume around atom i and r_α^{ij} denote the α component of the distance vector relating atom i and atom j. A point-wise measure of stress at atom i can be written as follows:

$$\Pi^i = \frac{1}{\Omega_i}\left[\frac{1}{2}\sum_i \sum_{i \neq j} \frac{U'}{r^{ij}} r_\alpha^{ij} r_\alpha^{ij} - \sum_i m v^i v^i\right] \tag{4.67}$$

From the philosophical standpoint a point-wise measure of stress has little rigorous scientific ground. For this reason, the point-wise measure shall be averaged over a representative number of atoms as follows:

$$\sigma = \frac{1}{N}\sum_i \Pi^i \tag{4.68}$$

While this measure of stress is of great interest, it cannot rigorously be related to a usual measure of stress defined at the scale of the continuum. Among others, it can be seen that by defining stress with (4.67) and (4.68) a deviation of stress would be prediction whether or not the system is subjected to rigid body motion.

4.5.2 Order: Centro-Symmetry

The centro-symmetry parameter, ρ , introduces a simple measure of atomic order within the system [23]. Let r_i denote the position of atom i within an FCC cell, the centro-symmetry parameter is then given by:

$$\rho = \sum_{i=1,6} |r_i + r_{i+6}|^2 \tag{4.69}$$

ρ is of great use to rapidly acknowledge the presence of stacking faults and partial dislocations. With this notation, it is clear that atoms in a perfect lattice configuration will have a centro-symmetry parameter equal to zero. For Au, surface atoms will have a centro-symmetry parameter equal to 24.9. Atoms in a stacking fault position and atoms halfway between HCP and FCC sites will have centro-symmetry parameters, respectively, equal to 8.3 and 2.1. As shown in Fig. 4.5, with this simple parameter, visualization of a partial dislocation loop (red atoms) and resulting stacking fault (yellow atoms) can rapidly be executed.

4.5.3 Boundaries Conditions

Similarly to any finite element and finite difference based simulations – used more readily in computational fluid mechanics – the boundary conditions imposed on the physical system are of critical importance. The extraneous

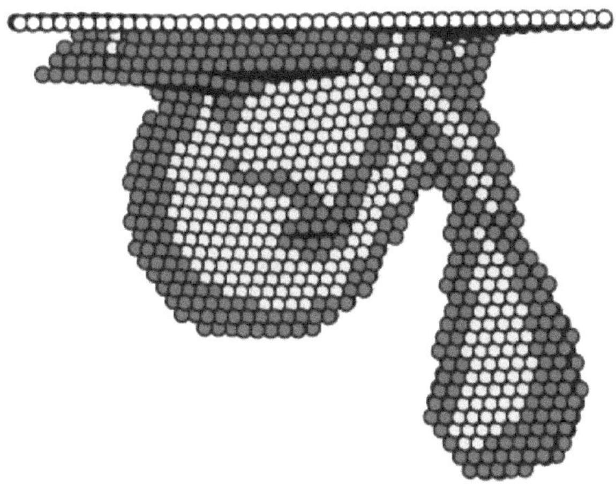

Fig. 4.5 Partial dislocation loop represented with the centro-symmetry parameter. Image extracted from [23]

difficulty in MD arises from the fact that all simulations essentially represent a transient response and that the simulation time is limited. The second difficulty in using MD is to obtain information relevant at higher scale arises from the limit in the size of the physical system. To overcome this limitation, periodic boundary conditions can be imposed on the system.

Depending on the problem studied, the use of periodic boundary conditions can aver helpful to eliminate simulation artifacts arising from free surfaces. For example, as will be shown in the following section, periodic boundary conditions are used to construct and simulate the behavior of bicrystal interfaces. As shown in Fig. 4.6, exhibiting a sketch of a two dimensional primary cell (central cell) repeated periodically, with the periodicity condition each atom, such as the red atom, for example, leaving the primary cell on a given side of the simulation box, will necessarily enter the primary cell from the side opposite to outlet side. Caution must be used when defining the boundary conditions. For example, a three-dimensional system subjected to periodic boundary conditions and represented by the NVE ensemble will not evolve due to the incompatibility in the fully periodic and constant volume conditions.

Alternatively, the primary cell surfaces can be treated as free surfaces which naturally allow studying free boundary effects. Insightful information regarding the notion of surface stress and more generally the domain of application of the Shuttleworth equation can be obtained from the use of free surfaces (see Chapter 8).

In addition to the periodic/free surface condition which can be imposed on each of the primary cell external surface, pressure, displacement or velocities can be imposed to either predict the system's equilibrium configuration or its transient response. These boundary conditions are to be selected with great care. For example, consider the bar of length L, shown in Fig. 4.7, and assume it

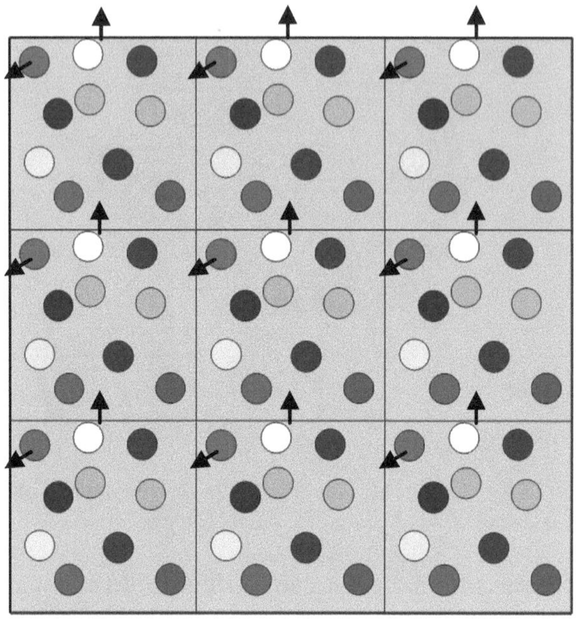

Fig. 4.6 Sketch of a two dimensional primary cell periodically repeated in all directions

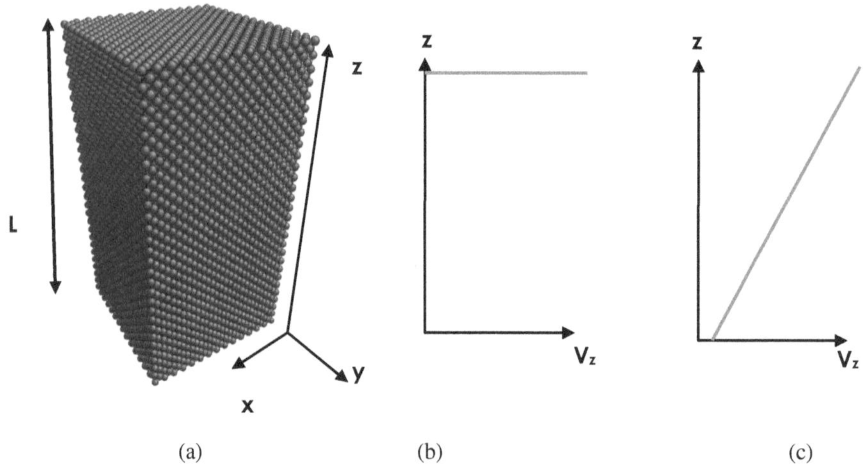

Fig. 4.7 Sketch of a bar (**a**) with two possible boundary conditions (**b**) and (**c**)

is modeled in the NVT ensemble. Assume all surfaces are free surfaces. If, as shown in Fig. 4.7(b), the bottom surface of the bar is constrained to a have a null velocity in the **z** direction and the top surface has a fixed velocity in the **z** direction set to $\mathbf{V_Z}$, the strain rate in the direction of loading will be:$\dot{\varepsilon}_Z = \frac{V_z \Delta t}{L \Delta t} = \frac{V_z}{L}$. Inside the bar, the strain in the **z**direction will not be homogeneous. Such a test would thus correspond more to a Shock test than to a typical tensile test. Alternatively, if as shown in Fig. 4.7(c), a ramp velocity was imposed along the z direction the strain in the z direction will be homogeneous. However, the strain rate imposed on the bar will not be constant. Clearly, it can be seen that boundary conditions (b) and (c) will lead to very different results and the appropriate choice of boundary condition is function of the study. Note that in the "thought" simulation presented in the above, the bar is still free to follow a rigid body motion in the **xy**plane. These could be prevented by fixing one or more atoms or by setting the overall system's momentum to zero. The two methods are obviously not equivalent.

4.6 Numerical Algorithms

Several numerical algorithms (Euler, trapezoidal, Runge Kutta, etc.) can be employed to solve the system's of equation of motion. The objective is clearly to use a numerical scheme that satisfies the following three requirements: (1) time efficiency, (2) precision, and (3) stability. For complete review on the subject of numerical integration the reader is referred to books dedicated to the subject [3]. Among the large number of schemes available, the most often used procedures presenting the most suitable compromise with respect to the three requirements mentioned in the above are typically second order methods such as the velocity Verlet (and similarly the leapfrog) method and predictor-corrector methods. Indeed, first-order methods (e.g., Euler implicit and explicit methods) suffer from poor stability and higher-order methods (e.g., Runge Kutta) are more computationally intensive.

4.6.1 Velocity Verlet and Leapfrog Algorithms

The velocity Verlet and leapfrog method correspond to two different formulations of the same algorithm. They will produce the exact same trajectory. The algorithms can be derived fairly easily by recalling Taylor's expansion. Given a function f of a variable x_n, where n corresponds to a given time step, if x_{n+1} is in the neighborhood of x_n and the function f is at least C^3, then Taylor's expansion of the function f in the neighborhood of x_n can be written as follows:

$$f(x_{n+1}) = f(x_n) + \Delta t \cdot f'(x_n) + \frac{(\Delta t)^2}{2} \cdot f''(x_n) + \frac{(\Delta t)^3}{6} \cdot f'''(x_n) + O(f''''(x_n)) \qquad (4.70)$$

Applying Equation (4.70) to the case of the position vector, r_i^n, at time step $n+1$ and $n-1$ and recalling the equilibrium equation – under its different possible form – one readily obtains:

$$r_i^{n+1} = r_i^n + \Delta t \cdot \dot{r}_i^n + \frac{(\Delta t)^2}{2} \cdot \ddot{r}_i^n + \frac{(\Delta t)^3}{6} \cdot \dddot{r}_i^n + \mathrm{O}(\ddddot{r}_i^n) \tag{4.71}$$

$$r_i^{n-1} = r_i^n - \Delta t \cdot \dot{r}_i^n + \frac{(\Delta t)^2}{2} \cdot \ddot{r}_i^n - \frac{(\Delta t)^3}{6} \cdot \dddot{r}_i^n + \mathrm{O}(\ddddot{r}_i^n) \tag{4.72}$$

In the above, superscript n denotes the *nth* time step. Adding (4.71) and (4.72) and using the force balance, one obtains the velocity Verlet scheme:

$$r_i^{n+1} = 2r_i^n - 2r_i^{n-1} + (\Delta t)^2 \cdot \ddot{r}_i^n \tag{4.73}$$

Here $\ddot{r}_i^n$ the acceleration of particle $i$at time step n is calculated from the expression of the force balance (e.g., Equations (4.1) or (4.61) or (4.64)). If the knowledge of the velocity is required, use of (4.70) leads to the following approximation:

$$\dot{r}_i^n = \frac{r_i^{n+1} - r_i^{n+1}}{2\Delta t} \tag{4.74}$$

As shown by Equation (4.73), the Verlet algorithm is second order. With this scheme, the energy drift is insignificant. The algorithm in the above can reformulated in the form of the leapfrog scheme where the velocities at time step $n+1/2$ are calculated from the accelerations at step n (obtained from the knowledge of the forces). Therefore, one has:

$$\dot{r}_i^{n+1/2} = \dot{r}_i^{n-1/2} + \Delta t \cdot \ddot{r}_i^n \tag{4.75}$$

and

$$r_i^{n+1} = r_i^n + \Delta t \cdot r_i^{n+1/2} \tag{4.76}$$

4.6.2 Predictor-Corrector

As shown in the above, the Verlet and leapfrog algorithms (1) can be easily implemented, (2) do not induce substantial energy drift, and (3) require the evaluation of forces only once per particle and per time-step. Therefore, these open algorithms are very frequently used in MD. However, closed methods such as the Gear algorithms, can improve the accuracy of the calculation without requiring substantial additional computational time. The idea here is to compute the position vector in two steps: (1) prediction via the use of Taylor's expansion, and (2) correction to minimize the prediction error. Complete derivations of the Gear method can be found in book by Gear [24].

First, let us introduce vector x_n containing all information given by a Taylor expansion of the position at time $n+1$:

$$x_i^n = \begin{pmatrix} r_i^n \\ \Delta t \cdot \dot{r}_i^n \\ \frac{(\Delta t)^2}{2} \cdot \ddot{r}_i^n \\ \frac{(\Delta t)^3}{6} \cdot \dddot{r}_i^n \end{pmatrix} \tag{4.77}$$

This representation is referred to as the Nordsieck (or N-) representation. Other representations such as the C and F representationS can also be used. For the sake of simplicity, only the N-representation is used here. Using (4.70) and applying it to the case of the position, and its first-, second-, and third-order derivation, one obtains a prediction of x_i^{n+1} which is denoted y_i^{n+1} and given by:

$$y_i^{n+1} = A x_i^n \tag{4.78}$$

with

$$A = \begin{pmatrix} 1 & 1 & 1 & 1 \\ 0 & 1 & 2 & 3 \\ 0 & 0 & 1 & 3 \\ 0 & 0 & 0 & 1 \end{pmatrix} \tag{4.79}$$

This method is referred to as the four-values Gear method because it uses the position and its first three time derivatives. While the Gear method can be expanded to an *nth*-value method. Typically, MD simulations use the four- or five-values method. The prediction step is defined by Equations (4.78) and (4.79). From these predictions, the difference between the predicted force and acceleration at step $n+1$ can be estimated. The second step, corresponding to the correction step, minimizes the error between predicted force and acceleration via use of a correction vector **a**:

$$x_i^{n+1} = y_i^{n+1} + a\frac{(\Delta t)^2}{2}\left(\tilde{F}_i\left(r_i^{n+1}\right) - \tilde{\ddot{r}}_i^{n+1}\right) \tag{4.80}$$

with

$$a = \begin{pmatrix} 1/6 \\ 5/6 \\ 1 \\ 1/3 \end{pmatrix} \tag{4.81}$$

Here, the symbol ∼ denotes a value obtained from the prediction step. Several different "formulations" based on the Gear method can be used – depending on the desired accuracy and stability – which will result in different values of the correction vector $\boldsymbol{a}$.

The predictor corrector method will necessarily be more computationally expansive than the simpler Verlet or leapfrog methods. However, the solution accuracy will be improved. This can be easily seen by numerically integrating a harmonic oscillator. This is left as an exercise for the reader.

4.7 Applications

As mentioned in the introduction to this chapter, the molecular dynamic method is particularly suitable for studies at the nanoscale. In the case of NC materials, a major part of the understanding of their plastic response at the atomic scale – the scale transition from the atomistic scale to the macroscopic scale remaining an open challenge – has been found via atomistic simulations. In this section some of the most interesting findings based on MD, based on the concepts introduced throughout this chapter, will be presented. Considering the vast number of MD studies performed on NC material, it is clear that this section cannot present all existing work. In addition to the following subsections, chapters 5 and 9 present MD and quasi-continuum simulations particularly dedicated to bicrystal simulations. In what follows, the construction and simulations methods to represent bicrystal interfaces responses and to model the mechanism of grain boundary migration and of dislocation/interface and dislocation/dislocation interaction will be presented.

4.7.1 Grain Boundary Construction

As shown in work by Spearot [21] and by Rittner and Seidman [25], bicrystal interfaces can be constructed by an ingenious use of molecular statics simulations. Although this technique was not presented in as much detail as molecular dynamics, its principle is very similar. Molecular statics rely on the use of the interatomic potential to find the equilibrium structure configuration via the use of conjugate gradient method. As shown in Chapter 5, five macroscopic degrees of freedom are required to describe the geometry of a grain boundary. The procedure shown here is based on the following steps. Consider, as shown in Fig. 4.8, two lattice blocks A and B with different crystallographic orientations. For example, both crystals A and B can share the same [001] axis parallel to the **y** axis such that the interface will correspond to a pure tilt grain boundary. If the misorientation angle between the interface plane and the [100] axis – as in the present case – is the same for both crystals, the grain boundary will correspond to a symmetric tilt grain boundary.

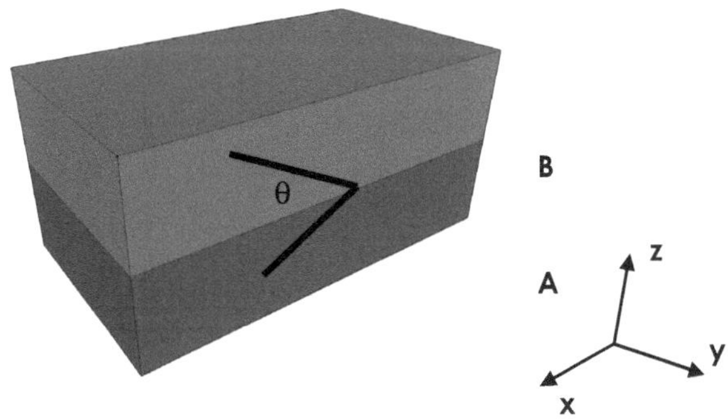

Fig. 4.8 Schematic of the bicrystal geometry block construction

A grain boundary interface can then be produced by first considering periodicity of the system with respect to the **yz** and **xz** planes. The top and bottom surfaces – parallel to the **xy** planes – are free surfaces such that any relaxation during the energy minimization procedure can occur. However, these surfaces must remain planar. With this configuration, realistic grain boundary structures can be generated by performing an energy minimization while simultaneously removing atoms located at distances smaller than an assigned critical value – the overlap distance – to other atoms. As shown by Spearot [21], with this approach the final configuration studied will depend on the overlap distance. The procedure described in the above will ensure that a minimum energy configuration can be obtained. In order to assess whether the previous configuration corresponds to a local energy minimum or to a global minimum, several neighboring initial configurations need to be considered.

As shown in Fig. 4.9, presenting (a) an HRTEM image of aΣ5(210) 53.1grain boundary in Al – composed of B' structural units – and (b) its corresponding prediction, realistic grain boundary structures can be generated via atomistic simulations. In Chapter 5, applications of this grain boundary construction

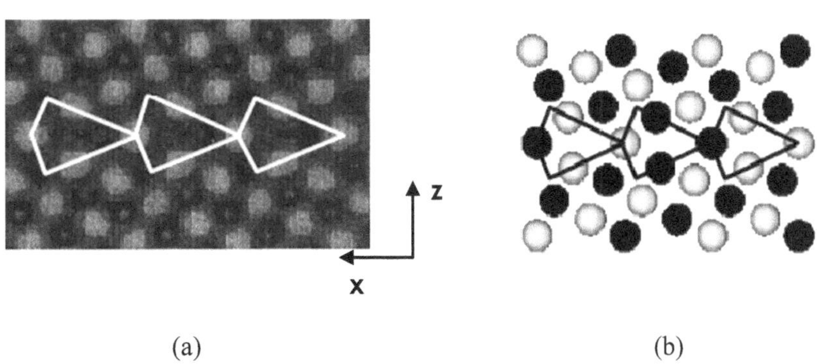

Fig. 4.9 HRTEM (**a**) and molecular statics predictions (**b**) of the structure of a symmetric tilt [001]Σ5(210) 53.1grain boundary in Al. Images extracted from [21]

method to calculate grain boundary energies and elastic and plastic response will be shown.

4.7.2 Grain Growth

Owing to their small grain size, NC polycrystals subjected to both complex boundary conditions (e.g., high external temperature and velocity gradient) can be studied. These studies provide great insight on the activity of particular mechanisms such as that of grain growth, to which this section is dedicated. Figure 4.10 shows a numerical model of a polycrystalline NC Pd structure containing 25 grains with average grain size 15 nm following a log-normal distribution. All grains are columnar two-dimensional grains. The two-dimensional structure is repeated periodically in both planar directions. The insert in Fig. 4.9 represents the primary cell. All grains share the same [001] crystallographic orientation (e.g., the outer plane axis). Therefore, all grain boundaries will be tilt grain boundaries (nonsymmetric). The crystallographic orientations are chosen such that all grain boundaries are large-angle type.

The structure in the above was created as follows. First, 25 seeds were randomly "planted" in the **x-y** plane. The geometrical template is then obtained from a Voronoi construction. In the simplest manner, the latter is based on the fact that for any given point **r** of a set of point, there exists one point closer to **r**. A boundary can then be constructed between the two points. A polycrystalline

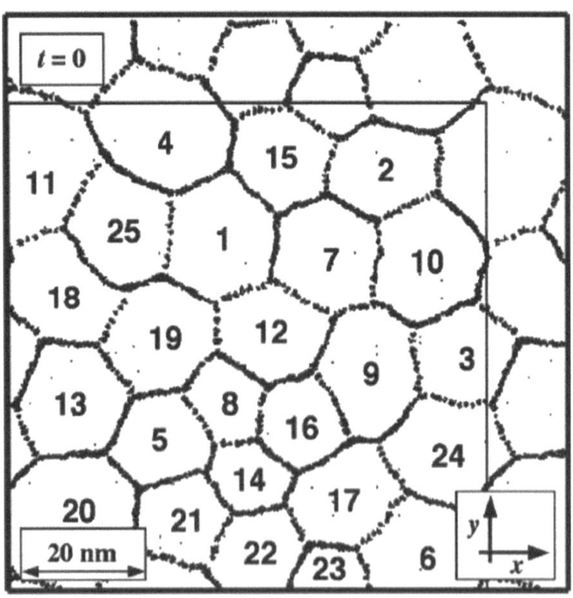

Fig. 4.10 Two-dimensional columnar NC polycrystals containing 25 grains [26]

structure can be reconstructed by repeating this procedure which was extremely simplified here. A Monte-Carlo algorithm was used to adjust the Voronoi tessellation such that a log-normal grain size distribution can be produced. Clearly, in the case of a 25-grain microstructure, it is rather difficult to ensure a proper log-normal distribution.

The second step consists of selecting a crystallographic orientation for each grain – which as mentioned in the above share the same [001] orientation. A random set of orientations, later refined with a Monte Carlo algorithm, is selected such as to yield only large-angle grain boundaries. Grain boundaries are then created by a procedure similar to that presented in previous subsection – consisting of removing atoms which are closer than the overlap distance to other atoms – is used.

From the above microstructure, the mechanism of grain growth can be simulated. As shown in Fig. 4.11(a), substantial grain growth – which as

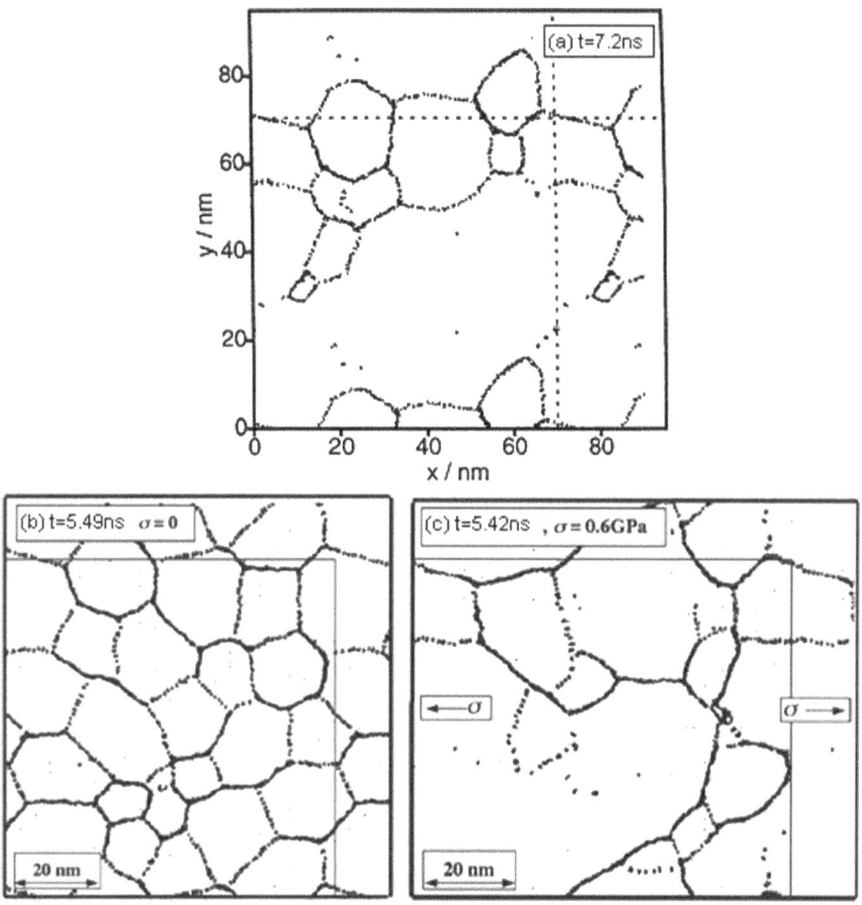

Fig. 4.11 Predicted microstructure after (**a**) temperature constraint at 1400 K ,(**b**) temperature constraint at 1200 K, and (**c**) temperature constraint at 1200 K and 600 MPa applied tensile stress

shown by Haslam et al. is initiated by grain rotation and followed by curvature driven grain boundary migration – can be engendered by subjecting the polycrystals to a high temperature of 1400 K for several picoseconds.

In order to evaluate the effect of an applied stress on grain growth, the same microstructure was subjected to different boundary conditions: (1) a constant high temperature with T = 1200 K and (2) a constant temperature with T = 1200 K and externally applied stress $\sigma = 600\ MPa$ as depicted in Fig. 4.11(c). Interestingly, it can be seen that, in contrast with previous simulation at 1400 K, in the absence of an external stress, no substantial grain growth was initiated after exposing the sample to 1200 K for several picoseconds. If a moderate stress is imposed to the polycrystal in addition to the external temperature, it can be seen that grain growth is greatly enhanced.

Although these simulations are typically not used to provide quantitative explanation for the numerically observed effect of stress, it can motivate interesting discussions notably on the dependence of mobility on stress. After careful analysis, the authors first discarded the role of elastic anisotropy and then suggested that stress activated grain boundary sliding and rotation may enhance grain boundary mobility and diffusion.

4.7.3 Dislocation in NC Materials

The size effect in the activity of dislocations is remarkable for there is a critical grain size below which commonly used models for coarse grained materials – based on the statistical storage and dynamic recovery of dislocations – do not apply anymore. Therefore, below the aforementioned critical grain size, dislocation activity can be studied via discrete approaches rather than statistical approaches. A more detailed discussion on the matter will be presented in Chapter 6.

In what follows, two example studies pertaining to the understanding of (1) the process of dislocation nucleation from grain boundaries and subsequent propagation [27] and (2) the interaction between mobile dislocations and twin boundaries – for which nucleation was too revealed by MD simulations discussed in Chapter 6 – will be presented.

4.7.3.1 Dislocation Nucleation and Propagation

Using a procedure, in essence, similar to that presented in previous subsection – based on Voronoi tessellation – more complex fully three-dimensional structures can be numerically generated as shown in Fig. 4.12. As shown, grains G0, G1, G2, G3, and G4 – G1 is not explicitly shown and corresponds to the plane where dislocation activity can be observed – are part of a primary cell composed of 15 grains with average grain size 12 nm. The potential used is the EAM potential of Mishin et al. fitted for Al. All grain boundaries are high-angle grain

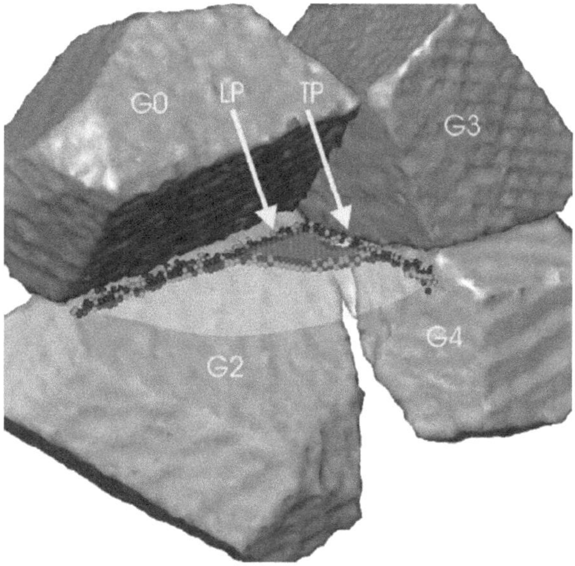

Fig. 4.12 Three-dimensional polycrystalline Al structure showing the emission of a leading and a trailing dislocation from different sites

boundaries. Also, as shown in Fig. 4.12, grain boundaries are nonperfectly planar (e.g., presence of ledges).

Note that the choice of Al for the material system studied is motivated by the fact that owing to its ratio of the unstable stacking fault energy over the stable stacking fault energy which approaches unity, both leading and trailing partial dislocations can typically be observed during the time of a simulation. This is generally not the case for Cu or Ni.

When the NC structure described in the above is subjected to a 1.6 GPa tensile stress, a trailing partial dislocation is nucleated from a ledge located along the G2/G1 grain boundary near a triple line. Following the emission of the leading partial dislocation, the ledge disappears. As shown in MD simulation on bicrystal interfaces, the ledge does not necessarily disappear following an emission event. Also, ledges can be generated by the atomic rearrangement following the emission of a dislocation from a perfectly planar grain boundary. Interestingly, simultaneously to the emission event a stress concentration arises along grain boundary G4/G1. Following the leading partial dislocation during its motion shows that as it travels along grain boundaries G2/G1 and G4/G1 it can be pinned at ledges. Upon depinning dislocation debris can remain attached to the ledge. Pinning/depinning events were shown to be thermally activated. Finally, after several picoseconds, a trailing partial dislocation is emitted from grain boundary G4/G1 in the neighborhood of the previously mentioned stress concentration.

4.7.3.2 Dislocation Twin Boundary Interaction

With a similar method to that used to create the two-dimensional columnar microstructure presented in the above (e.g., grain growth simulations), a primary cell containing four grains of equal grain size can be created (d = 30–100 nm). Using Al as the material system and orienting grains such that all grain boundaries are pure tilt and large-angle type, a tensile stress

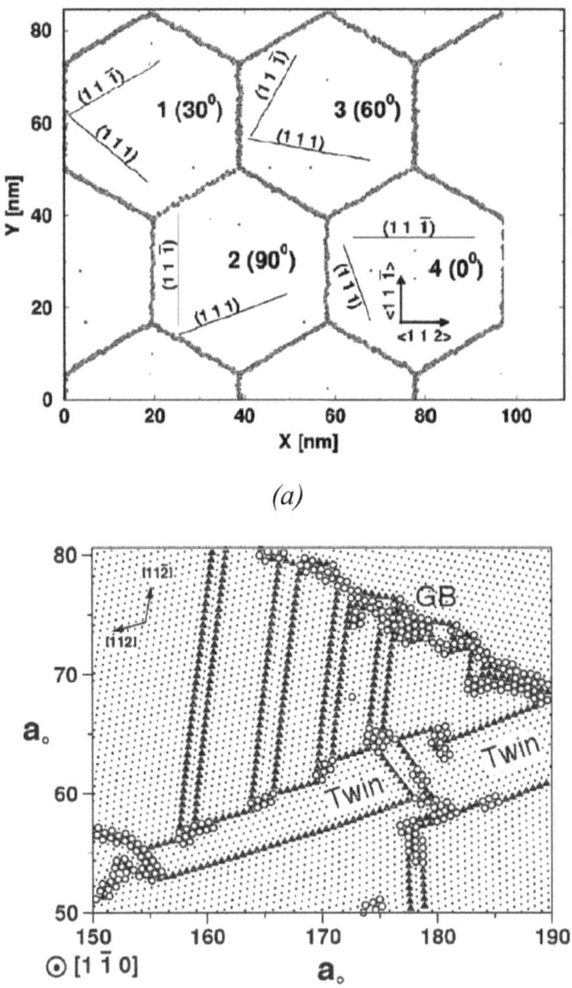

Fig. 4.13 (**a**) Microstructure containing four grains with same grain size d = 45 nm, (**b**) detwinning caused by the interaction between slip dislocations and twin interfaces. Images extracted from [29]

$\sigma = 2.2 - 2.5GPA$ can be applied – by use of the Parrinello-Rahman method [28] – parallel to the **x** axis while maintaining the system's temperature to 300 K.

Similarly to the fully three-dimensional simulations described in previous section, the emission of dislocations from grain boundaries is predicted. After some time, freshly nucleated dislocations will necessarily interact with one another. The reaction products must necessarily obey the conservation of the Burgers vector rule and can serve as visual illustration of the use of Thompson tetrahedron. Additionally, nonintuitive, processes were revealed by these simulations. As discussed in Chapter 6, twin domains can be formed in NC materials via emission of partial dislocations on parallel slip planes. An example of these reoriented domains is shown in Fig. 4.12(b). Upon meeting a twin interface, a slip dislocation can either (1) penetrate the interface or (2) dissociate into a partial dislocation with Burgers vector parallel to the twin plane and into a dislocation with Burgers vector perpendicular to the interface. A stair rod lock may not necessarily result from the dissociation event. Each possible reaction depends on the orientation of the dislocation line and of its Burgers vector with respect to the twin plane. Depending on the sign of the incoming dislocation, in the event where dissociation occurs, either a positive or a negative twinning dislocation will be generated which will lead to either growth or shrinkage of the twin domain (e.g., detwinning). As shown in Fig. 4.13(b) when numerous dislocations are emitted from the same grain boundary region, the twin domain can detwin substantially and even be cut through.

4.8 Summary

In this chapter, the fundamental of molecular dynamics simulations were presented. First, general considerations were discussed (e.g., equations of motion, Hamiltonian). Following this, the complexity associated with the description of interaction forces between atoms was discussed. Among others, the embedded atom method was introduced. Then, the concepts of statistical ensembles, such as the canonical, microcanonical, and isobaric isothermal ensemble, were recalled. Among others, the relation between each ensemble statistical distribution and thermodynamic quantities (e.g., S, G, H) was illustrated.

Second, molecular dynamic methods, consisting of augmenting the equations of motion to introduce temperature or pressure external constraints while allowing to sample the entire hyperspace of acceptable microstates, were introduced (e.g., Nosé Hoover molecular method, Melchionna et al.) Third, the most common numerical algorithms (e.g., velocity, Verlet, leapfrog, and Gear predictor-corrector) used in atomistic simulations were presented. Applicable boundary conditions were discussed.

Finally, several illustrations of the newly introduced concepts were depicted. For example, the construction of symmetric tilt grain boundaries via the combined use of energy minimization and constrained nonoverlapping was shown.

Additionally, dislocation activity and grain boundary instabilities were discussed from simulations on two-dimensional columnar and fully three-dimensional polycrystals.

References

1. Burghaus, U., J. Stephan, L. Vattuone, and J.M. Rogowska, *A Practical Guide to Kinetic Monte Carlo Simulations and Classical Molecular Dynamics Simulations.* Nova Science, New York, (2005)
2. Rapaport, D.C., *The Art of Molecular Dynamics Simulation.* Cambridge University Press, New York, (1995)
3. Chapra, C. and R.P. Canale, *Numerical Methods for Engineers.* McGraw-Hill, New York, (2005)
4. Kubo, R., *Statistical Mechanics.* North Holland, Amsterdam, (1988)
5. Wannier, G.H., *Statistical Physics.* Dover Publications, New York, (1966)
6. Hoover, W.G., ed. *Proceedings of the International School of Physics – Enrico Fermi-Molecular Dynamics Simulations of Statistical Mechanical Systems.* North Holland: Amsterdam, (1985)
7. Hildebrand, F.B., *Methods of Applied Mathematics*, Dover Publications, New York, (1992)
8. Hohenberg, P. and W. Kohn, Physical Review **136**, (1964)
9. Kohn, W. and L.J. Sham, Physical Review **140**, (1965)
10. Daw, M.S. and M.I. Baskes, Physical Review B **29**, (1984)
11. Foiles, S.M., M.I. Baskes, and M.S. Daw, Physical Review B **33**, (1986)
12. Finnis, M.W. and J.E. Sinclair, Philosophical Magazine A **50**, (1984)
13. Daw, M.S., Physical Review B **39**, (1989)
14. Baskes, M.I., Physical Review B **46**, (1992)
15. Chandler, D., *Introduction to Modern Statistical Mechanics.* Oxford University Press, New York, (1987)
16. Holian, B.L., H.A. Posch, and W.G. Hoover, Physical Review A **42**, (1990)
17. Hoover, W.G., A.J.C. Ladd, and B. Moran, Physical Review Letters **48**, 1818–1820 (1982)
18. Nose, S., Molecular Physics **52**, (1984)
19. Nose, S., Molecular Physics **57**, (1986)
20. Melchionna, S., G. Ciccotti, and B.L. Holian, Molecular Physics **78**, (1993)
21. Spearot, D., *Atomistic Calculations of Nanoscale Interface Behaviors in FCC Metals.* Georgia Institute of Technology, Atlanta, GA, (2005)
22. Tsai, D.H., Journal of Chemical Physics (1979)
23. Kelchner, C.L., S.J. Plimpton, and J.C. Hamilton, Physical Review B **58**, (1998)
24. Gear, C.W., *Numerical Initial Value Problems in Ordinary Differential Equations.* Upper Saddle River, NJ: Prentice Hall, (1971)
25. Rittner, J.D. and D.N. Seidman, Physical Review B **54**, (1996)
26. Haslam, A.J., S.R. Phillpot, D. Wolf, D. Moldovan, and H. Gleiter, Materials Science and Engineering A **318**, (2001)
27. Van Swygenhoven, H., P.M. Derlet, and A.G. Froseth, Acta Materialia **54**, (2006)
28. Parrinello, M. and A. Rahman, Journal of Applied Physics **52**, (1981)
29. Yamakov, V., D. Wolf, S.R. Phillpot, and H. Gleiter, Acta Materialia (2003)

Chapter 5
Grain Boundary Modeling

All structures tend to reach a minimum energy configuration; a perfect single crystal, for example, is the illustration of such configuration. However, the former structure with very low internal energy may not be suitable for all domains of application. Indeed, depending on the desired performance, the introduction of defects into a perfect microstructure can prove advantageous. Doped silicon plates and doped ceramics are good examples of the possible ameliorations resulting from the presence of defects in a material. Similarly to dopants, grain boundaries can lead to improved materials response. In general, grain boundaries provide barriers to the motion of dislocations within a grain – this in turns leads to a more pronounced hardening – and can also act as barrier to crack propagation, which can improve the materials' ductility.

As mentioned in Chapter 2, the volume fraction of grain boundaries is significantly higher in NC materials than in coarse-grain materials. Recall that grain boundary volume fractions as high as 50% were reported in early work on the matter. Clearly, the response of NC materials is affected by the amount and type of grain boundaries composing its microstructure. In Chapter 2, it was also seen that, along a given direction, a grain boundary can exhibit a changing character. That is, some regions of a grain boundary can exhibit no organization of their atomic arrangements while other regions may be well defined. Disordered regions can be considered as amorphous regions which typically exhibit an elastic perfect plastic response. However, the response of ordered regions of grain boundaries is less well known. This chapter discusses only ordered grain boundaries, for much can be learned from them.

Grain boundary modeling, in terms of geometry, elastic stress field, and excess energy, has motivated a large body of research over the past century. Clearly, it is out of the scope of this chapter to review all studies related to grain boundaries. The objective here is to recall key results related to grain boundaries. In particular, a short background will be given prior to describing continuum mechanic–based models. While at first sight continuum models may appear as obsolete compared to numerical models, the complementarity of the two approaches will be demonstrated with applications.

M. Cherkaoui, L. Capolungo, *Atomistic and Continuum Modeling of Nanocrystalline Materials*, Springer Series in Materials Science 112,
DOI 10.1007/978-0-387-46771-9_5,

5.1 Simple Grain Boundaries

Let A and B denote two separate crystal with similar structure. Upon joining crystal A and B, one creates a bicrystal containing a grain boundary. The geometry of the grain boundary does not depend only on the relative orientation of the two crystals. In general, a grain boundary has five macroscopic degrees of freedom and three microscopic degrees of freedom. Macroscopically, three degrees of freedom (e.g., rotations) are used to orient crystal B with respect to crystal A. Once the two crystals have been oriented, the plane defining the grain boundary must still be assigned. This choice consumes two degrees of freedom (rotations with respect to the two grains) [1]. In this case, there is no third degree of freedom assigned to the plane orientation since rotating a plane with respect to its normal does not affect the plane orientation. Microscopically, the three remaining degrees of freedom correspond to the translation vector of the two crystals composing the bicrystal.

It is easy to conceive that modeling of a general grain boundary – with its eight degrees of freedom assigned randomly – would be a gargantuan task. Instead, two types of grain boundaries have been subject to modeling efforts. These grain boundaries are referred to tilt and twist angle grain boundaries.

Let $\boldsymbol{u}$ be the unit vector representing the axis of relative orientation of the two crystals. The orientation of crystal B with respect to crystal A can then be given by vector $\boldsymbol{\omega} = \theta u$. Here, θ denotes the rotation angle. Each of the three components of ω represents one of the three degrees of freedom necessary to orient B with respect to A. The boundary orientation – the last two degrees of freedom – can be assigned with unit vector $\boldsymbol{n}$ denoting the normal to the grain boundary plane. With the above geometrical consideration, twin and tilt grain boundaries can be defined. As given by Read [2] – who pioneered the area of grain boundary engineering – a tilt grain boundary is such that the axis of relative orientation of the crystals, $\boldsymbol{u}$, lies in the grain boundary plane. In other words, $\boldsymbol{u}$ is perpendicular to $\boldsymbol{n}$. On the contrary, a twist grain boundary is such that $\boldsymbol{u} = \boldsymbol{n}$. In other words, the axis of relative orientation of the crystals is perpendicular to the grain boundary plane. A schematic of the simplest tilt grain boundary is given in Fig. 5.1(a). More tortuous geometries

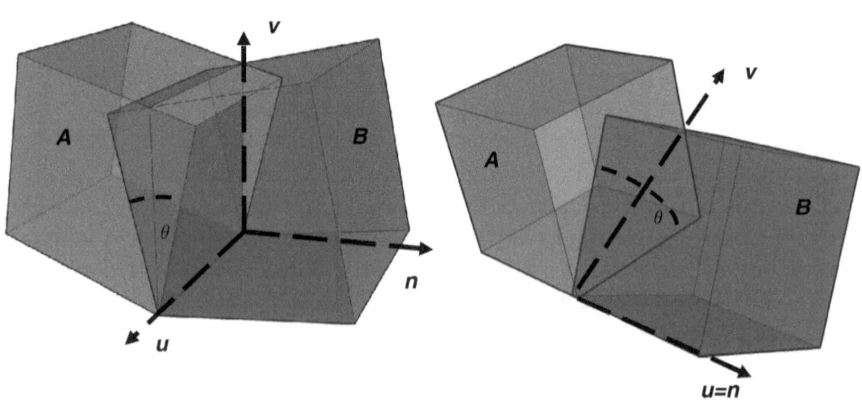

Fig. 5.1 (**a**) Simple tilt grain boundary, (**b**) simple twist grain boundary

could obviously be designed. Crystal A appears in red while crystal B appears in blue, and the grain boundary is defined as the region of intersection of both crystals (in dark red). The grain boundary plane is defined with vectors ***u*** and ***v***. Fig 5.1(b) shows a schematic of a simple twist grain boundary.

5.2 Energy Measures and Numerical Predictions

As expected, the energy of a given grain boundary is dependent on each of its degrees of freedom. In the case of tilt and twist grain boundaries, the interphase energy is thus dependent on the misorientation angle between the two crystals and on the grain boundary plane.

Grain boundary free energies can be measured experimentally via use of Herring's formula, which was originally derived from a variational approach (e.g., virtual displacements). Consider the junction of three crystals, as shown in Fig. 5.2, and let OA, OB, and OC represent respectively the interface between crystal 1 and 2, crystal 2 and 3, and crystal 3 and 1, respectively. Also let γ_1, γ_2 ,γ_2 and ϕ_1, ϕ_2, ϕ_3 denote the corresponding free energies and the angle formed by the interfaces, respectively. Let us note here that for the sake of simplicity the twists' contributions are not accounted for. The equilibrium configuration of the tricrystal is then given by [3]:

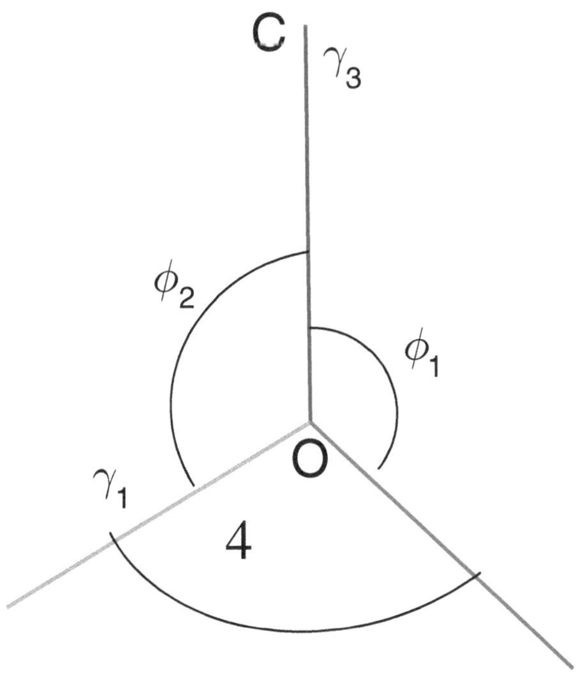

Fig. 5.2 Junction of three grain boundaries

$$\frac{\gamma_1}{\sin\phi_1} = \frac{\gamma_2}{\sin\phi_2} = \frac{\gamma_3}{\sin\phi_3} \tag{5.1}$$

Using Herring's formula, Gjostein and Rhines [4] systematically measured the interface free energy of simple tilt and twist grain boundaries with misorientation angles ranging from ~0 to ~70° in the case of copper. Figure 5.3 presents the measured data in the case of <001> pure tilt boundaries. The dashed line does not have any physical significance and simply serves as a guideline.

From Fig. 5.3, it can be seen that the energy of a simple tilt grain boundary increases with increasing misorientation angle, with a maximum at 43° after which the energy decreases. A similar trend was obtained in the case of pure twist grain boundaries. The experimental measures presented in the above do not exhibit the presence of "metastable" misorientations which would translate by the presence of cusps – additionally to the $\theta = 0$ cusp – in Fig. 5.3. However, the existence of such metastable configurations was clearly shown in several experiments. For this purpose, Chan and Baluffi [5] used the crystallite rotation method on Au [001] twist grain boundaries. It consists of first sintering small crystallites (~80 nm in diameter) onto a specimen at predetermined twist orientations and then subjecting the specimen to an anneal in situ so as to observe grain rotation towards relaxed configurations. It was found that all lattice oriented with $\theta < 34°$ tended to reorient towards $\theta = 0$, all lattice initially in the $35 \leq \theta \leq 40$ range reoriented towards $\theta = 36.9°$, and all crystallites orientated with $\theta < 40°$ reoriented towards $\theta = 45°$. From these anneal experiments it can be concluded that these particular twist orientation (e.g., $\theta = 0, 36.9$ and 45) appear more energetically favorable than other random orientations. Yet, these energy cusps were not rigorously found in measures shown in the above (Fig. 5.3).

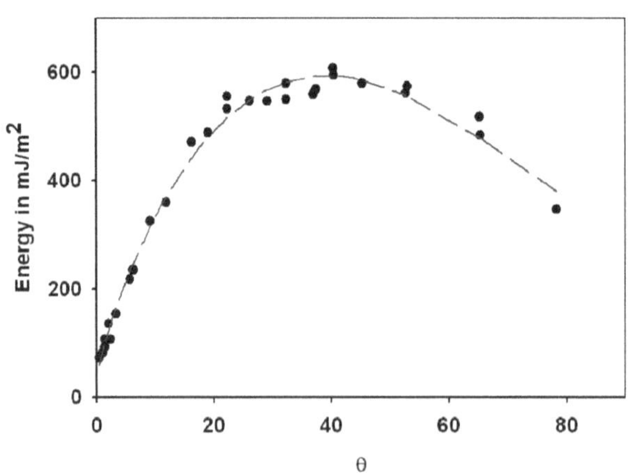

Fig. 5.3 Energy evolution with misorientation angle for copper pure tilt <001> grain boundaries. Experimental data from Gjostein and Rhines [4]

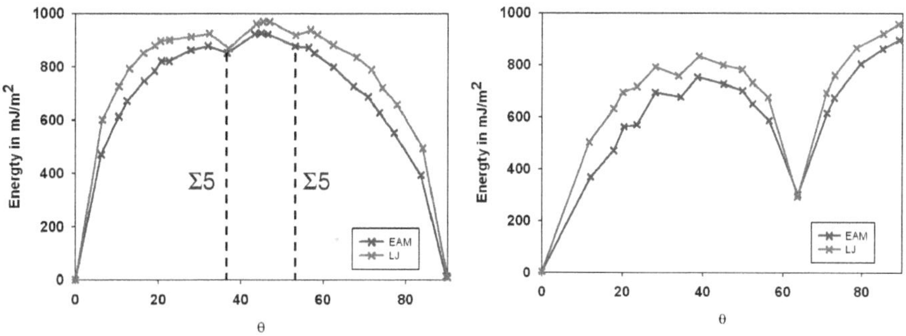

(a) (b)

Fig. 5.4 Energy evolution with misorientation angle for symmetric tilt grain boundaries on planes perpendicular to (**a**) <001> and (**b**) <112>. Data reproduced from Wolf [6]

On the other hand, molecular statics simulations were used to assess of the presence of energy cusps. These simulations were part of an extensive molecular based set of simulations by D. Wolf [6, 7]. Figure 5.4(a) and (b) presents the predictions of the evolution of symmetric tilt grain boundary free energy with misorientation angles for boundary planes perpendicular to the <001> and <112> orientations for Cu, respectively. Crosses refer to calculated date while lines serve as a guide to the eye. In order to overcome limitations related to the use of a particular potential, the author used both the Lennard Jones (LJ) potential and the embedded atom method (EAM) potential (which, as was discussed in Chapter 4, is more adequate to model FCC structures). In the case of the <001> symmetric tilt grain boundary, the presence of energy cusps – corresponding to misorientations from which the energy increases at an infinite rate as the misorientation angle is slightly changed – can be observed at the (310) 36.87° and (210) 53.13° misorientations. Similarly, energy cusps are predicted in the case of <112> symmetric tilt grain boundaries.

The existence and particular orientations corresponding to energy cusps both in symmetric tilt and twist grain boundaries has stimulated a large body of research aiming at understanding the correlation between grain boundary energy and it structure.

5.3 Structure Energy Correlation

Figure 5.1 clearly does not show the details of the atomic structure of the grain boundary nor does it explain the particular energy dependence on misorientation angles. Based on molecular simulations which revealed particular atomic arrangements within both tilt and twist grain boundaries, to be presented in what follows, several models resulting from physical considerations were introduced to easily describe and predict important atomic features of grain boundaries.

Prior to introducing these models let us discuss important geometrical and physical features of grain boundaries. First, grain boundaries can typically be sorted as low angle grain boundaries and large angle grain boundaries. Low-angle grain boundaries exhibit well-organized structures characterized by the discernible presence of dislocation arrays. Read and Shockley [8] first introduced a two-dimensional continuum model based on the dislocation arrangements allowing the prediction of the evolution of low angle grain boundaries as a function of misorientation angle [2]. This model will be presented next. Typically, the distinction between low- and large-angle grain boundaries is established on the following basis. As one fictitiously increases the misorientation angle of a given low-angle grain boundary, the dislocation density with the grain boundary increases (according to Frank formula). In the limit case, the number of dislocations composing the grain boundary will be such that the core of each dislocation will intersect. This limit case defines the onset of the domain of misorientation of large-angle grain boundaries. Typical values range between 20° and 25° of misorientation. With the argument in the above, one expects significant structural differences between low- and large-angle grain boundaries. However, this does not mean that large-angle grain boundaries necessarily lack structure. The following two subsections will present structure energy correlation models for both low-angle grain boundaries and large-angle grain boundaries.

In terms of statistical distribution of low- and large-angle grain boundaries, it was shown in work by Warrington and Boon [9] that, in polycrystals with random grain boundary distribution, the probability of low-angle grain boundaries should equal 0.000825. Deviation from this number would indicate that the grain boundary distribution is not given by a random distribution. In connection with Chapters 1 and 2, it can clearly be seen that the grain boundary distribution in NC materials is clearly not random.

5.3.1 Low-Angle Grain Boundaries: Dislocation Model

As mentioned above, low-angle grain boundaries are often assimilated, and several experimental studies concur with this conceptualization, as particular arrangements of dislocations. Let us clarify this concept by considering, as in the original work of Read [2] and Read Shockley [8], the formation of a low-angle tilt grain boundary. Read proposed the following representation of a low-angle grain boundary: in Fig. 5.5(a) two crystals (red and blue) are not yet connected by a grain boundary, upon creating the low-angle grain boundary (Fig. 5.5(b)), dislocations are present geometrically to ensure the degree of misorientation between the two crystals. In our case, disregarding other dislocation pairs which may be present in an actual grain boundary and would disappear after an anneal process, the minimum set of dislocations present in the grain boundary is an array of edge dislocation all parallel to the [001] axis.

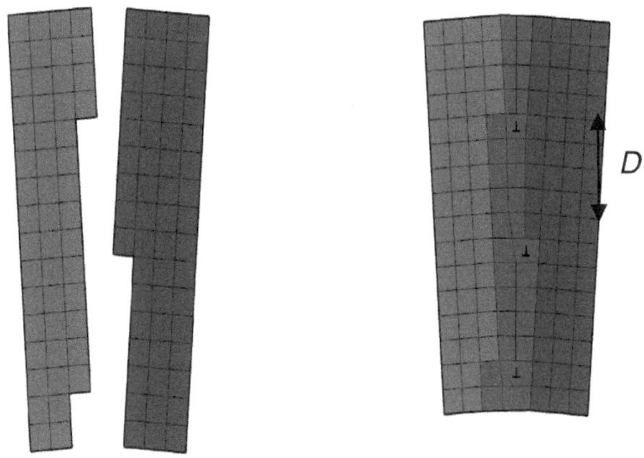

Fig. 5.5 Dislocation modeling of a low-angle symmetric tilt grain boundary; (**a**) two crystals and (**b**) two crystals joined by a grain boundary

The misfit between the two crystals is accommodated by both atomic misfit and elastic deformations. Note that in order to reduce the elastic energy within the grain boundary, the [100] planes of crystal A and B end at alternating intervals.

In the case of a simple symmetric tilt-angle grain boundary, the mean dislocation spacing D is given by Frank's formula, which is obtained with the following simple geometrical consideration: if D represents the average spacing between dislocations, each with Burgers vector **b**leading to $\mathbf{b}/2$ net displacement on each side of the grain boundary median plane (parallel to (001)), then recalling the misorientation angle on each side of the median plane is $\theta/2$ (see Fig. 5.1(a)), one obtains:

$$D = \frac{\mathbf{b}}{2\sin(\theta/2)} \tag{5.2}$$

Rigorous extensions of this simple law in the case of grain boundaries containing two or more different dislocation types can be found in Hirth and Lothe [10]. From this structural representation of low-angle grain boundaries, Read and Shockley developed a model entirely based on dislocation theory and predicting the energy vs. misorientation angle. While the model was limited to a two-dimensional representation, it could very well be extended to be fully three dimensional.

Read and Shockley and later Read proposed two derivations of their model, the first one being based solely on mathematical considerations while the second one is based on a more physical reasoning. For the sake of simplicity, only the intuitive derivation will be shown in detail and the mathematical derivation will only be briefly summarized.

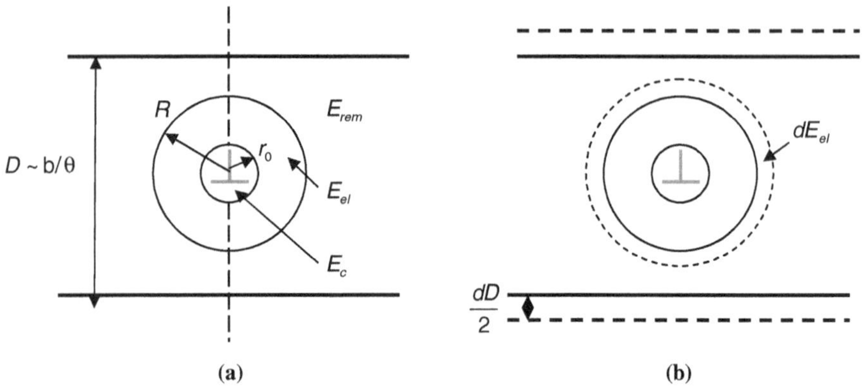

Fig. 5.6 (**a**) Schematic of the strip-divided grain boundary dislocation based representation, and (**b**) change in the grain boundary representation with a change in misorientation angle

First Proof: Physical Considerations

Let us consider the case of a simple tilt-grain boundary, as shown in Fig. 5.5(b). This grain boundary is composed of an array of edge dislocations. An equivalent representation of this simple grain boundary is presented in Fig. 5.6(a), where the grain boundary is divided in strips of length *D*– which in the small angle approximation is given by $D\sim\, \theta/2$-, the average dislocation spacing, positioned such that each strip, of infinite width, contains an edge dislocation positioned in its center.

The energy of a given strip is the sum of three contributions: the core energy and the edge dislocation, encompassed in the circle of radius r_0; the elastic deformation energy of the dislocation, which is encompassed in the circular area in between r_0 and R, proportional to D and which value will not affect the model's prediction as long as $R < D$; and the remaining energy of the strip. Let us name these terms, by order of citation, E_c, E_{el}, and E_{rem}. Therefore, the free energy of a given strip of grain boundary is given by:

$$E^{el}_{GB} = E_c + E_{el} + E_{rem} \tag{5.3}$$

Let us now decrease the misorientation angle by an amount –dθ (see Fig. 5.6 (b)). Then using Frank's formula in the above, the following relations are obtained:

$$-\frac{d\theta}{\theta} = \frac{dD}{D} = \frac{dR}{R} \tag{5.4}$$

Following the change in misorientation, the grain boundary energy will change by an amount dE^{el}_{GB} which will be the sum of the energy changes of each term in equation. The core energy is not expected to change significantly. Similarly, it can be shown that the term E_{rem} will not change when θ is changed. Indeed, as can be seen in Fig. 5.6(b), the area represented by E_{rem} increases with D^2 and the energy density varies as $1/D^2$.

In view of the argument above, the elementary change in the grain boundary energy following an elementary change in misorientation angle is given by the change in the elastic deformation energy of the dislocation. Formally, this is given by the energy in the ring encompassed in the radii R and $R+dR$. In linear elasticity, dE_{Gb}^{el} corresponds to the work done by the dislocation on a fictitious cut of the ring, and one readily obtains:

$$dE_{GB}^{el} = \frac{1}{2}\tau dR \cdot b \qquad (5.5)$$

Here τ denotes the shear stress on the cut of the ring and is given by $\tau \sim \tau_0 \frac{b}{R}$. For further details, consult Hirth and Lothe and Read. Using the above relation, Equation (5.4) and integrating the result, one obtains the following expression of energy of a low-angle grain boundary.

$$E_{GB}^{el} = E_0\theta(A - \ln\theta) \qquad (5.6)$$

$E_0\theta A$ is a constant energy per dislocation – including the energy of misfit in the core region – which is proportional to the density (i.e., to 1/D). $-E_0\theta\ln\theta$ is a term directly dependent on the elastic energy of a dislocation.

Second Proof: Mathematical Considerations

In Read and Shockley's original work, a more complex proof of relation (5.6) was given in the case of a simple grain boundary making an arbitrary angle ϕ about the common cube axis of the grain. In other words, a second degree of freedom is added and it corresponds to the orientation of the grain boundary plane. As mentioned above, such arbitrary grain boundary can be described by a multiple array of two different types of dislocations. Let us now summarize the methodology used to derive Equation (5.6).

- First, recall that the grain boundary energy can be written as the sum of a core part, inelastic by essence, and an elastic energy part. Rigorously, the core energy can be obtained by molecular simulations. Fortunately, in the case of low-angle grain boundaries, closed-form solutions can be found analytically.
- A longitudinal (x-axis) and a vertical axis (y-axis) is assigned to the grain boundary as well as corresponding dislocation densities (Frank's rule). The latter are calculated by assuming that lattices' planes are equivalent to dislocation flux lines.
- Choosing any "y" dislocation, the corresponding work term, which represents the elastic energy of such dislocation, can be calculated by considering the effect of all other "x" and "y" dislocations on its slip system. Similarly to Equation (5.5), the work terms are equal to half the lattice constant multiplied by the integral of the shear stress on the slip system. The energy

engendered by the "x" dislocations on the "y" dislocation is supposed not to depend on the position of the "y" dislocation and on the set of "x" considered. The same procedure is performed for an "x" dislocation.

- Finally, the interface energy per unit length is the sum, on the two types of slip systems (corresponding to the "x" and "y" dislocations), of the energy of a slip system multiplied by the number of slip systems. Although each term is diverging, the sum of the two terms converges. After some algebra one obtains equation (5.6).

In the case of this two degree of freedom grain boundary, the terms E_0 and A are given by:

$$E_0 = \frac{Ga}{4\pi(1-\nu)}(\cos\phi - \sin\phi) \tag{5.7}$$

and

$$A = A_0 - \frac{\sin 2\phi}{2} - \frac{\sin\phi \ln(\sin\phi) + \cos\phi \ln(\cos\phi)}{\sin\phi + \cos\phi} \tag{5.8}$$

where

$$A_0 = \ln\left(\frac{a}{2\pi r_0}\right) \tag{5.9}$$

G, φ, a and v represent the shear modulus, the orientation of the grain boundary, the inverse of the plane flux density, and Poisson's coefficient, respectively. r_0 is the lower bound used for the integration of the shear stress. This bound represents the smallest distance at which the material is elastically deformed.

The model above was applied to pure symmetric tilt grain boundaries in copper. Figure 5.7 presents a comparison between experimental data (dots) and the mode predictions (line). The model parameters E_0 and A were chosen to obtain a best fit of the low-angle grain boundary region. It was shown elsewhere that these parameters should be changed to obtain a better fit for larger grain boundary misorientations (e.g., $\theta > \sim 6°$). Regardless of the set of parameters chosen, the grain boundary dislocation model leads to adequate predictions only at low grain boundary misorientations.

5.3.2 Large-Angle Grain Boundaries

In order to circumvent the limitations of the grain boundary dislocation model, several models were developed to correlate the grain boundary energy with its macroscopic degrees of freedom (recall here that we focus primarily on pure tilt and twist grain boundaries). One of the objectives of these models also resides in

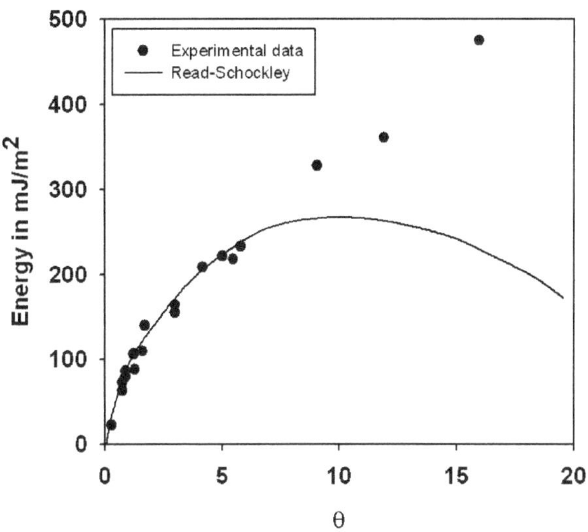

Fig. 5.7 Comparison between experimental measures and predictions given by the Read and Shockley dislocation model with low-angle parameters

providing a rationale behind the presence of energy cusps shown in Fig. 5.4. In this section, the coincident site lattice model will be recalled as well as the structural unit model and the disclination model.

5.3.2.1 CSL Model

The first of these models, referred to as the coincident site lattice (CSL) model, fathered by Bollman [11], introduces a measure of fit/misfit between the two crystals with their respective lattice. This geometrical model has been widely accepted in the community and is now often used to quickly describe the grain boundary structure. This model does not allow quantitative evaluations of grain boundary energies but presents a first explanation for the presence of metastable grain boundaries, identified by the presence of cusps. The argument here is that lower-energy grain boundaries are composed of a structure in which a "best-fit" of the two interpenetrating lattices of crystal A and B is obtained.

In the CSL model, the atomic arrangement within a given grain boundary is considered to result from the rigid junction of the two bodies followed by relaxation to improve lattice matching. The match of the two lattices at the median plane of the grain boundary is formally given by the CSL content. The CSL content, which describes the frequency of atoms positioned such that they are located in the continuity of both lattices from crystal A and B, is quantified with Σ. A coincident site is simply an atom in the grain boundary region which is in perfect continuity of both lattice A and B. This atom is a region of perfect match and is necessarily unstrained. Therefore, it is expected to be at a lower energy level. Due to the periodicity of the lattice, if one coincident site exists then an infinity of similar points also exist, which leads to a coincident site lattice.

Σ is given by the ratio of the volumes of the primitive unit cells of the CSL and the original crystal lattice. The lower Σ, the higher CSL content, and the better the match. In the BCC and FCC systems, $\Sigma = 3$ corresponds to a twin boundary. Clearly $\Sigma = 1$ corresponds to the perfect lattice case. To illustrate the evaluation of the CSL content, let us consider the following case of a $\Sigma = 5$ grain boundary presented in Fig. 5.8. Let us superimpose the lattices of crystal A and of crystal B, in red and blue, respectively. Crystal B is rotated by an arbitrary angle θ. A local frame is attached to each crystal. Clearly it can be seen that, with the given misorientation of the two crystals, several coincident sites can be found. Such points can be found at the origin of the blue and red frames and can also be found at the blue and red circles. The CSL content can be calculated in the frames of both crystal A and of crystal B. In the frame associated with crystal A (red frame), Σ' is given by the area of a unit cell of the CSL; this area is delimited by the x-axis and the vector relating two coincident sites (bold red vectors). One obtains $\Sigma' = 3^2 + 1^2 = 10$. Performing the same operation in the frame associated with the blue frame, one obtains $\Sigma = 2^2 + 1^2 = 5$. Bollmann introduced the following rule in calculating Σ: if Σ is even then $\Sigma \leftarrow \frac{\Sigma}{2}$, otherwise $\Sigma \leftarrow \Sigma$.

Upon rotating the two crystals A and B about the [001] direction, it can clearly be seen that all possible matching patterns are described in the range

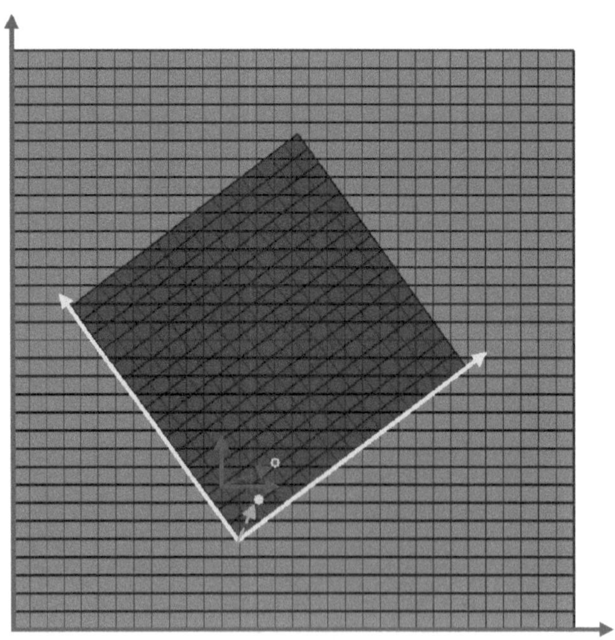

Fig. 5.8 Geometry of the coincident site lattice model

$0 < \theta < 45$. Therefore, in referring to a grain boundary by use of the CSL notation, one cannot solely mention the rotation axis and the CSL content. Either the misorientation angle or the grain boundary plane must be specified to avoid confusion. For example, let us consider the case of symmetric tilt grain boundaries rotated about [001]. In this case, both 36.87° and 53.13° misorientation correspond to a $\Sigma = 5$ grain boundary. As shown in Fig. 5.5 there seems to be a correspondence between CSL boundaries and energy cusps. Indeed, it can be seen that in that molecular simulations predict energy cusps corresponding to both the $\Sigma 5$ 36.87° and 53.13° grain boundaries. The concept of the O-lattice will present a rationale for such correspondence.

Introduction to the O Lattice

Geometry

Clearly, the CSL model is discontinuous with respect to θ. In other words, not all grain boundaries form a coincident site lattice. Also, a limitation inherent to the discontinuity of the CSL model stems from the fact that, as a coincident site is slightly moved out of its best fit position, the CSL model breaks down. In order to overcome such limitation, the CSL model was generalized, which lead to the concept of the O-lattice. The objective here is simply to introduce such a concept, for more details the reader is referred to Bollmann's book [11] and to the review by Balluffi et al. [12]

Prior to introducing the mathematics behind the O-lattice, let us present the idea behind it. A crystal lattice is composed of lattice point and also of "voids" present within each elementary cell of the crystal. An O-lattice point is simply a point of match between the two crystals. Here, the word *point* is meant in its general sense – it can either be a lattice point or a point where no atom is located.

Mathematically, the position of O-points engenders the existence of an O-lattice and can be assessed with the following reasoning. First, let the matrix $\boldsymbol{R}$ denote the transformation from lattice A to lattice B. Then any geometrical point of crystal A, which is given in term of its internal coordinates within a crystal cell and with the coordinates of the cell, is related to one of crystal B as follows:

$$x^B = R \cdot x^4 \tag{5.10}$$

Any point of the same class as x^A(i.e., having similar internal coordinates within a cell but different cell coordinate) can be related to x^A via a simple lattice translation given by vector t^A;

$$x^{A'} = x^A + t^A \tag{5.11}$$

An O-point, denoted x^O must necessarily respect the conditions given by Equations (5.10) and (5.11). Therefore it is given by:

$$\left(I - R^{-1}\right)x^O = t^A \tag{5.12}$$

With the relation above one can find the coordinates of the O-lattice. It can be seen that a coincident site is a particular O-point located at the corner of a cell. The solution of Equation (5.12) is left as an exercise for it is treated in great detail by Bollman [11]. However, let us note that the solution of Equation (5.12) for all possible cases shows that O-points are bounded in cells whose boundaries, defined by grain boundary dislocations, correspond to regions of worst fit between the lattices.

Significance

As mentioned earlier, the concepts of the coincident site lattice and its generalization (e.g., the O-lattice) were initially introduced to predict energetically favorable grain boundary orientation without the actual knowledge of the grain boundary energy. O-points define best matching points between the lattices defining the grain boundary. Therefore, one expects that the better the match, the lower the grain boundary energy. The question of finding minimum energy grain boundaries is thus equivalent to finding the periodicity of O-lattice points with the idea that smaller periods lead to more favorable grain boundaries. Two cases must then be considered.

First, when the O-lattice cells are much larger than the crystal lattice cell, then one can imagine that grain boundary relaxation is initiated at O-points and stops at cell walls. In that case, the periodicity of O-points is less relevant. Second, in the case where the O-lattice cells are of comparable size to that of the crystal lattice, then Bollman introduces the concept of pattern elements which are defined as subpattern of the grain boundary. The idea is that if a grain boundary is periodic it must be composed of a limited number of pattern elements. This idea is important because, as will be presented in an upcoming section, it is in direct connection with structural unit models. The number of pattern elements is equal to the number of different O-points with different internal coordinates.

Following the procedure introduced by Bollmann, the periodicity of the O-points can be calculated. Minimum energy grain boundaries then correspond to lower periods. In the O-lattice model, the presence of cusps in the energy vs. misorientation profiles result from the fact that a grain boundary whose misorientation is in the neighborhood of a minimum energy misorientation grain boundary will keep minimum energy periodicity of the grain boundary pattern, in order to remain in a relatively low energy state, with the presence of grain boundary dislocations. The presence of such dislocations is similar to the construction shown in Fig. 5.5 for low-angle grain boundaries. The Burgers vector of such grain boundary dislocations can be calculated via geometric arguments similar to that presented in Equations (5.10), (5.11), and (5.12). Finally, the presence of such grain boundary dislocation has been reported in several experimental studies [13].

5.3.2.2 Structural Units Models

Regardless of their agreement (or disagreement) with experimental measures, the O-lattice theory and the CSL model do not allow the evaluation of the relative grain boundary energy as a function of misorientation angle (in the case of pure tilt and twist grain boundaries). Nonetheless, these models bring out two interesting ideas.

First, grain boundaries exhibiting periodicity should be composed of a finite number of subpatterns. As will be seen in this section, molecular statics simulations will confirm this first point.

Second, departure from favorable misorientation is expected to be coupled with the presence of grain boundary dislocations (referred to as secondary dislocations). This second point was already discussed in Read and Shockley's original model. Indeed, in the dislocation model, the dislocation spacing is supposed uniform and the distance between dislocation is assumed to be a multiple of the distance between atomic planes. When this is not the case, additional dislocations are present within the grain boundary, as predicted by the O-lattice theory, the additional energy arising from the presence of such dislocations follows an equation similar to (5.6) where the misorientation angle is replaced by its deviation from an orientation considered in the dislocation model. In that case, energy cusps are expected when the spacing between the added dislocations is a multiple of the atomic planes spacing.

The structural unit model [14–16] is based on the two ideas presented above and is subsequently to be considered as an extension of the Read and Shockley dislocation model. The geometry of tilt boundaries was first investigated via molecular statics simulations on high Σ tilt grain boundaries. These simulations lead to the following postulates:

- Within a misorientation range, all tilt boundaries, with same median plane, are composed of a mixture of two structural patterns referred to as structural units.
- The grain boundaries limiting the misorientations range are composed of either a single type of fundamental structural unit or of multiple fundamental structural units. In that case, the delimiting grain boundary is referred to as multiple unit reference structure.
- Within two limiting grain boundaries are two structural units of the limiting grain boundary. The sequence of a structural unit is such that the minority units have the maximum spacing possible.

For example, it was shown that, in FCC metals, [001] symmetric tilt boundaries have the following delimiting ranges with following fundamental structural units (see Table 5.1):

Table 5.1 Delimiting grain boundaries for symmetric tilt [001] orientations

Range	Delimiting fundamental structural unit and corresponding Σ notation	
0°→36.87°	D, Σ1(110)	C, Σ5(310)
36.87°→53.13°	C, Σ5(310)	B or B', Σ5(210)
53.13°→90°	B or B', Σ5(310)	A, Σ1(100)

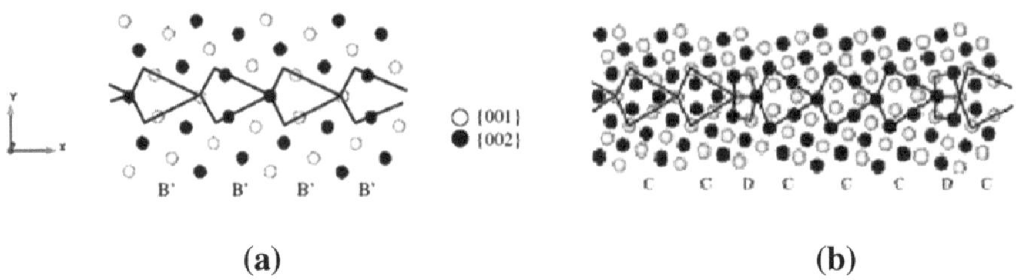

Fig. 5.9 Structures of (**a**) $\Sigma5(310)$ 36.87°, and (**b**) of a $\Sigma13(510)$ 67.38° symmetric tilt boundary. Images extracted from [17, 18]

It was shown that both $\Sigma5$ grain boundaries have two metastable states B and B' and C and C'. For the sake of simplicity these will not be recalled here. For the sake of illustration the structure of a $\Sigma5(210)$ 36.87°symmetric tilt boundary and of a $\Sigma13(510)$ 67.38° are shown in Fig. 5.9 [17]. In the following chapter, it will be shown that the presence of particular structural units (E structural units) can significantly affect the response of NC materials.

With the structural unit representation of grain boundaries and Read and Shockley's dislocation model, grain boundary energies as a function of misorientation angle can be predicted. Similarly to the dislocation model, the grain boundary energy can be written as the sum of a core energy term and an elastic energy term. The former will be calculated via the use of both molecular statics, giving the energy of particular structural units, and the structural unit model. The latter is the result of the presence of additional structural dislocations in the minority unit, which provides the misorientation away from the delimiting grain boundary. Therefore, the elastic contribution to the energy is given by Equation (5.6). The total grain boundary energy (per unit area) is written as follows.

$$E_{GB} = E^{el}_{GB} + E^{co}_{GB} \tag{5.13}$$

The calculation of core contribution, E^{co}_{GB}, is obtained via use of the structural unit model. Let us consider the case of a grain boundary composed of n structural units of type C and m structural units of type D. Also let $n > m$ such that C is the majority unit. In the case where $n = 5$ and $m = 2$, the grain boundary would then be of the form shown in Fig. 5.10.

The core energy of the grain boundary is then given by the sum of the contributions of segment CC and of segments CD (see Fig. 5.10). Let d_C, d_{CD} and E^{C}_{Co}, E^{CD}_{Co} denote the distance between two C units, the distance between a C unit and a D unit, and their respective energies with unit area. The core energy of the grain boundary is thus given by:

$$E^{co}_{GB} = \frac{\left[(n-m)d_C E^{co}_C + m d_{CD} E^{co}_{CD}\right]}{mD} \tag{5.14}$$

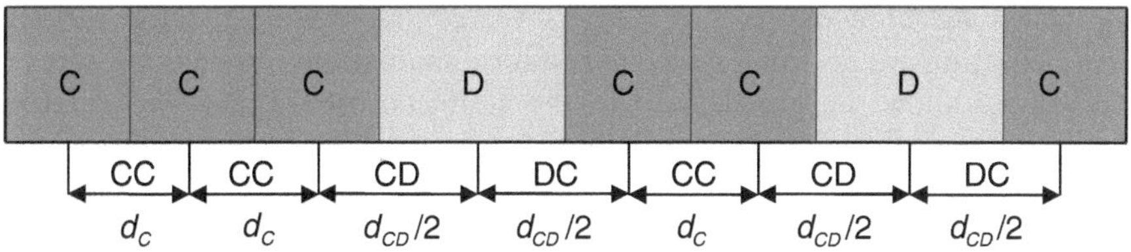

Fig. 5.10 Schematic of the construction of a grain boundary with the structural unit model

Recall that D denotes the average spacing between grain boundary dislocation and, therefore, with the previously mentioned argument, the average distance between minority structural units. Therefore, one obtains the following relation between D, d_C and d_{CD}:

$$mD = (n - m)d_C + md_{CD} \tag{5.15}$$

With (5.13), (5.14), (5.15), and Frank's formula the grain boundary core energy can be written as follows:

$$E_{GB}^{co} = E_C^{co} + d_{CD}\left(E_{CD}^{co} - E_C^{co}\right)\frac{\theta}{b} \tag{5.16}$$

The structural unit model has the same limitation as the grain boundary dislocation model in the sense that the core radius is unknown and must then be calculated to obtain a best fit. Figure 5.11 presents the model prediction (line),

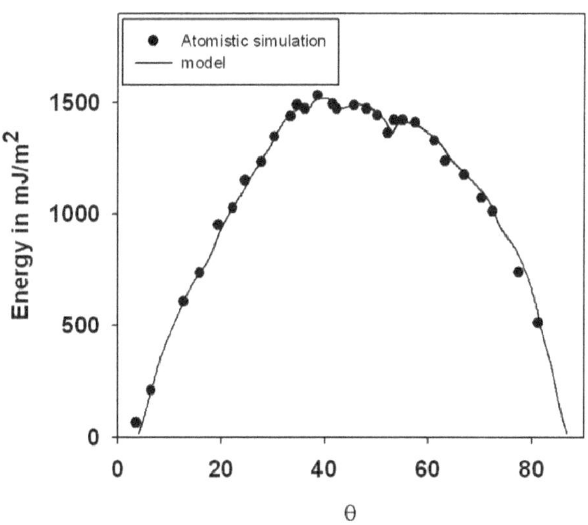

Fig. 5.11 Prediction of the evolution of the interface energy of the [001] tilt boundary with misorientation angle. Points represent molecular statics predictions and the solid line represent the structural unit model prediction

for which the core radius and Burgers vector are recalculated for various misorientation ranges and molecular statics simulations predictions (dots). The model allows an excellent fit of the atomistic predictions. Particularly, energy cusps are predicted in agreement with molecular simulations.

5.3.2.3 Disclination Models

Let us now introduce a third type of model based on the concept of disclinations. This model was first introduced by J.C.M. Li [19] and then applied by Shih and Li [20] to predict the energy dependence on misorientation in between energy cusps. The disclination model relies on the following idea: since grain boundaries are regions of intersections of two crystals with different rotational orientations, instead of describing their geometry with an assembly of linear defects, namely dislocation, a one-dimensional rotational defect, referred to as Volterra dislocation or as disclination, is used.

In what follows, the concept of disclination and disclination dipoles will first be briefly introduced, then the disclination model will be introduced.

Introduction to Disclination and Disclination Dipoles

Similarly to a dislocation, a disclination is a linear defect [21]. However, instead of translating the lattice in a manner similar to a dislocation, it leads to a lattice rotation. In other words, disclinations can be perceived as rotational defects bounding the surface of a cut to a continuum medium. This is illustrated in Fig. 5.12(a) and (b) presenting a wedge disclination. If $\boldsymbol{u}$ denotes the displacement between the two undeformed faces of the cut, then it is related to the disclination's strength –denoted with its Frank vector $\boldsymbol{w}$ which is equivalent to Burgers vector for dislocations – via the following relation:

$$u = (r - r_0) \times w \tag{5.17}$$

Here, r and r_0 denote the core radius and the distance between the rotation axis and the longitudinal axis of the cylinder. For ease of comprehension, one can consider that a disclination corresponds to the addition or to the subtraction of matter at the surface of a cut. A disclination is said to be positive if matter is subtracted to the medium and negative otherwise. Also, similarly to dislocations which can have either an edge or a screw character, a disclination can have a wedge (Fig. 5.12(a)) or a twist character. In that case, its Frank vector is perpendicular to the cylinder's radius. In what follows we are only interested in wedge disclinations.

Geometrically, it can be seen that a wedge disclination is equivalent to a wall of edge dislocations. Indeed as shown in Fig. 5.12(b) a wedge disclination of strength w leads to the same displacements as a wall of edge dislocations, with Burgers vector denoted b', equally spaced such that the distance between two dislocations is related to the disclinations' strength as follows:

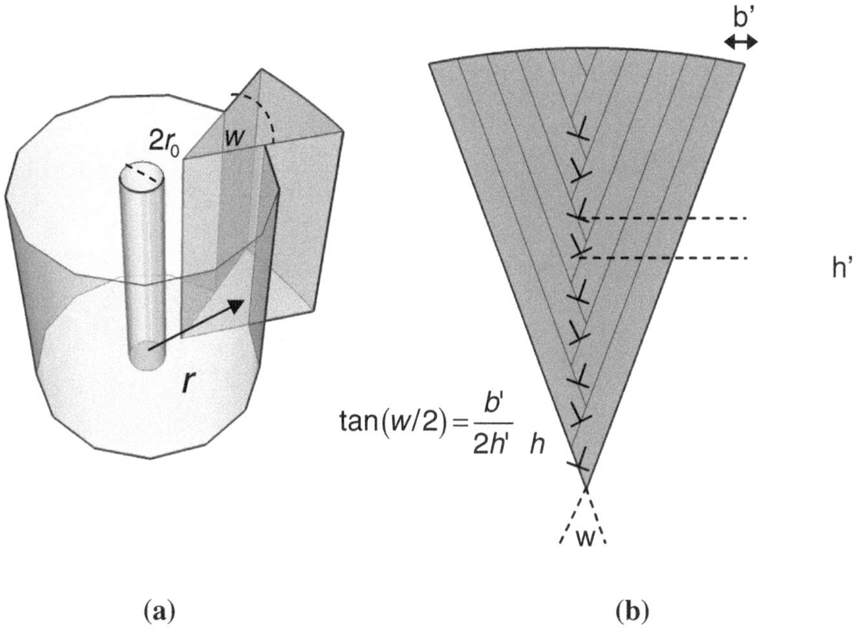

Fig. 5.12 Schematic of a wedge disclination (**a**) and (**b**) equivalent dislocation wall representation

$$\tan(w/2) = \frac{b'}{2h'} \tag{5.18}$$

As mentioned by Li, this representation falls apart when w is a symmetry operation. Note that disclination models are equivalent to dislocations arrangements solely in terms of stress field or strain field (but not both).

The geometrical equivalent representation between disclinations and dislocation walls suggests that grain boundaries could be equivalently represented by a disclination model. Moreover, as extensively presented by Romanov [21] any disclination can be equivalently represented by an arrangement of dislocations. Conversely, for any dislocation, an equivalent disclination arrangement can be found. For further detail the reader is referred to Romanov [21] and Romanov and Vladimirov [22].

As will be seen later, among the many equivalent dislocation/disclination representations, the equivalent representation of interest here is that of an edge dislocation. It can be shown that an edge dislocation can be equivalently represented as a single line two-rotation axis dipole; that is, two parallel disclinations of same strength but opposite signs separated by a small distance. Let us now see the advantage of disclination arrangements.

The stress field of a wedge disclination can be obtained without too much strain via elasticity theory since the displacements are known. The derivation becomes more difficult in the case of twist disclinations. Huang and Mura [23] obtained the following expression of the stress field, in units of $\mu w/(2\pi(1-\nu))$, of a wedge disclination (only the nonvanishing terms are presented):

$$\sigma_{xx} = \frac{1}{2}\ln\left(\frac{R^2}{x^2+y^2}\right) - \frac{y^2}{x^2+y^2}, \sigma_{yy} = \frac{1}{2}\ln\left(\frac{R^2}{x^2+y^2}\right) - \frac{x^2}{x^2+y^2} \tag{5.19}$$

$$\sigma_{zz} = \nu\left[\ln\left(\frac{R^2}{x^2+y^2}\right) - 1\right], \sigma_{xy} = \frac{xy}{x^2+y^2} \tag{5.20}$$

The expressions in the above are written in the case of isotropic elasticity where ν is the Poisson ratio and μ is the shear Modulus. Also, w is parallel to the z-axis, R denotes the outer radius of the medium considered, and the position vector is given by the x and y coordinates. Clearly, it can be seen that the stress field (σ_{xx}, for example) rapidly diverges as $x^2 + y^2$ approaches R. However, it can be easily shown that the energy of the wedge dislocation remains bounded. Nonetheless, the diverging stress field of a single disclination may be considered as an argument preventing the use of disclination theory.

Consider now a single line two-rotation axis dipole where the disclinations are separated by a small distance δy. Then, the stress field of as disclination dipole can be estimated with a Taylor's expansion (only the first term is kept). Therefore, taking the derivative of (5.19) and (5.20) with respect to y, one obtains the following expression of the stress field of the dipole considered, in units of $\boldsymbol{\delta} y \cdot \boldsymbol{\mu} w/(2\pi(1-\nu))$:

$$\sigma_{xx} = -\frac{-y(y^2+3x^2)}{(x^2+y^2)^2}, \sigma_{yy} = \frac{y(x^2-y^2)}{(x^2+y^2)^2} \tag{5.21}$$

$$\sigma_{zz} = -\frac{2\nu y}{(x^2+y^2)^2}, \sigma_{xy} = \frac{x(x^2-y^2)}{(x^2+y^2)^2} \tag{5.22}$$

In the expression above, one recognizes the expression of the stress field of an edge dislocation with Burgers vector $w\delta y$, which shows the equivalence between the disclination dipole considered in the above and an edge dislocation. More importantly, it can be seen that, contrary to the stress field of a disclination, the stress field of a screened disclination (e.g., disclination dipole) is not diverging. Therefore, it would be safe to assume that grain boundaries in general, and at least low-angle grain boundaries, could be modeled with use of disclination dipoles.

Relationship to Excess Energy Between Cusps

The disclination grain boundary model, which predicts the evolution of energy vs. misorientation in between energy cusps, is based on the following geometrical representation of grain boundaries, which is somewhat similar to the subpattern and structural unit model. In between two energy cusps, with misorientations θ_1 and θ_2, the grain boundary is represented as an alternate assembly of single line double rotation axis dipoles of strength $w_1 = \theta_1$ and $w_2 = \theta_2$. If $2L_1$ and $2L_2$ denote the separation distances between two w_1 and two w_2 dipoles, respectively, then for a given misorientation θ such that $\theta_1 < \theta < \theta_2$ one has the following relation:

$$\theta = \frac{L_1 w_1 + L_2 w_2}{L_1 + L_2} \tag{5.23}$$

This subpatterning of the grain boundary clearly differs from the structural unit model. Indeed, the sequence of disclination dipoles is not given by the minority unit rule presented in the above. Also, in the present model each cusp orientation necessarily represents a delimiting grain boundary (e.g., θ_1 or θ_2). Since all dipoles parallel Frank's vector, the sequence of alternating dipoles w_1 and w_2 is equivalent elastically to an alternate sequence of dipoles $\pm\Delta w = \pm(w_1 - w_2)$ with separation $H = 2L_1 + 2L_1$.

The excess energy between two energy cusps then resumes to the energy of a dipole wall. Using an edge wall dislocation representation of the dislocation dipoles, the excess energy between cusps is then the sum of the energy of edge dislocation walls of length H. After some algebra, one obtains:

$$E = \frac{\mu(\Delta w)^2}{8\pi(1-\nu)} \frac{H}{4\pi^2} f(\lambda_1) \tag{5.24}$$

with

$$f(\lambda) = -16 \int_0^{\lambda} (\lambda - \psi) \ln(2 \sin \psi) d\psi \text{ And } \lambda_1 = \frac{2\pi L_1}{H} \tag{5.25}$$

Typically, this model leads to fairly good agreement with experimental data. While it is of relatively easy use, for there is only one parameter that needs to be determined and no simulations at the atomistic scale are necessary, unlike the disclination model presented above, a priori knowledge of energy misorientation cusps is required. Nonetheless, it will be seen in next section that disclination-inspired grain boundary models have the great advantage of allowing modeling of grain boundary dislocation emission.

5.4 Applications

So far, several models were introduced to describe the structure of simple grain boundaries and to predict their corresponding energy. Let us now show how these models combined with purely numerical simulations (molecular statics and dynamics) can be used to predict the occurrence and activity of mechanisms particularly relevant to nanocrystalline materials. We will first focus on the atomic motion within grain boundaries in the elastic domain and then show some results in the case of plasticity.

5.4.1 Elastic Deformation: Molecular Simulations and the Structural Unit Model

Let us now show the advantage of the structural unit model in understanding particular behaviors of grain boundaries in the elastic regime. For this purpose several bicrystal interfaces where constructed via molecular statics (the construction method is described in Chapter 4). The following seven bicrystal interfaces were subjected to an increasing tensile load perpendicular to the grain boundary plane (see Table 5.2):

Table 5.2 Grain boundary misorientations and structures

Interface	Angle	Structure
Σ13(510)	22.6°	CDD
Σ17(410)	28.9°	CD.CD
Σ5(310)	36.9°	C
Σ29(730)	46.4°	B'B'C
Σ5(210)	53.1°	B'B'
Σ17(530)	61.9°	AB'
Σ13(320)	67.4°	AB'.AB'

Upon applying the increasing tensile load to the different bicrystal interfaces, their excess energies (with respect to the bulk energy) were recorded. The predicted excess energy evolutions for each interface are presented in Fig. 5.13(a). It can be seen that grain boundaries containing mixtures of only C and B' structural units present a decrease in their excess energies. Therefore, in the case studied, B' and C structural units appear to be less efficient at storing elastic energy than other structural units. However, as shown in Fig. 5.13(a), grain boundaries containing a mixture of B' or C structural units with either A or D structural units, which are basically perfect lattice regions, exhibit an increase in excess energy upon applying a tensile load on the bicrystal. Surrounded by A or D structural units, a B' or C structural unit is more likely to be able to expand in the lateral direction which would enhance the grain boundary ability for energy storage.

Also, one can notice in Fig. 5.13(a) the presence of sudden changes in the slope of the energy evolution of all grain boundaries containing C structural

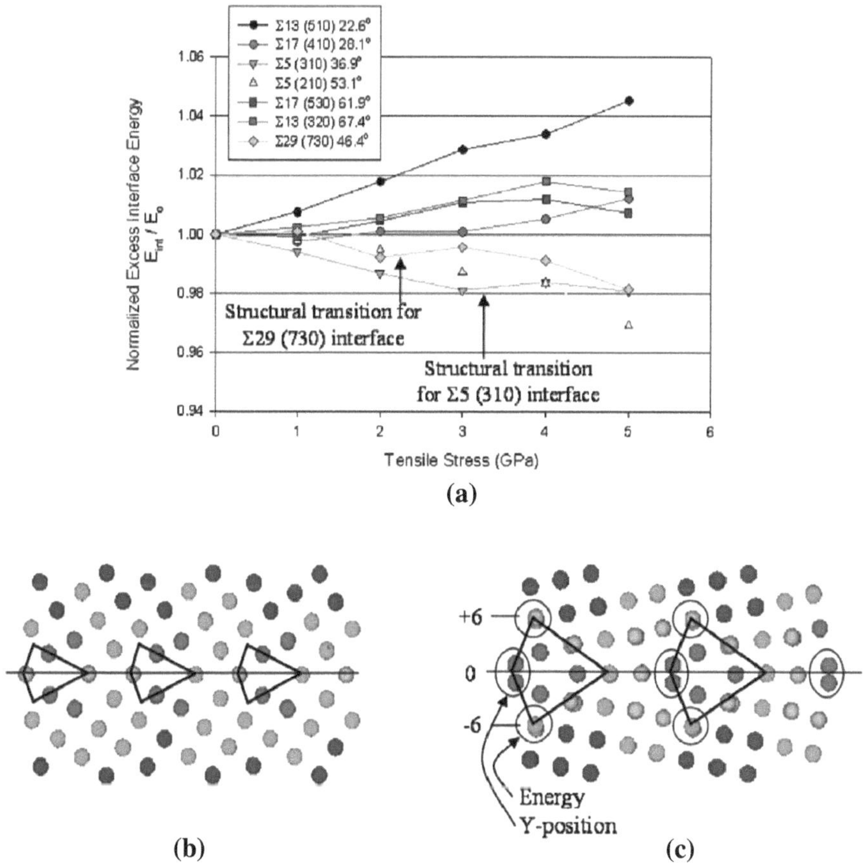

Fig. 5.13 (**a**) Evolution of bicrystals energy during tensile tests for several CSL orientations, (**b**) structure of a Σ13(320) AB'.AB' interface under 5 GPa tensile load, and (**c**) structure of a Σ13(510) CDD interface under 5 GPa tensile load

units. The structure of a Σ13(320) AB'.AB' and of a Σ13(510) CDD interface under 5 GPa tensile load are presented in Fig. 5.13(b) and (c). In Fig. 5.13(c) one can observe that the occurrence of an elastic transition mechanism which corresponds to the motion of atoms on each side of the grain boundary median plane. The excess energy of the grain boundary increases after the elastic transition has occurred. This mechanism may be a precursor to the grain boundary dislocation emission mechanism.

5.4.2 Plastic Deformation: Disclination Model and Dislocation Emission

Let us now present one of the many applications of disclination-based grain boundary models. For more detail the reader is referred to work by Gutkin [24, 25], Romanov [21], and others [26, 27]. The mechanism of dislocation

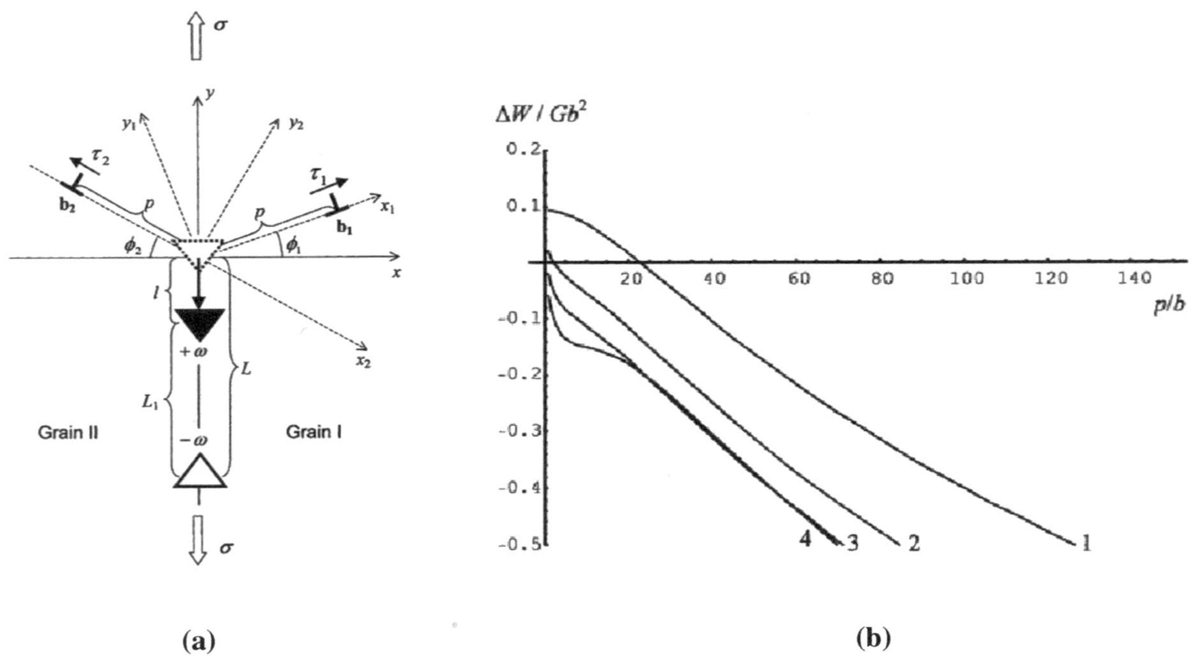

Fig. 5.14 Disclination-based model for grain boundary dislocation emission; (**a**) schematic [24] and (**b**) energy evolution as a function of emission distance p for 4 different angle configurations: (1) $\Phi_1 = \Phi_2 = 45°$, (2)$\Phi_1 = 30°$ and $\Phi_2 = 45°$, (3) $\Phi_1 = 20°$ and $\Phi_2 = 30°$, and (4)$\Phi_1 = \Phi_2 = 2°$

emission from grain boundaries has suggested particular interest in the NC community. While this mechanism is often studied via molecular simulations, disclination-based grain boundary representations are particularly suited to treat such problems for it has been shown that disclination motion is related to dislocation emission or absorption. Gutkin et al. [24] thus represent a grain boundary in a bicrystal as the concatenation of screened wedge disclinations of strength $\pm w = \pm(\theta_1 - \theta_2)$ (see Fig. 5.14(a)). Here, θ_1 and θ_2 represent two reference grain boundary misorientations. Note that the grain boundary representation used here is different from that of Li presented in previous section. In particular, the distance in between two disclinations of opposite sign is equivalent to an edge dislocation wall. It is assumed that the emission of two dislocations within each grain composing the bicrystal results from the motion of a wedge disclination of a distance l. It is thus suggested that grain boundary reorientation occurs as a result of grain boundary dislocation emission.

The mechanism of grain boundary dislocation emission by disclination motion is favorable if the energy after emission of the dislocation is lower than the system's initial energy. The initial energy of the system is simply the sum of a wedge disclination dipole's energy and of an edge dislocation wall. The energy after emission of a dislocation is the sum of the dipole's energy, in its new configuration, the dislocation wall energy, the energy of each emitted dislocation, and their interaction energies.

Disregarding any activation energy contribution, which can be obtained solely with molecular simulations, it can be seen in Fig. 5.14(b) that, depending on the angle at which dislocation are emitted, the process may be favorable. For example, emission of two dislocations symmetrically at a 45° angle with respect to the grain boundary longitudinal axis is not predicted to be favorable while an asymmetric emission with 20° and 30° orientations for dislocation 1 and 2, respectively, is a favorable process.

5.5 Summary

This chapter introduces a simple description of grain boundary geometry. In particular, symmetric tilt and twist grain boundaries are described as well as their excess energy evolution with misorientation angle. The presence of energy cusps is noted and motivates the introduction of four different structure/energy correlation models.

The first model introduced is that of Read and Shockley, which is valid in the low misorientation range, and based on the representation of grain boundaries as arrays of one or more types of dislocations. The coincident site lattice model and the O-lattice theory are then introduced. The fundamental novelty in these models is that while they do not allow quantitative prediction of grain boundary energies, a geometrical description of the degree of fit between two adjacent grains in the grain boundary region is introduced. The model suggests that in the neighborhood of grain boundary cusp orientations, grain boundaries shall exhibit the presence of secondary dislocations in order to keep a structure close to the metastable configuration (e.g., energy cusps).

Finally, the structural unit model and the disclination model are introduced. The first model, which is an extension of Read and Shockley's initial model, is based on the representation of grain boundaries as repeated sequences of particular atomic arrangements (structural units) whose energies are calculated by atomistic simulations. The disclination model, which is limited to predicting the excess energy between two non-necessary consecutive cusps orientations, is based on the existence of equivalent representations between dislocation arrangements and disclination arrangements, which is recalled as an introduction to disclination theory.

This chapter is concluded with two examples showing the use of molecular simulations, the CSL model and the disclination model, to investigate particular mechanisms associated with grain boundary–mediated deformation (e.g., elastic instabilities and grain boundary dislocation emission).

References

1. Gleiter, H., Materials Science and Engineering **52**, (1982)
2. Read, W.T., *Dislocations in Crystals.* McGraw-Hill, New York, (1953)
3. Herring, C., In: Gomer, R., and C.S. Smith, (eds.) *Structure and Properties of Solid Surfaces.* University of Chicago Press, Chicago, (1952)
4. Gjostein, N.A. and F.N. Rhines, Acta Metallurgica 7, (1959)
5. Chan, S.W. and R.W. Balluffi, Acta Metallurgica **33**(6), 1113–1119, (1985)
6. Wolf, D., Acta Metallurgica **38**, (1990)
7. Wolf, D., Acta Metallurgica **32**(5), 735–748, (1984)
8. Read, W.T. and W. Shockley, *Imperfections in Nearly Perfect Crystals.* New York: Wiley; London: Chapman & Hall, (1952)
9. Warrington, D.H. and M. Boon, Acta Metallurgica **23**(5), 599–607, (1975)
10. Hirth, J.P. and J. Lothe, *Theory of Dislocations.* Krieger Publishing Company, New York, (1982)
11. Bollmann, W., *Crystal Defects and Crystalline Interfaces.* Springer Verlag, New York, (1970)
12. Balluffi, R.W., A. Brokman, and A.H. King, Acta Metallurgica **30**, (1982)
13. Brandon, D.G., B. Ralph, S. Ranganathan, and M.S. Wald, Acta Metallurgica **12**(7), 813–821, (1964)
14. Balluffi, R.W. and P.D. Bristowe, Surface Science **144**, (1984)
15. Balluffi, R.W. and A. Brokman, Scripta Metallurgica **17**(8), 1027–1030, (1983)
16. Brokman, A. and R.W. Balluffi, Acta Metallurgica (1981)
17. Spearot, D.E., L. Capolungo, J. Qu, and M. Cherkaoui, Computational Materials Science **42**, 57–67, (2008)
18. Capolungo, L., D.E. Spearot, M. Cherkaoui, D.L. McDowell, J. Qu, and K.I. Jacob, Journal of the Mechanics and Physics of Solids **55**, (2007)
19. Li, J.C.M., Surface Science **31**, (1972)
20. Shih, K.K. and J.C.M. Li, Surface Science **50**, (1975)
21. Romanov, A.E., European Journal of Mechanics A/Solids **22**, (2003)
22. Romanov, A.E. and V.I. Vladimirov, *Dislocations in Solids.* Elsevier, City: Amsterdam, North Holland, (1992)
23. Huang, W. and T. Mura, Journal of Applied Physics **41**, (1970)
24. Gutkin, M.Y., I.A. Ovid'ko, and N.V. Skiba, Materials Science and Engineering **339**, (2003)
25. Gutkin, M.Y. and I.A. Ovid'ko, *Plastic Deformation in Nanocrystalline Materials.* Springer New York, (2004)
26. Hurtado, J.A., et al., Materials Science and Engineering **190**, (1995)
27. Mikaelyan, K.N., I.A. Ovid'ko, and A.E. Romanov, Materials Science and Engineering **288**, (2000)

Chapter 6
Deformation Mechanisms in Nanocrystalline Materials

Nanocrystalline (NC) materials have a particularly interesting microstructure characterized by large amounts of interfaces and, depending on the fabrication process, by the presence of defects (e.g., impurities, voids). This was discussed in Chapter 2. Prior to detailing the particular plastic deformation mechanisms associated with NC materials, let us recall some of the key features of the response of NC materials such as (1) the breakdown of the Hall-Petch law, (2) elastic pseudo perfect plastic response in quasi-static tests, and (3) increasing strain rate sensitivity parameter with decreasing grain size. All of these indicators clearly suggest that the activity of each probable deformation mechanism is likely to exhibit a pronounced size dependence.

As shown in Chapter 2, presenting the behavior of NC materials (e.g., tensile response, strain rate sensitivity, thermal stability), large discrepancies in the mechanical response of NC materials have been recorded experimentally. As a consequence, the nature of the active plastic deformation mechanisms – characteristic of NC materials regardless of the fabrication process – has been source of confusion. Fortunately, amelioration of sample qualities and extensive fundamental discussions – based on modeling, particularly on atomistic simulations (some key results from atomistic simulations were presented in chapters 4 and 5) – have allowed the field to reach maturity such that a general consensus was reached.

For the sake of completeness all mechanisms which have been supposed to be activated in NC materials will be discussed here. Note that among these some mechanisms have been found not to be likely to participate to the plastic deformation of NC materials.

6.1 Experimental Insight

In view of the peculiarities exhibited by NC materials – compared to that of conventional materials – detailed microstructure observation is necessary to provide insightful elements of explanation for all phenomena listed in Chapter 2.

M. Cherkaoui, L. Capolungo, *Atomistic and Continuum Modeling of Nanocrystalline Materials*, Springer Series in Materials Science 112,
DOI 10.1007/978-0-387-46771-9_6,

Transmission electron microscopy (TEM) (both ex situ and in situ) and high-resolution TEM studies are obviously the ideal candidates for such purpose.

Post-mortem observations [1] – typically performed after relatively large compressive stresses are applied to a sample – differ to that revealed by tensile in situ tests [2]. The difference clearly results from the loading conditions. These observations revealed key elements. For example, the microstructure of a 30 nm grain electrodeposited Ni was observed after compressive loads were imposed. Dislocation debris could be observed in some grains. However, the amount of stored dislocations was not sufficient to rationalize plastic deformation on the sole basis of dislocation activity. Similarly, slip traces could be observed in some grains. Yet, their occurrence was very limited compared to what would be observed in a conventional sample. It is thus likely that dislocation activity is reduced in NC materials with grain size in the neighborhood of ~30 nm. The occasional presence of cracks located at triple junctions was also revealed in these observations. However, this may be caused by the fabrication process (e.g., impurities in the sample).

In support of the aforementioned experiments, tensile tests on ~30 nm NC Cu also suggest a much reduced dislocation activity in nanograins. In particular, dislocation activity can be observed in grains as small as ~50 nm. However, no dislocation activity was observed in grains with size ~30 nm. Note that, as expected, the presence of growth twins within grains was shown to act as barrier to dislocation motion. However, near crack tips, in situ tensile tests on ~30 nm grain NC Ni show significant dislocation activity. In conventional materials, it is well known that crack tips can act as dislocation sources whose activation results from the large stress concentrations located at the tip.

In summary, most TEM observations discussed in the above show that dislocation activity in NC materials is reduced with grain size. As expected, it is also shown that dislocation activity is enhanced in regions of stress concentrations such as crack tip.

In situ tensile tests on 25 nm grain size electrodeposited Ni subjected to 0.001/s strain rate at 298 K and at 76 K have revealed unexpected features. As in previous observations, tensile tests at 298 K revealed the presence of very few full dislocations and growth twins. On the contrary, at 76 K the presence of deformation twins and stacking faults was observed in several nanograins. This is shown in Fig. 6.1, where grains exhibit lamellar domain decomposition (similar to what is observed in hexagonal close packed [hcp] metals and in some shape memory alloys). Several twins were shown not to cross the entire crystal. It was suggested that twins could be heterogeneously nucleated from activation of grain boundary partial dislocation emission. This appears to be a probable mechanism since, as opposed to hcp metals, twin planes and primary slip systems in the FCC are the same. This particular mechanism will be addressed in this chapter. Note that homogeneous defect nucleation (e.g., not directly arising from the presence of a defect) of dislocation dipoles and twins typically requires very large stress field on the order of the GPa.

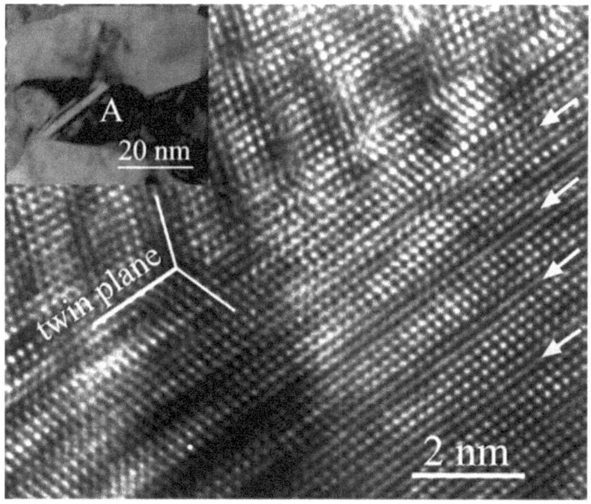

Fig. 6.1 HRTEM of a grain. Deformation twins are indicated by white lines and arrows indicate stacking faults

Further decreasing the grain size, tests at very low strain rate ($\dot{\varepsilon} = 10^{-8}/s$)n 10 nm grain Au thin films revealed no dislocation activity – similar experiments on 110 nm grain Au thin films showed significant dislocation activity – and showed that plastic deformation is driven by grain rotation. Unfortunately, the material density was not quantified. Nonetheless, these findings are of great interest. Indeed, grain rotation is typically observed in the case of superplastic deformation. For example, grain rotation was observed in Pb-62%Sn eutectic alloy with grain size in the order of ~10 μm tested in tension at 423 K [3].

6.2 Deformation Map

Previously discussed microscopies as well as the measured size effects in the response of NC materials are proof that plastic deformation in NC materials is not driven by the same mechanisms as in the case of conventional materials. Of interest here is the identification of each mechanism participating to plastic deformation of NC materials as a function of grain size. To this effect, a tentative simplified deformation map will be shown here. Note that as opposed to usual deformation maps (e.g., Ashby) – based on extensive experimental measures and describing the domain of activities of various deformation processes as a function of temperature and strain rate – the present one-dimensional map serves only the purpose of discussion, for such maturity level has not yet been reached in NC materials. Nonetheless, such an attempt, shown in Fig. 6.2, may serve as a basis to extend Ashby's map to a third dimension (e.g., grain size). Here, we restrict ourselves to quasi-static loading and to temperatures in the neighborhood of 298 K.

Three separate regimes are usually identified. The first regime describes polycrystalline materials with mean grain size ranging from several microns down to

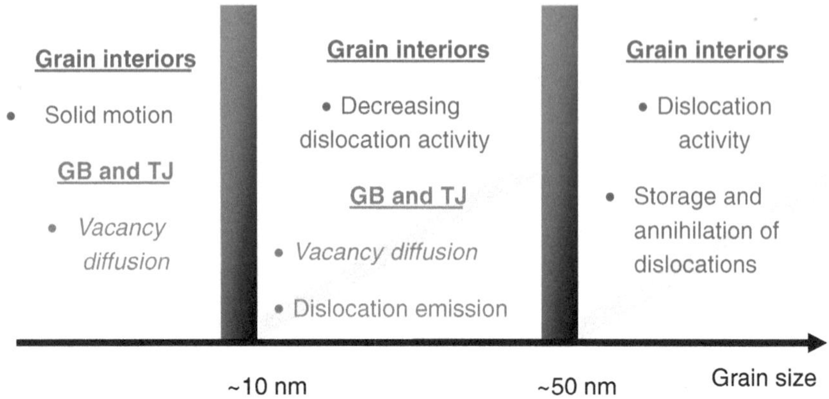

Fig. 6.2 Schematic of the deformation mode domains as a function of grain size

approximately ~50 nm, the second regime covers grain sizes between ~50 nm and ~10 nm, and the third regime corresponds to grain sizes smaller than ~10 nm. In conventional materials subjected to quasi-static loading well below the homologous melting temperature, plastic deformation results from the glide and interaction of dislocation loops. In the simplest form dislocation glide is modeled with a power law with an evolving critical resolved shear stress given by Taylor's stress. The dislocation density evolves via athermal storage of dislocations and dislocation annihilation operated by thermally activated mechanisms such as cross slip or climb. As presented in previous chapters, in this first regime, the volume fraction of grain boundary is very small (~0.01%), and grain boundaries act solely as barrier to dislocation motion – therefore contributing to the storage of dislocations.

In regime II, dislocation activity within grain interiors is reduced. This is likely to exacerbate the role of grain boundaries and triple junctions. Indeed, interfaces can act as dislocation sources and sinks such that the role of grain boundaries is not limited to that of dislocation barriers. As will be shown, low-angle grain boundaries, perfect planar large-angle grain boundaries, and grain boundary ledges can all act as dislocation sources. The macroscopic effect of this mechanism is, however, thought to be relatively small. Interestingly, in regime II, the presence of twins can be observed. This was discussed in Chapter 1. Note that this effect is also predicted by atomistic simulations. This is surprising, for in coarse face-centered cubic (FCC)structures, twinning is observed solely in dynamic loading and in materials with relatively low stacking fault energy. The presence of twins in NC materials may have a critical effect on the materials' response, for they can act as selective barrier to dislocation motion. For example, in the hcp system in which twinning is readily activated due to the lower crystal symmetry, molecular simulations on Ti have shown that a screw <a> basal dislocation, with line parallel to line of intersection of the $\left(10\bar{1}2\right)$ twin plane and the basal plane, passes through the twin boundary. On the contrary, a mixed <a> 60° dislocation dissociates into the $\left(10\bar{1}2\right)$ twin boundary upon meeting the fault [4].

Additionally, early work on NC materials suggested that, owing to an increase in the self-diffusivity of grain boundaries, the mechanism of Coble creep could be activated at room temperature. The possible activity of Nabarro-Herring creep was also documented. However, Nabarro-Herring creep represents the steady state vacancy diffusion through grain interiors and its activity was shown to be limited compared to that of Coble creep [5]. Recall that in most recent experiments it was shown that earlier creep tests on NC materials may have been influenced by artifacts such as crack nucleation and propagation.

Finally, in regime III (e.g., mean grain size smaller than ~10 nm), dislocation activity completely ceases and plastic deformation is mediated by grain boundaries through activation of mechanisms typically relevant in the case of superplastic responses: grain boundary sliding and grain rotation. These mechanisms which represent the solid motion of grains could be accommodated by vacancy diffusion. If not, soon after plastic deformation initiates, one expects to observe crack nucleation resulting from displacement incompatibilities at the interface.

6.3 Dislocation Activity

In pure metals with conventional grain size, plastic flow is dependent on dislocation activity. It is out of the scope of this chapter to present a detailed review on dislocation activity in conventional metals. Such reviews can be found elsewhere [6]. As stage II is initiated – corresponding to multislip while stage I corresponds to single slip and is typically not relevant to aggregates – mobile dislocation will become sessile upon interacting with obstacles such as grain boundaries and other dislocations. The increased stored dislocation activity leads to an increase in the critical resolved shear stress of active slip modes through Taylor's relation:

$$\tau_{crss} = \alpha b M \mu \sqrt{\rho} \tag{6.1}$$

Here, ρ denotes the stored dislocation density. Note that Taylor's relation has found substantial experimental support. More recent three-dimensional dislocation dynamics experiments proposed to correct the deviation reported by the previous authors as follows [7]:

$$\tau^{\alpha}_{sub} = k_{sub} \mu b^{\alpha} \sqrt{\rho_{sub}} \log\left(\frac{1}{b^{\alpha}\sqrt{\rho_{sub}}}\right) \tag{6.2}$$

Stage II is quickly followed by stage III corresponding to a regime of dynamic recovery of stored dislocations. The recovery mechanism (e.g., climb, cross-slip) is thermally activated. Stage III is characterized by the formation of subgranular structures aided by dislocation cross-slip. The latter was shown to be necessary for the formation of substructures. The resulting substructures are cells, exhibiting a low dislocation density, bounded by regions of high-dislocation density referred to as cell walls. In the FCC system, cells walls are typically misoriented by a few degrees with respect to the slip system (and to their perpendicular plane). Later deformation stages are characterized by the refinement of the cell wall structure leading to recrystallization.

Moving from the conventional polycrystalline aggregate to a NC material, both tensile tests – characterized by a pseudo elastic perfect plastic response- and TEM observations show that dislocation activity is reduced. A rationale for such size effect can be found by considering (1) the stability of dislocations within the nanosized grains and, (2) the size effect in the activation of dislocation sources. These two elements are discussed in what follows. Consider a simple two dimensional representation of the aggregate where grains are represented by square (see Fig. 6.3(a)) [8, 9]. Let d denote the grain size and d_s denote the size of the region in which an isolated dislocation is stable.

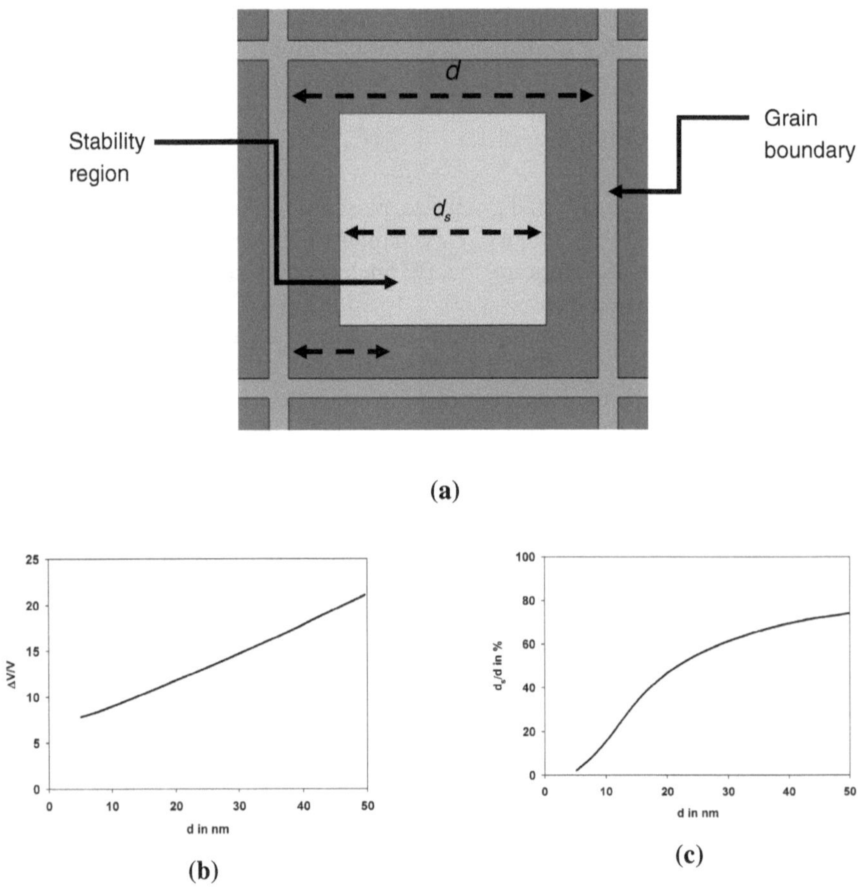

Fig. 6.3 (**a**) NC aggregate two dimensional representation (pink region: region of stability; gray region: grain boundaries); (**b**) excess volume measures in amorphous Ni-P alloy as a function of grain size [11]; (**c**) predictions of the evolution of the stability region of a single isolated dislocation with grain size [9]

As discussed in chapters 2 and 5 and shown in a comprehensive experimental study [10], grain boundaries exhibit an excess volume with respect to the crystal lattice. Let ΔV denote the excess volume at grain boundaries. It corresponds to the difference in the volume occupied by the same number of atoms in a grain boundary and in a lattice region. This excess volume in the grain boundary region is independent of the fabrication process. Experiments on amorphous NC NI-P alloys have shown that ΔV increases with decreasing grain size (see Fig. 6.3(b)). Atoms in the neighborhood of grain boundaries are expected to be displaced from an equilibrium position at distances far from the grain boundary. Denoting x the distance from grain boundaries (see Fig. 6.3(a)), the normal displacement, δ, of an atom can be expressed as $\delta = \frac{A}{x^2}$ where A is a function depending on the excess volume – obtained by simple consideration of the compatibility of displacement at the grain boundary grain interior interface – which increases with ΔV. Using Hooke's law it can be shown that the compressive stress, τ, resulting from the presence of excess volume, will decrease with increasing distance from the grain boundary as follows: $\tau = -\frac{4\mu A}{x^3}$. In a given crystal, an isolated dislocation is at equilibrium if the stress resulting from the surrounding lattice on the dislocation line, namely the Peierls stress, balances the stress related to the grain boundary excess volume. Figure 6.3(c) shows the evolution of the region of stability within a crystal as a function of grain size. Interestingly, it can be shown that the very simple argument presented in the above shows that as the grain size is decreased, the region of stability of an isolated dislocation decreases drastically. Note the model presented in the above does not consider the case of defects within crystal or within grain boundaries [8, 9]. As shown in Chapter 2, such defects are expected to favor the stability of dislocations. For example, if impurity atoms were added to grain boundaries, their excess volume is expected to decrease, which would result in smaller back stresses and, in turn, increase the size of the stability domain.

Consider now the size effect in the activation of dislocation sources. In conventional metals, mobile dislocations are typically nucleated by Frank and Read sources via the well-documented bow mechanism. Using a line tension model, the critical activation stress of a Frank Read source can be expressed as the ratio of the product of the norm of the dislocation Burgers vector, b, by the shear modulus, μ, divided by the length of the source L (e.g., the distance between the two pinning points):

$$\tau_c = \frac{\mu \cdot b}{L} \tag{6.3}$$

Within a given grain, the length of a Frank Read source is obviously limited by the grain size. Therefore, with Equation (6.3) the critical activation stress of a Frank Read source increases with a decrease in the grain size. To show appreciation for such effect the evolution of the critical activation stress with source

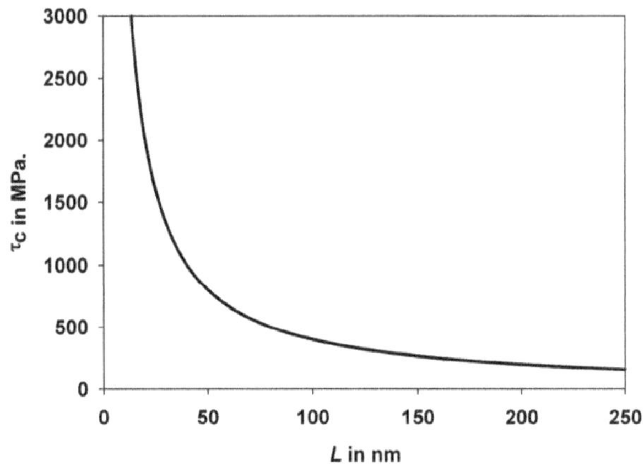

Fig. 6.4 Activation stress of a Frank Read source as a function of source length

length is shown in Fig. 6.4, with $b = 1$ nm and $\mu = 40GPa$.. Interestingly, it can be seen that the critical activation stress of, say, 20 nm, reaches very large value in the order of several GPa (1.6 GPa in the case of a 20 nm long source).

As shown in the above, a Frank Read source is not likely to be activated in small-grain NC materials. Moreover, since grain interiors in NC materials are usually defect free, the presence of Frank Read sources is expected to be rare. Recall that, on average, one would expect a NC sample with small grain size to have one dislocation line per grain initially in the microstructure (see Chapter 1).

Alternatively, one could imagine a process of homogeneous nucleation of a dislocation within a nanosized grain. Such process requires overcoming very large energetic barriers. This requires large local stresses to be present in the region of nucleation of the dislocation. Interestingly, the nucleation barrier may be decreased if – during the nucleation process – the dislocations Burgers vector grows simultaneously as the dislocation loop [12]. The energy change of a system during the process, ΔW, can be approximated as:

$$\Delta W = W_e + W_s - A \tag{6.4}$$

Where W_e, W_s, A denote the elastic strain energy of a dislocation with evolving Burgers vector **s** (the Burgers vector of a perfect dislocation is denoted 2.**b**), the stacking fault energy resulting from the growth of the dislocation Burgers vector, and A is the work done by the dislocation loop as it grows with **s**. Each term in Equation (6.4) can be obtained rather simply with linear elastic theory. The process is energetically favorable if there is a minimum energy path resulting in a decrease in the system's energy. As shown in Fig. 6.5, such a path could exist. However, very large stresses are required (larger than 2 GPa). Therefore, the mechanism of homogeneous nucleation of a dislocation within a crystal is not active in a typical tensile test (similar to those performed in situ and described in previous section). Note that this mechanism may be activated during shock experiments.

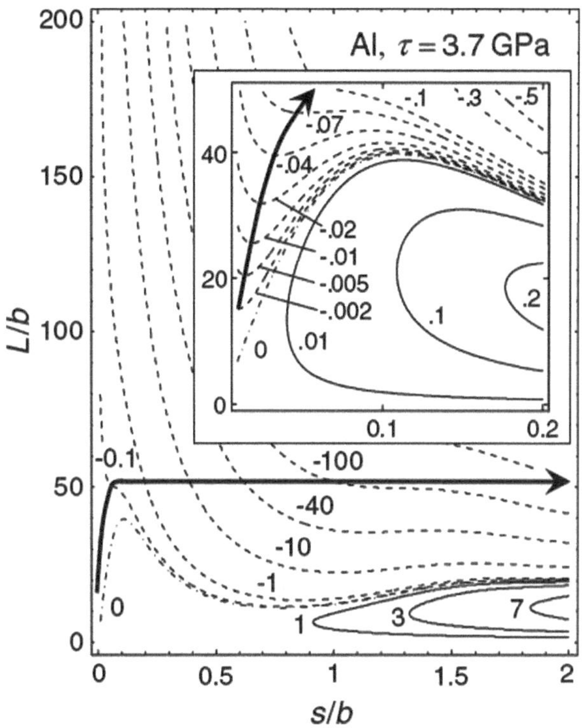

Fig. 6.5 Map of the system's energy during homogenous nucleation of a dislocation in Al under 3.7 GPa applied resolved shear stress [12]. The *horizontal axis* denotes the ratio of the dislocation Burgers vector over the Burgers vector of a perfect dislocation (2b), and the *vertical axis* denotes the ratio of the dislocation length L over b

Recall that dislocation activity ceases solely in grains with diameter smaller than ~10 nm. In a sample with average grain size 30 nm, some dislocation activity is still observed. As discussed in the above both nucleation of dislocation via activation of a Frank Read source or via a homogeneous process are not expected to operate. Therefore, another mechanism of creation of dislocations is to be expected. As shown in Chapter 5, grain boundaries are likely to act as dislocation sources when the grain size is small.

6.4 Grain Boundary Dislocation Emission

The mechanism of emission of dislocations from grain boundaries has received particular attention in several studies dedicated to NC materials. From the modeling perspective, the grand challenge is to predict the activation of the mechanism as well as its overall contribution to the plastic deformation. Interestingly, the conceptualization of such atypical mechanism was introduced long before the first NC sample could be produced. Indeed, in 1963, J.C.M. Li pioneered the area by suggesting that particular defects in grain boundaries, namely ledges, could act as dislocation donors [13]. In this view, a grain boundary ledge (e.g., a step) could separate itself from the grain boundary,

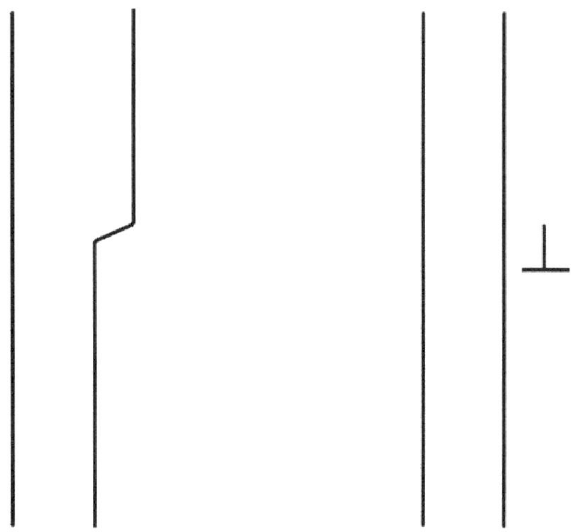

Fig. 6.6 Schematic dislocation emission process from a grain boundary ledge donor as imagined by J.C.M. Li

thus leading to a newly created dislocation within the crystal and to a perfect planar grain boundary. This process is schematically presented in Fig. 6.6. More recent studies showed that grain boundary ledges can operate as sources of dislocations and not simple donors. In other words, the ledge can remain attached to the grain boundary following the emission of the dislocation within the grain interior. Regardless of the details of the emission process at the atomistic scale, Li's model – in which no criterion describing the process was presented – first introduced the concept of grain boundary–assisted deformation.

Since then, several molecular simulations (both in static and dynamic), quasi-continuum simulations (presented in Chapter 9) and continuum models (see examples in Chapter 5) have shown that, in NC materials, grain boundaries can act as dislocation sources. Moreover, TEM observations concur on the matter. Prior to discussing the details of the mechanism of emission of dislocations by grain boundaries, let us discuss both the possible size effect in the activation and in the contribution of such mechanism. A typical intragranular dislocation source (e.g., Frank and Read sources) requires very large local stresses within the crystal's interior and the presence of these sources is expected to be extremely rare in NC materials. This shall favorize the emission of dislocations from grain boundaries (or triple junctions) for there are no obstacles, such as pile-ups, within the grains to prevent the activation process. Recall that TEM observations did not exhibit sufficient stored dislocation densities for typical hardening mechanism – from dislocation storage and thermally activated annihilation – to be considered active in NC materials with grain size smaller than $\sim$30 nm. Therefore, following the emission of a dislocation from interfaces, it is likely that a dislocation could travel from its emission site to the grain boundary opposite to the source. Following this, the dislocation is likely to be absorbed within the grain boundary. In conventional materials, numerous dislocations would be required for such mechanism to lead to significant plastic deformation. If we denote the grain size with d and the dislocation Burgers vector b, then the resolved shear strain resulting from the movement of a dislocation across a grain is on the order of b/d. Clearly, in NC materials fewer dislocations would be required to generate significant plastic deformation. For example, a single dislocation in a Cu grain of size 20 nm could result in $\sim$3% resolved shear strain.

Two types of nucleation processes can be distinguished: (1) emission of a dislocation from a low-angle grain boundary and (2) emission of a dislocation from a large-angle grain boundary. In the former case, grain boundaries can act as pinning points to dislocations and can act as typical Frank and Read sources. Note that here, too, the size effect discussed previously and shown in Fig. 6.4 will apply. Therefore, as the grain size is decreased such sources shall become more difficult to activate. Also, as discussed in Chapter 2, low-angle grain boundaries are typically not dominant in NC materials compared to large-angle grain boundaries.

As a consequence, most of the modeling effort (principally based on use of molecular simulations) has been focused on large-angle grain boundaries. Precisely, molecular dynamics simulations on two-dimensional columnar structures and fully three-dimensional structures and on bicrystal interfaces were used to investigate the atomistic details and activation of the mechanism of emission of dislocation from large-angle grain boundaries [14–18]. Additionally, quasi-continuum simulations on bicrystal interfaces were also used to that end. Several interesting features of the emission process were found [19]:

6.4.1 Dislocation Geometry

All simulations revealed that dislocations emitted from grain boundaries are emitted on the primary slip systems. Molecular simulations of a $\Sigma5(210)$ pure tile grain boundary in Cu subjected to a tensile test (normal to the interface plane) show that the nucleation event is localized on one of the primary $\{111\}<112>$ slip systems. The leading partial dislocation is connected to the interface by an intrinsic stacking fault (see Fig. 6.7(a) and (b)). The details of the process are as follows: (1) a leading partial dislocation is emitted from grain boundaries and (2) a trailing partial dislocation is emitted from the grain boundary. Note that this second point has been subject to some controversy. In particular, simulations on Al predict the emission of both the leading and the trailing partial dislocation. However, the same simulations on Cu predict solely the emission of the leading partial dislocation. Thus, a stacking fault would be left within the grain interior. However, note that TEM observations do not reveal the sufficient presence of stacking faults (although these are seen during in situ tensile tests at liquid nitrogen temperature) to conclude that solely a leading partial dislocation is emitted in NC materials.

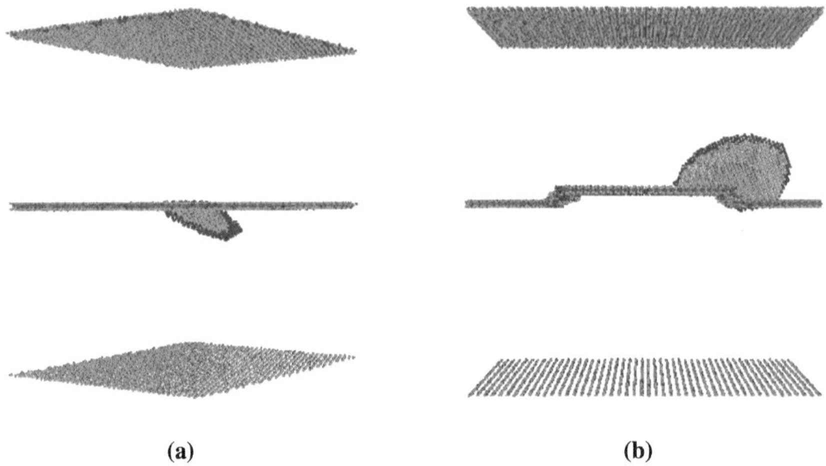

Fig. 6.7 Emission of a partial dislocation from (a) a perfect planar Σ5(210) grain boundary and (b) a stepped Σ5(210) grain boundary

6.4.2 Atomistic Considerations

Quasi-continuum simulations on the response of bicrystal interfaces with different misorientations (see Chapter 9) have shown that, in grain boundaries containing E structural units subjected to a shear stresses parallel to the grain boundary plane, atomic shuffling is active prior to the emission of a dislocation [20]. Similar shuffling process was also predicted in fully three-dimensional simulations following the emission of a leading partial dislocation. The mechanism of shuffling is clearly a relaxation process which is thus expected to decrease the local stress within the grain boundary prior to the emission of a dislocation. When such a process is activated, dislocation emission from a grain boundary would thus be expected to be slightly delayed compared to that of an emission process without pre-emission shuffling. As expected, the structure of the grain boundary will be affected by the dislocation emission process. So far, atomistic simulations have shown that, depending on the type of grain boundary, the emission of dislocations from the interface may have different influences on the grain boundary microstructure. For example, simulations on a Σ5(310) tilt grain boundary have shown that a grain boundary ledge can be created following the emission of a dislocation [15]. Molecular simulations have shown that a grain boundary ledge localized at a triple junction can be annihilated following the emission of the leading and trailing partial dislocation. Conversely, bicrystal simulations on stepped interfaces did not predict the annihilation of the ledge (see Fig. 6.7 (b)).

6.4.3 Activation Process

Grain boundary dislocation emission is a thermally activated process (i.e., stress alone can activate the mechanism). In other words, at a given temperature, sufficient energy must be provided to the grain boundary such that the energy barrier (e.g., free enthalpy of activation) preventing the activation of the mechanism can be overcome. Let ΔG denotes the activation, the free enthalpy of activation of a thermally activated mechanism is typically written as follows:

$$\Delta G = \Delta G_0\left(1 - \left(\frac{\tau}{\tau_c}\right)^p\right)^q \tag{6.5}$$

Here, ΔG_0 and τ_c denote the activation barrier at zero Kelvin (e.g., the necessary amount of energy to be brought to the system without additional energy brought by thermal activation), and the critical resolved shear stress sufficient to activate the process at zero Kelvin. Parameters p and q describe the shape of the energy barrier profile. The following constraint is imposed on p and q: $0 < p < 1$ and $1 < q < 2$. Supposing that a Boltzmann distribution can be used to describe the statistics of the emission process and with Equation (6.5) the probability of successful emission is given by an Arrhenius type of law: $P = \exp\left(-\frac{\Delta G}{kT}\right)$. Here, T and k denote the temperature and the Boltzmann constant. The free enthalpy of activation and the critical resolved shear stress, which are clearly dependent on the grain boundary geometry, can be estimated from molecular simulations on bicrystal interfaces. Precisely, the critical resolved shear stress, τ_c, can be obtained by applying an increasing tensile stress on an interface until a dislocation is emitted. Also, the free enthalpy of activation, ΔG_0 can be estimated as the difference in the interface excess energy between the initial relaxed state and the state at which a dislocation is emitted. By performing such simulations on two bicrystal interfaces – namely, a perfect planar and a stepped $\Sigma 5(210)$ interface – interesting results can be obtained. First, it can be found, as expected, that a step in the interface decreases the critical activation stress. In the particular case discussed above, the critical resolved shear stresses of the planar and stepped interfaces are 2580 and 2450 MPa. Figure 6.8 presents the evolution of the bicrystal energy (**a**) and of the interface excess energy (**b**) during tensile loading, as predicting from molecular dynamics simulations at 10 K. As expected, the total energy of the system increases with increasing strain (or time step in the present case). Also, it can be seen that the emission of a dislocation results in a sharp decrease in the systems energy. Interestingly, defects, such as steps, within grain boundaries can have a significant influence on the free activation enthalpy (see Fig. 6.8b). In the present example, calculations of free enthalpy of activation for the planar and stepped interfaces give 173.2 mJ/m^{2}and 103.8 mJ/m^2, respectively.

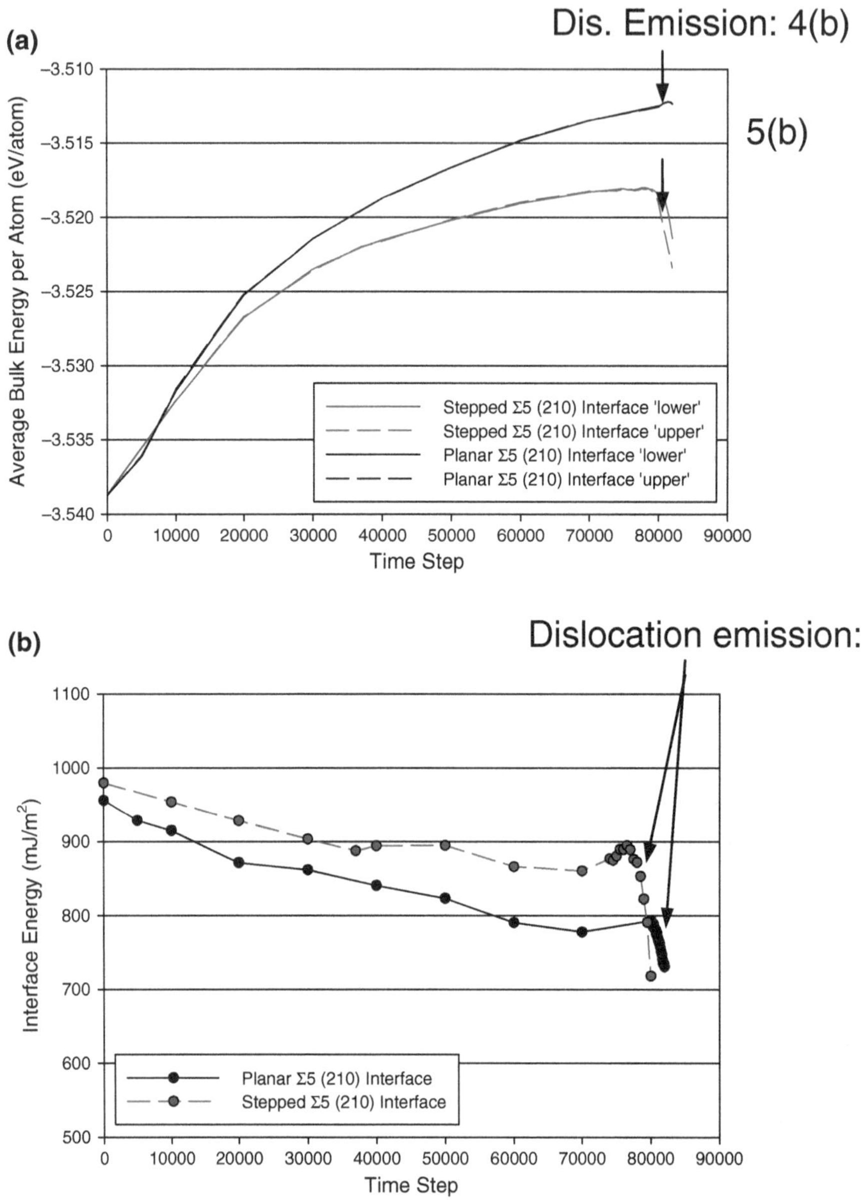

Fig. 6.8 (**a**) evolution of the bulk energy with time and (**b**) evolution of the interface energy with time, during tensile loading of a stepped and perfect planar Σ5(210) interface

6.4.4 Stability

As shown above, large local stresses are required within the grain boundaries for the emission of the first leading partial dislocation to be favorable. Within the crystals, it can be shown that fairly large stresses are required to drive both the leading and the trailing partial through the entire grain. As shown in original work by Asaro et al. [21, 22], the stress required to drive an extended dislocation can be written as the sum of the contribution of the stress required to engender a strain in the order of **b**/d (e.g., the shear strain engendered by the movement of a dislocation across a grain) and of the stress to drive two partial dislocations connected by a stacking fault across the grain. In the former case, a minimum energy path must be selected. Developing the mathematics, one obtains the following expression of the stress, τ, required to move two dislocations across a grain of size d.

$$\tau = \frac{\alpha - 1}{\alpha}\Gamma + \frac{1}{3}\frac{b}{d} \tag{6.6}$$

Here, α denotes the ratio of the grain size over the equilibrium distance between two partial dislocations. Note that this model does not describe the emission process but the motion of the emitted dislocations. Nonetheless, interesting size effects can be found. As described by Equation (6.6), as the grain size is decreased, the required shear stress will increase. For example, for 50 nm NC Cu one obtains $\tau = 211\,MPa.$ while for 10 nm NC Cu one has $\tau = 421\,MPa.$. Similarly, as the stacking fault energy increases, the critical stress will increase.

6.5 Deformation Twinning

Prior to describing the particulars of the mechanism of twinning in NC, let us briefly present a general introduction to deformation twinning in conventional materials. Comprehensive reviews on the matter can be found elsewhere [23]. Twinning corresponds to the mirror reorientation of atoms about a twinning plane. This mechanism is particularly interesting for it contributes to the plastic deformation in several different manners. First, depending on the system considered (e.g., hcp, FCC) it can engender a shear within the parent crystal. This can be easily seen in the case of an hcp crystal with a non perfect c/a ratio. Second, due to the lattice reorientation within the twin domains, slip systems which were not favorably oriented may become activated. Finally, it will affect the hardening response of the material due to the interaction of mobile dislocations with twin boundaries.

Twinning is typically active in materials with low symmetry – such as the hcp system – at low temperatures or at relatively large strain rates. As stated previously, deformation twinning was observed during tensile test at liquid nitrogen temperature on NC samples. No similar observation was made at room temperature. Clearly, the three contributions of twinning mentioned in the above are expected to be attenuated in NC materials due to the much reduced dislocation activity.

In conventional materials, and especially in the hcp system in which deformation twinning can be significant, the particulars of the nucleation of the mechanisms are not yet well known. Yet, several models were introduced to provide greater understanding of the process. These models can be sorted into two categories: (1) homogeneous nucleation models and (2) heterogeneous nucleation models. The former are based on the idea, that without the presence of defects, serving as seeds to the twin nucleus, a twin nucleus could spontaneously be created due to local stress concentrations [24]. Obviously, as in the case of spontaneous creation of a dislocation mentioned in the above, one would expect the need for very large stress fields to be present for such mechanism to be activated. Another twin nucleation mechanism introduced in conventional materials is, for example, that of nucleation from dissociation of a glide dislocation into one or two twinning dislocations (which may or not be zonal) [25, 26]. An illustration of such mechanism is presented in Fig. 6.9, where it can be seen that the result of the dissociation of a slip dislocation is a stair rod dislocation connected to one or more twinning dislocations.

Decreasing the grain size to the nanoregime, the process of dislocation dissociation is not likely to operate for the initial geometrical configuration is not expected to be found within a nanosized crystal. In NC materials, the process of twin nucleation is grain boundary mediated. Molecular simulations have shown – in the case of the FCC system – that it can be decomposed in three steps shown in Fig. 6.10. First, a partial dislocation is emitted from a grain boundary (see Fig. 6.10(a)). This dislocation remains connected to its source with a Stacking fault. Then, a second twinning dislocation is emitted from the same grain boundary. Interestingly, as shown in Fig. 6.10(b), this dislocation is emitted on a non-neighboring plane. This is the critical aspect step of the twin

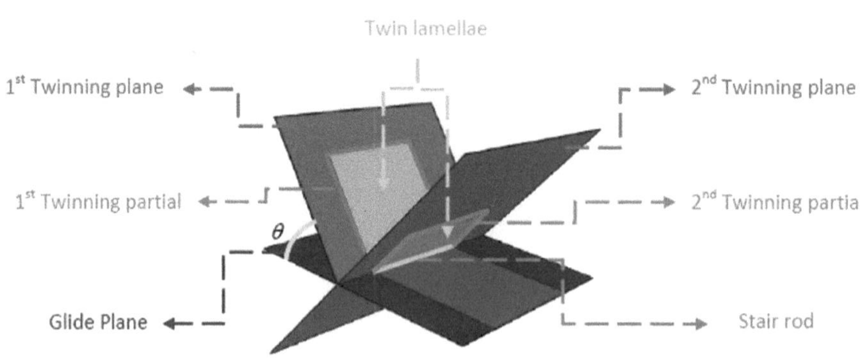

Fig. 6.9 Schematic representation of a twin nucleation process from nonplanar dissociation of a glide dislocation. The resulting defects are (1) a stair rod dislocation, (2) one or two twinning dislocation loop, and (3) one or two twin boundaries

Fig. 6.10 Molecular simulation of the twin nucleation process in NC Cu: (**a**) emission of a partial dislocation, (**b**) emission of a second partial dislocation on a non neighboring plane, and (**c**) transformation of the stacking faults into a twin nucleus via nucleation of a dislocation loop within the textured grain

nucleation process. Indeed, it can be seen that, after these first two steps have occurred, four consecutive planes are faulted. Finally the fault is transformed into a twin nucleus via nucleation of a third dislocation (or of a pair of antiparallel dislocations) within the faulted region. As a result, a two layer thick twin nucleus is created.

Finally, looking at the resolved shear stresses in the twin planes during the nucleation process, it was shown that large local shear stresses (in the order of ~3 GPa) are required to activate mechanism of twin nucleation.

6.6 Diffusion Mechanisms

In Chapter 2, it was shown that grain boundaries exhibit higher self-diffusivities than perfect lattice. This is especially true in the case of amorphous (or less organized) grain boundary regions. Also, early experiments on NC materials suggested that diffusion creep mechanisms (e.g., Nabarro Herring creep, Coble creep, and triple junction creep) could be activated at room temperature. Since then, it was shown that, similarly to conventional materials, diffusion creep is not likely to be activated at room temperature in NC materials. Nonetheless, vacancy diffusion shall remain easier to activate in NC materials. In particular, it may assist other grain boundary mediated mechanisms. For example, grain boundary sliding could benefit from diffusion mechanisms. Similarly, the penetration of a dislocation emitted from a grain boundary source will create a mass transfer within the grain boundary dislocation sink. This process, too, may be assisted by vacancy diffusion. Therefore, for the sake of completeness, let us briefly describe the pioneering vacancy diffusion models introduced by Nabarro-Herring [27] and by Coble [28].

In conventional metals, deformation maps were established to predict the range of activity, as function of temperature and resolved shear stress, of all plastic deformation mechanisms. An example is presented in Fig. 6.11 corresponding to 10 μm grain size Ni [29]. As shown, during a creep experiment, Coble creep and Nabarro-Herring creep are expected to operate under relatively low shear stresses. At a given temperature, an increase in the applied stress from the regime of activity of Coble creep, can lead to the activation of grain boundary sliding accommodated by vacancy diffusion (controlled grain boundaries or the lattice diffusion). Clearly, the deformation map shown in Fig. 6.11 will depend on the grain size.

In general, creep mechanisms are described with a phenomenological law. It is typically dependent on grain size, temperature, and stress and given, in its most generic form, by:

$$\dot{\varepsilon} = \frac{A \cdot D \cdot G \cdot b}{kT} \left(\frac{b}{d}\right)^{p} \left(\frac{\sigma}{G}\right)^{n} \tag{6.7}$$

Here A, D, G, b, k, T,and d denote a numerical constant, the diffusion coefficient, the shear modulus, the magnitude of Burger's vector, Boltzmann's constant, the temperature, and the grain size, respectively. σ denotes the applied stress, p and n are the size and stress exponents, respectively. The type of creep mechanism (e.g., controlled by lattice or by grain boundary diffusion) can usually be identified from the size and stress exponents. For example, a stress exponent equal to 1 and a size

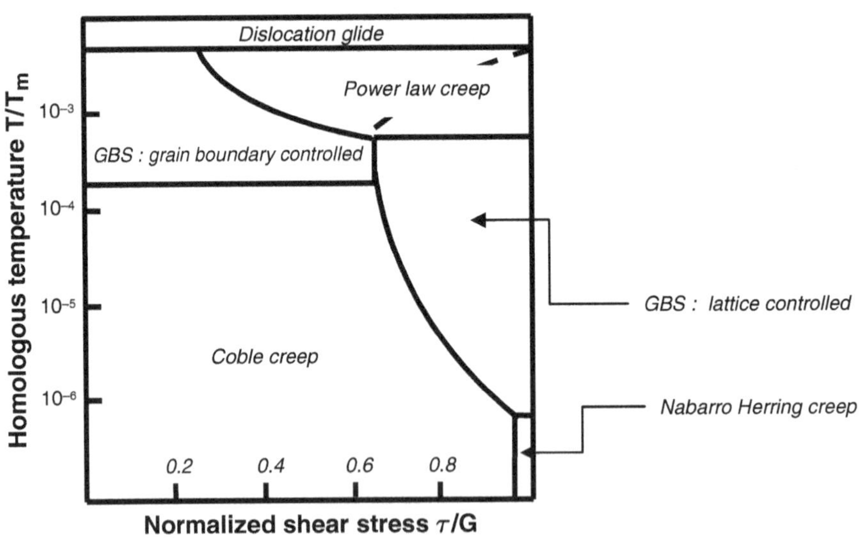

Fig. 6.11 Deformation map of pure Ni with 10 μm grain size

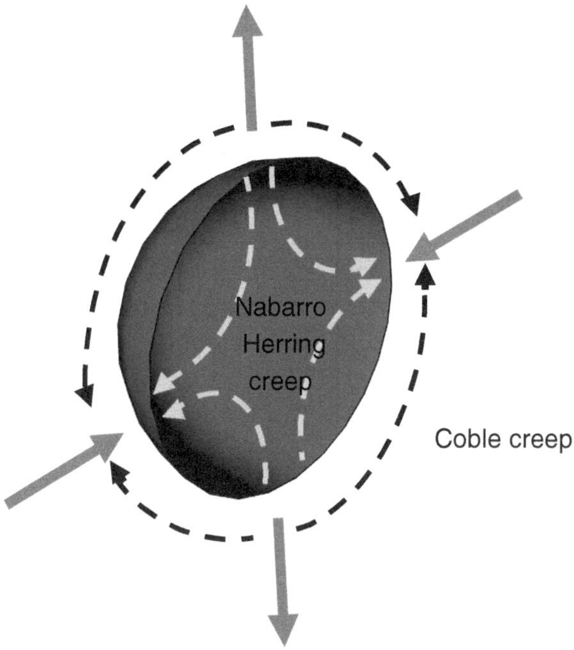

Fig. 6.12 Vacancy diffusion paths during Coble creep and Nabarro-Herring creep

exponent equal to 2 correspond to the mechanism of Nabarro-Herring creep. Let us now find the fundamentals leading to Equation (6.7). Of particular interest here are the mechanisms of steady state vacancy diffusion.

Within a polycrystalline aggregate, vacancy diffusion can occur via two different competing paths: grain interiors or grain boundaries. The former type of diffusion, controlled by the lattice self-diffusivity, is referred to as Nabarro-Herring creep, and the latter as Coble creep. An illustration of the two different diffusion paths is presented in Fig. 6.12.

6.6.1 Nabarro-Herring Creep

Interestingly, Herring's model (1950) was first introduced to explain the quasi-viscous behavior of metallic wires in traction under small load and high temperatures. The idea was to describe the transport of matter by diffusion within the grain interior. Under and applied stress on a spherical grain (as represented by the blue arrows in Fig. 6.12), the flux of matter, J, is driven by a chemical potential gradient$\nabla(\mu - \mu_h)$;

$$J = -\left(\frac{n_L D}{kT}\right)\nabla(\mu - \mu_h) \tag{6.8}$$

Here, n_L, D, k and T denote the number of sites per unit volume, the self-diffusion coefficient, the Boltzmann constant and the temperature. The chemical potential is dependent on the vacancy concentration. Taking mass conservation into account and minimizing the spherical crystal's free energy, the chemical potential can be related to the externally imposed stress and to the grain size. Depending on whether or not shear stresses are relaxed at interfaces, the chemical potential is given by:

$$\begin{aligned} \mu - \mu_h &= \mu_0 - \tfrac{\Omega_0}{d^2} \sum_{i,j} \sigma_{ij} x_i x_j && \text{in the relaxed case} \\ \mu - \mu_h &= \mu_0 - \tfrac{5\Omega_0}{2d^2} \sigma_{xx} \left(x^2 - y^2\right) && \text{in the nonrelaxed case} \end{aligned} \tag{6.9}$$

where x_r r = i, j are the coordinates, d is the grain size, and Ω_0 is the atomic volume. Finally, the creep law is obtained with the geometrical relation between the rate of displacement, the strain rate, and the normal flux. Finally, Nabarro-Herring's creep law reads:

$$\dot{\varepsilon}_{NH} = \frac{A_{NH} D_L G b}{kT} \left(\frac{b}{d}\right)^2 \left(\frac{\sigma}{G}\right) \tag{6.10}$$

σ is the stress, d the grain size, G the shear modulus, b the magnitude of Burger's vector, D_L is the crystals diffusion coefficient, k denotes Boltzmann's constant, T is the temperature, and A_{NH} is a numerical constant.

6.6.2 Coble Creep

As mentioned in the above, Coble creep represents the vacancy diffusion along grain boundaries. The derivation of the Coble creep law is similar to that introduced by Herring [27]. Vacancy concentration gradients are expressed as a function of the applied stress of temperature as follows:

$$\Delta C = \frac{C_0 \cdot \sigma \cdot \Omega}{kT} \tag{6.11}$$

Here, C_0, σ, Ω, k et T denote the initial equilibrium vacancy concentration, the stress normal to the grain boundary, the atomic volume, Boltzmann's constant, and the temperature, respectively. Then the flux is calculated such that mass conservation is ensured. In particular, vacancy creation on each facet is assumed uniform. Applying this to a spherical grain signifies that the rate of creation of vacancies is equal to the rate of annihilation of vacancies. Such condition is met at 60 degrees on a hemisphere (e.g., both the top and bottom regions have the same area). In the steady state regime and with Fick's law, the flux is given by:

$$J = D_v N w \frac{2\Delta C}{\pi d} d \sin 60 \tag{6.12}$$

w and D_v denote the average grain boundary thickness and the coefficient of diffusion of grain boundaries. As in the case of Nabarro herring creep, the strain rate engendered by vacancy diffusion is obtained by geometrical considerations (e.g., the rate of change of the crystal's volume is consistent with the volume change due to diffusion $Ja_0^3 = \pi d^2 \cdot \frac{\partial d}{dt}$) :

$$\dot{\boldsymbol{\varepsilon}} = \frac{A_{co} w \cdot \boldsymbol{\Omega} \cdot D_{GB}}{kT} \frac{\boldsymbol{\sigma}}{d^3} \tag{6.13}$$

Note that the size exponent of the Coble creep law is 3 while that of Nabarro-Herring creep is 2. Therefore, one expects Coble creep to dominate Nabarro-Herring at small grain sizes. Using micromechanical scale transitions techniques (presented in Chapter 7) and a two–phase representation of the aggregate (inclusion phase represent grain interiors), it can easily be shown that the activation of Coble creep in NC materials would soften the materials response [30].

6.6.3 Triple Junction Creep

Triple junctions, which exhibit a much less organized structure than pure tilt grain boundaries, for example, are naturally expected to provide shortcuts for vacancy diffusion. Using a similar method as presented above, a creep law accounting for triple junction creep can be established [31]:

$$\dot{\varepsilon} = K_{tl} \frac{D_{tl} \cdot \Omega \cdot \sigma}{kT} \frac{w^2}{d^4} \tag{6.14}$$

Here, K_{tl}, D_{tl}, Ω, w, k, σ and T denote a numerical constant, the coefficient of diffusion of triple junctions, the atomic volume, the average grain boundary thickness, Boltzmann's constant, the stress, and the temperature in Kelvin, respectively. Interestingly, Equation (6.14) suggests that such triple junction creep mechanism could become significant in the NC regime since the size exponent is equal to 4.

6.7 Grain Boundary Sliding

6.7.1 Steady State Sliding

Plastic deformation resulting from imperfectly bounded interfaces has been extensively studied to rationalize the superplastic deformation of aggregates with conventional grain size (see Fig. 6.11). Note that all models to be discussed in this subsection were established to model the creep response of conventional materials. The first notable contribution to the domain is that of Zener, who introduced the concept of viscous grain boundaries [32]. In this first model, the idea was to predict the reduction in elastic constants due to shear stress relaxation at grain boundaries during creep experiment. The solution of the problem was found by calculating the strain energy of a material (elastic isotropy was assumed) in which no shear stress is transmitted at grain boundary interfaces. The challenge resulted in introducing a displacement field respecting such condition. The solution can be written as the sum of a compatible field, of a field neutralizing the shearing stress and, of a dilatation filed. Finally the following relation, valid only in the case of isotropic elasticity, was obtained:

$$E_r = \frac{1}{2}\frac{(7+5\nu)}{(7+\nu-5\nu^2)}E \tag{6.15}$$

Here, E_r and E denote the "relaxed" Young's modulus and Young's modulus of the reference material.

Since then, more refined models have been introduced to model the mechanism of grain boundary sliding in the plastic regime. The simplest approach is based on the same phenomenological law given by Equation (6.7). It describes the creep response of polycrystalline materials driven by grain boundary sliding accommodated by steady state vacancy diffusion. Such mechanism is referred to as Lifschitz sliding [29]. The following expression is used when the vacancy path is similar to that of Coble creep (e.g., interfaces, see Fig. 6.12)

$$\dot{\boldsymbol{\varepsilon}}_{jg} \approx 2.E5 \cdot D_{jg}\frac{Gb}{kT}\left(\frac{b}{d}\right)^3\left(\frac{\boldsymbol{\sigma}}{G}\right)^2 \tag{6.16}$$

Alternatively, when the vacancy path is similar to that of Nabarro-Herring one has:

$$\dot{\boldsymbol{\varepsilon}}_{c} \approx 8.E6 \cdot D_{c}\frac{Gb}{kT}\left(\frac{b}{d}\right)^2\left(\frac{\boldsymbol{\sigma}}{G}\right)^2 \tag{6.17}$$

Here, $\dot{\varepsilon}_r$, D_r r $= jg$, c denote the average viscoplastic strain rate and the diffusion coefficients of grain boundaries and grain cores, respectively. b, k, T, d, G, and σ denote the magnitude of Burger's vector, Boltzmann constant, the temperature, the grain size, the shear modulus, and the stress, respectively. Note the similarities between the two expressions in the above and the expressions of Coble creep and Nabarro-Herrring creep, respectively.

The mechanism of grain boundary sliding accommodated by vacancy diffusion along grain boundaries was later described in great detail by Raj and Ashby [33]. The authors considered the effect of the grain boundary shape.

The reasoning is based on the fact that steady state grain boundary sliding driven by vacancy diffusion must respect (1) the equilibrium condition at the interface (e.g., null jump of the traction vector), (2) mass must be conserved during the process (e.g., null divergence of the flux of vacancies which is similar to the reasoning of Herring), and (3) the change of volume of the system must be consistent with that given by the vacancy flux. Representing the grain boundary shape with Fourier series, a solution of the problem can be found for any periodic grain boundary shape. For example, in the case of a sinusoidal grain boundary of wavelength λ and amplitude $h/2$, one obtains the following expression of the rate of relative displacement of the interface:

$$\dot{U} = \frac{8}{\pi} \frac{\lambda}{h^2} \frac{\Omega \tau_a}{kT} D_L \left(1 + \frac{\pi \delta}{\lambda} \frac{D_L}{D_B} \right) \tag{6.18}$$

Here, τ_a, D_L, and D_B are the shear stress applied at the interface and the lattice and grain boundary diffusivity, respectively. Alternatively, Langdon based his reasoning on the assumption that grain boundary sliding is driven by dislocation climb or glide [34]. The former is the controlling mechanism. Using an Arrhenius type of law to predict the rate of climb (as given in Friedel's work), the following expression of grain boundary sliding controlled by thermally activated climb:

$$\dot{\varepsilon} = \frac{\beta b^2 \sigma^2}{kTGd} D \tag{6.19}$$

Here, β, D, and d represent a numerical constant, the lattice self-diffusivity, and the grain size. Interestingly, the size exponent in Equation (6.18) is lower than obtained with reasoning based on vacancy diffusion.

6.7.2 Grain Boundary Sliding in NC Materials

All mechanisms presented in the previous subsection are based on either purely empirical or phenomenological descriptions of the mechanism of grain boundary sliding. In all cases the sliding process was assumed to be accommodated by a diffusing species (e.g.s vacancies, dislocations). During a tensile test, such steady state diffusion process is not likely to occur and the sliding process may not be accommodated by mass transfer. This would clearly lead to crack creation.

As in the case of the emission of dislocation by grain boundaries, atomistic simulations are the tool of choice to investigate the process of grain boundary sliding. In a comprehensive study on the response of bicrystal interfaces (both symmetric and asymmetric) subjected to pure shear constraints it was shown that, depending on the grain boundary microstructure, the mechanism of grain boundary sliding can be activated [19, 20]. The mechanism was shown to be preceded by atom shuffling. Figure 6.13 presents a plot of the evolution of the

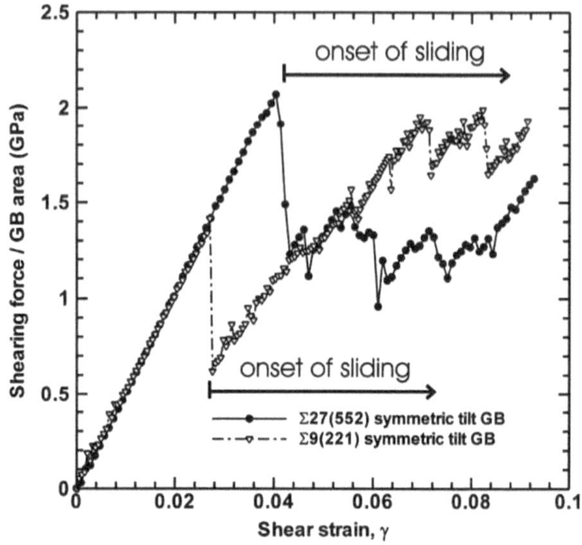

Fig. 6.13 Evolution of the shear stress as a function of the shear strain during the deformation in pure shear of two symmetrical bicrystal interfaces [19]

grain boundary strength as a function of shear strain during the sliding process. It can be seen that the process appears to be similar to a stick-slip mechanism. The grain boundary strength evolves quasi-periodically. In the course of one period, it increases with strain until a critical relative displacement is reached at which the strength decreases sharply.

In simulations presented above, the effect of triple junctions is not considered and any interface decohesion process cannot be simulated. To that end, an elastic-plastic constitutive response for imperfect interfaces was introduced and later implemented in finite element simulations [35]. While the atomistic particular cannot be reproduced with such approach, qualitative trends can be obtained (especially regarding the predictions of the ductility of NC materials). The idea here is to describe the grain boundary sliding process as the follow-up of two different regimes. Note that both tangential and normal displacement jumps are allowed at the interface. A normal displacement jump leads to the interface decohesion. The yield surface is defined by a normal and a tangential strength, $s^{(i)}$ where superscript $i$refers to the normal or to the tangential component, which evolves with strain rate. In the first regime (hard regime), where the relative displacements are smaller than a critical value (~1 nm), the interface is assumed to have a strength increasing with strain. Its evolution is given by:

$$\dot{s}^{(i)} = h_0^{(i)}\left(1 - \frac{s^{(i)}}{s^{*(i)}}\right)\dot{\gamma} \tag{6.20}$$

$h_0^{(i)}$, $s^{*(i)}$ are a numerical constant and the "intrinsic" grain boundary strength, which is on the order of several GPa (as shown in Fig. 6.13). In the second regime – when the relative displacement of two grain is large than a critical value – the interface strength is assumed to decrease proportionally to the shear strain rate until failure is reached (at relative displacement ~1.1 nm). At this point, the interface is fully incoherent. The interface strength is written as follows:

$$\dot{s}^{(i)} = -h^{(i)}_{0soft}\dot{\gamma} \tag{6.21}$$

The model presented in the above can predict the activity, and consequence, of grain boundary sliding not accommodated by a diffusion mechanism. While the approach is elegant, it generally leads to underestimated materials ductility. For example, in the case of 30 nm pure NC Ni it predicts strain to failure <4%.

6.8 Summary

The mechanisms relevant to the plastic deformation of NC materials, and in particular their possible size effects, are reviewed in this chapter. First, in situ and ex situ TEM observations are discussed. It is shown that, in NC materials, dislocation activity is severely reduced with grain size. Several arguments and models are presented to rationalize such size effects (e.g., difficulty in nucleating dislocations, low dislocation stability).

Second, the mechanism of emission of dislocations from grain boundaries is presented from both the atomistic and the continuum point of view. It is shown that grain boundaries, and more favorably grain boundary ledges and triple junctions, are sources of extended dislocations. Also, the critical driving stress allowing the motion of dislocation across grains is shown to be increasing with decreasing grain sizes.

Third, the mechanism of nucleation of twins within NC materials is introduced. Interestingly, it is shown that that the mechanism can be decomposed in three steps corresponding to the emission of partial dislocations from grain boundaries on non-neighboring planes.

Finally, fundamental models originally introduced to describe the creep response of polycrystalline materials deforming by grain boundary sliding and/or diffusion are recalled. Then, simulations investigating the grain boundary sliding process in NC materials are introduced. Details of atomic motion are presented. It is shown that grain boundary sliding lis ikely to be one of the most important mechanisms controlling the deformation of NC materials.

References

1. Kumar, K.S., S. Suresh, M.F. Chisholm, J.A. Horton, and P. Wang, Acta Materialia **51**, (2003)
2. Wu, X., Y.T. Zhu, M.W. Chen, and E. Ma, Scripta Materialia **54**, (2006)
3. Vevecka, A. and T.G. Langdon, Materials Science and Engineering A **187**, (1994)

4. Serra, A., D.J. Bacon, and R.C. Pond, Metallurgical and Materials Transactions; A; Physical Metallurgy and Materials Science **33**, (2002)
5. Kim, H.S., M.B. Bush, and Y. Estrin, Materials Science and Engineering A **276**, (2000)
6. Kocks, U.F. and H. Mecking, Progress in Materials Science **48**, (2003)
7. Madec, R., B. Devincre, and L.P. Kubin, Scripta Materialia **47**, (2002)
8. Qin, W., Z.H. Chen, P.Y. Huang, and Y.H. Zhuang, Journal of Alloys and Compounds **292**, (1999)
9. Qin, W., Y.W. Du, Z.H. Chen, and W.L. Gao, Journal of Alloys and Compounds **337**, (2002)
10. Van Petegem, S., F. Dalla Torre, D. Segers, and H. Van Swygenhoven, Scripta Materialia **48**, (2003)
11. Lu, K., R. Luck, and B. Predel, Materials Science and Engineering A **179–180**, (1994)
12. Gutkin, M.Y. and I.A. Ovid'ko, Acta Materialia **56**(7), 1642–1649, (2008)
13. Li, J.C.M., Transactions of the Metallurgical Society of AIME **227**, (1963)
14. Capolungo, L., D.E. Spearot, M. Cherkaoui, D.L. McDowell, J. Qu, and K.I. Jacob, Journal of the Mechanics and Physics of Solids **55**, (2007)
15. Spearot, D.E., K.I. Jacob, and D.L. McDowell, Acta Materialia **53**, (2005)
16. Van Swygenhoven, H., Materials Science and Engineering: **483–484**, 33–39, (2008)
17. Wolf, D., V. Yamakov, S.R. Phillpot, A. Mukherjee, and H. Gleiter, Acta Materialia **53**, (2005)
18. Yamakov, V., D. Wolf, M. Slalzar, S.R. Phillpot, and H. Gleiter, Acta Materialia **49**, (2001)
19. Warner, D.H., F. Sansoz, and J.F. Molinari, International Journal of Plasticity **22**, (2006)
20. Sansoz, F. and J.F. Molinari, Acta Materialia **53**(7), 1931–1944, (2005)
21. Asaro, R.J., P. Krysl, and B. Kad, Philosophical Magazine letters **83**, (2003)
22. Asaro, R.J. and S. Suresh, Acta Materialia **53**, (2005)
23. Christian, J.W. and S. Mahajan, Progress in Materials Science **39**, (1995)
24. Man Hyong, Y. and W. Chuan-Tseng, Philosophical Magazine **13**, (1966)
25. Mendelson, S., Materials Science and Engineering **4**, (1969)
26. Mendelson, S., Scripta Metallurgica **4**, (1970)
27. Herring, C., Journal of Applied Physics **21**, (1950)
28. Coble, R.L., Journal of Applied Physics **34**, (1963)
29. Luthy, H., R.A. White, and O.D. Sherby, Materials Science and Engineering **39**, (1979)
30. Capolungo, L., C. Jochum, M. Cherkaoui, and J. Qu, International Journal of Plasticity **21**, (2005)
31. Wang, N., Z. Wang, K.T. Aust, and U. Erb, Acta Metallurgica et Materialia **43**, (1995)
32. Zener, C., Physical Review **60**, (1941)
33. Raj, R. and M.F. Ashby, Metallurgical Transactions **2**, (1971)
34. Langdon, T.G., Philosophical Magazine **22**, (1970)
35. Wei, Y.J. and L. Anand, Journal of the Mechanics and Physics of Solids **52**, (2004)

Chapter 7
Predictive Capabilities and Limitations of Continuum Micromechanics

7.1 Introduction

As discussed in Chapter 3, the grain size dependence of mechanical response of nanocrystalline (NC) materials is caused by their local deformation mechanisms (e.g., Coble creep, twinning, grain boundary dislocation emission, grain boundary sliding) that rely on the typical nanoscale structure of grain boundaries and their extremely high-volume fraction. Although these deformation mechanisms have been highlighted by experimental observations and molecular dynamics simulations, it is rarely possible to directly relate their individual contributions to the macroscopic response of the material. This is primarily due to the fact that the scale and boundary conditions involved in molecular simulations are several orders of magnitude different from those in real experiments or of typical polycrystalline domains of interest.

Modeling the local mechanisms and reporting their effect on the overall behavior of NC materials are challenging problems that require the use of multiple approaches that rely on the classical continuum micromechanics where appropriate length scales can be introduced by mean of molecular dynamic simulations.

Micromechanical framework has no intrinsic length scale. To capture the size dependence in mechanical behavior of NC materials, appropriate length scale has to be introduced in the concept of continuum micromechanics. Within this context, most of the models rely on a generic idea that grain boundaries provide the effective action of the deformation mechanisms, which are different from the lattice dislocation mechanisms occurred in conventional coarse-grain polycrystalline materials. In other words, grain boundaries play the role of obstacles of lattice dislocations to strengthen a conventional polycrystalline material, whereas they serve as softening structural elements that carry the plastic flow in NC materials.

To deal with the deformation responses of NC materials, theoretical models adopting this generic idea introduce a length scale in continuum micromechanics to carry properly the deformation mechanisms generated by grain boundaries and their competitions with the conventional lattice dislocation motion. The length scale is introduced by assigning a finite thickness to grain boundaries along with appropriate continuum models describing grain boundary defects.

M. Cherkaoui, L. Capolungo, *Atomistic and Continuum Modeling of Nanocrystalline Materials*, Springer Series in Materials Science 112,
DOI 10.1007/978-0-387-46771-9_7, © Springer Science+Business Media, LLC 2009

Molecular dynamic is a reliable tool in describing the active role of grain boundaries and developing the associated continuum models to be incorporated in classical continuum mechanics to capture the size dependence of the deformation response of NC materials. In general, theoretical models adopting this philosophy can be divided into two basic categories that may rely on the classical concepts of continuum micromechanics:

1. Models invoking the concept of two-phase composite with grain interiors and grain boundaries playing the role of constitutive phases. Models falling into this category have proven to be an effective way to capture a characteristic length scale to describe the deformation response of NC materials [8, 9, 32]
2. Models adopting a crystal plasticity type description that deal with nanoscale effects of grain boundaries on conventional lattice dislocation motion, competition between various deformation mechanisms, and the effect grain orientations and grain size distribution on the overall response of NC materials. Note that classical strain gradient crystal plasticity models may fall into this category if the role of grain boundary is properly identified. In general, models in this category rely on a double-scale transition method; a scale transition from the atomic scale to the mesoscopic scale (nano single crystal) must first be performed to describe the role of grain boundaries, followed by a second scale transition from the mesoscopic scale to the macroscopic scale (polycrystalline aggregate). The second scale transition can be replaced by appropriate finite element calculations.

The main focus of this chapter is to provide an overview of the main concepts of continuum micromechanics. The principles of micromechanics that rely on the elementary inclusion problem of Eshelby are developed. The methodology is first illustrated in the case of linear problems that carry the overall elastic responses of heterogeneous materials. Secondly, focus is placed on extensions of linear solutions to the case of nonlinear problems describing the plastic flow of composite materials. Then, attention is paid to the continuum description of the elementary lattice dislocation motion within the concept of crystal plasticity in conventional polycrystalline materials. The chapter will end with an illustrative example of the contribution of Jiang and Weng [32] describing the deformation response of NC materials. Other models belonging to both categories are developed and discussed in Chapter 9.

7.2 Continuum Micromechanics: Definitions and Hypothesis

Most of engineering materials are heterogeneous in nature. They generally consist of different constituents or phases, which are distinguishable at a specific scale. Each constituent may show different physical properties (e.g., elastic moduli, thermal expansion, yield stress, electrical conductivity, heat conduction, etc.) and/or material orientations, and may be heterogeneous at smaller scales. Therefore, the continuum micromechanics based methodologies lie in the definition of the source of heterogeneities of the constituents, from which the overall physical properties of the heterogeneous material are derived (e.g., elastic moduli, thermal expansion, yield stress, electrical conductivity, heat conduction, etc.). Continuum micromechanics have been applied with a certain success to derive the overall physical properties of a class of heterogeneous materials such as composite materials, polycrystalline materials, and porous and cellular materials.

Throughout this chapter, our concerns are the mechanical properties of heterogeneous materials. However, there are large studies in the literature devoted to nonmechanical properties of heterogeneous materials using adaptable continuum micromechanics techniques.

In the present section, the general concepts governing the continuum micromechanics are addressed. The main concern is the definition of the representative volume element (RVE), and how the RVE represents statistically the heterogeneous material to make a link between the local stress and strain fields to the global ones.

7.2.1 Definition of the RVE: Basic Principles

As discussed above, the main feature of continuum micromechanics is to derive macroscopic mechanical properties from the corresponding microscopic ones. For such a purpose, the definition of the general framework of continuum micromechanics requires certain conditions to be fulfilled:

1. Since the microstructure of heterogeneous materials is generally complex, but at least to some extent random, a realistic link between macro and micro quantities is performed only under certain approximations. Typically, these approximations are based on the ergodic condition. Basically, the ergodic condition assumes that the heterogeneous material being statistically homogeneous. In other words, one can select randomly within the heterogeneous material sufficiently large volumes, called mesodomains (or homogeneous equivalent medium), so that appropriate averaging schemes over these domains give rise to the same mechanical properties, corresponding to the overall or effective mechanical properties. [*Remark* – Not all of heterogeneous materials can be treated as statistically homogeneous. Some cases may require nonstandard analysis.]
2. The definition of macro and micro mechanical properties (or in a large sense the physical properties) requires an appropriate separation between different length scales. In fact, in the framework of micromechanics, the stress and strain fields are split into contributions corresponding to different length scales. It is assumed that these length scales are sufficiently different in terms of the order of magnitude, so that for each pair of them, the fluctuations of stress and strain field (micro or local quantities) at the smaller length scale

influence the overall behavior (or macroscopic) at the larger length only via their averages, and, conversely, fluctuations of stress and strain fields as well as the compositional gradients (macro or global quantities) are not significant at the smaller length scale. Therefore, at this scale, these macro fields are locally uniform and can be described as uniform applied stresses and strains. This scale separation leading to the definition of micro and macro fields corresponds to the macrohomogeneity condition. [*Remark* – The macrohomogeneity condition requires the fulfilment of the ergodic condition to ensure the averaging procedures to make a link between local fields and global ones.]

As can be seen in the next two sections, the ergodic and macrohomogeneity conditions allow the definition of the RVE with appropriate boundary conditions. Such a definition gives, within the framework of continuum micromechanics, a rigorous and systematic way to derive overall mechanical properties by taking appropriate information at the microscale. [*Remark* – The reader should not confuse the length scale that is required for the definition of microscopic quantities and an intrinsic length scale, which in general doesn't appear explicitly in the framework of continuum micromechanics.]

7.2.1.1 Ergodic Condition

Ergodicity is a mathematics term, meaning "space filling." Ergodic theory has its origin from the work of Boltzmann in statistical physics. Ergodic theory in statistical mechanics refers to where time and space distribution averages are equal.

Let us see how the ergodic condition works in the case of heterogeneous materials and how it lies in the statistical homogeneous presentation of such materials.

Consider a heterogeneous medium defined by a finite volume, V. Suppose that the volume V is partitioned into n-disjoint random set or phases. Each phase $I(I = 1, 2, \ldots, N)$ is supposed to occupy a set of subvolume $V^I(\mathbf{r}')$ $(I = 1, 2, \ldots, N)$, where $\mathbf{r}'$ is the position vector with respect to a reference medium. Let define on V a probability density function, $p(\mathbf{r}')$.

The characteristic function, $\theta^I(\mathbf{r}, \mathbf{r}')$, for the phase, I, reads

$$\theta^I(\mathbf{r}, \mathbf{r}') = \begin{cases} 1, & \text{if } \mathbf{r} \in V^I(\mathbf{r}') \\ 0, & \text{otherwise} \end{cases} \tag{7.1}$$

with the property that

$$\sum_{I=1}^{N} \theta^I(\mathbf{r}, \mathbf{r}') = 1 \tag{7.2}$$

Therefore, the probability $\Im_1^I(\mathbf{r})$ to find the phase I at a chosen point, $\mathbf{r}$, is expressed by

$$\Im_1^I(\mathbf{r}) = \langle \theta^I(\mathbf{r},\mathbf{r}') \rangle = \iiint\limits_V \theta^I(\mathbf{r},\mathbf{r}')p(\mathbf{r}')dV \tag{7.3}$$

$\Im_1^I(\mathbf{r})$ is referred to as the one-point correlation function for the characteristic function, θ^I.

Generally, the probability to find a phase, I, at n-different, $\mathbf{r}_i (i = 1, 2, \ldots, n)$, defines the n-point correlation function $\Im_n^I(\mathbf{r}_i)_{i=1,2,\ldots,n}$ given by

$$\begin{aligned}\Im_n^I(\mathbf{r}_i)_{i=1,2,\ldots,n} &= \langle \theta^I(\mathbf{r}_1,\mathbf{r}')\theta^I(\mathbf{r}_2,\mathbf{r}')\ldots\theta^I(\mathbf{r}_n,\mathbf{r}')\rangle \\ &= \iiint\limits_V \theta^I(\mathbf{r}_1,\mathbf{r}')\theta^I(\mathbf{r}_2,\mathbf{r}')\ldots\theta^I(\mathbf{r}_n,\mathbf{r}')p(\mathbf{r}')dV\end{aligned} \tag{7.4}$$

A heterogeneous medium is defined as statistically homogenous if the probability to find a certain phase at a particular material point of the heterogeneous medium is independent on the position of this point within the finite volume representing the heterogeneous medium. In other words, the n-point correlation function, $\Im_n^I(\mathbf{r}_i)_{i=1,2,\ldots,n}$, is invariant under translation, so that

$$\begin{aligned}\forall \mathbf{r}' \in V \quad \Im_n^I(\mathbf{r}_i)_{i=1,2,\ldots,n} &= \Im_n^I(\mathbf{r}_i + \mathbf{r}')_{i=1,2,\ldots,n} \\ &= \Im_n^I(\mathbf{r}_{12}, \mathbf{r}_{13}, \ldots \mathbf{r}_{1n}) \text{ if } \mathbf{r}' = -\mathbf{r}_1\end{aligned} \tag{7.5}$$

where $\mathbf{r}_{1j} = \mathbf{r}_1 - \mathbf{r}_j (j = 2, 3, \ldots, n)$

The ergodic hypothesis suggests that the complete probabilistic information on the microstructure is obtained within a volume sufficiently large corresponding to the "ergodic media," known also as the infinite medium. If this medium further satisfies the macrohomogeneity conditions, it will correspond to the RVE.

Under the ergodic conditions, the probability function writes

$$p(\mathbf{r}) = \frac{1}{V} \tag{7.6}$$

and therefore, the n-point correlation function reads

$$\Im_n^I = \lim_{V\to\infty} \frac{1}{V} \iiint\limits_V \theta^I(\mathbf{r})\theta^I(\mathbf{r},\mathbf{r}_{12})\ldots\theta^I(\mathbf{r},\mathbf{r}_{1n})dV \tag{7.7}$$

For one-point correlation function, one has

$$\Im_1^I = \frac{1}{V}\iiint\limits_V \theta^I(\mathbf{r})dV = \frac{1}{V}\iiint\limits_{V^1} dV = f^I \tag{7.8}$$

which is the volume fraction of the phase I.

7.2.1.2 Macrohomogeneity Condition and Resulting Properties

Under the ergodic condition, the definition of a statistically homogeneous representative volume element is completed by the introduction of two length scales:

1. A local scale or "microscale" with a characteristic length, *d*, which corresponds to smallest constituent whose physical properties, orientation, and shape are judged to have direct first-order effects on the overall physical properties of the statistically homogeneous representative volume element. [*Remark* – The choice of the microscale is generally adapted to the problem under analysis. An appropriate choice should be guided by systematic "multiscale" experimental observations. Generally speaking, an optimum choice would be the one that includes a good balance between the definitions of the microscale that have a first-order effect on the overall properties, and the simplicity of the resulting model (solution of the field equations, time consuming simulations, etc.).]
2. A macroscopic scale that should be large enough to fulfil, on one hand, the ergodic condition, and on the other hand, the definition of macro fields that are locally uniform and that can be described as uniform applied stresses and strains. Typically, if we denote by, *D*, the characteristic length of the macro element, it must be orders of magnitudes larger than the typical dimension of the micro constituent, *d*; i.e., $d/dDD << 1$ (e.g., in characterizing a mass of compacted fine powders in powder-metallurgy, with grain of submicron size, a macro element of a dimension of 100 microns would fulfil the macrohomogeneity condition, whereas in characterizing an earth dam as a continuum, with aggregates of many centimetres in size, the absolute dimension of the macro element would be of the order of tens meters).

The macro element corresponds then to the RVE, defined, in the following, by its volume, V, and its boundary, ∂V. If we denote by $\bar{\boldsymbol{\varepsilon}}$ and $\bar{\boldsymbol{\sigma}}$ the associated macro strain and stress fields, respectively, the macrohomogeneity condition is expressed such that the definition of local or micro fields, $\boldsymbol{\varepsilon}(\mathbf{r})$ and $\boldsymbol{\sigma}(\mathbf{r})$, satisfies the following relationships:

$$\bar{\boldsymbol{\varepsilon}} = \langle \boldsymbol{\varepsilon}(\mathbf{r}) \rangle_V = \frac{1}{V} \iiint_V \boldsymbol{\varepsilon}(\mathbf{r}) dV \text{ and } \bar{\boldsymbol{\sigma}} = \frac{1}{V} \iiint_V \boldsymbol{\sigma}(\mathbf{r}) dV = \langle \boldsymbol{\sigma}(\mathbf{r}) \rangle_V \tag{7.9}$$

In other words, for any points, $\mathbf{r}, \mathbf{r}' \in V$, one has

$$\text{if} \|\mathbf{r} - \mathbf{r}'\| \leq \alpha d \text{ then } \bar{\boldsymbol{\varepsilon}}(\mathbf{r}) = \bar{\boldsymbol{\varepsilon}}(\mathbf{r}') \text{ and } \bar{\boldsymbol{\sigma}}(\mathbf{r}) = \bar{\boldsymbol{\sigma}}(\mathbf{r}') \tag{7.10}$$

with α being orders of magnitudes.

On the other hand, the macrohomogeneity condition assumes also that the fluctuations of stress and strain field (micro or local quantities) at the smaller length scale influence the overall behaviour (or macroscopic) at the larger length only via their averages. This is formally expressed by the following:

$$\boldsymbol{\varepsilon}(\mathbf{r}) = \bar{\boldsymbol{\varepsilon}} + \boldsymbol{\varepsilon}'(\mathbf{r}) \quad \text{and} \quad \boldsymbol{\sigma}(\mathbf{r}) = \bar{\boldsymbol{\sigma}} + \boldsymbol{\sigma}'(\mathbf{r}) \tag{7.11}$$

with

$$\frac{1}{V}\iiint_V \boldsymbol{\varepsilon}'(\mathbf{r})dV = \frac{1}{V}\iiint_V \boldsymbol{\sigma}'(\mathbf{r})dV = 0 \tag{7.12}$$

where, $\boldsymbol{\varepsilon}'(\mathbf{r})$ and $\boldsymbol{\sigma}'(\mathbf{r})$ stand for the fluctuating part of the local strain and stress fields, respectively.

[*Remark* – The above relation between local and macroscopic stress and strain fields are not fulfilled in some cases of heterogeneous materials, where sufficient length scale separation is not possible, like free surfaces of heterogeneous materials, macroscopic interfaces adjoined by at least one heterogeneous material, marked compositional or load gradient (e.g., heterogeneous beams under bending loads). In such situations, special homogenization techniques like strain gradient theory are used.]

7.2.2 Field Equations and Averaging Procedures

An introduction of an appropriate RVE allows a link between different scales leading to the definition of overall (or macroscopic) mechanical properties from those given at a suitable microscale. Such a link is fundamentally based on averaging techniques, and on appropriate constitutive laws defined at different scales. [*Remark* – As it will be discussed below, the definition of the overall mechanical properties from those given at the microscale is generally estimation.]

In this section, attention is focused on developing averaging theorems devoted to heterogeneous materials with arbitrary constituents, so that no specific indications on the constitutive law are given (i.e., the constituent may behave linearly or nonlinearly, rate-dependent or rate independent). However, any limitations of such averaging schemes are discussed throughout this chapter.

7.2.2.1 Field Equations and Boundary Conditions

In what follows, we assume the usual conditions which define the framework of continuum micromechanics to be satisfied: a random homogeneous material is assumed to obey macrohomogeneity requirements, which implies that the pertinent scale lengths of the body differ by one order of magnitude at least from each other. This separation of the scales allows the RVE to be the homogeneous equivalent medium.

As a continuum, the "selected" RVE is regarded as a "structural" element subjected at it boundary ∂V to an overall mechanical loading, that consist on forces and displacements. Within the framework of continuum micromechanics, the formulation of boundary-value problems disregards body forces, and do not include the inertia terms for a broad range of problems. The main concerns are to derive the overall average properties of the RVE (elastic moduli, yield stresses, electrical conductivity, etc.), when it is subjected to the boundary data corresponding to the uniform fields in the homogeneous equivalent medium which the RVE is assigned to represent. In other words, an RVE may be viewed as a heterogeneous material under prescribed boundary data which correspond to the uniform macroscopic fields. A general procedure consists in estimating the overall average strain increment as a function of the corresponding prescribed incremental surface forces or, conversely, the average stress increment, as a function of the prescribed incremental surface displacements. The prescribed incremental surface tractions may be taken as spatially uniform, or, in the converse case, the prescribed incremental surface displacements may be assumed as spatially linear. [*Remark* – Whether boundary displacements or boundary tractions are regarded as prescribed, a viable micromechanical approach should produce equivalent overall constitutive parameters for the corresponding macro-element. For example, if the instantaneous moduli and compliance are being calculated, then the resulting instantaneous modulus tensor obtained for the prescribed incremental surface displacements should be the inverse of the instantaneous compliance tensor obtained for the prescribed incremental surface tractions of the RVE.]

Consider an RVE with volume, V, bounded by a regular surface, ∂V. A typical point in V is identified by its position vector, $\mathbf{r}$, with components $x_i (i = 1, 2, 3)$, relative to a fixed rectangular Cartesian coordinate system. The unit base vectors of this coordinate system are denoted by $\vec{e}_i (i = 1, 2, 3)$, so that the position vector $\mathbf{r}$ reads $\mathbf{r} = x_i \vec{e}_i$, we adopt throughout this book the Einstein convention, where repeated indices are summed.

As discussed above, the boundary conditions applied to the RVE are of two distinct types:

1. Surface tractions $\mathbf{t}^{\text{o}}(\mathbf{r})$, which are in equilibrium with a certain uniform stress field $\boldsymbol{\sigma}^{\text{o}}$, that is

$$t_i^{\text{o}}(\mathbf{r}) = \sigma_{ij}^{\text{o}} n_j(\mathbf{r}) \quad \mathbf{r} \in \partial V \tag{7.13}$$

where $\mathbf{n}$ denotes the outer unit normal vector of ∂V.

2. Spatially linear displacement field $\mathbf{u}^{\text{o}}(\mathbf{r})$, derived from a certain uniform strain field $\boldsymbol{\varepsilon}^{\text{o}}$, that is

$$u_i^{\text{o}}(\mathbf{r}) = \varepsilon_{ij}^{\text{o}} x_j \quad \mathbf{r} \in \partial V \tag{7.14}$$

The application of this loading will give rise to displacement, $\mathbf{u}(\mathbf{r})$, strain, $\boldsymbol{\varepsilon}(\mathbf{r})$, and stress, $\boldsymbol{\sigma}(\mathbf{r})$, fields at each point $\mathbf{r}$ of the volume V. Under the prescribed surface data, the RVE must be in equilibrium and its overall deformation compatible. The governing field equations at any point in V include the balance of linear and angular momenta (in absence of body forces and with quasistatic conditions),

$$\sigma_{ij,j}(\mathbf{r}) = 0, \;\; \sigma_{ij}(\mathbf{r}) = \sigma_{ji}(\mathbf{r}) \tag{7.15}$$

as well as the compatibility conditions under small strain hypothesis

$$\varepsilon_{ij}(\mathbf{r}) = \frac{1}{2}\left(u_{i,j}(\mathbf{r}) + u_{j,i}(\mathbf{r})\right) \tag{7.16}$$

where the comma followed by an index denotes partial differentiation with respect to the corresponding coordinate variable.

When the self-equilibrating traction vector, $\mathbf{t}^{o}(\mathbf{r})$, is prescribed on the boundary, ∂V, of the RVE, then

$$\sigma_{ij}(\mathbf{r})n_j = t_i^{o}(\mathbf{r}) \quad \text{on } \partial V \tag{7.17}$$

On the other hand, when the displacements, $\mathbf{u}^{o}$, are assumed prescribed on the boundary of the RVE, it follows that

$$u_i(\mathbf{r}) = u_i^{o}(\mathbf{r}) = \varepsilon_{ij}^{o} x_j \quad \text{on } \partial V \tag{7.18}$$

Note that any stress field, $\boldsymbol{\sigma}^{*}(\mathbf{r})$, fulfilling the field equations (7.15) and the boundary conditions (7.17) is called statistically admissible. Conversely, any displacement field, $\mathbf{u}^{*}(\mathbf{r})$, fulfilling the boundary conditions (7.18) and leading to a compatible strain field, $\boldsymbol{\varepsilon}^{*}(\mathbf{r})$, is called kinematically admissible.

The field equations and boundary conditions developed above are expressed in terms of the total stress and strain. This may be sufficient for certain problems in elasticity. However, in most engineering applications, the microstructure evolves in the course of deformation (plastic flow, void propagation, etc.), leading to change of material properties. Therefore, incremental formulations are required. Under such formulations, a rate problem is considered, where traction rates, $\dot{\mathbf{t}}^{o}(\mathbf{r})$, or velocity, $\dot{\mathbf{u}}^{o}(\mathbf{r})$, may be regarded as prescribed on the boundary of the RVE. Here the rates may be measured in terms of a monotone increasing parameter, since no inertia effects are included. For a rate-dependent material response, however, the actual time must be used.

In rate formulations, the basic field equations are simply obtained by substituting in Equations (7.15, 7.16, 7.17, and 7.18) the corresponding rate quantities, $\dot{\mathbf{u}}(\mathbf{r})$, $\dot{\boldsymbol{\varepsilon}}(\mathbf{r})$, and $\dot{\boldsymbol{\sigma}}(\mathbf{r})$.

7.2.2.2 Volume Averages of Stress and Strain Fields

A fundamental issue in continuum micromechanics is that the average stress and strain, $\bar{\boldsymbol{\sigma}}$ and $\bar{\boldsymbol{\varepsilon}}$, over the RVE volume, must be expressed in terms of the prescribed boundary data only, so that the overall properties of the equivalent homogeneous medium, represented by the RVE, follows directly from the relation between $\bar{\boldsymbol{\sigma}}$ and $\bar{\boldsymbol{\varepsilon}}$. This is shown in this section both for prescribed surface tractions and linear displacements.

Traction Boundary Conditions

It follows from the field Equation (7.15) that

$$\sigma_{ij} = \sigma_{ik}\delta_{jk} = \sigma_{ik,k}x_j + \sigma_{ik}x_{j,k} = \left(\sigma_{ik}x_j\right)_{,k} \tag{7.19}$$

where δ is the Kronecker delta.

When traction boundary conditions $t_i^{\mathrm{o}}(\mathbf{r})$ are prescribed and with help of the Gauss theorem, the average stress, $\bar{\boldsymbol{\sigma}}$, is expressed by

$$\bar{\sigma}_{ij} = \frac{1}{V}\iiint\limits_{V} (\sigma_{ik}x_j)_{,k} \; dV = \frac{1}{V}\iint\limits_{\partial V} \sigma_{ik}x_j n_k dS = \frac{1}{V}\iint\limits_{\partial V} t_i^{\mathrm{o}} x_j dS \tag{7.20}$$

If further $t_i^{\mathrm{o}}(\mathbf{r}) = \sigma_{ij}^{\mathrm{o}} n_j(\mathbf{r})$, one has

$$\begin{aligned} \bar{\sigma}_{ij} &= \frac{1}{V}\iint\limits_{\partial V} t_i^{\mathrm{o}} x_j dS = \sigma_{ik}^{\mathrm{o}} \left\{ \frac{1}{V}\iint\limits_{\partial V} n_k(\mathbf{r}) x_j dS \right\} \\ &= \sigma_{jk}^{\mathrm{o}} \left\{ \frac{1}{V}\iiint\limits_{V} x_{j,k} dV \right\} = \sigma_{jk}^{\mathrm{o}}\delta_{jk} = \sigma_{ij}^{\mathrm{o}} \end{aligned} \tag{7.21}$$

Displacement Boundary Conditions

From the compatibility conditions (7.16) and thanks to the Gauss theorem, the average strain, $\bar{\boldsymbol{\varepsilon}}$, reads in view of prescribed boundary displacements $\mathbf{u}^{\mathrm{o}}(\mathbf{r})$

$$\begin{aligned} \bar{\varepsilon}_{ij} &= \frac{1}{V}\iiint\limits_{V} \varepsilon_{ij}(\mathbf{r}) dV = \frac{1}{2V}\iiint\limits_{V} (u_{i,j}(\mathbf{r}) + u_{j,i}(\mathbf{r})) dV \\ &= \frac{1}{2V}\iint\limits_{\partial V} (n_i u_j^{\mathrm{o}} + n_j u_i^{\mathrm{o}}) dS \end{aligned} \tag{7.22}$$

If further $\mathbf{u}^{\mathrm{o}}(\mathbf{r})$ is linear, one has

$$
\begin{aligned}
\bar{\varepsilon}_{ij} &= \frac{1}{2}\left[\varepsilon^{\mathrm{o}}_{jk}\left\{\frac{1}{V}\iint\limits_{\partial V} n_j x_k dS\right\} + \varepsilon^{\mathrm{o}}_{jk}\left\{\frac{1}{V}\iint\limits_{\partial V} n_i x_k dS\right\}\right] \\
&= \frac{1}{2}\left[\varepsilon^{\mathrm{o}}_{jk}\delta_{jk} + \varepsilon^{\mathrm{o}}_{jk}\delta_{jk}\right] = \varepsilon^{\mathrm{o}}_{ij}
\end{aligned}
\tag{7.23}
$$

However, if we denote by $\bar{\mathbf{u}}$ the average displacement field, whose components are expressed by

$$
\bar{u}_i = \frac{1}{V}\iiint\limits_V u_i(\mathbf{r})dV \tag{7.24}
$$

the condition of incompressible materials is required to express $\bar{\mathbf{u}}$ in terms of the prescribed displacement boundary conditions only. In fact,

$$
\begin{aligned}
\bar{u}_i &= \frac{1}{V}\iiint\limits_V u_i(\mathbf{r})dV = \frac{1}{V}\iiint\limits_V u_j(\mathbf{r})x_{i,j}\,dV \\
&= \frac{1}{V}\iiint\limits_V (u_j(\mathbf{r})x_i)_{,j}\,dV - \frac{1}{V}\iiint\limits_V u_{j,j}(\mathbf{r})x_i\,dV
\end{aligned}
\tag{7.25}
$$

and with the help of Gauss theorem, it follows that

$$
\bar{u}_i = \frac{1}{V}\iint\limits_{\partial V} u^{\mathrm{o}}_j n_j x_i dS - \frac{1}{V}\iiint\limits_V u_{j,j}(\mathbf{r})x_i dV \tag{7.26}
$$

In the case of incompressible material $u_{j,j}(\mathbf{r}) = 0$ and then

$$
\bar{u}_i = \frac{1}{V}\iint\limits_{\partial V} u^{\mathrm{o}}_j n_j x_i dS \tag{7.27}
$$

Note that the average strain, $\bar{\boldsymbol{\varepsilon}}$ defined by (7.22) and (7.23) is unchanged by adding a rigid-body translation or rotation. In fact, let define by $\mathbf{u}^{\mathrm{r}}$ a rigid translation associated with an antisymmetric infinitesimal rotation tensor, $\boldsymbol{\omega}^{\mathrm{r}}$. This will produce at each material point of the RVE an additional displacement given by

$$
u^{\mathrm{r}}_i + \omega^{\mathrm{r}}_{ik}x_k \tag{7.28}
$$

The corresponding additional average displacement gradient writes

$$\iiint\limits_V \left(u^{\mathrm{r}}_{i,j} + (\omega^{\mathrm{r}}_{ik} x_k)_{,j}\right) dV = \left\{ \frac{1}{V} \iint\limits_{\partial V} n_j ds \right\} u^{\mathrm{r}}_i + \left\{ \frac{1}{V} \iint\limits_{\partial V} n_j x_k ds \right\} \omega^{\mathrm{r}}_{ik} \qquad (7.29)$$

with

$$\begin{aligned} &\frac{1}{V} \iint\limits_{\partial V} n_j dS = \frac{1}{V} \iint\limits_{\partial V} \delta_{ji} n_i dS = \frac{1}{V} \iiint\limits_V \delta_{ji,i} dV = 0 \\ &\left\{ \frac{1}{V} \iint\limits_{\partial V} n_j x_k dS \right\} \omega^{\mathrm{r}}_{ik} = \left\{ \frac{1}{V} \iiint\limits_V x_{j,k}\, dV \right\} \omega^{\mathrm{r}}_{ik} = \delta_{jk}\, \omega^{\mathrm{r}}_{ik} = \omega^{\mathrm{r}}_{ij} \end{aligned} \qquad (7.30)$$

$$\text{Therefore,} \quad \iiint\limits_V \left(u^{\mathrm{r}}_{i,j} + \left(\omega^{\mathrm{r}}_{ik} x_k\right)_{,j}\right) dV = \omega^{\mathrm{r}}_{ij} \qquad (7.31)$$

Equation (7.31) shows then that a rigid-body rotation or translation does not affect the macroscopic strain, $\bar{\boldsymbol{\varepsilon}}$.

7.2.2.3 Hill Lemma

For any stress and strain fields, $\boldsymbol{\sigma}(\mathbf{r})$ and $\boldsymbol{\varepsilon}(\mathbf{r})$ at a given point of the RVE under prescribed boundary traction or boundary displacement, one has the following result

$$\langle \sigma_{ij} \varepsilon_{ij} \rangle - \langle \sigma_{ij} \rangle \langle \varepsilon_{ij} \rangle = \frac{1}{V} \iint\limits_{\partial V} \{ u_i - x_j \langle u_{i,j} \rangle \} \{ \sigma_{ik} n_k - \langle \sigma_{ik} \rangle n_k \} dS \qquad (7.32)$$

The proof of (7.32) is straightforward if one develops the following quantities

$$\sigma_{ij} \varepsilon_{ij} = \sigma_{ij} u_{i,j} = \left(\sigma_{ij} u_i\right)_{,j} - \sigma_{ij,j} u_i = \left(\sigma_{ij} u_i\right)_{,j} \qquad (7.33)$$

Then the average is given, in view of the Gauss theorem

$$\langle \sigma_{ij} \varepsilon_{ij} \rangle = \frac{1}{V} \iint\limits_{\partial V} \sigma_{ij}\, n_j\, u_i\, dS = \frac{1}{V} \iint\limits_{\partial V} t^{\mathrm{o}}_i u_i dS \qquad (7.34)$$

Let now develop the quantity within the surface integral in (7.32). One has

$$\begin{aligned} &\{ u_i - x_j \langle u_{i,j} \rangle \} \{ \sigma_{ik} n_k - \langle \sigma_{ik} \rangle n_k \} = \\ &\quad u_i \sigma_{ik} n_k - u_i n_k \langle \sigma_{ik} \rangle - \sigma_{ik} n_k x_j \langle u_{i,j} \rangle + x_j n_k \langle u_{i,j} \rangle \langle \sigma_{ik} \rangle \end{aligned} \qquad (7.35)$$

and then the surface integral writes

$$\frac{1}{V}\iint_{\partial V} u_i\sigma_{ik}n_k dS - \left\{\frac{1}{V}\iint_{\partial V} u_i n_k dS\right\}\langle\sigma_{ik}\rangle - \left\{\frac{1}{V}\iint_{\partial V} \sigma_{ik}n_k x_j dS\right\}\langle u_{i,j}\rangle$$
$$-\left\{\frac{1}{V}\iint_{\partial V} x_j n_k dS\right\}\langle u_{i,j}\rangle\langle\sigma_{ik}\rangle \tag{7.36}$$

with

$$\frac{1}{V}\iint_{\partial V} u_i n_k dS = \frac{1}{V}\iiint_V u_{i,k} dV = \langle u_{i,k}\rangle, \frac{1}{V}\iint_{\partial V} u_i\sigma_{ik}n_k dS$$
$$= \frac{1}{V}\iint_{\partial V} t_i^{\mathrm{o}} u_i dS = \langle\ \sigma_{ij}\,\varepsilon_{ij}\ \rangle \tag{7.37}$$
$$\frac{1}{V}\iint_{\partial V} \sigma_{ik}\, n_k\, x_j\, dS = \langle\sigma_{ij}\rangle, \frac{1}{V}\iint_{\partial V} x_j\, n_k\, dS = \frac{1}{V}\iiint_V x_{j,k} dV = \delta_{jk}$$

Finally,

$$\frac{1}{V}\iint_{\partial V} \{u_l - x_j\langle u_{l,j}\rangle\}\{\sigma_{lk}n_k - \langle\sigma_{lk}\rangle n_k\}dS$$
$$= \langle\sigma_{ij}\varepsilon_{ij}\rangle - \langle u_{i,k}\rangle\langle\sigma_{ik}\rangle - \langle\sigma_{ij}\rangle\langle u_{i,j}\rangle + \delta_{jk}\langle u_{i,j}\rangle\langle\sigma_{ik}\rangle \tag{7.38}$$
$$= \langle\sigma_{ij}\varepsilon_{ij}\rangle - \langle\varepsilon_{ik}\rangle\langle\sigma_{ik}\rangle - \langle\sigma_{ij}\rangle\langle\varepsilon_{ij}\rangle + \langle\varepsilon_{ik}\rangle\langle\sigma_{ik}\rangle = \langle\sigma_{ij}\varepsilon_{ij}\rangle - \langle\varepsilon_{ij}\rangle\langle\sigma_{ij}\rangle$$

Generally, for statically admissible stress fields, $\boldsymbol{\sigma}^*(\mathbf{r})$, and kinematically admissible strain fields (derived from a kinematically displacement fields), $\boldsymbol{\varepsilon}^*(\mathbf{r})$. The surface integral in (7.32) vanishes, since:

for statistically admissible stress $\boldsymbol{\sigma}_{ij}^*(\mathbf{r})n_j = t_i^{\mathrm{o}}(\mathbf{r}) = \langle\boldsymbol{\sigma}_{ij}(\mathbf{r})\rangle n_j$ on ∂V, and for kinematically admissible displacement $u_i^*(\mathbf{r}) = u_i^{\mathrm{o}}(\mathbf{r}) = \langle\varepsilon_{ij}(\mathbf{r})\rangle x_j$ on ∂V. Therefore, equation (7.32) is reduced to the following result

$$\langle\sigma_{ij}^*\varepsilon_{ij}^*\rangle = \langle\sigma_{ij}^*\rangle\langle\varepsilon_{ij}^*\rangle \tag{7.39}$$

corresponding to the Hill's Lemma. It is also known as Hill's macrohomogeneity condition or Mandel-Hill condition.

Such conditions imply that the volume averaged strain energy density of a heterogeneous material can be obtained from the volume averages of the stresses and strains, provided the micro- and macroscales are sufficiently different. Accordingly, homogenization can be interpreted as finding a homogeneous comparison material that is energetically equivalent to a given microstructured material. [*Remark* – Averaging schemes involving surface integral formulations, hold only for perfect interfaces between the constituents of the heterogeneous medium. Otherwise, correction terms involving the displacements jumps across failed interfaces must be added.]

For rate problems, the averaging schemes developed above are simply extended by substituting in the different equations the corresponding rate quantities, $\dot{\mathbf{u}}(\mathbf{r})$, $\dot{\boldsymbol{\varepsilon}}(\mathbf{r})$, and $\dot{\boldsymbol{\sigma}}(\mathbf{r})$, and by adapting rate boundary conditions. In general, in the case of perfect interfaces, one has the following relationships

$$\begin{aligned} \bar{\dot{\boldsymbol{\sigma}}} &= \langle \dot{\boldsymbol{\sigma}}(\mathbf{r}) \rangle = \frac{d}{dt} \langle \boldsymbol{\sigma}(\mathbf{r}) \rangle = \dot{\bar{\boldsymbol{\sigma}}} \\ \bar{\dot{\boldsymbol{\varepsilon}}} &= \langle \boldsymbol{\varepsilon}(\mathbf{r}) \rangle = \frac{d}{dt} \langle \boldsymbol{\varepsilon}(\mathbf{r}) \rangle = \dot{\bar{\boldsymbol{\varepsilon}}} \end{aligned} \tag{7.40}$$

Note relationships (7.40) must be corrected and adapted to other stress and strain measures in the case of finite deformations.

7.2.3 Concluding Remarks

As discussed through the ergodic and macrohomogeneity requirements, the main feature of continuum micromechanical framework is that the material content of the RVE cannot be described in a deterministic way. Even with the ergodic hypothesis, the statistical description of the spatial distribution of the constituent phases cannot be performed completely. Therefore, continuum micromechanical framework does not allow more than bounding or estimating the overall material properties of the considered heterogeneous material. An approach is the more accurate, the more completely the available information on the phase distribution is used. This is why a rigorous and systematic "multiscale" experimental analysis is required to develop pertinent micromechanics approaches.

Such limitations in describing completely the spatial distribution of the constituent phases lead basically to two groups of micromechanics modeling.

The first group comprises methods that describe the microstructure on the basis of "limited" statistical information using, generally, one-point correlation function. Within this group, one can define two major methodologies:

- Mean field approaches and related methods: The local fields within each constituent are approximated by their phase averages, i.e., piecewise uniform stress and strain fields are employed. Such descriptions typically use information on the microscopic topology, the inclusion shape and orientation, and, to some extent, on the statistics of the phase distribution. Mean field approaches tend to be formulated in terms of the phase concentration tensors, they pose relatively low computational requirements, they have been highly successful in describing the thermoelastic response of inhomogeneous materials, and their use for modelling nonlinear inhomogeneous materials is a subject of active research.

- Variational bounding methods (see below): Variational principles are used to obtain upper and lower bounds on the overall elastic tensors, elastic moduli, secant moduli, and other physical properties of inhomogeneous materials. Many analytical bounds are obtained on the basis of phase-wise constant stress (polarization) fields. Bounds aside from their intrinsic interest| are important tools for assessing other models of inhomogeneous materials. In addition, in many cases one of the bounds provides good estimates for the physical property under consideration, even if the bounds are rather slack. Many bounding methods are closely related to mean field approaches.

The second group of approximations is based on studying discrete microstructures and includes basically periodic microfield approaches or unit cell methods. The real inhomogeneous material is approximated by an infinitely extended model material with a periodic phase arrangement. The corresponding periodic microfields are usually evaluated by analyzing unit cells (which may describe microgeometries ranging from rather simplistic to highly complex) via analytical or numerical methods. Unit cell methods are typically used for performing materials characterization of inhomogeneous materials in the nonlinear range, but they can also be employed as micromechanically based constitutive models. The high resolution of the microfields provided by periodic microfield approaches can be very useful for studying the initiation of damage at the microscale. However, because they inherently give rise to periodic configurations of damage, periodic microfield approaches are not well suited for investigating phenomena such as the interaction of the microstructure with macroscopic cracks. Periodic microfield approaches can give detailed information on the local stress and strain fields within a given unit cell, but they tend to be computationally expensive.

7.3 Mean Field Theories and Eshelby's Solution

In real situations of heterogeneous materials, the description of phase distribution and spatial variations of resulting stress and strain microfields is beyond the capabilities of major approaches in continuum micromechanics. For convenience, most of homogenization techniques use appropriate approximations which lead to the mean field concepts. Such approximations lie in the description of spatial distribution of phases and microgeometries on the basis of statistical information using one-point correlation function, which lead to the introduction of phase volume fractions. Basically, mean field theories rely on the fact that microfields (stresses and strains) are approximated by their phase averages $\bar{\boldsymbol{\varepsilon}}^{\alpha}$ and $\bar{\boldsymbol{\sigma}}^{\alpha}$ (α is a given phase of the material), in other words, to build up such methodologies, the assumption of piecewise uniform properties is required. Relationships between microscopic quantities and macroscopic ones can be formally written in the case of linear problems as

$$\bar{\varepsilon}_{ij}^{\alpha} = A_{ijkl}\,\bar{\varepsilon}_{kl},\ \bar{\sigma}_{ij}^{\alpha} = B_{ijkl}\,\bar{\sigma}_{kl} \tag{7.41}$$

where **A** and **B** define the strain and stress concentration tensors, respectively. Typically, these fourth-order tensors can capture appropriate information on the microscopic topology, the inclusion shape, and orientation, and, to some extent, on the statistics of the phase distribution.

In the following, different mean field theories are discussed in details. These approaches are formulated in terms of phase concentration tensors (stress and strain), which require simple computational schemes. The methodology has been used extensively with a certain success to describe the thermoelastic behavior of composite materials; its adoption in nonlinear inhomogeneous materials is a subject of active research. A few attempts will be addressed in this chapter.

7.3.1 Eshelby's Inclusion Solution

A large proportion of the mean field approaches used in continuum mechanics of heterogeneous materials are based on the elementary Eshelby's inclusion problem [15]. The Eshelby's framework deals with the problem of stress and strain distribution in homogeneous infinite elastic media (elastic constant $\mathbf{L}^{\mathrm{o}}$ and compliance $\mathbf{M}^{\mathrm{o}}$) containing an ellipsoidal subregion with volume V^{I} called inclusion that spontaneously changes its shape and/or size or in other words, undergoes a certain transformation so that it no longer fits into its previous space in the surrounding medium. Eshelby's results show that if an elastic homogeneous ellipsoidal inclusion in an infinite linear elastic matrix is subjected to a uniform strain $\boldsymbol{\varepsilon}^{\mathrm{t}}$ describing the spontaneous change in shape and/or size, uniform stress and strain are induced in the constrained inclusion. The resulting uniform strain $\boldsymbol{\varepsilon}^{I}$ is related to the stress-free strain $\boldsymbol{\varepsilon}^{\mathrm{t}}$ as follows.

$$\varepsilon_{ij}^{\mathrm{I}} = S_{ijkl}^{I}\,\varepsilon_{kl}^{\mathrm{t}} \tag{7.42}$$

[*Remark* – The stress-free strain $\boldsymbol{\varepsilon}^{\mathrm{t}}$ is also called "unconstrained strain," eigenstrain," or "transformation strain.". The concept of Eshelby's inclusion has been used extensively to describe real metallurgical situations of practical interest like solid-solid phase transformation, thermal misfit during temperature change, plastic strains, diffusion, etc. However, this concept is conditioned by the uniformity of $\boldsymbol{\varepsilon}^{\mathrm{t}}$ so that Equation (7.42) holds and, therefore, its adoption in homogenization procedures requires a certain number of hypotheses, which in turn depends on the scale of description.]

Recall that Equation (7.42) holds only if the inclusion is ellipsoidal in shape. $\mathbf{S}^I$ is called the Eshelby's fourth-order tensor, which depends on material properties $\mathbf{L}^{\mathrm{o}}$ and aspect ratios of the ellipsoid. $\mathbf{S}^I$ is expressed in terms of the Green's functions $\mathbf{G}$ by

$$S^I_{ijmn} = -\frac{1}{2} L^{\mathrm{o}}_{klmn} \iiint\limits_{V^I} \left(G_{ik,lj}(\mathbf{r}-\mathbf{r}') + G_{jk,li}(\mathbf{r}-\mathbf{r}') \right) dV' \quad \text{if } \mathbf{r} \in V^I \tag{7.43}$$

On the other hand, the elastic constitutive law of the inclusion

$$\sigma^I_{ij} = L^{\mathrm{o}}_{ijkl} \left(\varepsilon^I_{kl} - \varepsilon^{\mathrm{t}}_{kl} \right) \tag{7.44}$$

could be written as

$$\sigma^I_{ij} = L^{\mathrm{o}}_{ijkl} \varepsilon^I_{kl} + \tau^{\mathrm{t}}_{ij} \text{ with } \tau^{\mathrm{t}}_{ij} = -L^{\mathrm{o}}_{ijkl} \varepsilon^{\mathrm{t}}_{kl} \tag{7.45}$$

The quantity $\boldsymbol{\tau}^{\mathrm{t}}$ is called polarization, it could be interpreted as the resulting stress in the inclusion after the spontaneous transformation, if the inclusion is not allowed to deform elastically.

Alternatively, the Eshelby's solution (7.42) is expressed in terms of polarization as follows

$$\boldsymbol{\varepsilon}^I_{ij} = -P^I_{ijkl} \tau^{\mathrm{t}}_{kl} \text{ with } P^I_{ijmn} = S^I_{ijkl} M^{\mathrm{o}}_{klmn} \tag{7.46}$$

The fourth-order tensor $\mathbf{P}^I$, called the Hill's polarization tensor, has more interesting properties than the Eshelby's tensor. It is shown easily that $\mathbf{P}^I$ is symmetric, positive-defined and its inverse "greater" than the elastic constant $\mathbf{L}^{\mathrm{o}}$ in the way of associated quadratic forms, in other words, for any second-order tensor $\boldsymbol{\alpha} > 0$, one has

$$\alpha_{ij} \left(P^I_{ijkl} \right)^{-1} \alpha_{kl} > \alpha_{ij} L^{\mathrm{o}}_{ijkl} \alpha_{kl} > 0 \tag{7.47}$$

which can be written in an abridged manner

$$\left(P^I_{ijkl} \right)^{-1} > L^{\mathrm{o}}_{ijkl} > 0 \tag{7.48}$$

since $\mathbf{P}^I$ and $\mathbf{L}^{\mathrm{o}}$ are symmetric and positive-defined, (7.48) is equivalent to

$$0 < P^I_{ijkl} < D^{\mathrm{o}}_{ijkl} \tag{7.49}$$

and according to (7.46), which equivalent to $\mathbf{S}^I = \mathbf{P}^I : (\mathbf{M}^{\mathrm{o}})^{-1}$, the inequality (7.49) leads to the following properties resulting from the Eshelby's theory:

- The total strain $\boldsymbol{\varepsilon}^I$ of the inclusion is "smaller" than the eigenstrain $\boldsymbol{\varepsilon}^{\mathrm{t}}$,
- The Eshelby's tensor $\mathbf{S}^I$ is "smaller" than unity.

If one deals with the stress $\boldsymbol{\sigma}^I = \mathbf{L}^{\mathrm{o}} : (\boldsymbol{\varepsilon}^I - \boldsymbol{\varepsilon}^{\mathrm{t}})$ in the inclusion, it could be expressed by

$$\sigma^I_{ij} = -Q^I_{ijkl}\,\varepsilon^{\mathrm{t}}_{kl} \text{ with } Q^I_{ijmn} = L^{\mathrm{o}}_{ijkl}\big(I_{klmn} - S^I_{klmn}\big) = L^{\mathrm{o}}_{ijmn} - L^{\mathrm{o}}_{ijkl}P^I_{klpq}L^{\mathrm{o}}_{pqmn} \quad (7.50)$$

The fourth-order tensor $\mathbf{Q}^I$ has been also introduced by Hill, it could be seen as dual of $\mathbf{P}^I$ and has the same properties as $\mathbf{P}^I$. Furthermore, the Eshelby's problem is also formulated in terms of polarization rather then the eigenstrain, so that one defines a dual Eshelby's tensor $\tilde{\mathbf{S}}^I$ linking the stress $\boldsymbol{\sigma}^I$ to the polarization $\boldsymbol{\tau}^{\mathrm{t}}$ as follows

$$\sigma^I_{ij} = \tilde{S}^I_{ijkl}\tau^{\mathrm{t}}_{kl} \text{ with } \tilde{S}^I_{ijpq} = Q^I_{ijkl}M^{\mathrm{o}}_{klpq} = I_{ijpq} - L^{\mathrm{o}}_{ijkl}S^I_{klmn}M^{\mathrm{o}}_{mnpq} \quad (7.51)$$

$\tilde{\mathbf{S}}^I$ has the same properties as the Eshelby's tensor but it rarely used.

7.3.2 *Inhomogeneous Eshelby's Inclusion: "Constraint" Hill's Tensor*

The elementary solution of Eshelby's problem could be extended to cover the particular case of Eshelby's inhomogeneous inclusion (Fig. 7.1). It corresponds to the case of a linear elastic infinite medium (elastic constant $\mathbf{L}^{\mathrm{o}}$ and compliance $\mathbf{M}^{\mathrm{o}}$) containing an ellipsoidal subregion I with elastic properties $\mathbf{L}^I = \mathbf{L}^{\mathrm{o}} + \Delta\mathbf{L}^I$ where $\Delta\mathbf{L}^I$ is symmetric, non-null, and not necessarily defined. This system is built up so that it is in equilibrium at a relaxed state. Suppose that, as in the previous section, the inclusion undergoes a spontaneous transformation characterized by a polarization $\boldsymbol{\tau}^{\mathrm{t}}$, which leads to an eigenstrain $\boldsymbol{\varepsilon}^{\mathrm{t}} = -\mathbf{S}^I : \boldsymbol{\tau}^{\mathrm{t}}$. For example, this configuration may describe the problem of a single inclusion made by a certain material whose properties are different from

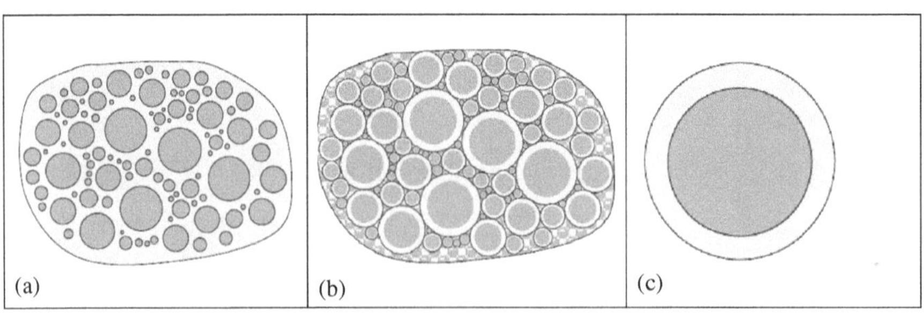

Fig. 7.1 The composite sphere assemblage model

those of the surrounding medium and the inclusion is subjected to a thermal expansion whereas the surrounding medium is insensitive to the temperature change. This system is more complicated than the previous one, however, by rewriting the inclusion constitutive law as

$$\sigma^I_{ij} = L^{\mathrm{o}}_{ijkl}\,\varepsilon^I_{kl} + \tau^{\mathrm{t}'}_{ij} \text{ with } \tau^{\mathrm{t}'}_{ij} = \tau^{\mathrm{t}}_{ij} + \Delta L^I_{ijkl}\,\varepsilon^I_{kl} \tag{7.52}$$

the problem will be equivalent formally to the previous elementary problem expressed in terms of polarization. [*Remark* – The application of Eshelby's solution to the present problem is subjected to the condition of uniform polarization $\boldsymbol{\tau}^{\mathrm{t}'}$ inside the inclusion. Such a condition is not necessarily satisfied since the strain $\boldsymbol{\varepsilon}^I$ may fluctuate inside the inclusion.]

By assuming a uniform strain $\boldsymbol{\varepsilon}^I$, the Eshelby's solution leads to

$$\varepsilon^I_{ij} = -P^I_{ijkl}\tau^{\mathrm{t}'}_{kl} = -P^I_{ijkl}\left(\tau^{\mathrm{t}}_{kl} + \Delta L^I_{klmn}\varepsilon^I_{mn}\right) \tag{7.53}$$

where one should be noticed that the tensor $\mathbf{P}^I$ depends on the elastic constants $\mathbf{L}^{\mathrm{o}}$ of the infinite medium. It result from (7.53) that

$$\left(\left(P^I_{ijkl}\right)^{-1} - L^{\mathrm{o}}_{ijkl} + L^I_{ijkl}\right)\varepsilon^I_{kl} = -\,\tau^{\mathrm{t}}_{ij} \tag{7.54}$$

which leads to the definition of the following tensor

$$H^I_{ijmn} = \left(P^I_{ijmn}\right)^{-1} - L^{\mathrm{o}}_{ijmn} = L^{\mathrm{o}}_{ijkl}\left\{\left(S^I_{klmn}\right)^{-1} - I_{klmn}\right\} \tag{7.55}$$

known as the constraint Hill's tensor. As Eshelby's tensor $\mathbf{S}^I$, $\mathbf{H}^I$ depends on the shape of the inclusion and the elastic constant $\mathbf{L}^{\mathrm{o}}$ of the infinite medium. $\mathbf{H}^I$ is also symmetric, positive-defined, and its dimensions are the ones of elastic constants. Equation (7.54) is therefore equivalent to

$$\varepsilon^I_{ij} = -\left(H^I_{ijkl} + L^I_{ijkl}\right)^{-1}\tau^{\mathrm{t}}_{kl} \tag{7.56}$$

If one deals with the dual relationships expressed in terms of homogeneous stress in the inclusion by using the eigenstrain $\boldsymbol{\varepsilon}^{\mathrm{t}}$ instead of polarization, one has

$$\sigma^I_{ij} = -\left(\tilde{H}^I_{ijkl} + M^I_{ijkl}\right)^{-1}\varepsilon^{\mathrm{t}}_{kl} \text{ with } \tilde{H}^I_{ijkl} = \left(H^I_{ijkl}\right)^{-1} = \left(Q^I_{ijkl}\right)^{-1} - M^{\mathrm{o}}_{ijkl} \tag{7.57}$$

Remark – Interpretation of the Hill Tensor – Equations (7.56) and (7.57) – expressed in the particular case of rigid inclusion provides a mechanical explanation of the constraint Hill tensor. In fact, if $\mathbf{L}^I \gg \mathbf{L}^{\mathrm{o}}$, one has also $\mathbf{L}^I \gg \mathbf{H}^I > 0$ and therefore from (7.56) it follows

$$\varepsilon_{ij}^{I} \approx -\left(M_{ijkl}^{I}\right)^{-1} \tau_{kl}^{\mathrm{t}} = \varepsilon_{ij}^{\mathrm{t}} \tag{7.58}$$

which means that under the condition of stiff elastic properties, the inclusion "imposes" the entire transformation strain to the surrounding medium. On the other hand, the stress in the inclusion results from (7.57) using the property $0 < \mathbf{M}^I \ll \tilde{\mathbf{H}}^I$

$$\sigma_{ij}^{I} \approx -\left(\tilde{H}_{ijkl}^{I}\right)^{-1} \varepsilon_{kl}^{\mathrm{t}} = -H_{ijkl}^{I} \varepsilon_{kl}^{\mathrm{t}} \tag{7.59}$$

It results from (7.58) and (7.59) that under the condition of rigid inclusion, the constitutive law of the inclusion $\boldsymbol{\sigma}^I = \mathbf{L}^I : (\boldsymbol{\varepsilon}^I - \boldsymbol{\varepsilon}^{\mathrm{t}})$ is undetermined since $\mathbf{L}^I$ is too high whereas $\boldsymbol{\varepsilon}^I - \boldsymbol{\varepsilon}^{\mathrm{t}}$ is too small. However, the stress $\boldsymbol{\sigma}^I$ is given by (7.59), which provides a certain mechanical interpretation of the Hill's tensor as a reaction of the surrounding infinite medium to the deformation imposed by the inclusion. This is the reason why $\mathbf{H}^I$ is called the constraint tensor. The negative sign in (7.59) could be interpreted, for example, by the fact that the surrounding infinite medium will act by compressive stresses if the inclusion is subjected to a pure dilatation in an isotropic infinite medium.

In conclusion, the constraint Hill's tensor describes the forces exerted by an infinite medium on a subregion in response to its homogeneous deformation. Since $\mathbf{H}^I$ is independent of the elastic constants of the inclusion, this property should be generalized to any elastic behavior of the inclusion. In fact, the combination of Equations (7.56) and (7.57), leads to a general definition of $\mathbf{H}^I$

$$\sigma_{ij}^{I} = -H_{ijkl}^{I} \varepsilon_{kl}^{I} \tag{7.60}$$

7.3.3 Eshelby's Problem with Uniform Boundary Conditions

The previous classical and inhomogeneous Eshelby's problems are developed under the conditions of vanishing far fields. The concept of Eshelby's inclusion can be extended to cases where a uniform mechanical strain $\boldsymbol{\varepsilon}^{\mathrm{o}}$ or external stress $\boldsymbol{\sigma}^{\mathrm{o}}$ is applied to a perfectly bonded inhomogeneous elastic inclusion in an infinite matrix. The strain $\boldsymbol{\varepsilon}^I$ in the inclusion will the superposition of the applied strain $\boldsymbol{\varepsilon}^{\mathrm{o}}$ and of the additional term resulting from the transformation strain $\boldsymbol{\varepsilon}^{\mathrm{t}}$, such that

$$\boldsymbol{\varepsilon}_{ij}^{I} = \boldsymbol{\varepsilon}_{ij}^{\mathrm{o}} + S_{ijkl}^{I} \boldsymbol{\varepsilon}_{kl}^{\mathrm{t}} \tag{7.61}$$

or by introducing the polarization

$$\boldsymbol{\varepsilon}_{ij}^{I} = \boldsymbol{\varepsilon}_{ij}^{\mathrm{o}} - P_{ijkl}^{I}\tau_{kl}^{\mathrm{t}} = \boldsymbol{\varepsilon}_{ij}^{\mathrm{o}} - \left(H_{ijkl}^{I} + L_{ijkl}^{\mathrm{o}}\right)^{-1}\tau_{kl}^{\mathrm{t}} \tag{7.62}$$

The stresses result simply from (7.60) as

$$\sigma_{ij}^{I} = \sigma_{ij}^{\mathrm{o}} - H_{ijkl}^{I}\left(\varepsilon_{kl}^{I} - \varepsilon_{kl}^{\mathrm{o}}\right) \tag{7.63}$$

or by introducing the polarization with the "dual" Eshelby's tensor $\tilde{\mathbf{S}}^{I}$

$$\sigma_{ij}^{I} = \sigma_{ij}^{\mathrm{o}} + \tilde{S}_{ijkl}^{I}\tau_{kl}^{\mathrm{t}} \tag{7.64}$$

[*Remark* – Equations (7.61) and (7.64) are usually expressed as $\boldsymbol{\sigma}^{I} = \boldsymbol{\sigma}^{\mathrm{o}} + \boldsymbol{\sigma}^{\mathrm{d}}$ and $\boldsymbol{\varepsilon}^{I} = \boldsymbol{\varepsilon}^{\mathrm{o}} + \boldsymbol{\varepsilon}^{\mathrm{d}}$ where $\boldsymbol{\sigma}^{\mathrm{d}} = \tilde{\mathbf{S}}^{I} : \boldsymbol{\tau}^{\mathrm{t}}$ and $\boldsymbol{\varepsilon}^{\mathrm{d}} = \mathbf{S}^{I} : \boldsymbol{\varepsilon}^{\mathrm{t}}$ are defined to be the disturbance stress and strain, respectively generated by the eigenstrain.

Consider an inhomogeneous problem consisting in an infinite elastic medium (properties $\mathbf{L}^{\mathrm{o}}$) containing an ellipsoidal inclusion (properties $\mathbf{L}^{I}$). No eigenstrain is assumed to occur in the inclusion. When this system is subjected to a far field homogeneous strain $\boldsymbol{\varepsilon}^{\mathrm{o}}$, the inclusion deforms homogeneously and its mechanical state is expressed by the following equations

$$\sigma_{ij}^{I} = \sigma_{ij}^{\mathrm{o}} - H_{ijkl}^{I}\left(\varepsilon_{kl}^{I} - \varepsilon_{kl}^{\mathrm{o}}\right), \sigma_{ij}^{\mathrm{o}} = L_{ijkl}^{\mathrm{o}}\,\varepsilon_{kl}^{\mathrm{o}}, \sigma_{ij}^{I} = L_{ijkl}^{I}\,\varepsilon_{kl}^{I} \tag{7.65}$$

describing the interaction between local and far fields, the constitutive law at far field state and inclusion constitutive law, respectively. The combination of these equations leads to the following equations

$$\varepsilon_{ij}^{I} = \left(H_{ijkl}^{I} + L_{ijkl}^{I}\right)^{-1}\left(H_{klmn}^{I} + L_{klmn}^{\mathrm{o}}\right)\varepsilon_{mn}^{\mathrm{o}} = \left(H_{klmn}^{I} + L_{klmn}^{I}\right)^{-1}\left(P_{klmn}^{I}\right)^{-1}\varepsilon_{mn}^{\mathrm{o}} \tag{7.66}$$

and

$$\sigma_{ij}^{I} = \left(\tilde{H}_{ijkl}^{I} + M_{ijkl}^{I}\right)^{-1}\left(\tilde{H}_{klmn}^{I} + M_{klmn}^{\mathrm{o}}\right)\sigma_{mn}^{\mathrm{o}} = \left(\tilde{H}_{klmn}^{I} + M_{klmn}^{I}\right)^{-1}\left(Q_{klmn}^{I}\right)^{-1}\sigma_{mn}^{\mathrm{o}} \tag{7.67}$$

which involve the polarization tensors $\mathbf{P}^{I}$ and $\mathbf{Q}^{I}$. Recall that as for the Hill's tensor $\mathbf{H}^{I}$, these tensors depend on shape of the inclusion and the elastic properties of the infinite medium.

The problem of inhomogeneous inclusion is equivalent to the elementary Eshelby's problem with an applied homogeneous far field and a polarization $\boldsymbol{\tau}^{\mathrm{t}} = \left(\mathbf{L}^{I} - \mathbf{L}^{\mathrm{o}}\right) : \boldsymbol{\varepsilon}^{I}$ subjected by the inclusion. The polarization is usually expressed in terms of the applied far field by introducing a tensor $\mathbf{T}^{I}$ as follows

$$\tau_{ij}^{\mathrm{t}} = T_{ijkl}^{I}\,\varepsilon_{kl}^{\mathrm{o}} \tag{7.68}$$

$\mathbf{T}^I$ depends on the elastic constant of the inclusion and surrounding medium as well as on he shape of the inclusion. It is expressed by

$$\begin{aligned} T^I_{ijpq} &= \left(L^I_{ijkl} - L^{\mathrm{o}}_{ijkl}\right)\left(H^I_{klmn} + L^I_{klmn}\right)^{-1}\left(H^I_{mnpq} + L^{\mathrm{o}}_{mnpq}\right) \\ &= \left(L^I_{ijkl} - L^{\mathrm{o}}_{ijkl}\right) - \left(L^I_{ijkl} - L^{\mathrm{o}}_{ijkl}\right)\left(H^I_{klmn} + L^I_{klmn}\right)^{-1}\left(L^I_{mnpq} - L^{\mathrm{o}}_{mnpq}\right) \\ &= \left(P^I_{ijkl}\right)^{-1}\left\{P^I_{klmn} - \left(H^I_{klmn} + L^I_{klmn}\right)^{-1}\right\}\left(P^I_{mnpq}\right)^{-1} \end{aligned} \tag{7.69}$$

It follows from (7.69) that $\mathbf{T}^I$ is symmetric but not necessarily positive-defined. It can be checked easily that $\mathbf{T}^I$ is increasing function of $\mathbf{C}^I$ and when $\mathbf{C}^I$ varies from 0 to $+\infty$ at a given $\mathbf{C}^{\mathrm{o}}$, $\mathbf{T}^I$ varies from $\left(\mathbf{P}^I - \mathbf{M}^{\mathrm{o}}\right)^{-1} < 0$ to $\left(\mathbf{P}^I\right)^{-1} > 0$ with $\mathbf{T}^I = 0$ for $\mathbf{L}^I = \mathbf{L}^{\mathrm{o}}$. Therefore, it follows that

$$\left(\mathbf{P}^I - \mathbf{M}^{\mathrm{o}}\right)^{-1} \leq \mathbf{T}^I \leq \left(\mathbf{P}^I\right)^{-1},\ \mathbf{L}^I \geq \mathbf{L}^{\mathrm{o}} \Rightarrow \mathbf{T}^I \geq 0,\ \mathbf{L}^I \leq \mathbf{L}^{\mathrm{o}} \Rightarrow \mathbf{T}^I \leq 0 \tag{7.70}$$

The dual relationship of (7.68) expresses the eigenstrain in inclusion in terms of homogeneous applied stresses as

$$\boldsymbol{\varepsilon}^{\mathrm{t}}_{ij} = \tilde{T}^I_{ijpq}\sigma^{\mathrm{o}}_{kl} \text{ with } \tilde{T}^I_{ijpq} = -M^{\mathrm{o}}_{ijkl}T^I_{klmn}M^{\mathrm{o}}_{mnpq} \tag{7.71}$$

As a summary, the elementary problem of Eshelby and its extension to inhomogeneous cases with or without applied homogeneous far fields are based on the homogeneity of stress and strain fields in the inclusion. This fundamental property results from the assumption of an ellipsoidal inclusion, the concept of infinite medium and the linearity of the constitutive laws. In addition to the Eshelby's tensor, the analysis results in the introduction of other tensors of practical interest like the polarization tensors $\mathbf{P}^I$ and $\mathbf{Q}^I$. In particular, the Hill's tensor is fundamental in describing the constraint effect of the infinite medium on the deformation of the inclusion (Equation (7.60)) with given elastic properties. [*Remark* – The above analysis could be generalized to the case to a linear thermoelastic behavior of the inclusion, in other word, when the infinite medium is subjected to an eigenstrain.]

The Eshelby's solution and its extension constitute the basic framework for different mean field theories to derive the overall elastic properties of inclusion-matrix composites. This is the purpose of the next sections.

7.3.4 Basic Equations Resulting from Averaging Procedures

Consider a RVE with multiple phases of inhomogeneities, $\alpha = 1, 2, \ldots, n$. The elastic tensor and compliance tensors in the matrix are denoted by $\mathbf{L}^{\mathrm{M}}$ and $\mathbf{M}^{\mathrm{M}}$, respectively. The elastic tensors and compliance tensors in the constituent phases are denotes by $\mathbf{L}^{\alpha}$ and $\mathbf{M}^{\alpha}$ where $\alpha = 1, 2, \ldots, n$. V^{M} and V^{α} are the volumes of the matrix and inhomogeneity α, respectively, and $V = V^{\mathrm{M}} \cup \sum_{1}^{n} V^{\alpha}$ is the volume of the RVE. Further, we denote by $f^{\alpha} = V^{\alpha}/V$ and $f^{\mathrm{M}} = V^{\mathrm{M}}/V$ the volume fractions of the α – phase and the matrix, respectively.

Define the average stress and average strain in the matrix and in the inclusions as follows:

$$\begin{aligned} \bar{\sigma}_{ij}^{\mathrm{M}} &= \frac{1}{V^{\mathrm{M}}} \iiint\limits_{V^{\mathrm{M}}} \sigma_{ij}(\mathbf{r}) dV, & \bar{\boldsymbol{\varepsilon}}_{ij}^{\mathrm{M}} &= \frac{1}{V^{\mathrm{M}}} \iiint\limits_{V^{\mathrm{M}}} \varepsilon_{ij}(\mathbf{r}) dV \\ \bar{\sigma}_{ij}^{\alpha} &= \frac{1}{V^{\alpha}} \iiint\limits_{V^{\alpha}} \sigma_{ij}(\mathbf{r}) dV, & \bar{\boldsymbol{\varepsilon}}_{ij}^{\alpha} &= \frac{1}{V^{\alpha}} \iiint\limits_{V^{\alpha}} \varepsilon_{ij}(\mathbf{r}) dV \end{aligned} \tag{7.72}$$

By definition let denote by

$$\begin{aligned} \bar{\sigma}_{ij} &= \frac{1}{V} \iiint\limits_{V} \sigma_{ij}(\mathbf{r}) dV \\ &= \frac{1}{V} \left[\frac{V^{\mathrm{M}}}{V^{\mathrm{M}}} \iiint\limits_{V^{\mathrm{M}}} \sigma_{ij}(\mathbf{r}) dV + \sum_{1}^{n} \frac{V^{\alpha}}{V^{\alpha}} \iiint\limits_{V^{\alpha}} \sigma_{ij}(\mathbf{r}) dV \right] \\ &= f^{\mathrm{M}} \bar{\sigma}_{ij}^{\mathrm{M}} + \sum_{1}^{n} f^{\alpha} \bar{\sigma}_{ij}^{\alpha} \end{aligned} \tag{7.73}$$

Similarly one has

$$\begin{aligned} \bar{\boldsymbol{\varepsilon}}_{ij} &= \frac{1}{V} \iiint\limits_{V} \boldsymbol{\varepsilon}_{ij}(\mathbf{r}) dV \\ &= \frac{1}{V} \left[\frac{V^{\mathrm{M}}}{V^{\mathrm{M}}} \iiint\limits_{V^{\mathrm{M}}} \boldsymbol{\varepsilon}_{ij}(\mathbf{r}) dV + \sum_{1}^{n} \frac{V^{\alpha}}{V^{\alpha}} \iiint\limits_{V^{\alpha}} \boldsymbol{\varepsilon}_{ij}(\mathbf{r}) dV \right] \\ &= f^{\mathrm{M}} \bar{\boldsymbol{\varepsilon}}_{ij}^{\mathrm{M}} + \sum_{1}^{n} f^{\alpha} \bar{\boldsymbol{\varepsilon}}_{ij}^{\alpha} \end{aligned} \tag{7.74}$$

so that the elastic constitutive laws of each phase are expressed as

$$\begin{aligned} \bar{\sigma}_{ij}^{\mathrm{M}} &= L_{ijkl}^{\mathrm{M}} \bar{\varepsilon}_{kl}^{\mathrm{M}} \text{ and } \bar{\varepsilon}_{ij}^{\mathrm{M}} = M_{ijkl}^{\mathrm{M}} \bar{\sigma}_{kl}^{\mathrm{M}} \\ \bar{\boldsymbol{\sigma}}_{ij}^{\alpha} &= L_{ijkl}^{\alpha} \bar{\boldsymbol{\varepsilon}}_{kl}^{\alpha} \text{ and } \bar{\boldsymbol{\varepsilon}}_{ij}^{\alpha} = M_{ijkl}^{\alpha} \bar{\boldsymbol{\sigma}}_{kl}^{\alpha} \end{aligned} \tag{7.75}$$

On the other hand, the overall elastic stiffness and compliance tensors $\bar{\mathbf{L}}$ and $\bar{\mathbf{M}}$ are defined such that

$$\bar{\sigma}_{ij} = \bar{L}_{ijkl}\,\bar{\varepsilon}_{kl} \text{ and } \bar{\varepsilon}_{ij} = \bar{M}_{ijkl}\,\bar{\sigma}_{kl} \tag{7.76}$$

Therefore, it follows from Equations (7.73) and (7.75)

$$f^{\mathrm{M}}\bar{\sigma}_{ij}^{\mathrm{M}} = \bar{\sigma}_{ij} - \sum_{1}^{n} f^{\alpha}\bar{\sigma}_{ij}^{\alpha} = \bar{L}_{ijkl}\,\bar{\varepsilon}_{kl} - \sum_{1}^{n} f^{\alpha} L_{ijkl}^{\alpha}\,\bar{\varepsilon}_{kl}^{\alpha} \tag{7.77}$$

On the other hand, from Equations (7.74) and (7.75) leads

$$f^{\mathrm{M}}\bar{\sigma}_{ij}^{\mathrm{M}} = f^{\mathrm{M}} L_{ijkl}^{\mathrm{M}}\,\bar{\varepsilon}_{kl}^{\mathrm{M}} = L_{ijkl}^{\mathrm{M}}\left(\bar{\boldsymbol{\varepsilon}}_{kl} - \sum_{1}^{n} f^{\alpha}\bar{\boldsymbol{\varepsilon}}_{kl}^{\alpha}\right) \tag{7.78}$$

Combining Equations (7.77) and (7.78) yields

$$\left(\bar{L}_{ijkl} - L_{ijkl}^{\mathrm{M}}\right)\bar{\boldsymbol{\varepsilon}}_{kl} = \sum_{1}^{n} f^{\alpha}\left(L_{ijkl}^{\alpha} - L_{ijkl}^{\mathrm{M}}\right)\bar{\boldsymbol{\varepsilon}}_{kl}^{\alpha} \tag{7.79}$$

If the boundary conditions are given in terms of displacement, Equation (7.79) is equivalent to

$$\left(\bar{L}_{ijkl} - L_{ijkl}^{\mathrm{M}}\right)\boldsymbol{\varepsilon}_{kl}^{\mathrm{o}} = \sum_{1}^{n} f^{\alpha}\left(L_{ijkl}^{\alpha} - L_{ijkl}^{\mathrm{M}}\right)\bar{\boldsymbol{\varepsilon}}_{kl}^{\alpha} \tag{7.80}$$

Following similar steps, one can show that

$$\left(\bar{M}_{ijkl} - M_{ijkl}^{\mathrm{M}}\right)\bar{\sigma}_{kl} = \sum_{1}^{n} f^{\alpha}\left(M_{ijkl}^{\alpha} - M_{ijkl}^{\mathrm{M}}\right)\bar{\sigma}_{kl}^{\alpha} \tag{7.81}$$

If the traction boundary conditions are prescribed, Equation (7.80) leads to

$$\left(\bar{M}_{ijkl} - M_{ijkl}^{\mathrm{M}}\right)\sigma_{kl}^{\mathrm{o}} = \sum_{1}^{n} f^{\alpha}\left(M_{ijkl}^{\alpha} - M_{ijkl}^{\mathrm{M}}\right)\bar{\sigma}_{kl}^{\alpha} \tag{7.82}$$

Equations (7.79) and (7.82) constitute ones of the basic tools of the mean field theory. It is required to generate different approaches for the overall elastic properties of heterogeneous materials. In the following, three different attempts are developed and discussed.

7.4 Effective Elastic Moduli for Dilute Matrix-Inclusion Composites

Mean field theories for dilute matrix-inclusion composites are based on the concept of Eshelby's inclusion to derive the effective properties of composites where the volume fractions of inhomogeneities or inclusions are sufficiently small. Under such conditions, Eshelby's theory constitutes a good approximation of stress and strain fields in homogeneous inclusions embedded in a matrix. The procedure uses in general the Eshelby's equivalent inclusion method. However, equivalent solution is derived by adopting the Green's function techniques.

7.4.1 Method Using Equivalent Inclusion

This method is based on the Eshelby's theory for homogeneous inclusions, basically Equation (7.42), by introducing the concept of equivalent homogeneous inclusions, which consists in replacing an actual perfectly bonded inhomogeneous inclusion (which has different elastic properties than the matrix) with a fictitious equivalent homogeneous inclusion on which an appropriate fictitious equivalent eigenstrain $\boldsymbol{\varepsilon}^{\mathrm{t}}$ is made to act. This equivalent eigenstrain must be chosen in such a way that the same strain and stress fields $\bar{\boldsymbol{\varepsilon}}^{\alpha}$ and $\bar{\boldsymbol{\sigma}}^{\alpha}$ are obtained in the constrained state of the actual inhomogeneous inclusion and the equivalent homogeneous inclusion with prescribed eigenstrain.

The conditions of equal stresses and strains in the actual inclusion (with elastic constant $\mathbf{L}^{\alpha}$) and the equivalent homogenous one (with elastic constant $\mathbf{L}^{\mathrm{M}}$) under an applied far field strain $\boldsymbol{\varepsilon}^{\mathrm{o}}$ are expressed by

$$\bar{\sigma}_{ij}^{\alpha} = L_{ijkl}^{\alpha}\,\bar{\varepsilon}_{kl}^{\alpha} = L_{ijkl}^{\mathrm{M}}\left(\bar{\varepsilon}_{kl}^{\alpha} - \varepsilon_{kl}^{\mathrm{t}}\right) \text{ with } \bar{\varepsilon}_{ij}^{\alpha} = \varepsilon_{ij}^{\mathrm{o}} + S_{ijkl}^{\alpha}\,\varepsilon_{kl}^{\mathrm{t}} \tag{7.83}$$

from which one gets

$$\bar{\varepsilon}_{ij}^{\alpha} = \left(L_{ijkl}^{\mathrm{M}} - L_{ijkl}^{\alpha}\right)^{-1} L_{klmn}^{\mathrm{M}}\,\varepsilon_{mn}^{\mathrm{t}} \tag{7.84}$$

or

$$\bar{\boldsymbol{\varepsilon}}_{ij}^{\alpha} = \tilde{A}_{ijkl}^{\alpha}\boldsymbol{\varepsilon}_{kl}^{\mathrm{t}} \text{ with } \tilde{A}_{ijmn}^{\alpha} = \left(L_{ijkl}^{\mathrm{M}} - L_{ijkl}^{\alpha}\right)^{-1} L_{klmn}^{\mathrm{M}} \tag{7.85}$$

Substituting $\bar{\boldsymbol{\varepsilon}}^{\alpha}$ in (7.85) by using (7.83), leads to the following equation

$$\boldsymbol{\varepsilon}_{ij}^{\mathrm{t}} = \left(\tilde{A}_{ijkl}^{\alpha} - S_{ijkl}^{\alpha}\right)^{-1}\boldsymbol{\varepsilon}_{kl}^{\mathrm{o}} = \left(\tilde{A}_{ijkl}^{\alpha} - S_{ijkl}^{\alpha}\right)^{-1} M_{klmn}^{\mathrm{M}}\sigma_{mn}^{\mathrm{o}} \tag{7.86}$$

from which one can deduce the average stress in the inclusion by using its constitutive law

$$\bar{\sigma}^{\alpha}_{ij} = L^{\alpha}_{ijkl}\tilde{A}^{\alpha}_{klmn}\left(\tilde{A}^{\alpha}_{mnpq} - S^{\alpha}_{mnpq}\right)^{-1} M^{\mathrm{M}}_{pqrs}\sigma^{\mathrm{o}}_{rs} \tag{7.87}$$

Equation (7.87) allows us to determine the stress concentration tensor defined by Equation (7.41), which is expressed as

$$\bar{B}^{\alpha}_{ijrs} = L^{\alpha}_{ijkl}\,\tilde{A}^{\alpha}_{klmn}\left(\tilde{A}^{\alpha}_{mnpq} - S^{\alpha}_{mnpq}\right)^{-1} M^{\mathrm{M}}_{pqrs} \tag{7.88}$$

Finally, from the definition (7.82) of the overall compliance tensor $\bar{\mathbf{M}}$ of the inclusion-matrix composite, it follows that

$$\bar{M}_{ijmn} = \left\{ I_{ijkl} + \sum_{\alpha=1}^{n} f^{\alpha}\left(\tilde{A}^{\alpha}_{ijkl} - S^{\alpha}_{ijkl}\right)^{-1} \right\} M^{\mathrm{M}}_{klmn} \tag{7.89}$$

Recall that the Eshelby's tensor $\mathbf{S}^{\alpha}$ in (7.89) depends on the shape of the inclusion and the elastic properties of the matrix.

Under prescribed stress boundary conditions described by a homogeneous far field stress $\boldsymbol{\sigma}^{\mathrm{o}}$, the conditions of equal stresses and strains in the actual inclusion (with elastic compliance $\mathbf{M}^{\alpha}$) and the equivalent homogenous one (with elastic constant $\mathbf{M}^{\mathrm{M}}$) give

$$\bar{\varepsilon}^{\alpha}_{ij} = M^{\alpha}_{ijkl}\,\bar{\sigma}^{\alpha}_{ij} = M^{\mathrm{M}}_{ijkl}\left(\bar{\sigma}^{\alpha}_{kl} - \tau^{\alpha}_{kl}\right) \text{ with } \bar{\sigma}^{\alpha}_{ij} = \sigma^{\mathrm{o}}_{ij} + \tilde{S}^{\alpha}_{ijkl}\,\tau^{\alpha}_{kl} \tag{7.90}$$

from which one gets

$$\bar{\sigma}^{\alpha}_{ij} = \left(M^{\mathrm{M}}_{ijkl} - M^{\alpha}_{ijkl}\right)^{-1} M^{\mathrm{M}}_{klmn}\tau^{\alpha}_{mn} \tag{7.91}$$

or

$$\bar{\sigma}^{\alpha}_{ij} = \tilde{B}^{\alpha}_{ijmn}\tau^{\alpha}_{mn} \text{ with } \tilde{B}^{\alpha}_{ijmn} = \left(M^{\mathrm{M}}_{ijkl} - M^{\alpha}_{ijkl}\right)^{-1} M^{\mathrm{M}}_{klmn} \tag{7.92}$$

Therefore, from (7.90) and (7.92), one has the following expression:

$$\tau^{\alpha}_{ij} = \left(\tilde{B}^{\alpha}_{ijkl} - \tilde{S}^{\alpha}_{ijkl}\right)^{-1}\sigma^{\mathrm{o}}_{kl} = \left(\tilde{B}^{\alpha}_{ijkl} - \tilde{S}^{\alpha}_{ijkl}\right)^{-1} M^{\mathrm{M}}_{klmn}\varepsilon^{\mathrm{o}}_{mn} \tag{7.93}$$

and thanks to the constitutive law of the inclusion, it follows that

$$\bar{\boldsymbol{\varepsilon}}^{\alpha}_{ij} = M^{\alpha}_{ijkl}\tilde{B}^{\alpha}_{klmn}\left(\tilde{B}^{\alpha}_{mnpq} - \tilde{S}^{\alpha}_{mnpq}\right)^{-1} L^{\mathrm{M}}_{pqrs}\,\boldsymbol{\varepsilon}^{\mathrm{o}}_{rs} \tag{7.94}$$

Equation (7.94) gives the average strain in the inclusion in terms of the overall resulting strain. Therefore, one can express the strain concentration tensor defined by (7.41) as

$$\bar{A}^{\alpha}_{ijrs} = M^{\alpha}_{ijkl}\,\tilde{B}^{\alpha}_{klmn}\left(\tilde{B}^{\alpha}_{mnpq} - \tilde{S}^{\alpha}_{mnpq}\right)^{-1} L^{\mathrm{M}}_{pqrs} \tag{7.95}$$

Finally, from the definition (7.80) of the overall elastic tensor $\bar{\mathbf{L}}$ of the inclusion-matrix composite, it follows that

$$\bar{L}_{ijmn} = \left\{ I_{ijkl} + \sum_{\alpha=1}^{n} f^{\alpha}\left(\tilde{B}^{\alpha}_{ijkl} - \tilde{S}^{\alpha}_{ijkl}\right)^{-1} \right\} L^{\mathrm{M}}_{klmn} \tag{7.96}$$

Direct expressions of $\bar{\mathbf{L}}$ and $\bar{\mathbf{M}}$ results directly from Equations (7.66) and (7.67), which give the stress and strain concentration tensors in terms of Hill's and polarization tensors as

$$\bar{A}^{\alpha}_{ijmn} = \left(H^{I}_{ijkl} + L^{I}_{ijkl}\right)^{-1}\left(P^{I}_{klmn}\right)^{-1} \text{ and } \bar{B}^{\alpha}_{ijrs} = \left(\tilde{H}^{I}_{ijkl} + M^{I}_{ijkl}\right)^{-1}\left(Q^{I}_{klmn}\right)^{-1} \tag{7.97}$$

and therefore, it results from (7.80) and (7.82) that

$$M_{ijpq} = M^{\mathrm{M}}_{ijpq} + \sum_{\alpha=1}^{n} f^{\alpha}\left(M^{\alpha}_{ijkl} - M^{\mathrm{M}}_{ijkl}\right)\left(\tilde{H}^{\alpha}_{klmn} + M^{\alpha}_{klmn}\right)^{-1}\left(Q^{\alpha}_{mnpq}\right)^{-1} \tag{7.98}$$

and

$$\bar{L}_{ijpq} = L^{\mathrm{M}}_{ijpq} + \sum_{\alpha=1}^{n} f^{\alpha}\left(L^{\alpha}_{ijkl} - L^{\mathrm{M}}_{ijkl}\right)\left(H^{\alpha}_{klmn} + L^{\alpha}_{klmn}\right)^{-1}\left(P^{\alpha}_{mnpq}\right)^{-1} \tag{7.99}$$

where $\mathbf{H}^{\alpha}$, $\tilde{\mathbf{H}}^{\alpha}$, $\mathbf{P}^{\alpha}$ and $\mathbf{Q}^{\alpha}$ depend in addition to the shape of inclusion on the elastic properties of the matrix.

Equations (7.98) and (7.99) are derived under the assumption that the inclusions are dilutely dispersed in the matrix and therefore do not "feel" any effects due to their neighbors. In other words, in addition to the polarization imposed the surrounding medium, the inclusions are loaded by the unperturbed applied stress $\boldsymbol{\sigma}^{\mathrm{o}}$ or applied strain $\boldsymbol{\varepsilon}^{\mathrm{o}}$, so that the concentration tensors are independent of the inclusion volume fraction. As a result, the above analysis is valid only for vanishing small inclusion volume fractions.

Another feature of dilute description is that the obtained results from prescribed stress boundary conditions (Equation (7.99)) are different from those obtained under prescribed strain boundary conditions (Equation (7.98)). In fact, it can be shown easily that

$$\bar{\mathbf{L}} : \bar{\mathbf{M}} = \boldsymbol{I} + \mathrm{O}\left(f^{\alpha^2}\right) \neq \boldsymbol{I} \tag{7.100}$$

This is why the mean field theory based on the dilute conditions is known to be not consistent.

In this description, the Eshelby's tensor and related tensors depend only on the materials properties of the matrix and on the aspect ratio of the inclusion, i.e., expression for the Eshelby's tensor of ellipsoidal inclusions are independent of the material symmetry and properties of the inclusions. Analytical results for isotropic matrix containing dilute spherical inclusion are developed in the following. The obtained results will show clearly the nonconsistency of the dilute description.

7.4.2 Analytical Results for Spherical Inhomogeneities and Isotropic Materials

From the definition (7.89) of $\bar{\mathbf{M}}$

$$M_{ijmn} = \left\{ I_{ijkl} + \sum_{\alpha=1}^{n} f^{\alpha} \left(\tilde{A}^{\alpha}_{ijkl} - S^{\alpha}_{ijkl} \right)^{-1} \right\} M^{\mathrm{M}}_{klmn} \text{ with } \tilde{A}^{\alpha}_{ijmn}$$

$$= \left(L^{\mathrm{M}}_{ijkl} - L^{\alpha}_{ijkl} \right)^{-1} L^{\mathrm{M}}_{klmn}$$

we will develop successively the different terms involved in this expression.

Introduction of the E-basis orthogonal decomposition

$$E^{1}_{ijkl} = \frac{1}{3}\delta_{ij}\,\delta_{kl},\, E^{2}_{ijkl} = \frac{1}{3}\delta_{ij}\,\delta_{kl} + \frac{1}{2}\left\{\delta_{ik}\delta_{jl} + \delta_{il}\delta_{jk}\right\} \tag{7.101}$$

with $\mathbf{E}^1 : \mathbf{E}^1 = \mathbf{E}^1, \mathbf{E}^2 : \mathbf{E}^2 = \mathbf{E}^2, \mathbf{E}^1 : \mathbf{E}^2 = \mathbf{E}^2 : \mathbf{E}^1 = 0, \mathbf{E}^1 + \mathbf{E}^2 = \mathbf{I}$

one has

$$L^{\mathrm{M}}_{ijkl} = 3\,K^{\mathrm{M}} E^{1}_{ijkl} + 2\mu^{\mathrm{M}} E^{2}_{ijkl} \text{ and } M^{\mathrm{M}}_{ijkl} = \frac{1}{3\,K^{\mathrm{M}}} E^{1}_{ijkl} + \frac{1}{2\mu^{\mathrm{M}}} E^{2}_{ijkl} \tag{7.102}$$

Therefore

$$L^{\mathrm{M}}_{ijkl} - L^{\alpha}_{ijkl} = 3\left(K^{\mathrm{M}} - K^{\alpha}\right) E^{1}_{ijkl} + 2\left(\mu^{\mathrm{M}} - \mu^{\alpha}\right) E^{2}_{ijkl} \tag{7.103}$$

and

$$\left(L^{\mathrm{M}}_{ijkl} - L^{\alpha}_{ijkl} \right)^{-1} = \frac{1}{3(K^{\mathrm{M}} - K^{\alpha})} E^{1}_{ijkl} + \frac{1}{2(\mu^{\mathrm{M}} - \mu^{\alpha})} E^{2}_{ijkl} \tag{7.104}$$

Combining Equations (7.102) and (7.104) yields

$$
\begin{aligned}
\tilde{A}^{\alpha}_{ijmn} &= \left(L^{\mathrm{M}}_{ijkl} - L^{\alpha}_{ijkl}\right)^{-1} L^{\mathrm{M}}_{klmn} = \left\{\frac{1}{3(K^{\mathrm{M}} - K^{\alpha})} E^{1}_{ijkl} + \frac{1}{2(\mu^{\mathrm{M}} - \mu^{\alpha})} E^{2}_{ijkl}\right\} \\
&\quad \times \left\{3\,K^{\mathrm{M}} E^{1}_{klmn} + 2\mu^{\mathrm{M}} E^{2}_{klmn}\right\} \\
&= \frac{K^{\mathrm{M}}}{(K^{\mathrm{M}} - K^{\alpha})} E^{1}_{ijmn} + \frac{\mu^{\mathrm{M}}}{(\mu^{\mathrm{M}} - \mu^{\alpha})} E^{2}_{ijmn}
\end{aligned}
\tag{7.105}
$$

On the other hand the Eshelby's tensor is given by (J. Qu and M. Cherkaoui, 2006)

$$
S^{\alpha}_{ijkl} = s^{\mathrm{M}}_{1} E^{1}_{ijkl} + s^{\mathrm{M}}_{2} E^{2}_{ijkl} \text{ with } s^{\mathrm{M}}_{1} = \frac{1 + \nu^{\mathrm{M}}}{3(1 - \nu^{\mathrm{M}})} \text{ and } s^{\mathrm{M}}_{2} = \frac{2(4 - 5\nu^{\mathrm{M}})}{15(1 - \nu^{\mathrm{M}})} \tag{7.106}
$$

which leads to

$$
\tilde{A}^{\alpha}_{ijkl} - S^{\alpha}_{ijkl} = \left\{\frac{K^{\mathrm{M}}}{(K^{\mathrm{M}} - K^{\alpha})} - s^{\mathrm{M}}_{1}\right\} E^{1}_{ijkl} + \left\{\frac{\mu^{\mathrm{M}}}{(\mu^{\mathrm{M}} - \mu^{\alpha})} - s^{\mathrm{M}}_{2}\right\} E^{2}_{ijkl} \tag{7.107}
$$

and

$$
\left(\tilde{A}^{\alpha}_{ijkl} - S^{\alpha}_{ijkl}\right)^{-1} = \frac{1}{\frac{K^{\mathrm{M}}}{(K^{\mathrm{M}} - K^{\alpha})} - s^{\mathrm{M}}_{1}} E^{1}_{ijkl} + \frac{1}{\frac{\mu^{\mathrm{M}}}{(\mu^{\mathrm{M}} - \mu^{\alpha})} - s^{\mathrm{M}}_{2}} E^{2}_{ijkl} \tag{7.108}
$$

Hence

$$
\begin{aligned}
\bar{M}_{ijmn} &= \left\{ E^{1}_{ijkl} + E^{2}_{ijkl} + \sum_{\alpha=1}^{n} \frac{f^{\alpha}}{\frac{K^{\mathrm{M}}}{(K^{\mathrm{M}} - K^{\alpha})} - s^{\mathrm{M}}_{1}} E^{1}_{ijkl} \right. \\
&\quad \left. + \sum_{\alpha=1}^{\mathrm{n}} \frac{f^{\alpha}}{\frac{\mu^{\mathrm{M}}}{(\mu^{\mathrm{M}} - \mu^{\alpha})} - s^{\mathrm{M}}_{2}} E^{2}_{ijkl} \right\} \left(\frac{1}{3\,K^{\mathrm{M}}} E^{1}_{klmn} + \frac{1}{2\mu^{\mathrm{M}}} E^{2}_{klmn}\right)
\end{aligned}
\tag{7.109}
$$

or

$$
\begin{aligned}
\bar{M}_{ijkl} &= \frac{1}{3\,K^{\mathrm{M}}} \left\{1 + \sum_{\alpha=1}^{n} \frac{f^{\alpha}}{\frac{K^{\mathrm{M}}}{(K^{\mathrm{M}} - K^{\alpha})} - s^{\mathrm{M}}_{1}}\right\} E^{1}_{ijkl} \\
&\quad + \frac{1}{2\mu^{\mathrm{M}}} \left\{1 + \sum_{\alpha=1}^{\mathrm{n}} \frac{f^{\alpha}}{\frac{\mu^{\mathrm{M}}}{(\mu^{\mathrm{M}} - \mu^{\alpha})} - s^{\mathrm{M}}_{2}}\right\} E^{2}_{ijkl}
\end{aligned}
\tag{7.110}
$$

Finally, from the isotropic expression $\bar{\mathbf{M}} = \frac{1}{\bar{K}}\mathbf{E}^1 + \frac{1}{2\bar{\mu}}\mathbf{E}^2$, one concludes that

$$\frac{\bar{K}}{K^{\mathrm{M}}} = 1 + \sum_{\alpha=1}^{n} \frac{f^{\alpha}}{\frac{K^{\mathrm{M}}}{(K^{\mathrm{M}}-K^{\alpha})} - s_1^{\mathrm{M}}}$$
$$\frac{\bar{\mu}}{\mu^{\mathrm{M}}} = 1 + \sum_{\alpha=1}^{\mathrm{n}} \frac{f^{\alpha}}{\frac{\mu^{\mathrm{M}}}{(\mu^{\mathrm{M}}-\mu^{\alpha})} - s_2^{\mathrm{M}}} \tag{7.111}$$

The same procedure is followed for the overall elastic constant $\bar{\mathbf{L}}$

$$\bar{L}_{ijmn} = \left\{ I_{ijkl} + \sum_{\alpha=1}^{n} f^{\alpha}\left(\tilde{B}^{\alpha}_{ijkl} - \tilde{S}^{\alpha}_{ijkl}\right)^{-1} \right\} L^{\mathrm{M}}_{klmn} \text{ with } \tilde{B}^{\alpha}_{ijmn}$$
$$= \left(M^{\mathrm{M}}_{ijkl} - M^{\alpha}_{ijkl}\right)^{-1} M^{\mathrm{M}}_{klmn}$$

The E-basis orthogonal decomposition gives

$$\begin{aligned}\tilde{B}^{\alpha}_{ijmn} &= \left(\frac{1}{3}\left(\frac{1}{K^{\mathrm{M}}} - \frac{1}{K^{\alpha}}\right)E^1_{ijkl} + \frac{1}{2}\left(\frac{1}{\mu^{\mathrm{M}}} - \frac{1}{\mu^{\alpha}}\right)E^2_{ijkl}\right)^{-1}\left(\frac{1}{3\,K^{\mathrm{M}}}E^1_{klmn} + \frac{1}{2\mu^{\mathrm{M}}}E^2_{klmn}\right)\\
&= \left(\frac{3\,K^{\mathrm{M}}K^{\alpha}}{K^{\alpha} - K^{\mathrm{M}}}E^1_{ijkl} + \frac{2\mu^{\mathrm{M}}\mu^{\alpha}}{\mu^{\alpha} - \mu^{\mathrm{M}}}E^2_{ijkl}\right)\left(\frac{1}{3\,K^{\mathrm{M}}}E^1_{klmn} + \frac{1}{2\mu^{\mathrm{M}}}E^2_{klmn}\right)\\
&= \frac{K^{\alpha}}{K^{\alpha} - K^{\mathrm{M}}}E^1_{ijmn} + \frac{\mu^{\alpha}}{\mu^{\alpha} - \mu^{\mathrm{M}}}E^2_{ijmn}\end{aligned} \tag{7.112}$$

Under the conditions of spherical inclusions and isotropic elastic matrix, the dual Eshelby's tensor is expressed by

$$\tilde{S}^{\alpha}_{ijkl} = \left(1 - s_1^{\mathrm{M}}\right)E^1_{ijkl} + \left(1 - s_2^{\mathrm{M}}\right)E^2_{ijkl} \tag{7.113}$$

Therefore

$$\left(\tilde{B}^{\alpha}_{ijkl} - \tilde{S}^{\alpha}_{ijkl}\right) = \left(\frac{K^{\mathrm{M}}}{K^{\alpha} - K^{\mathrm{M}}} + s_1^{\mathrm{M}}\right)E^1_{ijkl} + \left(\frac{\mu^{\mathrm{M}}}{\mu^{\alpha} - \mu^{\mathrm{M}}} + s_2^{\mathrm{M}}\right)E^2_{ijkl} \tag{7.114}$$

and

$$\left(\tilde{B}^{\alpha}_{ijkl} - \tilde{S}^{\alpha}_{ijkl}\right)^{-1} = \frac{1}{\frac{K^{\mathrm{M}}}{K^{\alpha}-K^{\mathrm{M}}} + s_1^{\mathrm{M}}}E^1_{ijkl} + \frac{1}{\frac{\mu^{\mathrm{M}}}{\mu^{\alpha}-\mu^{\mathrm{M}}} + s_2^{\mathrm{M}}}E^2_{ijkl} \tag{7.115}$$

Hence

$$\bar{L}_{ijmn} = \left\{ \left(1 + \sum_{\alpha=1}^{n} \frac{f^\alpha}{\frac{K^M}{K^\alpha - K^M} + s_1^M} \right) E^1_{ijkl} + \left(1 + \sum_{\alpha=1}^{n} \frac{1}{\frac{\mu^M}{\mu^\alpha - \mu^M} + s_2^M} \right) E^2_{ijkl} \right\}$$

$$\left(3\, K^M E^1_{ijkl} + 2\mu^M E^2_{ijkl} \right)$$

$$= 3\, K^M \left(1 + \sum_{\alpha=1}^{n} \frac{f^\alpha}{\frac{K^M}{K^\alpha - K^M} + s_1^M} \right) E^1_{ijkl} + 2\mu^M \left(1 + \sum_{\alpha=1}^{n} \frac{1}{\frac{\mu^M}{\mu^\alpha - \mu^M} + s_2^M} \right) E^2_{ijkl} \tag{7.116}$$

Finally, from the definition $\bar{\mathbf{L}} = 3\bar{K}\mathbf{E}^1 + 2\bar{\mu}\,\mathbf{E}^2$, it results

$$\frac{\bar{K}}{K^M} = 1 - \sum_{\alpha=1}^{n} \frac{f^\alpha}{\frac{K^M}{K^M - K^\alpha} - s_1^M}$$
$$\frac{\bar{\mu}}{\mu^M} = 1 - \sum_{\alpha=1}^{\mathrm{n}} \frac{1}{\frac{\mu^M}{\mu^M - \mu^\alpha} - s_2^M} \tag{7.117}$$

Clearly, the results (7.117) are different from those obtained under prescribed traction boundary conditions (Equation (7.111)). As stated above, the results agree to the first order of the volume fraction. Therefore, the dilute description provides non consistent results in the homogenization scheme.

7.4.3 Direct Method Using Green's Functions

Before to develop consistent homogenization schemes, Green's function techniques are adopted to deal with the dilute homogenization scheme by a direct methodology, which gives equivalent solution as the above procedure. The problem consists in deriving the overall elastic properties of a dilute inclusion-matrix composite. For such a purpose, consider an infinite medium with elastic constant $\mathbf{L}^M$ and volume V containing an inclusion with elastic properties $\mathbf{L}^\alpha$ and volume V^α. The infinite medium is subjected to a homogeneous stress or strain boundary conditions described by $\boldsymbol{\sigma}^o$ and $\boldsymbol{\varepsilon}^o$, respectively.

For any material point of the infinite medium, the local elastic properties are given by

$$l_{ijkl}(\mathbf{r}) = L^M_{ijkl} + \left(L^\alpha_{ijkl} - L^M_{ijkl} \right) \theta^\alpha(\mathbf{r}) \tag{7.118}$$

with

$$\theta^{\alpha}(\mathbf{r}) = \begin{cases} 1 & \text{if} \quad \mathbf{r} \in V^{\alpha} \\ 0 & \text{if} \quad \mathbf{r} \notin V^{\alpha} \end{cases} \tag{7.119}$$

The governing equation are the elastic constitutive law at any material writes

$$\boldsymbol{\sigma}_{ij}(\mathbf{r}) = l_{ijkl}(\mathbf{r})\varepsilon_{kl}(\mathbf{r}) = \frac{1}{2} l_{ijkl}(\mathbf{r})\left(u_{k,l}(\mathbf{r}) + u_{l,k}(\mathbf{r})\right) = l_{ijkl}(\mathbf{r})u_{k,l}(\mathbf{r}) \tag{7.120}$$

and equilibrium equations in absence of body forces

$$\boldsymbol{\sigma}_{ij,j}(\mathbf{r}) = 0 \tag{7.121}$$

Combining (7.120) and (7.121) with (7.118)

$$L^{\mathrm{M}}_{ijkl}\, u_{k,lj}(\mathbf{r}) + \left[\left(L^{\alpha}_{ijkl} - L^{\mathrm{M}}_{ijkl}\right)\theta^{\alpha}(\mathbf{r})u_{k,l}(\mathbf{r})\right]_{,j} = 0 \tag{7.122}$$

which could be considered as a Navier-Stocks type problem with body forces

$$f_i = \left[\left(L^{\alpha}_{ijkl} - L^{\mathrm{M}}_{ijkl}\right)\theta^{\alpha}(\mathbf{r})u_{k,l}(\mathbf{r})\right]_{,j} \tag{7.123}$$

The solution of partial derivative Equation (7.122) is given in terms of Green's functions as

$$u_i(\mathbf{r}) = u_i^{\mathrm{o}}(\mathbf{r}) + \iiint\limits_V G_{ki}(\mathbf{r}-\mathbf{r}')\left[\left(L^{\alpha}_{klmn} - L^{\mathrm{M}}_{klmn}\right)u_{m,n}(\mathbf{r}')\theta^{\alpha}(\mathbf{r}')\right]_{,l'} dV' \tag{7.124}$$

with $u_i(\mathbf{r}) = u_i^{\mathrm{o}}(\mathbf{r})$ if $\mathbf{r} \to \infty$. Applying the divergence theorem (7.124) gives

$$u_i(\mathbf{r}) = u_i^{\mathrm{o}}(\mathbf{r}) + \iiint\limits_V G_{ki,l}(\mathbf{r}-\mathbf{r}')\left(L^{\alpha}_{klmn} - L^{\mathrm{M}}_{klmn}\right)u_{m,n}(\mathbf{r}')\theta^{\alpha}(\mathbf{r}')dV' \tag{7.125}$$

and therefore the displacement gradient writes

$$u_{i,j}(\mathbf{r}) = u^{\mathrm{o}}_{i,j} + \iiint\limits_{V^{\alpha}} G_{ki,lj}(\mathbf{r}-\mathbf{r}')\left(L^{\alpha}_{klmn} - L^{\mathrm{M}}_{klmn}\right)\varepsilon_{mn}(\mathbf{r}')dV' \tag{7.126}$$

or

$$\varepsilon_{ij}(\mathbf{r}) = \varepsilon^{\mathrm{o}}_{ij} - \iiint\limits_{V^{\alpha}} \Gamma_{ijkl}(\mathbf{r}-\mathbf{r}')\tau^{\mathrm{t}}_{kl}(\mathbf{r}')dV' \tag{7.127}$$

with

$$\tau^{\mathrm{t}}_{kl}(\mathbf{r}') = \left(C^{\alpha}_{klmn} - C^{\mathrm{M}}_{klmn}\right)\varepsilon_{mn}(\mathbf{r}') \tag{7.128}$$

defines the polarization, and

$$\Gamma_{ijkl}(\mathbf{r}-\mathbf{r}') = -\frac{1}{2}\left(G_{ki,lj}(\mathbf{r}-\mathbf{r}') + G_{kj,li}(\mathbf{r}-\mathbf{r}')\right) \tag{7.129}$$

is the modified Green's tensor.

Then the average strain in the inclusion is calculated from (7.127)

$$\bar{\varepsilon}^{\alpha}_{ij} = \frac{1}{V^{\alpha}}\iiint\limits_{V^{\alpha}} \varepsilon_{ij}(\mathbf{r})dV = \varepsilon^{\mathrm{o}}_{ij} - \iiint\limits_{V^{\alpha}} \Gamma_{ijkl}(\mathbf{r}-\mathbf{r}')dV\left\{\frac{1}{V^{\alpha}}\iiint\limits_{V^{\alpha}} \tau_{kl}(\mathbf{r}')dV'\right\} \tag{7.130}$$

In (7.130) we denote by

$$\bar{\tau}^{t}_{kl} = \frac{1}{V^{\alpha}}\iiint\limits_{V^{\alpha}} \tau_{kl}(\mathbf{r}')dV' \tag{7.131}$$

the average polarization in the inclusion and since $\mathbf{r} \in V^{\alpha}$ in the averaging procedure (7.130), it results from Eshelby's results that

$$\iiint\limits_{V^{\alpha}} \Gamma_{ijkl}(\mathbf{r}-\mathbf{r}')dV = P^{\alpha}_{ijkl} \tag{7.132}$$

is uniform and defines the polarization tensor $\mathbf{P}^{\alpha}$.

Finally, the average strain in the inclusion is expressed by

$$\bar{\varepsilon}^{\alpha}_{ij} = \varepsilon^{\mathrm{o}}_{ij} - P^{\alpha}_{ijkl}\,\bar{\tau}^{\mathrm{t}}_{kl} \tag{7.133}$$

Equation (7.133) is equivalent to the one defined by (7.62) and therefore the effective properties of the dilute inclusion-matrix are estimated by using the procedure developed in the previous section.

7.5 Mean Field Theories for Nondilute Inclusion-Matrix Composites

As discussed above, the mean field theory resulting directly from the inhomogeneous Eshelby's solutions is valid only for low volume fractions of inclusions. As the volume fractions of inhomogeneities increase, the interaction with their neighbors must explicitly be taken into consideration. These interactions between individual inclusions, on one hand, give rise to inhomogeneous stress and strain fields within each inhomogeneity. This is defined as to be intraparticle fluctuations. On the other hand, the interactions cause the levels of the average stresses and strains in individual inhomogeneities to differ. This is defined as to be interparticle fluctuations. Within the framework of mean field theories these interactions are accounted for by combining the concept of Eshelby's inclusion with appropriate approximations. In the following, the well-known approaches are developed in detail, namely the self-consistent approximation and the Mori-Tanaka type estimates.

7.5.1 The Self-Consistent Scheme

The self-consistent mean field theory is based on the concept of inhomogeneous Eshelby's inclusion solutions, in which the infinite medium have the elastic properties of the unknown overall or effective properties of the inclusion-matrix composite taken into consideration. Therefore, the self-consistent scheme deals with the problem of stress and strain distribution in homogeneous infinite elastic media with properties $\bar{\mathbf{L}}$ and $\bar{\mathbf{M}}$ containing an ellipsoidal subregion with volume V^α and elastic properties $\mathbf{L}^\alpha$ and $\mathbf{M}^\alpha$.

One of major differences between the self-consistent scheme and dilute mean theory procedure lies in the treatment of strain and stress boundary conditions prescribed at the boundary ∂V of the infinite medium. As it is noticed below, this difference ensures the consistency of the self-consistent scheme.

Consider the prescribed homogeneous stress boundary condition

$$t_i^{\mathrm{d}} = \sigma_{ij}^{\mathrm{o}} n_j \tag{7.134}$$

with $\bar{\boldsymbol{\sigma}} = \boldsymbol{\sigma}^{\mathrm{o}}$. In the self-consistent mean field theory, the resulting far field strain $\boldsymbol{\varepsilon}^{\mathrm{o}}$ is expressed by

$$\varepsilon_{ij}^{\mathrm{o}} = \bar{M}_{ijmn}\sigma_{mn}^{\mathrm{o}} = \bar{M}_{ijmn}\bar{\sigma}_{mn} \tag{7.135}$$

Therefore

$$\varepsilon_{ij}^{\mathrm{o}} = \bar{M}_{ijmn}\bar{\sigma}_{mn} = \bar{\varepsilon}_{ij} \tag{7.136}$$

Similarly, under strain displacement boundary conditions

$$u_i^{\mathrm{d}} = \varepsilon_{ij}^{\mathrm{o}} x_j \text{ with } \mathbf{r}(x_j) \in \partial V \tag{7.137}$$

with $\bar{\boldsymbol{\varepsilon}} = \boldsymbol{\varepsilon}^{\mathrm{o}}$. The resulting far field stress is

$$\sigma_{ij}^{\mathrm{o}} = \bar{L}_{ijmn}\,\varepsilon_{mn}^{\mathrm{o}} = \bar{L}_{ijmn}\,\bar{\varepsilon}_{mn} = \bar{\sigma}_{ij} \tag{7.138}$$

As stated above, another major difference between the self-consistent method and dilute suspension scheme is that to derive the effective properties, the Eshelby's equivalent principle is applied with respect to the homogenized overall properties instead of those of the matrix.

Within the self-consistent scheme, the concept of equivalent inclusion ensures that

$$L^{\alpha}_{ijkl}\,\bar{\varepsilon}^{\alpha}_{kl} = \bar{L}_{ijkl}\left(\bar{\varepsilon}^{\alpha}_{kl} - \varepsilon^{\mathrm{t}}_{kl}\right) \text{ with } \bar{\varepsilon}^{\alpha}_{ij} = \varepsilon^{\mathrm{o}}_{ij} + S^{\alpha}_{ijkl}\,\varepsilon^{\mathrm{t}}_{kl} \tag{7.139}$$

or

$$M^{\alpha}_{ijkl}\,\bar{\sigma}^{\alpha}_{ij} = \bar{M}_{ijkl}\left(\bar{\sigma}^{\alpha}_{kl} - \tau^{\alpha}_{kl}\right) \text{ with } \bar{\sigma}^{\alpha}_{ij} = \sigma^{\mathrm{o}}_{ij} + \tilde{S}^{\alpha}_{ijkl}\,\tau^{\alpha}_{kl} \tag{7.140}$$

Therefore the average strain and stress inside the $\alpha-$ th phase inclusion is expressed in terms of the resulting eigenstrain and polarization as

$$\begin{aligned} \bar{\varepsilon}^{\alpha}_{ij} &= \tilde{A}^{\alpha}_{ijkl}\,\varepsilon^{\mathrm{t}}_{kl} \text{ with } \tilde{A}^{\alpha}_{ijmn} = \left(\bar{L}_{ijkl} - L^{\alpha}_{ijkl}\right)^{-1}\bar{L}_{ijkl} \\ \bar{\sigma}^{\alpha}_{ij} &= \tilde{B}^{\alpha}_{ijmn}\tau^{\alpha}_{mn} \text{ with } \tilde{B}^{\alpha}_{ijmn} = \left(\bar{M}_{ijkl} - M^{\alpha}_{ijkl}\right)^{-1}\bar{M}_{klmn} \end{aligned} \tag{7.141}$$

Substituting (7.141) in (7.139) and (7.140), one can relate the average strain and stress inside the α-th phase inclusion to the applied boundary conditions as

$$\begin{aligned} \bar{\varepsilon}^{\alpha}_{ij} &= \tilde{A}^{\alpha}_{ijkl}\left(\tilde{A}^{\alpha}_{klmn} - S^{\alpha}_{lmn}\right)^{-1}\varepsilon^{\mathrm{o}}_{mn} \\ \bar{\sigma}^{\alpha}_{ij} &= \tilde{B}^{\alpha}_{ijkl}\left(\tilde{B}^{\alpha}_{klmn} - \tilde{S}^{\alpha}_{klmn}\right)^{-1}\sigma^{\mathrm{o}}_{mn} \end{aligned} \tag{7.142}$$

which allows us to determine the strain and stress concentration tensors as

$$\begin{aligned} \bar{A}^{\alpha}_{ijmn} &= \tilde{A}^{\alpha}_{ijkl}\left(\tilde{A}^{\alpha}_{klmn} - S^{\alpha}_{lmn}\right)^{-1} \\ \bar{B}^{\alpha}_{ijmn} &= \tilde{B}^{\alpha}_{ijkl}\left(\tilde{B}^{\alpha}_{klmn} - \tilde{S}^{\alpha}_{klmn}\right)^{-1} \end{aligned} \tag{7.143}$$

Note that $\bar{\boldsymbol{\varepsilon}} = \boldsymbol{\varepsilon}^{\mathrm{o}}$ and $\bar{\boldsymbol{\sigma}} = \boldsymbol{\sigma}^{\mathrm{o}}$ are used to set up the expression (7.143) for stress and strain concentration tensors.

Since by definition $\boldsymbol{\sigma}^{\alpha} = \mathbf{L}^{\alpha} : \boldsymbol{\varepsilon}^{\alpha}$ and $\boldsymbol{\varepsilon}^{\alpha} = \mathbf{M}^{\alpha} : \boldsymbol{\sigma}^{\alpha}$, one can derive the following relationship between stress and strain concentration tensors

$$\begin{aligned} \bar{B}^{\alpha}_{ijpq} &= L^{\alpha}_{ijkl}\,\bar{A}^{\alpha}_{klmn}\,\bar{M}_{mnpq} \\ \bar{A}^{\alpha}_{ijpq} &= M^{\alpha}_{ijkl}\,\bar{B}^{\alpha}_{klmn}\,\bar{L}_{mnpq} \end{aligned} \tag{7.144}$$

In the case of prescribed stress boundary conditions, combination of (7.143) with the basic averaging Equation (7.82) yields to the following self-consistent estimate of the overall compliance of a nondilute inclusion-matrix composite

$$\begin{aligned}\bar{M}_{ijrs} &= M^{\mathrm{M}}_{ijrs} + \sum_{\alpha=1}^{n} f^{\alpha}\left(M^{\alpha}_{ijkl} - M^{\mathrm{M}}_{klmn}\right)\bar{B}^{\alpha}_{klrs} \\ &= M^{\mathrm{M}}_{ijrs} + \sum_{\alpha=1}^{n} f^{\alpha}\left(M^{\alpha}_{ijkl} - M^{\mathrm{M}}_{ijkl}\right)L^{\alpha}_{klmn}\,\bar{A}^{\alpha}_{mnpq}\,\bar{M}_{pqrs}\end{aligned} \tag{7.145}$$

The same procedure is followed under prescribed displacement boundary conditions where Equation (7.79) combined with (7.143) yields to a self-consistent estimate of the overall stiffness

$$\begin{aligned}\bar{L}_{ijrs} &= L^{\mathrm{M}}_{ijrs} + \sum_{\alpha=1}^{n} f^{\alpha}\left(L^{\alpha}_{ijkl} - L^{\mathrm{M}}_{ijkl}\right)\bar{A}^{\alpha}_{klrs} \\ &= L^{\mathrm{M}}_{ijrs} + \sum_{\alpha=1}^{n} f^{\alpha}\left(L^{\alpha}_{ijkl} - L^{\mathrm{M}}_{ijkl}\right)M^{\alpha}_{klmn}\,\bar{B}^{\alpha}_{mnpq}\,\bar{L}_{pqrs}\end{aligned} \tag{7.146}$$

From (7.145) and (7.146) we should now check the consistency of the self-consistent estimate, in other words

$$\bar{\mathbf{M}} : \bar{\mathbf{L}} = \mathbf{I}$$

or

$$\bar{\mathbf{M}} = \bar{\mathbf{L}}^{-1} \text{ and } \bar{\mathbf{M}}^{-1} = \bar{\mathbf{L}}$$

should be fulfilled.

Consider

$$\begin{aligned}\mathbf{M}^{\mathrm{M}} &= \mathbf{M}^{\mathrm{M}} : \bar{\mathbf{L}} : \bar{\mathbf{L}}^{-1} = \mathbf{D}^{\mathrm{M}} : \left\{\mathbf{L}^{\mathrm{M}} + \sum_{\alpha=1}^{n} f^{\alpha}\left(\mathbf{L}^{\alpha} - \mathbf{L}^{\mathrm{M}}\right) : \bar{\mathbf{A}}^{\alpha}\right\} : \bar{\mathbf{L}}^{-1} \\ &= \bar{\mathbf{L}}^{-1} + \sum_{\alpha=1}^{n} f^{\alpha}\mathbf{M}^{\mathrm{M}} : \left(\mathbf{L}^{\alpha} - \mathbf{L}^{\mathrm{M}}\right) : \bar{\mathbf{A}}^{\alpha} : \bar{\mathbf{L}}^{-1}\end{aligned} \tag{7.147}$$

Note that in (7.147) we used the expression (7.146) of $\bar{\mathbf{L}}$. Since

$$\mathbf{M}^{\mathrm{M}} : \left(\mathbf{L}^{\alpha} - \mathbf{L}^{\mathrm{M}}\right) = \mathbf{M}^{\mathrm{M}} : \mathbf{L}^{\alpha} - \mathbf{I} = -\left(\mathbf{M}^{\alpha} - \mathbf{M}^{\mathrm{M}}\right) : \mathbf{L}^{\alpha} \tag{7.148}$$

(7.148) is equivalent to

$$\mathbf{M}^{\mathrm{M}} = \bar{\mathbf{L}}^{-1} - \sum_{\alpha=1}^{n} f^{\alpha}\left(\mathbf{M}^{\alpha} - \mathbf{M}^{\mathrm{M}}\right) : \mathbf{L}^{\alpha} : \bar{\mathbf{A}}^{\alpha} : \bar{\mathbf{L}}^{-1} \tag{7.149}$$

Which leads to

$$\bar{\mathbf{L}}^{-1} = \mathbf{M}^{\mathrm{M}} + \sum_{\alpha=1}^{n} f^{\alpha}\left(\mathbf{M}^{\alpha} - \mathbf{M}^{\mathrm{M}}\right) : \mathbf{L}^{\alpha} : \bar{\mathbf{A}}^{\alpha} : \bar{\mathbf{L}}^{-1} \tag{7.150}$$

Comparison of (7.151) with the expression (7.145) of $\bar{\mathbf{M}}$ allows us to conclude that $\bar{\mathbf{L}}^{-1} = \bar{\mathbf{M}}$. Similar procedure can be followed to show that $\bar{\mathbf{M}}^{-1} = \bar{\mathbf{L}}$. [*Remark* – The self-consistent estimate of overall or effective properties of inclusion-matrix composites (Equations (7.145) and (7.146)) are implicit in nature since the stress and strain concentration tensors depend on the effective properties through the Eshelby's tensor. Therefore, the use the self-consistent methodology is not straightforward and requires iterative procedures for numerical calculations.]

We propose in the following to develop analytical results in the case of a spherical inclusions and isotropic materials by using the E-basis orthogonal decomposition.

From

$$\bar{L}_{ijrs} = L^{\mathrm{M}}_{ijrs} + \sum_{\alpha=1}^{n} f^{\alpha}\left(L^{\alpha}_{ijkl} - L^{\mathrm{M}}_{ijkl}\right)\bar{A}^{\alpha}_{klrs} \text{ with } \bar{A}^{\alpha}_{ijmn} = \tilde{A}^{\alpha}_{ijkl}\left(\tilde{A}^{\alpha}_{klmn} - S^{\alpha}_{lmn}\right)^{-1}$$

one has successively

$$\begin{aligned}
L^{\mathrm{M}}_{ijkl} &= 3K^{\mathrm{M}}E^{1}_{ijkl} + 2\mu^{\mathrm{M}}E^{2}_{ijkl} \text{ and } L^{\alpha}_{ijkl} = 3K^{\alpha}E^{1}_{ijkl} + 2\mu^{\alpha}E^{2}_{ijkl} \\
\tilde{A}^{\alpha}_{ijmn} &= \left(\bar{L}_{ijkl} - L^{\alpha}_{ijkl}\right)^{-1}\bar{L}_{klmn} = \left\{\frac{1}{3(\bar{K}-K^{\alpha})}E^{1}_{ijkl} + \frac{1}{2(\bar{\mu}-\mu^{\alpha})}E^{2}_{ijkl}\right\} \\
&\quad \left\{3\bar{K}E^{1}_{klmn} + 2\bar{\mu}E^{2}_{klmn}\right\} \\
&= \frac{\bar{K}}{(\bar{K}-K^{\alpha})}E^{1}_{ijmn} + \frac{\bar{\mu}}{(\bar{\mu}-\mu^{\alpha})}E^{2}_{ijmn}
\end{aligned}$$

and

$$\begin{aligned}
\bar{A}^{\alpha}_{ijmn} &= \tilde{A}^{\alpha}_{ijkl}\left(\tilde{A}^{\alpha}_{klmn} - S^{\alpha}_{lmn}\right)^{-1} = \left(\frac{\bar{K}}{(\bar{K}-K^{\alpha})}E^{1}_{ijkl} + \frac{\bar{\mu}}{(\bar{\mu}-\mu^{\alpha})}E^{2}_{ijkl}\right) \\
&\quad \left(\frac{1}{\frac{\bar{K}}{(\bar{K}-K^{\alpha})} - \bar{s}_1}E^{1}_{klmn} + \frac{1}{\frac{\bar{\mu}}{(\bar{\mu}-\mu^{\alpha})} - \bar{s}_2}E^{2}_{klmn}\right)
\end{aligned}$$

which yields

$$\bar{A}^{\alpha}_{ijmn} = \frac{\bar{K}}{\bar{K} - (\bar{K}-K^{\alpha})\bar{s}_1}E^{1}_{ijmn} + \frac{\bar{\mu}}{\bar{\mu} - (\bar{\mu}-\mu^{\alpha})\bar{s}_2}E^{2}_{ijmn}$$

and

$$\bar{L}_{ijmn} = 3\left(K^{\mathrm{M}} + \sum_{\alpha=1}^{n} \frac{\bar{K}(K^{\alpha} - K^{\mathrm{M}})}{\bar{K} - (\bar{K} - K^{\alpha})\bar{s}_1}\right) E^1_{ijmn} + 2\left(\mu^{\mathrm{M}} + \sum_{\alpha=1}^{n} \frac{\bar{\mu}(\mu^{\alpha} - \mu^{\mathrm{M}})}{\bar{\mu} - (\bar{\mu} - \mu^{\alpha})\bar{s}_2}\right) E^2_{ijmn}$$

From the definition $\bar{C} = 3\bar{E}^1 + 2\bar{\mu}E^2$ one can conclude that

$$\frac{\bar{K}}{K^{\mathrm{M}}} = 1 + \sum_{\alpha=1}^{n} \frac{\left(\frac{K^{\alpha}}{K^{\mathrm{M}}} - 1\right)}{1 - \left(1 - \frac{K^{\alpha}}{\bar{K}}\right)\bar{s}_1} \tag{7.151}$$

$$\frac{\bar{\mu}}{\mu^{\mathrm{M}}} = 1 + \sum_{\alpha=1}^{n} \frac{\left(\frac{\mu^{\alpha}}{\mu^{\mathrm{M}}} - 1\right)}{1 - \left(1 - \frac{\mu^{\alpha}}{\bar{\mu}}\right)\bar{s}_2}$$

where $\bar{s}_1 = \dfrac{1+\bar{\nu}}{3(1-\bar{\nu})}$ and $\bar{s}_2 = \dfrac{2(4-5\bar{\nu})}{15(1-\bar{\nu})}$

7.5.2 Interpretation of the Self-Consistent

The self-consistent mean field theory could be derived directly by using the integral equation based on Green's function for an infinite medium. Let consider a RVE (volume V and boundary ∂V) of the inhomogeneous composite subjected to displacement or traction boundary conditions. Let assume displacement boundary conditions. The problem governing equations are as follow:

The quasistatic equilibrium without applied body forces

$$\boldsymbol{\sigma}_{ij,j}(\mathbf{r}) = 0 \tag{7.152}$$

The boundary conditions

$$u^{\mathrm{o}}_i(\mathbf{r}) = \varepsilon^{\mathrm{o}}_{ij} x_j \text{ if } \mathbf{r} \in \partial V \tag{7.153}$$

Kinematic relations

$$\varepsilon_{ij}(\mathbf{r}) = \frac{1}{2}\left(u_{i,j}(\mathbf{r}) + u_{j,i}(\mathbf{r})\right) \tag{7.154}$$

The local constitutive law

$$\sigma_{ij} = l_{ijkl}(\mathbf{r})\varepsilon_{ij}(\mathbf{r}) \tag{7.155}$$

At this stage, the methodology consist in introducing an unknown reference with properties $\mathbf{L}^{\mathrm{o}}$ so that the local elastic constant $\mathbf{l}(\mathbf{r})$ are decomposed into a homogeneous part $\mathbf{L}^{\mathrm{o}}$ and a fluctuating part $\delta\mathbf{l}(\mathbf{r})$ such that

$$l_{ijkl}(\mathbf{r}) = L^{o}_{ijkl} + \delta l_{ijkl}(\mathbf{r}) \tag{7.156}$$

Then by following the same procedure as (7.123), one gets a system of partial derivative equations

$$L^{o}_{ijkl}\, u_{k,lj}(\mathbf{r}) + \left[\delta l_{ijkl}(\mathbf{r}) u_{k,l}(\mathbf{r})\right]_{,j} = 0 \tag{7.157}$$

which is transformed to integral equations by using the Green's functions

$$\varepsilon_{ij}(\mathbf{r}) = \varepsilon^{o}_{ij} \iiint\limits_{V'} \Gamma_{ijkl}\,(\mathbf{r}-\mathbf{r}')\,\delta l_{ijkl}\,(\mathbf{r}')\,\varepsilon_{ij}(\mathbf{r}')dV' \tag{7.158}$$

Originally, the self-consistent mean field theory has its great interest in the properties of the modified Green tensor Γ , which can be divided for any homogeneous medium with elastic moduli $\mathbf{L}^{o}$ into a local part Γ^{loc} and nonlocal part Γ^{nloc} such as

$$\Gamma_{ijkl}(\mathbf{r}) = \Gamma^{\mathrm{loc}}_{ijkl}\,\delta(\mathbf{r}) + \Gamma^{\mathrm{nloc}}_{ijkl}\,(\mathbf{r}) \tag{7.159}$$

Substituting (7.159) in (7.158), and using the properties of the Dirac function $\delta(\mathbf{r})$, the integral equation becomes

$$\varepsilon_{ij}(\mathbf{r}) = \varepsilon^{o}_{ij} - \Gamma^{\mathrm{loc}}_{ijkl}\,\delta c_{klmn}(\mathbf{r})\varepsilon_{mn}(\mathbf{r}) - \iiint\limits_{V^{\alpha}} \Gamma^{\mathrm{nloc}}_{ijkl}\,(\mathbf{r}-\mathbf{r}')\,\delta l_{klmn}\,(\mathbf{r}')\varepsilon_{mn}(\mathbf{r}')\,dV' \tag{7.160}$$

where the integral form in (7.160) is generally difficult to estimate due to high and stochastic fluctuations of the field $\delta\mathbf{l}(\mathbf{r}') : \boldsymbol{\varepsilon}(\mathbf{r}')$. To overcome this difficulty, the self-consistent mean field theory for elastic materials came out with an original idea, which consists in choosing the reference medium $\mathbf{L}^{o}$ so that the mean value of the field $\delta\mathbf{l}(\mathbf{r}) : \boldsymbol{\varepsilon}(\mathbf{r})$ vanishes and therefore the integral form in (7.160) could be neglected. This condition of vanishing mean value of the fluctuating field is also known as the consistency condition. In fact, this condition writes

$$\iiint\limits_{V'} \delta l_{klmn}(\mathbf{r}')\varepsilon_{mn}(\mathbf{r}')dV' = \iiint\limits_{V'} \left(l_{klmn}(\mathbf{r}') - L^{o}_{klmn}\right)\varepsilon_{mn}(\mathbf{r}')dV' = 0 \tag{7.161}$$

which is equivalent to

$$\iiint\limits_{V'} \sigma_{kl}(\mathbf{r}')dV' = L^{o}_{klmn} \iiint\limits_{V'} \varepsilon_{mn}(\mathbf{r}')dV' \text{ or } \bar{\sigma}_{kl} = L^{o}_{klmn}\bar{\varepsilon}_{mn} \tag{7.162}$$

Expression (7.162) shows an interesting property which consists in the typical choice of the reference medium $\mathbf{L}^{\mathrm{o}}$ to fulfill the consistency condition stated above. Clearly, it results from (7.162) that the properties of the reference medium should be the effective properties of the considered composite, i.e., $\mathbf{L}^{\mathrm{o}} = \bar{\mathbf{L}}$.

Under the self-consistent approximation, Equation (7.162) is reduced to

$$\varepsilon_{ij}(\mathbf{r}) = \varepsilon_{ij}^{\mathrm{o}} - \Gamma_{ijkl}^{\mathrm{loc}}\,\delta l_{klmn}(\mathbf{r})\varepsilon_{mn}(\mathbf{r}) = \varepsilon_{ij}^{\mathrm{o}} - \Gamma_{ijkl}^{\mathrm{loc}}(l_{klmn}(\mathbf{r}) - \bar{L}_{klmn})\varepsilon_{mn}(\mathbf{r}) \quad (7.163)$$

Since the effective properties the effective properties through the averaging schemes require the average strain in each individual α–th phase, it results from (7.163) that

$$\bar{\varepsilon}_{ij}^{\alpha} = \varepsilon_{ij}^{\mathrm{o}} - \Gamma_{ijkl}^{\mathrm{loc}}\left(L_{klmn}^{\alpha} - \bar{L}_{klmn}\right)\bar{\varepsilon}_{mn}^{\alpha} \quad (7.164)$$

or by introducing the polarization, one has

$$\bar{\varepsilon}_{ij}^{\alpha} = \varepsilon_{ij}^{\mathrm{o}} - P_{ijkl}^{\alpha}\,\bar{\tau}_{kl}^{\mathrm{t}} \quad (7.165)$$

where

$$\bar{\tau}_{kl}^{\mathrm{t}} = \left(L_{klmn}^{\alpha} - \bar{L}_{klmn}\right)\bar{\varepsilon}_{mn}^{\alpha}$$

and

$$P_{ijkl}^{\alpha} = \Gamma_{ijkl}^{\mathrm{loc}}$$

Here the polarization tensor $\mathbf{P}^{\alpha}$ depends on the inclusion shape and on the effective elastic constant of the inclusion-matrix composite. Finally, by adopting similar methodology as the previous section, equation (7.165) can be combined with appropriate averaging procedures to derive similar expressions as (7.145) and (7.146) for the effective properties.

7.5.3 Mori-Tanaka Mean Field Theory

7.5.3.1 Mori Tanaka's Two-Phase Model

The Mori-Tanaka two-phase model is also known as the two phase double inclusion method. The typical feature of this homogenization scheme is that the related elementary inclusion problem consists in two ellipsoidal domains, which are coaxial, similar in shape and made by the same material (with elastic constant $\mathbf{L}^{\mathrm{o}}$). One of the ellipsoids represents the infinite medium (volume V) and the other the inclusion or inhomogeneity (volume V^{I}).

First, assume that a uniform eigenstrain is prescribed in the inclusion, so that at each material point $\mathbf{r}$ one has

$$\varepsilon_{ij}^{\mathrm{t}}(\mathbf{r}) = \varepsilon_{ij}^{\mathrm{t}}\,\theta^{I}(\mathbf{r}) \quad (7.166)$$

where $\theta^I(\mathbf{r})$ is given by (7.119)

By combining the equilibrium equations

$$\sigma_{ij,j}(\mathbf{r}) = 0 \tag{7.167}$$

with the constitutive law

$$\sigma_{ij}(\mathbf{r}) = L^{\mathrm{o}}_{ijkl}\left(u_{k,l}(\mathbf{r}) - \varepsilon^{\mathrm{t}}_{ij}(\mathbf{r})\right) \tag{7.168}$$

one obtains the following system of partial derivative equations

$$L^{\mathrm{o}}_{ijkl}\,u_{k,lj}(\mathbf{r}) - \left[L^{\mathrm{o}}_{ijkl}\,\varepsilon^{\mathrm{t}}_{kl}\theta^I(\mathbf{r})\right]_{,j} = 0 \tag{7.169}$$

which is transformed to an integral equation by using the Green's function

$$\varepsilon_{ij}(\mathbf{r}) = \iiint_{V^I} \Gamma_{ijkl}(\mathbf{r}-\mathbf{r}')L^{\mathrm{o}}_{klmn}\varepsilon'_{mn}dV' \tag{7.170}$$

Note that Equation (7.170) is obtained under no prescribed boundary conditions.

If $\mathbf{r} \in V^I$ the integral $\iiint_{V'} \mathbf{\Gamma}(\mathbf{r}-\mathbf{r}')dV'$ is uniform and leads to definition of Eshelby's inclusion $\mathbf{S}^I$. Therefore, Equation (7.170) is equivalent to the elementary Eshelby's solution $\boldsymbol{\varepsilon}^I = \mathbf{S}^I : \boldsymbol{\varepsilon}^{\mathrm{t}}$ where

$$\mathbf{S}^I = \left\{\iiint_{V^I} \mathbf{\Gamma}(\mathbf{r}-\mathbf{r}')dV'\right\} : \mathbf{L}^{\mathrm{o}} \text{ if } \mathbf{r} \in V^I \tag{7.171}$$

However, at this point, we are interested by the average strain $\bar{\boldsymbol{\varepsilon}}^{V-V^I}$ in the region belonging to $V - V^I$. By definition

$$\begin{aligned}\bar{\boldsymbol{\varepsilon}}^{V-V^I}_{ij} &= \frac{1}{V-V^I}\iiint_{V-V^I} \varepsilon_{ij}(\mathbf{r})dV \\ &= \frac{1}{V-V^I}\iiint_{V-V^I}\left(\iiint_{V^I} \Gamma_{ijkl}(\mathbf{r}-\mathbf{r}')dV'\right)L^{\mathrm{o}}_{klmn}\,\varepsilon^{\mathrm{t}}_{mn}dV\end{aligned} \tag{7.172a}$$

where

$$\begin{aligned}\iiint_{V-V^I}\left(\iiint_{V^I} \Gamma_{ijkl}(\mathbf{r}-\mathbf{r}')dV'\right) = &\iiint_{V}\left(\iiint_{V^I} \Gamma_{ijkl}(\mathbf{r}-\mathbf{r}')dV'\right)dV \\ &- \iiint_{V^I}\left(\iiint_{V^I} \Gamma_{ijkl}(\mathbf{r}-\mathbf{r}')dV'\right)dV\end{aligned}$$

Based on the Eshelby's results (7.171), one can easily show that

$$\iiint_{V}\left(\iiint_{V^I}\Gamma_{ijkl}(\mathbf{r}-\mathbf{r}')dV'\right)dV = V^I S^V_{ijkl}$$

and

$$\iiint_{V^I}\left(\iiint_{V^I}\Gamma_{ijkl}(\mathbf{r}-\mathbf{r}')dV'\right)dV = V^I S^I_{ijkl}$$

Since the two ellipsoids have the same shape and made by the same materials their related Eshelby's tensors $\mathbf{S}^I$ and $\mathbf{S}^V$ are identical. Therefore Equation (7.172a, b) leads to

$$\bar{\varepsilon}^{V-V^I}_{ij} = \frac{V^I}{V-V^I}\left(S^V_{ijkl} - S^I_{ijkl}\right)L^{\mathrm{o}}_{klmn}\,\varepsilon^{\mathrm{t}}_{mn}dV = 0 \tag{7.172b}$$

The result (7.172a, b) is known as the Mori-Tanaka lemma, it comes as a direct consequence of the scalable property of Eshelby's tensor.

The Mori-Tanaka's two-phase model or double inclusion model is a straightforward application of Mori-Tanaka's lemma. The original idea consists in choosing an infinite ellipsoidal medium having the elastic properties of the matrix and containing an ellipsoidal subregion representing the inhomogeneity. Under prescribed displacement or traction boundary. The equivalent Eshelby's inclusion principle leads to

$$\bar{\sigma}^{\alpha}_{ij} = L^{\alpha}_{ijkl}\,\bar{\varepsilon}^{\alpha}_{kl} = L^{\mathrm{M}}_{ijkl}\left(\bar{\varepsilon}^{\alpha}_{kl} - \varepsilon^{\mathrm{t}}_{kl}\right) \tag{7.173}$$

where

$$\bar{\varepsilon}^{\alpha}_{ij} = \varepsilon^{\mathrm{o}}_{ij} + S^{\alpha}_{ijkl}\,\varepsilon^{\mathrm{t}}_{kl} \tag{7.174}$$

Equation (7.173) leads to

$$\bar{\varepsilon}^{\alpha}_{ij} = \tilde{A}^{\alpha}_{ijkl}\,\varepsilon^{\mathrm{t}}_{kl}\ \text{with}\ \tilde{A}^{\alpha}_{ijmn} = \left(L^{\mathrm{M}}_{ijkl} - L^{\alpha}_{ijkl}\right)^{-1}L^{\mathrm{M}}_{klmn} \tag{7.175}$$

Substituting (7.175) in (7.174) one has

$$\varepsilon^{\mathrm{t}}_{ij} = \left(\tilde{A}^{\alpha}_{klmn} - S^{\alpha}_{lmn}\right)^{-1}\varepsilon^{\mathrm{o}}_{mn} \tag{7.176}$$

and therefore

$$\bar{\varepsilon}_{ij}^{\alpha}=\left(I_{ijmn}+S_{ijkl}^{\alpha}\left(\tilde{A}_{klmn}^{\alpha}-S_{lmn}^{\alpha}\right)^{-1}\right)\varepsilon_{mn}^{o} \tag{7.177}$$

Te constitutive law (7.173) leads to the average stress in the inclusion as

$$\bar{\sigma}_{ij}^{\alpha}=L_{ijkl}^{M}\left(I_{klmn}+\left(S_{klmn}^{\alpha}-I_{klmn}\right)\left(\tilde{A}_{mnpq}^{\alpha}-S_{mnpq}^{\alpha}\right)^{-1}\right)\varepsilon_{pq}^{o} \tag{7.178}$$

On the other hand, the Mori-Tanaka lemma states that the average disturbance strain in the region $V-V^{I}$ representing the matrix is null and therefore $\bar{\boldsymbol{\varepsilon}}^{M}=\boldsymbol{\varepsilon}^{o}$ and $\bar{\boldsymbol{\sigma}}^{M}=\mathbf{L}^{M}:\boldsymbol{\varepsilon}^{o}$.

Let f be the volume fraction of the inclusion phase. We then have the following equations resulting from the averaging schemes

$$\begin{aligned}\bar{\sigma}_{ij}&=f^{M}\bar{\sigma}_{ij}^{M}+f\bar{\sigma}_{ij}^{\alpha}\\ \bar{\varepsilon}_{ij}&=f^{M}\bar{\varepsilon}_{ij}^{M}+f\bar{\varepsilon}_{ij}^{\alpha}\end{aligned} \tag{7.179}$$

Substituting (7.178) in (7.179) and using the constitutive law of each phase, one obtains

$$\begin{aligned}\bar{\sigma}_{ij}&=(1-f)L_{ijkl}^{M}\varepsilon_{kl}^{o}+fL_{ijkl}^{M}\left(I_{klmn}+\left(S_{klmn}^{\alpha}-I_{klmn}\right)\left(\tilde{A}_{mnpq}^{\alpha}-S_{mnpq}^{\alpha}\right)^{-1}\right)\varepsilon_{pq}^{o}\\ &=L_{ijkl}^{M}\left(I_{klmn}+f(S_{klmn}^{\alpha}-I_{klmn})\left(\tilde{A}_{mnpq}^{\alpha}-S_{mnpq}^{\alpha}\right)^{-1}\right)\varepsilon_{pq}^{o}\end{aligned} \tag{7.180}$$

and

$$\begin{aligned}\bar{\varepsilon}_{ij}&=(1-f)\varepsilon_{ij}^{o}+f\left(I_{ijmn}+S_{ijkl}^{\alpha}\left(\tilde{A}_{klmn}^{\alpha}-S_{lmn}^{\alpha}\right)^{-1}\right)\varepsilon_{mn}^{o}\\ &=\left(I_{ijmn}+fS_{ijkl}^{\alpha}\left(\tilde{A}_{klmn}^{\alpha}-S_{lmn}^{\alpha}\right)^{-1}\right)\varepsilon_{mn}^{o}\end{aligned} \tag{7.181}$$

or

$$\varepsilon_{ij}^{o}=\left(I_{ijmn}+fS_{ijkl}^{\alpha}\left(\tilde{A}_{klmn}^{\alpha}-S_{lmn}^{\alpha}\right)^{-1}\right)^{-1}\bar{\varepsilon}_{mn}^{\alpha} \tag{7.182}$$

Substituting (7.182) in (7.180) leads to

$$\bar{\boldsymbol{\sigma}}=\mathbf{L}^{M}:\left(\mathbf{I}+f(\mathbf{S}^{\alpha}-I):\left(\tilde{\mathbf{A}}^{\alpha}-\mathbf{S}^{\alpha}\right)^{-1}\right):\left(\mathbf{I}+f\mathbf{S}^{\alpha}:\left(\tilde{\mathbf{A}}^{\alpha}-\mathbf{S}^{\alpha}\right)^{-1}\right)^{-1}:\bar{\boldsymbol{\varepsilon}} \tag{7.183a}$$

or from the definition $\bar{\boldsymbol{\sigma}}=\bar{\mathbf{L}}:\bar{\boldsymbol{\varepsilon}}$, Equation (7.182) provides the following expression

$$\bar{\mathbf{L}} = \mathbf{L}^{\mathrm{M}} : \left(\mathbf{I} + f(\mathbf{S}^{\alpha} - I) : \left(\tilde{\mathbf{A}}^{\alpha} - \mathbf{S}^{\alpha}\right)^{-1}\right) : \left(\mathbf{I} + f\mathbf{S}^{\alpha} : \left(\tilde{\mathbf{A}}^{\alpha} - \mathbf{S}^{\alpha}\right)^{-1}\right)^{-1} \quad (7.183b)$$

which is a double inclusion Mori-Tanaka estimate of the overall or effective elastic properties of an inclusion-matrix two-phase composite.

7.5.3.2 Mori Tanaka's Mean Field Theory

Within the Mori-Tanaka mean field, the interactions between the inclusions are estimated by the following averaging scheme, which is based on the concept of ellipsoidal inclusion with an appropriate properties and appropriate boundary conditions. In this scheme, each individual inclusion is taken to be in interaction with an infinite medium having the properties of the matrix. However, a fundamental difference with the dilute mean field approach is that the boundary conditions (far field conditions) are given in terms of the average stress or average strain in the matrix. Therefore within this scheme, the equivalent Eshelby's inclusions principle is equivalent to

$$\bar{\sigma}_{ij}^{\alpha} = L_{ijkl}^{\alpha}\,\bar{\varepsilon}_{kl}^{\alpha} = L_{ijkl}^{\mathrm{M}}\left(\bar{\varepsilon}_{kl}^{\alpha} - \varepsilon_{kl}^{\mathrm{t}}\right) \text{ with } \bar{\varepsilon}_{ij}^{\alpha} = \bar{\varepsilon}_{ij}^{\mathrm{M}} + S_{ijkl}^{\alpha}\,\varepsilon_{kl}^{\mathrm{t}} \quad (7.183c)$$

and

$$\bar{\varepsilon}_{ij}^{\alpha} = M_{ijkl}^{\alpha}\,\bar{\sigma}_{ij}^{\alpha} = M_{ijkl}^{\mathrm{M}}\left(\bar{\sigma}_{kl}^{\alpha} - \tau_{kl}^{\alpha}\right) \text{ with } \bar{\sigma}_{ij}^{\alpha} = \bar{\sigma}_{ij}^{\mathrm{M}} + \tilde{S}_{ijkl}^{\alpha}\,\tau_{kl}^{\alpha} \quad (7.184)$$

from which one gets

$$\bar{\varepsilon}_{ij}^{\alpha} = \tilde{A}_{ijkl}^{\alpha}\,\varepsilon_{kl}^{\mathrm{t}} \text{ with } \tilde{A}_{ijmn}^{\alpha} = \left(L_{ijkl}^{\mathrm{M}} - L_{ijkl}^{\alpha}\right)^{-1} L_{klmn}^{\mathrm{M}} \quad (7.185)$$

and

$$\bar{\sigma}_{ij}^{\alpha} = \tilde{B}_{ijmn}^{\alpha}\tau_{mn}^{\alpha} \text{ with } \tilde{B}_{ijmn}^{\alpha} = \left(M_{ijkl}^{\mathrm{M}} - M_{ijkl}^{\alpha}\right)^{-1} M_{klmn}^{\mathrm{M}} \quad (7.186)$$

It results from (7.183a,b,c) and (7.185) that

$$\bar{\varepsilon}_{ij}^{\alpha} = \tilde{A}_{ijkl}^{\alpha}\left(\tilde{A}_{klmn}^{\alpha} - S_{lmn}^{\alpha}\right)^{-1}\bar{\varepsilon}_{mn}^{\mathrm{M}} \text{ or } \bar{\varepsilon}_{ij}^{\alpha} = \overset{\approx}{A}{}_{ijkl}^{\alpha}\,\bar{\varepsilon}_{kl}^{\mathrm{M}} \quad (7.187)$$

where

$$\overset{\approx}{A}{}_{ijmn}^{\alpha} = \tilde{A}_{ijkl}^{\alpha}\left(\tilde{A}_{klmn}^{\alpha} - S_{lmn}^{\alpha}\right)^{-1} \quad (7.188)$$

Similarly from (7.184) and (7.186) one has

$$\bar{\sigma}_{ij}^{\alpha} = \tilde{B}_{ijkl}^{\alpha}\left(\tilde{B}_{klmn}^{\alpha} - \tilde{S}_{klmn}^{\alpha}\right)^{-1}\bar{\sigma}_{mn}^{\mathrm{M}} \quad \text{or} \quad \bar{\sigma}_{ij}^{\alpha} = \tilde{\tilde{B}}_{ijkl}^{\alpha}\bar{\sigma}_{ij}^{\mathrm{M}} \tag{7.189}$$

where

$$\tilde{\tilde{B}}_{ijmn}^{\alpha} = \tilde{B}_{ijkl}^{\alpha}\left(\tilde{B}_{klmn}^{\alpha} - \tilde{S}_{klmn}^{\alpha}\right)^{-1} \tag{7.190}$$

From the averaging procedures (7.73) and (7.74)

$$\begin{aligned} \bar{\sigma}_{ij} &= f^{\mathrm{M}}\bar{\sigma}_{ij}^{\mathrm{M}} + \sum_{1}^{n} f^{\alpha}\bar{\sigma}_{ij}^{\alpha} \\ \bar{\varepsilon}_{ij} &= f^{\mathrm{M}}\bar{\varepsilon}_{ij}^{\mathrm{M}} + \sum_{1}^{n} f^{\alpha}\bar{\varepsilon}_{ij}^{\alpha} \end{aligned} \tag{7.191}$$

and based on (7.187) one may find that

$$\bar{\varepsilon}_{ij}^{\mathrm{M}} = \left[\left(1 - \sum_{1}^{n} f^{\alpha}\right) + \sum_{1}^{n} f^{\alpha}\tilde{\tilde{A}}_{ijkl}^{\alpha}\right]^{-1} \bar{\varepsilon}_{kl} \text{ or } \bar{\varepsilon}_{ij}^{\mathrm{M}} = \bar{A}_{ijkl}^{\mathrm{M}}\bar{\varepsilon}_{kl} \tag{7.192}$$

where

$$\bar{A}_{ijkl}^{\mathrm{M}} = \left[\left(1 - \sum_{1}^{n} f^{\alpha}\right) + \sum_{1}^{n} f^{\alpha}\tilde{\tilde{A}}_{ijkl}^{\alpha}\right]^{-1} \tag{7.193}$$

Similarly

$$\bar{\sigma}_{ij}^{\mathrm{M}} = \left[\left(1 - \sum_{1}^{n} f^{\alpha}\right) + \sum_{1}^{n} f^{\alpha}\tilde{\tilde{B}}_{ijkl}^{\alpha}\right]^{-1} \sigma_{kl}^{\mathrm{o}} \quad \text{or} \quad \bar{\sigma}_{ij}^{\mathrm{M}} = \bar{B}_{ijkl}^{\mathrm{M}}\,\varepsilon_{kl}^{\mathrm{o}} \tag{7.194}$$

with

$$\bar{B}_{ijkl}^{\mathrm{M}} = \left[\left(1 - \sum_{1}^{n} f^{\alpha}\right) + \sum_{1}^{n} f^{\alpha}\tilde{\tilde{B}}_{ijkl}^{\alpha}\right]^{-1} \tag{7.195}$$

Accordingly, equations (7.187) and (7.189) yield to

$$\bar{\varepsilon}_{ij}^{\alpha} = \tilde{\tilde{A}}_{ijkl}^{\alpha}\left[\left(1 - \sum_{1}^{n} f^{\alpha}\right) + \sum_{1}^{n} f^{\alpha}\tilde{\tilde{A}}_{klmn}^{\alpha}\right]^{-1} \bar{\varepsilon}_{mn} \tag{7.196}$$

$$\bar{\sigma}_{ij}^{\alpha} = \tilde{\tilde{B}}_{ijkl}^{\alpha}\left[\left(1 - \sum_{1}^{n} f^{\alpha}\right) + \sum_{1}^{n} f^{\alpha}\tilde{\tilde{B}}_{klmn}^{\alpha}\right]^{-1} \bar{\sigma}_{mn} \tag{7.197}$$

Then it results from (7.191) that

$$\begin{aligned}
\bar{\sigma}_{ij} &= f^{\mathrm{M}}\bar{\sigma}_{ij}^{\mathrm{M}} + \sum_{1}^{n} f^{\alpha}\bar{\sigma}_{ij}^{\alpha} \\
&= \left(1 - \sum_{1}^{n} f^{\alpha}\right) L_{ijkl}^{\mathrm{M}}\bar{\varepsilon}_{kl}^{\mathrm{M}} + \sum_{1}^{n} f^{\alpha} L_{ijkl}^{\alpha}\bar{\varepsilon}_{kl}^{\alpha} \\
&= \left(1 - \sum_{1}^{n} f^{\alpha}\right) L_{ijkl}^{\mathrm{M}}\bar{A}_{klmn}^{\mathrm{M}}\bar{\varepsilon}_{mn} + \sum_{1}^{n} f^{\alpha} L_{ijkl}^{\alpha}\tilde{\bar{A}}_{klmn}^{\alpha}\bar{A}_{mnpq}^{\mathrm{M}}\bar{\varepsilon}_{pq} \\
&= \left(\left(1 - \sum_{1}^{n} f^{\alpha}\right) L_{ijmn}^{\mathrm{M}} + \sum_{1}^{n} f^{\alpha} L_{ijkl}^{\alpha}\tilde{\bar{A}}_{klmn}^{\alpha}\right)\bar{A}_{mnpq}^{\mathrm{M}}\bar{\varepsilon}_{pq} \\
&= \bar{L}_{ijpq}\bar{\varepsilon}_{pq}
\end{aligned} \tag{7.198}$$

Similarly

$$\begin{aligned}
\bar{\varepsilon}_{ij} &= f^{\mathrm{M}}\bar{\varepsilon}_{ij}^{\mathrm{M}} + \sum_{1}^{n} f^{\alpha}\bar{\varepsilon}_{ij}^{\alpha} \\
&= \left(1 - \sum_{1}^{n} f^{\alpha}\right) M_{ijkl}^{\mathrm{M}}\bar{\sigma}_{kl}^{\mathrm{M}} + \sum_{1}^{n} f^{\alpha} M_{ijkl}^{\alpha}\bar{\sigma}_{kl}^{\alpha} \\
&= \left(1 - \sum_{1}^{n} f^{\alpha}\right) M_{ijkl}^{\mathrm{M}}\bar{B}_{klmn}^{\mathrm{M}}\bar{\sigma}_{mn} + \sum_{1}^{n} f^{\alpha} M_{ijkl}^{\alpha}\approx{B}_{klmn}^{\alpha}\bar{A}_{mnpq}^{\mathrm{M}}\bar{\sigma}_{pq} \\
&= \left(\left(1 - \sum_{1}^{n} f^{\alpha}\right) M_{ijmn}^{\mathrm{M}} + \sum_{1}^{n} f^{\alpha} M_{ijkl}^{\alpha}\approx{B}_{klmn}^{\alpha}\right)\bar{B}_{mnpq}^{\mathrm{M}}\bar{\sigma}_{pq} \\
&= \bar{M}_{ijpq}\bar{\sigma}_{pq}
\end{aligned} \tag{7.199}$$

Finally, expressions (7.198) and (7.199) provide the Mori-Tanaka estimates of the overall elastic properties of the composite as

$$\bar{L}_{ijpq} = \left(\left(1 - \sum_{1}^{n} f^{\alpha}\right) L_{ijmn}^{\mathrm{M}} + \sum_{1}^{n} f^{\alpha} L_{ijkl}^{\alpha}\approx{A}_{klmn}^{\alpha}\right)\bar{A}_{mnpq}^{\mathrm{M}} \tag{7.200}$$

$$\bar{M}_{ijpq} = \left(\left(1 - \sum_{1}^{n} f^{\alpha}\right) M_{ijmn}^{\mathrm{M}} + \sum_{1}^{n} f^{\alpha} M_{ijkl}^{\alpha}\approx{B}_{klmn}^{\alpha}\right)\bar{B}_{mnpq}^{\mathrm{M}} \tag{7.201}$$

By replacing $\tilde{\bar{\mathbf{A}}}^{\alpha}$, $\bar{\mathbf{A}}^{\mathrm{M}}$, $\tilde{\bar{\mathbf{B}}}^{\alpha}$, and $\bar{\mathbf{B}}^{\mathrm{M}}$ by their expressions, (7.200) and (7.201) are equivalent to

$$\bar{L}_{ijpq} = \left(\sum_{\alpha=0}^{n} f^{\alpha} L^{\alpha}_{ijkl} \overset{\approx\alpha}{A}_{klmn}\right)\left(\sum_{\alpha=0}^{n} f^{\alpha} \overset{\approx\alpha}{A}_{mnpq}\right)^{-1} \tag{7.202}$$

$$\bar{M}_{ijpq} = \left(\sum_{\alpha=0}^{n} f^{\alpha} L^{\alpha}_{ijkl} \overset{\approx\alpha}{B}_{klmn}\right)\left(\sum_{\alpha=0}^{n} f^{\alpha} \overset{\approx\alpha}{B}_{mnpq}\right)^{-1} \tag{7.203}$$

Note that expressions (7.202) and (7.203) are valid for a composite with $n+1$ phases. $\alpha = 0$ corresponds to the matrix and therefore $\overset{\approx}{\mathbf{A}}^{0} = \overset{\approx}{\mathbf{B}}^{0} = \mathbf{I}$.

In accordance with their derivations, Mori-Tanaka type theories describe composites consisting in aligned ellipsoidal inclusions embedded in a matrix, i.e., inhomogeneous materials with a distinct matrix-inclusion microtopology. Contrary to the self-consistent methodology, the resulting equations from the Mori-Tanaka mean field theory are explicit and therefore numerically implemented in a straightforward way.

In the previous sections, we have introduced several methods of estimating the effective stiffness (compliance) tensors for a given composite, namely, the Eshelby method (or the dilute concentration method), the Mori-Tanaka method and the self-consistent method. As stated at the beginning of the present chapter, none of these methods are able to capture a size effect. In fact, based on the assumptions made in deriving them, all these methods ignore the spatial distribution of the inhomogeneities, that is, they all assume uniform distribution. However, the shapes and orientations of the ellipsoidal inhomogeneities are taken into account through the Eshelby tensor $\mathbf{S}$. Note that the Eshelby tensor is shape-dependent, but not size-dependent. Thus, the effective modulus tensor predicted by these methods will not depend on the size of the inhomogeneities. Interactions among the inhomogeneities are taken into consideration differently by different methods. In general, the Eshelby method works only for very dilute concentration, whereas the self-consistent ans Mori-Tanaka estimates are applicable to somewhat higher concentration. However, these approaches fail to predict accurately the properties for composites with high contrast between the inhomogeneities, as for example, the case of porous materials and rigid inclusions. To overcome this deficiency, self-consistent multi-inclusion methods have been introduced by Christensen and Lo [11]. This approach is based on the pioneering work of Hashin and Shtrikman [23], known as the composite sphere assemblage model.

7.6 Multinclusion Approaches

7.6.1 The Composite Sphere Assemblage Model

The composite sphere model was introduced by Hashin in 1962. As shown in (Fig. 7.1), the topology of this model involves various sizes of spherical coated inclusions, in which a particle inclusion is surrounded by a concentric matric shell. The volume fraction of the particle and the matrix material are the same in each sphere, but the spheres can be of any size to fill an arbitrary volume. By homogenizing each individual composite sphere (Fig. 7.1c) subjected to volumetric boundary conditions, Hashin found an excat solution for the effective bulk modulus $\bar{K}$ of the composite material, expressed by

$$\bar{K} = K^{\mathrm{M}} + \frac{f(K^{1} - K^{\mathrm{M}})(3\,K^{\mathrm{M}} + 4\mu^{\mathrm{M}})f}{(3\,K^{\mathrm{M}} + 4\mu^{\mathrm{M}}) + (K^{1} - K^{\mathrm{M}})(1 - f)} \tag{7.204}$$

where the superscript (M) stands for the matrix matrial and (1) for the particulate phase, where f represents its volume fraction.

Contrary to the bulk modulus, only bounds have been found for the shear moduls by composite sphere assemblage model.

7.6.2 *The Generalized Self-Consistent Model of Christensen and Lo*

Christensen and Lo [11] succeeded in obtaining the exact shear stiffness for the composite shown in Fig. 7.1a, by considering a single composite sphere embedded in an infinitely extended equivalent homogeneous material. The generalized self-consistent model also known as the three-phase approach solves the elementary composite inclusion problem shown in Fig. 7.2. Under presecribe volumetric boundary conditions, the three-phase self-consistent model provides the same expression (Equation (7.204)) of the bulk modulus.

Figure 7.3 exhibits the variation of the effective bulk modulus, normalized by the matrix's bulk modulus, of a composite material composed of porosities

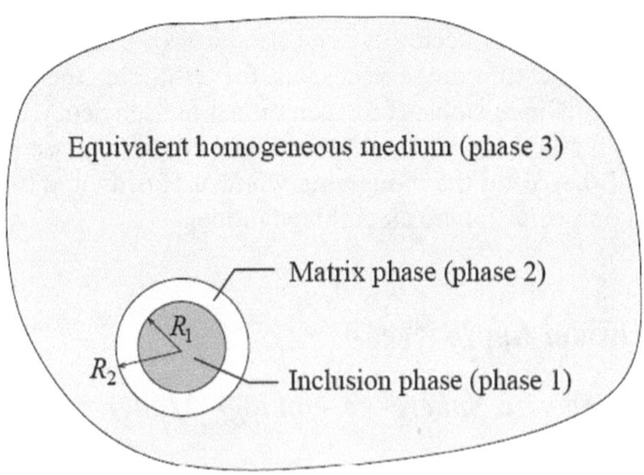

Fig. 7.2 Christensen and Lo model [11]

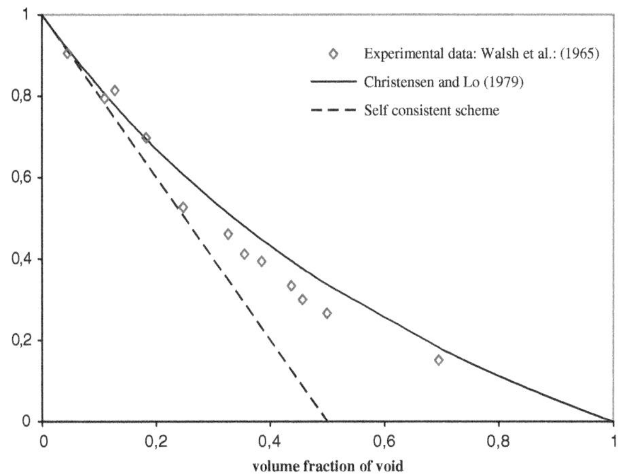

Fig. 7.3 Normalized effective bulk modulus versus void volume fraction: comaparisons between self-consistent model, Christensen and Lo, and experimental results

embedded in a polymer matrix. Predictions given by the typical self-consistent model (Equation (7.151) adapted to a two-phase composite material) and by the generalized self-consistent model (Equation (7.204)) are compared with Walsh et al. experimental results (1965).

Figure 7.3 shows that the typical self-consistent method underestimates experimental results. The variation the effective bulk modulus given by this model is almost linear up to a volume fraction of voids of 50% and gives a percolation threshold at 50% of the volume fraction of the spherical voids. Moreover, one can notice that the generalized self-consistent is in good agreement with experiments, even for a large amount of voids. This corroborates the discussion stated above about the limitations of the self-consistent method the corrections that may introduce the generalized self-consistent model.

In the case of simple shear, Christensen and Lo [11] have estimated the shear modulus $\bar{\mu}$ of a particulate composite. The prescribed simple shear deformation or simple shear stress boundary conditions yields to the following quadratic equation

$$A\left(\frac{\bar{\mu}}{\mu^{\mathrm{M}}}\right)^2 + B\left(\frac{\bar{\mu}}{\mu^{\mathrm{M}}}\right) + C = 0 \tag{7.205}$$

where A, B, and C are constant defined as follows:

$$\begin{aligned} A = {} & 8\left(\mu^1/\mu^{\mathrm{M}} - 1\right)(4 - 5\nu_m)\eta_1 c^{10/3} - 2\left[63\left(\mu^1/\mu^{\mathrm{M}} - 1\right)\eta_2 + 2\eta_1\eta_3\right] f^{7/3} \\ & + 252\left(\mu^1/\mu^{\mathrm{M}} - 1\right)\eta_2 f^{5/3} - 25\left(\mu^1/\mu^{\mathrm{M}} - 1\right)\left(7 - 12\nu^{\mathrm{M}} + 8\left(\nu^{\mathrm{M}}\right)^2\right)\eta_2 f \\ & + 4\left(7 - 10\nu^{\mathrm{M}}\right)\eta_2\eta_3 \end{aligned}$$

$$B = -4\left(\mu^{\mathrm{I}}/\mu^{\mathrm{M}} - 1\right)\left(4 - 5\nu^{\mathrm{M}}\right)\eta_1 f^{10/3} + 4\left[63\left(\mu^{\mathrm{I}}/\mu^{\mathrm{M}} - 1\right)\eta_2 + 2\eta_1\eta_3\right] f^{7/3}$$
$$- 504\left(\mu^{\mathrm{I}}/\mu^{\mathrm{M}} - 1\right)\eta_2 f^{5/3} + 150\left(\mu^{\mathrm{I}}/\mu^{\mathrm{M}} - 1\right)\left(3 - \nu^{\mathrm{M}}\right)\nu^{\mathrm{M}}\eta_2 f$$
$$+ 3\left(15\nu^{\mathrm{M}} - 7\right)\eta_2\eta_3$$

$$C = 4\left(\mu^{\mathrm{I}}/\mu^{\mathrm{M}} - 1\right)\left(5\nu^{\mathrm{M}} - 7\right)\eta_1 f^{10/3} - 2\left[63\left(\mu^{\mathrm{I}}/\mu^{\mathrm{M}} - 1\right)\eta_2 + 2\eta_1\eta_3\right] f^{7/3}$$
$$+ 252\left(\mu^{\mathrm{I}}/\mu^{\mathrm{M}} - 1\right)\eta_2 f^{5/3} + 25\left(\mu^{\mathrm{I}}/\mu^{\mathrm{M}} - 1\right)\left(\left(\nu^{\mathrm{M}}\right)^2 - 7\right)\eta_2 f - (7 + 5\nu)\eta_2\eta_3$$

with

$$\eta_1 = \left(\mu^{\mathrm{I}}/\mu^{\mathrm{M}} - 1\right)\left(49 - 50\nu^{\mathrm{I}}\nu^{\mathrm{M}}\right) + 35\left(\mu^{\mathrm{I}}/\mu^{\mathrm{M}}\right)\left(\nu^{\mathrm{I}} - 2\nu^{\mathrm{M}}\right) + 35\left(2\nu^{\mathrm{I}} - \nu^{\mathrm{M}}\right)$$
$$\eta_2 = 5\nu_1\left(\mu^{\mathrm{I}}/\mu^{\mathrm{M}} - 8\right) + 7\left(\mu^{\mathrm{I}} + \mu^{\mathrm{M}} + 4\right)$$
$$\eta_3 = \left(\mu^{\mathrm{I}}/\mu^{\mathrm{M}}\right)\left(8 - 10\nu^{\mathrm{M}}\right) + \left(7 - 5\nu^{\mathrm{M}}\right)$$

The generalized self-consistent predictions of the shear modulus of particulate composites were successful, even in the case of "asymptoty" configurations such as rigid particles and voided materials. As it will be shown in the next section, the three-phase model fall in general between the Hashin-Schtrickman bounds.

Based on Hermans work (1967), Christensen and Lo [11] developed the same model in the case of infinitely long parallel cylinders. The approach consists in considering a cylindrical three-phase model. In this model, the elementary cylindrical cell is surrounded by a third cylinder of large dimensions, composed of the equivalent homogenized material and having the homogenized effective properties of the composite (Fig. 7.4).

In the case of a composite with long fibers, the homogenized behavior is completely defined with five independent moduli: (1) longitudinal Young's

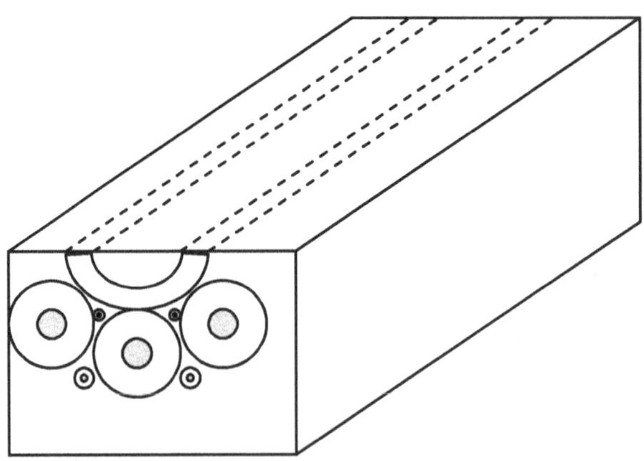

Fig. 7.4 Composite cylinders problem of Christensen and Lo

modulus, (2) Poisson's ratio, (3) shear modulus, (4) lateral hydrostatic bulk modulus, and (5) transverse shear modulus. The first four moduli were evaluated by Hashin and Rosen [22], and later by Hill [27] and Hashin [19]. They used Hashin and Shtrikman composite spheres model [23] and considered a cylindrical fiber, of radius a, surrounded y a matrix cylinder, of radius$b$. The two radii are related to the volume fraction of fibers by $f = a^2/b^2$. The fifth modulus was found by Christensen and Lo [11]. The five moduli are expressed as follows:

Longitudinal Young's modulus:

$$\bar{E}_L = fE^{\mathrm{I}} + (1-f)E^{\mathrm{M}} + \frac{f(1-f)(\nu^{\mathrm{I}} - \nu^{\mathrm{M}})^2}{\frac{f}{K^{\mathrm{I}}} + \frac{1}{\mu^{\mathrm{M}}} + \frac{1-f}{K^{\mathrm{I}}}} \tag{7.206}$$

Poisson's ratio:

$$\bar{\nu} = f\nu^{\mathrm{I}} + (1-f)\nu^{\mathrm{M}} + \frac{f(1-f)(\nu^{\mathrm{I}} - \nu^{\mathrm{M}})(1/K^{\mathrm{M}} - 1/K^{\mathrm{I}})}{\frac{f}{K^{\mathrm{I}}} + \frac{1}{\mu^{\mathrm{M}}} + \frac{1-f}{K^{\mathrm{I}}}} \tag{7.207}$$

Longitudinal shear modulus:

$$\bar{\mu}_L = \mu^{\mathrm{M}} \frac{\mu^{\mathrm{I}}(1+f) + \mu^{\mathrm{M}}(1-f)}{\mu^{\mathrm{I}}(1-f) + \mu^{\mathrm{M}}(1+f)} \tag{7.208}$$

Lateral hydrostatic bulk modulus:

$$\bar{K}_L = K^{\mathrm{M}} + \frac{f}{\frac{1}{K^{\mathrm{I}} - K^{\mathrm{M}}} + \frac{1-f}{K^{\mathrm{M}} + \mu^{\mathrm{M}}}} \tag{7.209a}$$

Transverse shear modulus:

$$\bar{\mu}_T = \mu^{\mathrm{M}} + \frac{f\mu^{\mathrm{M}}}{\frac{\mu^{\mathrm{M}}}{\mu^{\mathrm{I}} - \mu^{\mathrm{M}}} + \frac{(1-f)(K^{\mathrm{M}} + 2\mu^{\mathrm{M}})}{2\,K^{\mathrm{M}} + 2\mu^{\mathrm{M}}}} \tag{7.209b}$$

7.6.3 The n + 1 Phases Model of Herve and Zaoui

This model presented by Herve and Zaoui [25], is a generalized version of the three-phase model, for a composite which contains layered spherical inclusions. By analyzing this model (Fig. 7.5) with homogeneous boundary conditions prescribed at infinity the exact bulk and shear moduli are obtained. It is noted that the effective bulk modulus $\bar{K} = \bar{K}^{(n)}$ appears in recursive form and is successively obtained from the innermost sphere to the outermost sphere as

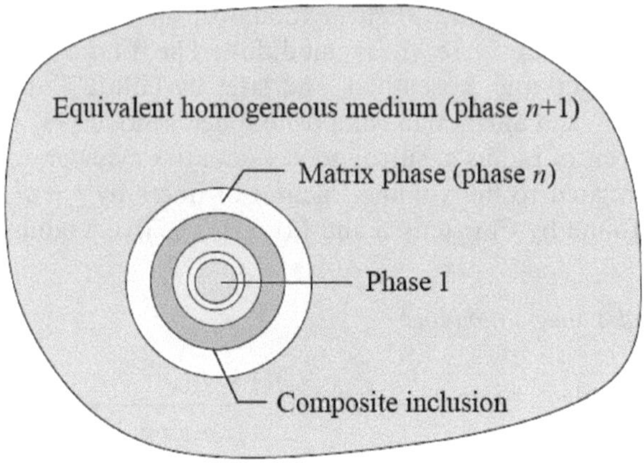

Fig. 7.5 The n + 1- phases model of Herve and Zaoui [25]

$$\bar{K}^{(i)} = \bar{K}^{(i)} + \frac{\left(R^{i-1}/R^{(i)}\right)^3 \left(\bar{K}^{(i-1)} - K^{(i)}\right)\left(3\,K^{(i)} + 4\mu^{(i)}\right)}{\left(3\,K^{(i)} + 4\mu^{(i)}\right) + \left(1 - \left(R^{i-1}/R^{(i)}\right)^3\right)\left(\bar{K}^{(i-1)} - K^{(i)}\right)} \tag{7.210}$$

where $R^{(i)}$,$R^{(i-1)}$ are the outer and inner raddii of phase i, respectively, $\bar{K}^{(i)}$ is the effective bulk modulus of the layered spherical inclusion which contains phase 1 to phase i. The effective shear modulus can be calculated from the quadratic equation

$$A\left(\frac{\bar{\mu}}{\mu^{(n)}}\right)^2 + 2B\left(\frac{\bar{\mu}}{\mu^{(n)}}\right) + C = 0 \tag{7.211}$$

where the constants A, B and C are given in Herve and Zaoui [25]. Similar to the three-phase model, this model is appropriate for very fine gradation of the inclusions. The application of this model can be found in Herve and Pellegrini [24], and Garboczi and Bentz [16]. When $n = 3$, the four-phase sphere model can be used to analyze the generalized self-consitent model of Christensen and Lo. This model has been used to study the mechanical properties of concrete and cement-mortar [21].

7.7 Variational Principles in Linear Elasticity

Obtaining the effective modulus tensor of a heterogeneous material is often a very difficult task. In most cases, only approximate solutions can be found. Several of these approximate estimates have been discussed in the previous sections. Although exact solutions to the effective moduli may not be found easily, for all practical purposes, knowing the bounds for these moduli is enough. In this section, some of these bounds are derived based on variation principles that rely on the concept of standard generalized materials.

The theory of standard generalized materials lies on the two concepts:

- The introduction of appropriate internal variables containing all necessary information about all system history at time t. The choice of internal variables depend on the considered material,
- The definition of thermodynamic potentials which are convex and corresponding to free energies and dissipative potentials.

The state variables of the system are the microscopic strain $\boldsymbol{\varepsilon}(\mathbf{r})$ and the internal variables $\boldsymbol{\alpha}$ describing irreversible processes.

The free energy ω leads to the definition, by mean of the constitutive laws, the stress from the strain and the thermodynamic forces $\mathbf{A}$ associated to dissipative mechanisms $\boldsymbol{\alpha}$ of the system. The definition of a dissipative potential φ links the rate of dissipative mechanisms to related driving forces.

The constitutive laws are given by

$$\begin{cases} \boldsymbol{\sigma} = \dfrac{\partial \omega}{\partial \boldsymbol{\varepsilon}}(\boldsymbol{\varepsilon}, \boldsymbol{\alpha}) \\ \mathbf{A} = -\dfrac{\partial \omega}{\partial \boldsymbol{\alpha}}(\boldsymbol{\varepsilon}, \boldsymbol{\alpha}) \end{cases} \tag{7.212}$$

and the so-called complementary laws

$$\mathbf{A} = \frac{\partial \varphi}{\partial \dot{\boldsymbol{\alpha}}}(\dot{\boldsymbol{\alpha}}) \tag{7.213}$$

If one defines ψ as the dual potential of φ , the complementary laws writes

$$\dot{\boldsymbol{\alpha}} = \frac{\partial \psi}{\partial \mathbf{A}}(\mathbf{A}) \tag{7.214}$$

In the case of an inhomogeneous material where the phases are assumed to be standard generalized, the homogenized material is also standard generalized. In other words, the homogenization scheme or the scale transitions methods should keep the characteristic of standard generalized, however, the number of internal variables describing the macroscopic potentials may be infinite.

In the following sections, the studied materials are supposed to be standard generalized.

7.7.1 Variational Formulation: General Principals

Let us consider an RVE with volume V and boundary ∂V representing an elastic heterogeneous material. The boundary conditions on ∂V are given in terms of combined loading: traction boundary condition t^{d} on ∂V^{T} and displacement boundary condition u^{d} on ∂V^{u} with $\partial V^{\mathrm{T}} \cup \partial V^{\mathrm{u}} = \partial V$.

The resolution of the elastic problem consists in finding the fields $\mathbf{u}$, $\boldsymbol{\varepsilon}$ and $\boldsymbol{\sigma}$ so that $\mathbf{u}$ and $\boldsymbol{\varepsilon}$ fulfil the conditions of kinematics

$$\begin{cases} \boldsymbol{\varepsilon} = \frac{1}{2}(\nabla \mathbf{u} +^{\mathrm{t}} \nabla \mathbf{u}) \text{ within } V \\ \mathbf{u} \text{ is continous in } V \\ \mathbf{u} = \mathbf{u}^{\mathrm{d}} \text{ on } \partial V^{\mathrm{u}} \end{cases} \tag{7.215}$$

Under the conditions (7.215) $\mathbf{u}$ and $\boldsymbol{\varepsilon}$ are defined to be kinematically admissibles.

$\boldsymbol{\sigma}$ fulfil the conditions of static in the presence of body forces $\mathbf{f}$

$$\begin{cases} \text{``div}\boldsymbol{\sigma} + \mathbf{f} = 0 \text{ inside } V \\ \boldsymbol{\sigma}.\mathbf{n} = \mathbf{t}^{\mathrm{d}} \text{ on } \partial V \end{cases} \tag{7.216}$$

Under the conditions (7.216) $\boldsymbol{\sigma}$ is called statically admissible.

$\boldsymbol{\varepsilon}$ and $\boldsymbol{\sigma}$ are related at each material point of V by the constitutive law

$$\boldsymbol{\sigma}(\mathbf{r}) = \frac{\partial \omega}{\partial \boldsymbol{\varepsilon}}(\boldsymbol{\varepsilon}(\mathbf{r}), \mathbf{r}) \tag{7.217}$$

or

$$\boldsymbol{\varepsilon}(\mathbf{r}) = \frac{\partial \phi}{\partial \varepsilon}(\sigma(\mathbf{r}), \mathbf{r}) \tag{7.218}$$

where ϕ is the dual of ω.

7.7.1.1 Extreme Variational Principle in Linear Elasticity

Minimum Potential Energy Principle

Consider an RVE representing an inclusion-matrix composite. The total potential energy of the elastic solid is

$$\Pi(\mathbf{u}) = \frac{1}{2} \iiint_V \sigma_{ij}(\mathbf{r}) \varepsilon_{ij}(\mathbf{r}) dV - \iiint_V f_i(\mathbf{r}) u_i(\mathbf{r}) dV - \iint_{\partial V^T} t_i^{\mathrm{d}} u_i(\mathbf{r}) dS \tag{7.219}$$

or by introducing the kinematic condition (7.215) with linear elastic constitutive law $\boldsymbol{\sigma}(\mathbf{r}) = \mathbf{l}(\mathbf{r}) : \boldsymbol{\varepsilon}(\mathbf{r})$

$$\Pi(\mathbf{u}) = \frac{1}{2} \iiint_V l_{ijkl}(\mathbf{r}) u_{i,j}(\mathbf{r}) u_{k,l}(\mathbf{r}) dV - \iiint_V f_i(\mathbf{r}) u_i(\mathbf{r}) dV - \iint_{\partial V^T} t_i^{\mathrm{d}} u_i(\mathbf{r}) dS \tag{7.220}$$

Consider a trial function $\mathbf{u}^*(\mathbf{r})$ which is kinematically admissible, and a test "virtual" $\delta\mathbf{u}^*(\mathbf{r})$ displacement with the condition $\delta\mathbf{u}^*(\mathbf{r}) = 0$ if $\mathbf{r} \in V$. We say that $\Pi(\mathbf{u}^*)$ reaches an extreme if the stationary condition is fulfilled, that is

$$\begin{aligned}\delta\Pi(\mathbf{u}^*) = &\iiint\limits_V l_{ijkl}(\mathbf{r}) u^*_{i,j}(\mathbf{r}) \delta u_{k,l}(\mathbf{r}) dV - \iiint\limits_V f_i(\mathbf{r}) \delta u_i(\mathbf{r}) dV \\ &- \iint\limits_{\partial V^{\mathrm{T}}} t_i^{\mathrm{d}} \delta u_i(\mathbf{r}) dS = 0\end{aligned} \tag{7.221}$$

Equation (7.220) is known as the virtual principle in solid mechanics, it also means the equilibrium conditions. In fact, the expression (7.221) is often called the weak formulation of Navier equation in computational mechanics. This can be readily shown by using the divergence theorem in (7.221) as

$$\begin{aligned}\delta\Pi(\mathbf{u}^*) &= \iiint\limits_V \sigma_{ij}(\mathbf{r}) \delta u^*_{i,j}(\mathbf{r}) dV - \iiint\limits_V f_i(\mathbf{r}) \delta u^*_i(\mathbf{r}) dV - \iint\limits_{\partial V^{\mathrm{T}}} t_i^{\mathrm{d}} \delta\, u^*_i(\mathbf{r})\, dS = 0 \\ &= \iiint\limits_V \Big(\big(\sigma_{ij}(\mathbf{r})\, \delta u^*_i(\mathbf{r})_{,j}\big) - \sigma_{ij,j}(\mathbf{r})\, \delta u^*_i(\mathbf{r}) \Big) dV \\ &\quad - \iiint\limits_V f_i(\mathbf{r}) \delta\, u^*_i(\mathbf{r}) dV - \iint\limits_{\partial V^{\mathrm{T}}} t_i^{\mathrm{d}}\, \delta u^*_i(\mathbf{r})\, dS \\ &= \iiint\limits_V \Big(\big(\sigma_{ij}(\mathbf{r})\, \delta u^*_i(\mathbf{r})_{,j}\big) - \sigma_{ij,j}(\mathbf{r})\, \delta u^*_i(\mathbf{r}) \Big) dV \\ &\quad - \iiint\limits_V f_i(\mathbf{r})\, \delta u^*_i(\mathbf{r})\, dV - \iint\limits_{\partial V^{\mathrm{T}}} t_i^{\mathrm{d}}\, \delta u^*_i(\mathbf{r})\, dS \\ &= \iint\limits_{\partial V} \sigma_{ij}(\mathbf{r}) n_j\, \delta\, u^*_i(\mathbf{r}) dS - \iiint\limits_V \big(\sigma_{ij,j}(\mathbf{r}) + f_i(\mathbf{r})\big)\, \delta u^*_i(\mathbf{r})\, dV \\ &\quad - \iint\limits_{\partial V^{\mathrm{T}}} t_i^{\mathrm{d}}\, \delta u^*_i(\mathbf{r}) dS = \iint\limits_{\partial V^{\mathrm{T}}} \big(\sigma_{ij}(\mathbf{r})\, n_j - t_i^{\mathrm{d}}\big) \delta\, u^*_i(\mathbf{r}) dS \\ &\quad - \iiint\limits_V \big(\sigma_{ij,j}(\mathbf{r}) + f_i(\mathbf{r})\big) \delta u^*_i(\mathbf{r}) dV - \iint\limits_{\partial V^{\mathrm{u}}} \sigma_{ij}(\mathbf{r}) n_j\, \delta\, u^*_i(\mathbf{r}) dS\end{aligned} \tag{7.222}$$

which leads to Navier equations

$$\sigma_{ij,j}(\mathbf{r}) + f_i(\mathbf{r}) = 0 \tag{7.223}$$

and the stress boundary conditions

$$\sigma_{ij}(\mathbf{r})n_j = t_i^{\mathrm{d}} = \sigma_{ij}^{\mathrm{o}} n_j \text{ if } \mathbf{r} \in \partial V^{\mathrm{T}} \tag{7.224}$$

Now examine the fluctuation $\Delta\Pi(\mathbf{u}^*)$ of the potential energy around an equilibrium configuration, it reads

$$\begin{aligned}
\Delta\Pi(\mathbf{u}^*) = {} & \Pi(\mathbf{u}^* + \delta\mathbf{u}) - \Pi(\mathbf{u}^*) \\
= {} & \frac{1}{2}\iiint\limits_V l_{ijkl}(\mathbf{r})\left(u_{i,j}^*(\mathbf{r}) + \delta u_{i,j}(\mathbf{r})\right)\left(u_{i,j}^*(\mathbf{r}) + \delta u_{i,j}(\mathbf{r})\right)dV \\
& - \iiint\limits_V f_i(\mathbf{r})\left(u_i^*(\mathbf{r}) + \delta u_i(\mathbf{r})\right)dV - \iint\limits_{\partial V^{\mathrm{T}}} t_i^{\mathrm{d}}\left(u_i^*(\mathbf{r}) + \delta u_i(\mathbf{r})\right)dS \\
& - \frac{1}{2}\iiint\limits_V l_{ijkl}(\mathbf{r})u_{i,j}^*(\mathbf{r})u_{k,l}^*(\mathbf{r})dV \\
& + \iiint\limits_V f_i(\mathbf{r})u_i^*(\mathbf{r})dV + \iint\limits_{\partial V^{\mathrm{T}}} t_i^{\mathrm{d}} u_i^*(\mathbf{r})dS
\end{aligned} \tag{7.225}$$

After few straightforward simplifications, (7.225) writes

$$\begin{aligned}
\Delta\Pi(\mathbf{u}^*) = {} & \iiint\limits_V l_{ijkl}(\mathbf{r})u_{i,j}^*(\mathbf{r})\delta u_{k,l}(\mathbf{r})dV - \iiint\limits_V f_i(\mathbf{r})\delta u_i(\mathbf{r})dV \\
& - \iint\limits_{\partial V^{\mathrm{T}}} t_i^{\mathrm{d}}\delta u_i(\mathbf{r})dS + \frac{1}{2}\iiint\limits_V l_{ijkl}(\mathbf{r})\delta u_{i,j}(\mathbf{r})\delta u_{k,l}(\mathbf{r})dV = \delta\Pi(\mathbf{u}^*) + \delta^2\Pi
\end{aligned} \tag{7.226}$$

where

$$\delta^2\Pi = \frac{1}{2}\iiint\limits_V l_{ijkl}(\mathbf{r})\delta u_{i,j}(\mathbf{r})\delta u_{k,l}(\mathbf{r})dV \tag{7.227}$$

and $\delta\Pi(\mathbf{u}^*)$ is given by (7.222).

From the equilibrium condition $\delta\Pi(\mathbf{u}^*) = 0$, one can conclude that $\Delta\Pi(\mathbf{u}^*) = \delta^2\Pi > 0$. This means that for all kinematically admissible field $\mathbf{u}^*$, the equilibrium solution is the one minimizing the total potential energy. In other words, the real solution in terms of kinematically admissible displacements renders the potential energy an absolute minimum. That is

$$\Pi(\mathbf{u}) \leq \Pi(\mathbf{u}^*) \text{ or } \Pi(\mathbf{u}) = \inf_{\mathbf{u}^*}\Pi(\mathbf{u}^*) \tag{7.228}$$

If we consider an RVE with prescribed displacement boundary conditions on its entire boundary ∂V, so that

$$u_i^{\mathrm{d}} = \varepsilon_{ij}^{\mathrm{o}} x_j \text{ if } \mathbf{r}(x_j) \in \partial V \ \left(\partial V^{\mathrm{u}} = \partial V; \partial V^{\mathrm{T}} = \emptyset\right) \tag{7.229}$$

it follows from the definition (7.219) that

$$\Pi(\mathbf{u}^*) = V\, W(\mathbf{u}^*) \tag{7.230}$$

where $W(\mathbf{u}^*)$ is the macroscopic elastic strain density expressed by

$$W(\mathbf{u}^*) = \frac{1}{2V} \iiint_V l_{ijkl}(\mathbf{r}) \varepsilon_{ij}^*(\mathbf{r}) \varepsilon_{kl}^*(\mathbf{r}) dV \tag{7.231}$$

Hence the minimum energy principle (7.228) becomes

$$W(\mathbf{u}) = \inf_{\mathbf{u}^*} W(\mathbf{u}^*) \tag{7.232}$$

Application: The Voigt bound

We will show in the following that a special choice of kinematically admissible displacement field leads to one of possible lower bounds known as the simple Voigt solution for composite materials. For such a purpose, consider a inclusion-matrix composite with n phases, where the α−th phase has homogeneous elastic properties $\mathbf{L}^{\alpha}$.

For the real solution characterized by the kinematically and statistically fields $\mathbf{u}(\mathbf{r})$ and $\boldsymbol{\sigma}(\mathbf{r})$, respectively, the elastic energy density reads (Hill's lemma, Section 7.1.)

$$W(\mathbf{u}) = \frac{1}{2} \iiint_V \sigma_{ij}(\mathbf{r}) \varepsilon_{ij}(\mathbf{r}) dV = \frac{1}{2} \bar{\sigma}_{ij}\, \bar{\varepsilon}_{ij} = \frac{1}{2} \bar{\sigma}_{ij}\, \varepsilon_{ij}^{\mathrm{o}} = \frac{1}{2}\, \varepsilon_{kl}^{\mathrm{o}}\, \bar{L}_{ijkl}\, \varepsilon_{ij}^{o} \tag{7.233}$$

On the other hand

$$\begin{aligned} W(\mathbf{u}^*) &= \frac{1}{2} \iiint_V \sigma_{ij}^*(\mathbf{r}) \varepsilon_{ij}^*(\mathbf{r}) dV = \frac{1}{2} \bar{\sigma}_{ij}\, \bar{\varepsilon}_{ij} = \frac{1}{2} \bar{\sigma}_{ij}\, \bar{\varepsilon}_{ij}^* \\ &= \frac{1}{2} \bar{\sigma}_{ij}\, \varepsilon_{ij}^{\mathrm{o}} = \frac{1}{2} \varepsilon_{ij}^{\mathrm{o}} \sum_{\alpha=0}^{n} f^{\alpha} L_{ijkl}^{\alpha}\, \bar{\varepsilon}_{kl}^{\alpha} \end{aligned} \tag{7.234}$$

If we choose in (7.234) $\bar{\boldsymbol{\varepsilon}}^{\alpha} = \boldsymbol{\varepsilon}^{\mathrm{o}}$ (which derives from a kinematically admissible field), the minimum energy principle (7.232) writes

$$\frac{1}{2}\varepsilon^{o}_{kl}\,\bar{L}_{ijkl}\,\varepsilon^{o}_{ij} \leq \frac{1}{2}\varepsilon^{o}_{ij}\sum_{\alpha=0}^{n} f^{\alpha}\,\bar{L}^{\alpha}_{ijkl}\,\bar{\varepsilon}^{\alpha}_{kl} \quad \text{or} \quad \frac{1}{2}\varepsilon^{o}_{kl}\,\bar{L}_{ijkl}\,\varepsilon^{o}_{ij} \leq \frac{1}{2}\varepsilon^{o}_{ij}\sum_{\alpha=0}^{n} f^{\alpha}\,\bar{L}^{\alpha}_{ijkl}\,\varepsilon^{o}_{kl} \tag{7.235}$$

which leads

$$\bar{L}_{ijkl} \leq \sum_{\alpha=0}^{n} f^{\alpha} C^{\alpha}_{ijkl} \tag{7.236}$$

where $\bar{L}^{\mathrm{v}}_{ijkl} = \sum_{\alpha=0}^{n} f^{\alpha} C^{\alpha}_{ijkl}$ is the Voigt solution

Minimum Complementary Potential Energy Principle

For a statistically admissible stress field $\boldsymbol{\sigma}^{*}(\mathbf{r})$, the complementary potential energy is expressed by

$$\begin{aligned}\Pi^{c}(\boldsymbol{\sigma}) &= \frac{1}{2}\iiint_{V} \sigma^{*}_{ij}(\mathbf{r})\varepsilon^{*}_{ij}(\mathbf{r})dV - \iint_{\partial V^{u}} u^{d}_{i}\sigma^{\mathrm{d}}_{ij}\sigma^{*}_{ij}(\mathbf{r})n_{j}dS \\ &= \frac{1}{2}\iiint_{V} m_{ijkl}(\mathbf{r})\sigma^{*}_{ij}(\mathbf{r})\sigma^{*}_{kl}(\mathbf{r})dV - \iint_{\partial V^{u}} u^{\mathrm{d}}_{i}\sigma^{*}_{ij}(\mathbf{r})n_{j}dS\end{aligned} \tag{7.237}$$

where $\mathbf{m}(\vec{r}) = \mathbf{l}^{-1}(\vec{r})$ is the local compliance tensor.

The stationary conditions $\delta\Pi^{c}(\boldsymbol{\sigma}^{*}) = 0$ of complementary energy is known as the virtual force principle in continuum mechanics, or the week form of compatibility condition in computational mechanics. Consider a virtual stress field $\delta\boldsymbol{\sigma}(\mathbf{r})$ with the boundary condition $\delta\boldsymbol{\sigma}(\mathbf{r}) = 0$ if $\mathbf{r} \in \partial V^{\mathrm{T}}$, the stationary concept reads

$$\delta\Pi^{c}(\boldsymbol{\sigma}^{*}) = \frac{1}{2}\iiint_{V} m_{ijkl}(\mathbf{r})\sigma^{*}_{ij}(\mathbf{r})\delta\sigma^{*}_{kl}(\mathbf{r})dV - \iint_{\partial V^{u}} u^{\mathrm{d}}_{i}\delta\sigma^{*}_{ij}(\mathbf{r})n_{j}dS = 0 \tag{7.238}$$

which can be rewritten as

$$\begin{aligned}\delta\Pi^{c}(\boldsymbol{\sigma}^{*}) = &\iiint_{V} \varepsilon^{*}_{ij}(\mathbf{r})\delta\sigma_{ij}(\mathbf{r})dV - \frac{1}{2}\iiint_{V}\left(u^{*}_{i,j}(\mathbf{r}) + u^{*}_{j,i}(\mathbf{r})\delta\sigma_{ij}(\mathbf{r})dV\right) \\ &+ \iiint_{V} u^{*}_{i,j}(\mathbf{r})\delta\sigma_{ij}(\mathbf{r})dV - \iint_{\partial V^{u}} u^{d}_{i}\delta\sigma_{ij}(\mathbf{r})n_{j}dS = 0\end{aligned} \tag{7.239}$$

Integration by parts with divergence theorem leads to

$$\delta\Pi^c(\boldsymbol{\sigma}) = \iiint\limits_V \left(\varepsilon^*_{ij}(\mathbf{r}) - \frac{1}{2}\left(u^*_{i,j}(\mathbf{r}) + u^*_{j,i}(\mathbf{r})\right)\right)\delta\sigma_{ij}(\mathbf{r})dV + \iint\limits_{\partial V^{\mathrm{u}}} u^*_i(\mathbf{r})\delta\sigma_{ij}(\mathbf{r})n_j dS - \iiint\limits_V u_i(\mathbf{r})\delta\sigma_{ij,j}(\mathbf{r})dV - \iint\limits_{\partial V^{\mathrm{u}}} u^{\mathrm{d}}_i\delta\sigma_{ij}(\mathbf{r})n_j dS = 0 \tag{7.240}$$

where $\iiint\limits_V u_i(\mathbf{r})\delta\sigma_{ij,j}(\mathbf{r})dV = 0$
since the field $\delta\boldsymbol{\sigma}(\mathbf{r})$ is statistically admissible.

Hence

$$\delta\Pi^c(\boldsymbol{\sigma}^*) = \iiint\limits_V \left(\varepsilon^*_{ij}(\mathbf{r}) - \frac{1}{2}\left(u^*_{i,j}(\mathbf{r}) + u^*_{j,i}(\mathbf{r})\right)\right)\delta\sigma_{ij}(\mathbf{r})dV + \iint\limits_{\partial V^{\mathrm{u}}} (u^*_i(\mathbf{r}) - u^{\mathrm{d}}_i)\delta\sigma_{ij}(\mathbf{r})n_j dS = 0 \tag{7.241}$$

which leads to the compatibility conditions

$$\varepsilon^*_{ij}(\mathbf{r}) - \frac{1}{2}\left(u^*_{i,j}(\mathbf{r}) + u^*_{j,i}(\mathbf{r})\right) \tag{7.242}$$

and the natural displacement boundary conditions

$$u^*_i(\mathbf{r}) = u^{\mathrm{d}}_i \text{ if } \mathbf{r} \in \partial V^{\mathrm{u}} \tag{7.243}$$

On the other hand, the extreme principle is shown by dealing with the perturbance of the complementary energy around an equilibrium position. That is

$$\begin{aligned}\Delta\Pi^c(\boldsymbol{\sigma}^*) &= \Pi^c(\boldsymbol{\sigma}^* + \delta\boldsymbol{\sigma}) - \Pi^c(\boldsymbol{\sigma}^*)\\ &= \frac{1}{2}\iiint\limits_V m_{ijkl}(\mathbf{r})\left(\sigma^*_{kl}(\mathbf{r}) + \delta\sigma_{kl}(\mathbf{r})\right)\left(\sigma^*_{ij}(\mathbf{r}) + \delta\sigma_{ij}(\mathbf{r})\right)dV\\ &\quad - \iint\limits_{\partial V^{\mathrm{u}}} u^{\mathrm{d}}_i\left(\sigma^*_{ij}(\mathbf{r}) + \delta\sigma_{ij}(\mathbf{r})\right)n_j dS\\ &\quad - \frac{1}{2}\iiint\limits_V m_{ijkl}(\mathbf{r})\sigma^*_{kl}(\mathbf{r})\sigma^*_{ij}(\mathbf{r})dV + \iint\limits_{\partial V^{\mathrm{u}}} u^{\mathrm{d}}_i\sigma^*_{ij}(\mathbf{r})n_j dS\\ &= \iiint\limits_V m_{ijkl}(\mathbf{r})\sigma^*_{kl}(\vec{r})\sigma^*_{ij}(\mathbf{r})dV - \iint\limits_{\partial V^{\mathrm{u}}} u^{\mathrm{d}}_i\delta\sigma_{kl}(\mathbf{r})n_j dS\\ &\quad + \frac{1}{2}\iiint\limits_V m_{ijkl}(\mathbf{r})\delta\sigma_{kl}(\mathbf{r})\delta\sigma_{ij}(\mathbf{r})dV\end{aligned} \tag{7.244}$$

or

$$\Delta\Pi^c(\boldsymbol{\sigma}^*) = \delta\Pi^c(\boldsymbol{\sigma}^*) + \delta^2\Pi^c \tag{7.245}$$

where

$$\delta^2\Pi^c = \frac{1}{2}\iiint_V m_{ijkl}(\mathbf{r})\delta\sigma_{kl}(\mathbf{r})\delta\sigma_{ij}(\mathbf{r})dV \tag{7.246}$$

Therefore, the stationary condition leads to $\Delta\Pi^c(\boldsymbol{\sigma}^*) = \delta^2\Pi^c > 0$. This means that for all statistically admissible fields $\boldsymbol{\sigma}^*(\mathbf{r})$, the equilibrium solution is the one minimizing the total complementary potential energy. In other words, the real solution in terms of statistically admissible stresses renders the complementary potential energy an absolute minimum. That is

$$\Pi^c(\boldsymbol{\sigma}) \leq \Pi^c(\boldsymbol{\sigma}^*) \text{ or } \Pi^c(\boldsymbol{\sigma}) = \inf_{\boldsymbol{\sigma}^*} \Pi^c(\boldsymbol{\sigma}^*) \tag{7.247}$$

If we consider an RVE with prescribed traction boundary conditions on its entire boundary ∂V, so that

$$\sigma^*_{ij}(\mathbf{r})n_j = t_i^{\mathrm{d}} = \sigma^{\mathrm{o}}_{ij}n_j \text{ if } \mathbf{r} \in \partial V\left(\partial V^{\mathrm{T}} = \partial V; \partial V^{\mathrm{u}} = \emptyset\right) \tag{7.248}$$

Then from (7.237), the complementary potential energy reads

$$\Pi^c(\boldsymbol{\sigma}^*) = \frac{1}{2}\iiint_V m_{ijkl}(\mathbf{r})\sigma^*_{kl}(\mathbf{r})\sigma^*_{kl}(\mathbf{r})dV = W^c(\boldsymbol{\sigma}^*)V \tag{7.249}$$

where

$$W^c(\boldsymbol{\sigma}^*) = \frac{1}{2V}\iiint_V m_{ijkl}(\mathbf{r})\sigma^*_{kl}(\mathbf{r})\sigma^*_{kl}(\mathbf{r})dV \tag{7.250}$$

is the macroscopic complementary elastic energy density.

Therefore, the minimum complementary energy principle is equivalent to

$$W^c(\boldsymbol{\sigma}) = \inf_{\boldsymbol{\sigma}^*} W^c(\boldsymbol{\sigma}^*) \tag{7.251}$$

Application: Reuss Bound

The purpose of this section is the deal with a special choice of statistically admissible stress field leading to one of possible upper bounds known as the simple Reuss solution for composite materials. For such a purpose, consider a inclusion-matrix composite with n phases, where the α−th phase has homogeneous elastic properties $\mathbf{L}^{\alpha}$.

For the real solution characterized by the kinematically and statistically fields $\mathbf{u}(\mathbf{r})$ and $\boldsymbol{\sigma}(\mathbf{r})$, respectively, the elastic energy density reads (Hill's Lemma)

$$W^{c}(\sigma)=\frac{1}{2}\iiint\limits_{V}\sigma_{ij}(\mathbf{r})\varepsilon_{ij}(\mathbf{r})dV=\frac{1}{2}\left(\iiint\limits_{V}\sigma_{ij}(\mathbf{r})dV\right)\left(\iiint\limits_{V}\varepsilon_{ij}(\mathbf{r})dV\right)=\frac{1}{2}\bar{\sigma}_{ij}\,\bar{\varepsilon}_{ij} \tag{7.252}$$

or

$$W^{c}(\boldsymbol{\sigma})=\frac{1}{2}\sigma^{\mathrm{o}}_{ij}\,\bar{L}_{ijkl}\,\sigma^{\mathrm{o}}_{kl} \tag{7.253}$$

Similarly

$$W^{c}(\boldsymbol{\sigma}^{*})=\frac{1}{2}\iiint\limits_{V}\sigma^{*}_{ij}(\mathbf{r})\varepsilon^{*}_{ij}(\mathbf{r})dV=\sigma^{\mathrm{o}}_{ij}\frac{1}{2V}\iiint\limits_{V}\varepsilon^{*}_{ij}(\mathbf{r})dV=\frac{1}{2}\sigma^{\mathrm{o}}_{ij}\sum_{\alpha=0}^{n}f^{\alpha}m^{\alpha}_{ijkl}\,\bar{\sigma}^{\alpha}_{kl} \tag{7.254}$$

If we choose in (7.234) $\bar{\boldsymbol{\sigma}}^{\alpha}=\boldsymbol{\sigma}^{\mathrm{o}}$ (homogeneous stress field, which is a statistically admissible field), the minimum energy principle (7.251) writes

$$\frac{1}{2}\sigma^{\mathrm{o}}_{ij}\bar{L}_{ijkl}\sigma^{\mathrm{o}}_{kl}\leq\frac{1}{2}\sigma^{\mathrm{o}}_{ij}\sum_{\alpha=0}^{n}f^{\alpha}L^{\alpha}_{ijkl}\sigma^{\mathrm{o}}_{kl} \tag{7.255}$$

or

$$\frac{1}{2}\sigma^{\mathrm{o}}_{ij}\bar{L}^{-1}_{ijkl}\sigma^{\mathrm{o}}_{kl}\leq\frac{1}{2}\sigma^{\mathrm{o}}_{ij}\sum_{\alpha=0}^{n}f^{\alpha}\left(L^{\alpha}_{ijkl}\right)^{-1}\sigma^{\mathrm{o}}_{kl} \tag{7.256}$$

which leads to

$$\left(\sum_{\alpha=0}^{n}f^{\alpha}\left(L^{\alpha}_{ijkl}\right)^{-1}\right)^{-1}\leq\bar{L}_{ijkl} \tag{7.257}$$

where $\bar{L}^{\mathrm{R}}_{ijkl}=\left(\sum\limits_{\alpha=0}^{n}f^{\alpha}\left(L^{\alpha}_{ijkl}\right)^{-1}\right)^{-1}$ is the Reuss solution for composite materials.

Finally combining Reuss and Voigt solution one has the following bounds for linear elastic properties of an n + 1-phases composite material

$$\left(\sum_{\alpha=0}^{n} f^{\alpha}\left(L_{ijkl}^{\alpha}\right)^{-1}\right)^{-1} \leq \bar{L}_{ijkl} \leq \sum_{\alpha=0}^{n} f^{\alpha} L_{ijkl}^{\alpha} \tag{7.258}$$

For isotropic materials one obtains respectively the following expressions for bulk and shear modulus

$$\frac{1}{\sum_{\alpha=0}^{n} \frac{f^{\alpha}}{K^{\alpha}}} \leq \bar{K} \leq \sum_{\alpha=0}^{n} f^{\alpha} K^{\alpha} \tag{7.259}$$

and

$$\frac{1}{\sum_{\alpha=0}^{n} \frac{f^{\alpha}}{\mu^{\alpha}}} \leq \bar{\mu} \leq \sum_{\alpha=0}^{n} f^{\alpha} \mu^{\alpha} \tag{7.260}$$

where the Voigt bound could be seen as an arithmetic average whereas the Reuss bound could be viewed as an harmonic average. It should be noticed that these bounds do not take into account any interaction between the phases. In addition, in the case of high contrast between the phases in terms of their elastic properties, the Voigt and Reuss solutions give large bounding of the composite overall elastic properties. Therefore, Voigt and Reuss bounds provide a restrictive utility for practical situations.

7.7.2 Hashin-Shtrikman Variational Principles

The Hashin-Shtrikman variation principle provides a powerful tool to narrow the gap between the Reuss bound and the Voigt bound. It is based on the principal of polarization previously introduced by Hill (see previous sections) in linear elasticity to describe the fluctuation of elastic constant by a stress polarization tensor. That is

$$\sigma_{ij}(\mathbf{r}) = l_{ijkl}(\mathbf{r})\varepsilon_{kl}(\mathbf{r}) = L_{ijkl}^{o}\,\varepsilon_{kl}(\mathbf{r}) + \left(l_{ijkl}(\mathbf{r}) - L_{ijkl}^{o}\right)\varepsilon_{kl}(\mathbf{r}) \tag{7.261}$$

where $\mathbf{L}^{o}$ describes the elastic constant of a reference homogeneous medium.

(7.261) can be written as

$$\sigma_{ij}(\mathbf{r}) = C_{ijkl}^{o}\,\varepsilon_{kl}(\mathbf{r}) + p_{ij}(\mathbf{r}) \tag{7.262}$$

where

$$p_{ij}(\mathbf{r}) = \left(l_{ijkl}(\mathbf{r}) - L_{ijkl}^{o}\right)\varepsilon_{kl}(\mathbf{r}) = \Delta l_{ijkl}(\mathbf{r})\varepsilon_{kl}(\mathbf{r}) \tag{7.263}$$

The Hashin-Shtrikman variational principal deals with two boundary problems:

1. The real composite material with trial fields $\mathbf{u}^*(\mathbf{r}), \boldsymbol{\varepsilon}^*(\mathbf{r})$, and $\boldsymbol{\sigma}^*(\mathbf{r})$ fulfilling the following conditions

$$\begin{cases} \varepsilon^*_{ij}(\mathbf{r}) = \frac{1}{2}\left(u^*_{j,i}(\mathbf{r}) + u^*_{j,i}(\mathbf{r})\right) \text{within } V \\ \mathbf{u}^* \text{ is continous in } V \\ \mathbf{u}^* = \mathbf{u}^{\mathrm{d}} \text{ on } \partial V^{\mathrm{u}} = \partial V(\partial V^{\mathrm{T}} = \emptyset) \\ \boldsymbol{\sigma}^*_{ij,j}(\mathbf{r}) = 0 \text{ and } \boldsymbol{\sigma}^*_{ij}(\mathbf{r}) = L^{\mathrm{o}}_{ijkl}\boldsymbol{\varepsilon}^*_{kl}(\mathbf{r}) + p^*_{ij}(\mathbf{r}) \end{cases} \tag{7.264}$$

The last conditions in (7.264) is equivalent to

$$L^{\mathrm{o}}_{ijkl}u^*_{kl,j}(\mathbf{r}) + p^*_{ij,j}(\mathbf{r}) = 0 \tag{7.265}$$

2. A comparison homogeneous solid with properties $\mathbf{L}^{\mathrm{o}}$ and therefore without any polarization

$$\begin{cases} \bar{\varepsilon}_{ij} = \frac{1}{2}\left(\bar{u}_{i,j} + \bar{u}_{j,i}\right) \text{ within } V \\ \bar{\mathbf{u}} \text{ is continous in } V \\ \bar{\mathbf{u}} = \mathbf{u}^{\mathrm{d}} \text{ on } \partial V^{\mathrm{u}} = \partial V(\partial V^{\mathrm{T}} = \emptyset) \\ \bar{\sigma}_{ij,j} = 0 \text{ and } \bar{\sigma}_{ij} = L^{\mathrm{o}}_{ijkl}\bar{\varepsilon}_{kl} \end{cases} \tag{7.266}$$

If we denote $W(\mathbf{u})$ and $W(\bar{\mathbf{u}})$ the elastic energy densities of the real composite and the comparison composite, respectively, the main purpose of Hashin-Shtrikman variational principle is to set up a lower and upper bounds for the difference $W(\mathbf{u}) - W(\bar{\mathbf{u}})$, where we denote by $\mathbf{u}$ the real solution of the problem.

For such a purpose, Hashin-Shtrikman introduced a functional $HS(\boldsymbol{\varepsilon}^*, \mathbf{p}^*)$ expressed by

$$\begin{aligned} HS(\boldsymbol{\varepsilon}^*, \mathbf{p}^*) =& \frac{1}{2}\iiint_V \Big(L^{\mathrm{o}}_{ijkl}\,\bar{\varepsilon}_{ij}\,\bar{\varepsilon}_{kl} - \Delta l^{-1}_{ijkl}(\mathbf{r})p^*_{ij}(\mathbf{r})p^*_{kl}(\mathbf{r}) \\ &+ p^*_{ij}(\mathbf{r})\tilde{\boldsymbol{\varepsilon}}(\mathbf{r}) + 2p^*_{ij}(\mathbf{r})\bar{\varepsilon}_{ij}\Big)dV \end{aligned} \tag{7.267}$$

where the following decomposition is introduced

$$\begin{aligned} &\tilde{\varepsilon}_{ij}(\mathbf{r}) = \varepsilon^*_{ij}(\mathbf{r}) - \bar{\varepsilon}_{ij} \\ &\tilde{u}_i(\mathbf{r}) = u^*_i(\mathbf{r}) - \bar{u}_i \text{ with } \tilde{u}_i(\mathbf{r}) = 0 \text{ if } \mathbf{r} \in \partial V \end{aligned} \tag{7.268}$$

Hashin-Shtrikman's principle states the following:

With the condition (7.265) which states the equilibrium in terms of polarization, the functional

- $HS(\boldsymbol{\varepsilon}^*, \mathbf{p}^*)$ is stationary, i.e., $\delta HS(\boldsymbol{\varepsilon}^*, \mathbf{p}^*) = 0$,
- $\delta^2 HS(\boldsymbol{\varepsilon}^*, \mathbf{p}^*) > 0$, if $\Delta \mathbf{l} < 0$, $HS(\boldsymbol{\varepsilon}^*, \mathbf{p}^*)$, $HS(\boldsymbol{\varepsilon}^*, \mathbf{p}^*)$ has a minimum,
- $\delta^2 HS(\boldsymbol{\varepsilon}^*, \mathbf{p}^*) < 0$, if $\Delta \mathbf{l} > 0$, $HS(\boldsymbol{\varepsilon}^*, \mathbf{p}^*)$ has a maximum.

The proof of such statements is straightforward developed in the following.

Let first determine the perturbance of $HS(\boldsymbol{\varepsilon}^*, \mathbf{p}^*)$ with respect to a virtual strain and polarization fluctuation $\delta\boldsymbol{\varepsilon}$ and $\delta\mathbf{p}$

$$\begin{aligned}\Delta HS(\boldsymbol{\varepsilon}^*, \mathbf{p}^*) &= HS(\boldsymbol{\varepsilon}^* + \delta\boldsymbol{\varepsilon}, \mathbf{p}^* + \delta\mathbf{p}) - HS(\boldsymbol{\varepsilon}^*, \mathbf{p}^*)\\ &= \frac{1}{2}\iiint_V \Big(-2\Delta l^{-1}_{ijkl}(\mathbf{r}) p^*_{ij}(\mathbf{r}) \delta p_{kl}(\mathbf{r}) + p^*_{ij}(\mathbf{r}) \delta\tilde{\varepsilon}_{ij}(\mathbf{r})\\ &\quad + \delta p_{ij}(\mathbf{r}) \tilde{\varepsilon}_{ij}(\mathbf{r}) + 2\delta p_{ij}(\mathbf{r}) \bar{\varepsilon}_{ij}\Big) dV\\ &\quad + \frac{1}{2}\iiint_V \Big(-\Delta l^{-1}_{ijkl}(\mathbf{r}) \delta p_{ij}(\mathbf{r}) \delta p_{kl}(\mathbf{r}) + \delta p_{ij}(\mathbf{r}) \delta\tilde{\varepsilon}_{ij}(\mathbf{r})\Big) dV\\ &= \delta HS(\boldsymbol{\varepsilon}^*, \mathbf{p}^*) + \delta^2 HS\end{aligned} \tag{7.269}$$

Now examine $\delta HS(\boldsymbol{\varepsilon}^*, \mathbf{p}^*)$, which can be written as

$$\begin{aligned}\delta HS(\boldsymbol{\varepsilon}^*, \mathbf{p}^*) = -\frac{1}{2}\iiint_V &\Big(2\Delta l^{-1}_{ijkl}(\mathbf{r}) p^*_{ij}(\mathbf{r}) \delta p_{kl}(\mathbf{r}) - 2\bar{\varepsilon}_{ij}\, \delta p_{ij}(\mathbf{r})\\ &- \tilde{\varepsilon}_{ij}(\mathbf{r}) \delta p_{ij}(\mathbf{r}) - p^*_{ij}(\mathbf{r}) \delta\tilde{\varepsilon}_{ij}(\mathbf{r})\Big)\, dV\end{aligned} \tag{7.270}$$

or

$$\begin{aligned}\delta HS(\varepsilon^*, p^*) = -\frac{1}{2}\iiint_V &\Big(2\Delta l^{-1}_{ijkl}(\mathbf{r}) p^*_{ij}(\mathbf{r}) \delta p_{kl}(\mathbf{r}) - 2\Big(\varepsilon^*_{ij}(\mathbf{r}) - \tilde{\varepsilon}_{ij}(\mathbf{r})\Big) \delta p_{ij}(\mathbf{r})\\ &- \tilde{\varepsilon}_{ij}(\mathbf{r}) \delta p_{ij}(\mathbf{r}) - p^*_{ij}(\mathbf{r}) \delta\tilde{\varepsilon}_{ij}(\mathbf{r})\Big)\, dV\end{aligned} \tag{7.271}$$

In addition, (7.271) can be rewritten as

$$\begin{aligned}\delta HS(\boldsymbol{\varepsilon}^*, \mathbf{p}^*) = -\frac{1}{2}\iiint_V &\Big(2\Big(\Delta l^{-1}_{ijkl}(\mathbf{r}) p^*_{ij}(\mathbf{r}) - \varepsilon^*_{ij}(\mathbf{r})\Big) \delta p_{kl}(\mathbf{r})\\ &+ \tilde{\varepsilon}_{ij}(\mathbf{r}) \delta p_{ij}(\mathbf{r}) - p^*_{ij}(\mathbf{r}) \delta\tilde{\varepsilon}_{ij}(\mathbf{r})\Big) dV\end{aligned} \tag{7.272}$$

Finally it results from the definition (7.263) of polarization which is equivalent to

$$\Delta l^{-1}_{ijkl}(\mathbf{r})p^*_{ij}(\mathbf{r}) - \varepsilon^*_{ij}(\mathbf{r}) = 0 \tag{7.273}$$

that

$$\delta HS(\boldsymbol{\varepsilon}^*, \mathbf{p}^*) = -\frac{1}{2}\iiint\limits_V \left(\tilde{\varepsilon}_{ij}(\mathbf{r})\delta p_{ij}(\mathbf{r}) - p^*_{ij}(\mathbf{r})\delta\tilde{\varepsilon}_{ij}(\mathbf{r})\right) dV \tag{7.274}$$

On the other hand, from the equilibrium condition

$$L^{\text{o}}_{ijkl}\,\tilde{\varepsilon}_{kl,j}(\mathbf{r}) + p^*_{ij,j}(\mathbf{r}) = 0$$

which is can be reorganized as

$$t_{ij}(\mathbf{r}) = C^{\text{o}}_{ijkl}\,\tilde{\varepsilon}_{kl}(\mathbf{r}) + p^*_{ij}(\mathbf{r}) \text{ with } t_{ij,j}(\mathbf{r}) = 0 \tag{7.275}$$

or by introducing the virtual field

$$\delta t_{ij}(\mathbf{r}) = C^{\text{o}}_{ijkl}\,\delta\tilde{\varepsilon}_{kl}(\mathbf{r}) + \delta p_{ij}(\mathbf{r}) \text{ with } \delta t_{ij,j}(\mathbf{r}) = 0 \tag{7.276}$$

Substituting (7.276) into (7.274) gives

$$\begin{aligned}\delta HS(\boldsymbol{\varepsilon}^*, \mathbf{p}^*) = -\frac{1}{2}\iiint\limits_V &\left(\tilde{\varepsilon}_{ij}(\mathbf{r})\left(\delta t_{ij}(\mathbf{r}) - L^{\text{o}}_{ijkl}\;\delta\;\tilde{\varepsilon}_{kl}(\vec{r})\right)\right.\\ &- \left.\left(t_{ij}(\mathbf{r}) - L^{\text{o}}_{ijkl}\,\tilde{\varepsilon}_{kl}(\mathbf{r})\right)\delta\tilde{\varepsilon}_{ij}(\mathbf{r})\right) dV\end{aligned} \tag{7.277}$$

or

$$\begin{aligned}\delta HS(\boldsymbol{\varepsilon}^*, \mathbf{p}^*) = -\frac{1}{2}\iiint\limits_V &\left(\left(\tilde{\varepsilon}_{ij}(\mathbf{r})\delta t_{ij}(\mathbf{r}) - t_{ij}(\mathbf{r})\delta\tilde{\varepsilon}_{ij}(\mathbf{r})\right)\right.\\ &\left.-\tilde{\varepsilon}_{ij}(\mathbf{r})L^{\text{o}}_{ijkl}\;\delta\;\tilde{\varepsilon}_{kl}(\mathbf{r}) + \delta\;\tilde{\varepsilon}_{ij}(\mathbf{r})L^{\text{o}}_{ijkl}\tilde{\varepsilon}_{kl}(\mathbf{r})\right) dV\end{aligned} \tag{7.278}$$

which is reduced to

$$\begin{aligned}\delta HS(\boldsymbol{\varepsilon}^*, \mathbf{p}^*) &= -\frac{1}{2}\iiint\limits_V \left(\tilde{\varepsilon}_{ij}(\mathbf{r})\;\delta\;t_{ij}(\mathbf{r}) - t_{ij}(\mathbf{r})\;\delta\;\tilde{\varepsilon}_{ij}(\mathbf{r})\right) dV\\ &= -\frac{1}{2}\iiint\limits_V \left(\tilde{u}_{i,j}(\mathbf{r})\;\delta\,t_{ij}(\mathbf{r}) - t_{ij}(\mathbf{r})\;\delta\;u_{i,j}(\mathbf{r})\right) dV\end{aligned} \tag{7.279}$$

or by partial derivative using the divergence theorem, one has

$$\begin{aligned}\delta HS(\boldsymbol{\varepsilon}^*,\mathbf{p}^*) =&\frac{1}{2}\iiint\limits_V \left(\tilde{u}_i(\mathbf{r})\delta t_{ij,j}(\mathbf{r}) - t_{ij,j}(\mathbf{r})\delta u_i(\mathbf{r})\right)dV \\ &-\frac{1}{2}\iint\limits_{\partial V} \left(\tilde{u}_i(\mathbf{r})\delta t_{ij}(\mathbf{r}) - t_{ij}(\mathbf{r})\delta u_i(\mathbf{r})\right)n_j dS = 0\end{aligned} \tag{7.280}$$

since

$$\tilde{u}_i(\mathbf{r}) = \delta u_i(\mathbf{r}) = 0 \text{ when } r \in \partial V \text{ and } t_{ij,j}(\mathbf{r}) = \delta t_{ij,j}(\mathbf{r}) = 0$$

Hence the stationary condition $\delta HS(\boldsymbol{\varepsilon}^*,\mathbf{p}^*) = 0$ of Hashin-Shtrikman's functional is proved. Now we examine the extremum condition by analysing $\delta^2 HS(\boldsymbol{\varepsilon}^*,\mathbf{p}^*)$, which is expressed by

$$\delta^2 HS = \frac{1}{2}\iiint\limits_V \left(-\Delta l^{-1}_{ijkl}(\mathbf{r})\delta p_{ij}(\mathbf{r})\delta p_{kl}(\mathbf{r}) + \delta p_{ij}(\mathbf{r})\delta\varepsilon_{ij}(\mathbf{r})\right) dV \tag{7.281}$$

Substituting (7.276) into (7.281) leads to

$$\begin{aligned}\delta^2 HS =&\frac{1}{2}\iiint\limits_V \Big(-\Delta l^{-1}_{ijkl}(\mathbf{r})\delta p_{ij}(\mathbf{r})\delta p_{kl}(\mathbf{r}) \\ &-L^{\mathrm{o}}_{ijkl}\,\delta\tilde{\varepsilon}_{kl}(\mathbf{r})\delta\varepsilon_{ij}(\mathbf{r}) + \delta t_{ij}(\mathbf{r})\delta\varepsilon_{ij}(\mathbf{r})\Big) dV\end{aligned} \tag{7.282}$$

where

$$\begin{aligned}&\iiint\limits_V \delta t_{ij}(\mathbf{r})\delta\varepsilon_{ij}(\mathbf{r})\, dV = \iiint\limits_V \delta t_{ij}(\mathbf{r})\delta u_{i,j}(\mathbf{r})\, dV \\ &= \iint\limits_{\partial V} \delta t_{ij}(\mathbf{r})n_j\delta u_i(\mathbf{r})dS - \iiint\limits_V \delta t_{ij,j}(\mathbf{r})\delta u_i(\mathbf{r})\, dV = 0\end{aligned} \tag{7.283}$$

Therefore (7.283) is reduced to

$$\delta^2 HS = \left(-\frac{1}{2}\right)\iiint\limits_V \left(\Delta l^{-1}_{ijkl}(\mathbf{r})\delta p_{ij}(\mathbf{r})\delta p_{kl}(\mathbf{r}) + L^{\mathrm{o}}_{ijkl}\,\delta\varepsilon_{kl}(\mathbf{r})\delta\varepsilon_{ij}(\mathbf{r})\right) dV \tag{7.284}$$

Clearly, (7.284) shows that if $\Delta\mathbf{l} > 0$, $\Delta HS(\boldsymbol{\varepsilon}^*,\mathbf{p}^*) = \delta^2 HS < 0$. Therefore, $HS(\boldsymbol{\varepsilon}^*,\mathbf{p}^*)$ achieves a maximum value. However, if $\Delta\mathbf{l} < 0$ the judgment is not systematic.

To clarify the condition under which $\Delta\mathbf{l} < 0$ we consider the following positive integral

$$T=\iiint_V L^{o-1}_{ijkl}(\mathbf{r})\delta p_{ij}(\mathbf{r})\delta p_{kl}(\mathbf{r})dV \tag{7.285}$$

and if we substitute $\delta p_{ij}(\mathbf{r})=\delta t_{ij}(\mathbf{r})-L^{o}_{ijkl}(\mathbf{r})\delta\varepsilon_{kl}(\mathbf{r})$ into it, it could be easily shown that

$$T=\iiint_V \left(L^{o-1}_{ijkl}(\mathbf{r})\delta t_{ij}(\mathbf{r})\delta t_{kl}(\mathbf{r})-2\delta t_{ij}(\mathbf{r})\delta\varepsilon_{ij}(\mathbf{r})+L^{o}_{ijkl}(\mathbf{r})\delta\varepsilon_{ij}(\mathbf{r})\delta\varepsilon_{kl}(\mathbf{r})\right)dV \tag{7.286}$$

and thanks to (7.283) , it results

$$T=\iiint_V \left(L^{o-1}_{ijkl}(\mathbf{r})\delta t_{ij}(\mathbf{r})\delta t_{kl}(\mathbf{r})+C^{o}_{ijkl}(\mathbf{r})\delta\varepsilon_{ij}(\mathbf{r})\delta\varepsilon_{kl}(\mathbf{r})\right)dV \tag{7.287}$$

Therefore, one can conclude from (7.285) and (7.287), that

$$\iiint_V L^{o-1}_{ijkl}(\mathbf{r})\delta p_{ij}(\mathbf{r})\delta p_{kl}(\mathbf{r})dV>\iiint_V L^{o}_{ijkl}(\mathbf{r})\,\delta\varepsilon_{ij}(\mathbf{r})\delta\varepsilon_{kl}(\mathbf{r})dV \tag{7.288}$$

which leads to the following inequality

$$\begin{aligned}\delta^2 HS(\boldsymbol{\varepsilon}^*,\mathbf{p}^*)&=\left(-\frac{1}{2}\right)\iiint_V\left(\Delta l^{-1}_{ijkl}(\mathbf{r})\delta p_{ij}(\mathbf{r})\delta p_{kl}(\mathbf{r})+L^{o}_{ijkl}\,\delta\varepsilon_{kl}(\mathbf{r})\delta\varepsilon_{ij}(\mathbf{r})\right)dV\\ &\geq\left(-\frac{1}{2}\right)\iiint_V\left(\Delta l^{-1}_{ijkl}(\mathbf{r})+L^{o-1}_{ijkl}\right)\delta p_{ij}(\mathbf{r})\delta p_{kl}(\mathbf{r})dV\end{aligned} \tag{7.289}$$

On the other hand, we can readily show that $\Delta\mathbf{l}^{-1}+\mathbf{L}^{o-1}=\mathbf{L}^{o-1}:\mathbf{l}:\Delta\mathbf{l}^{-1}$ and therefore it results from (7.289)

$$\delta^2 HS(\boldsymbol{\varepsilon}^*,\mathbf{p}^*)\geq\left(-\frac{1}{2}\right)\iiint_V \delta\mathbf{p}(\mathbf{r}):\mathbf{L}^{o-1}:\mathbf{l}(\mathbf{r}):\delta\mathbf{p}(\mathbf{r})dV \tag{7.290}$$

Now the analysis becomes straightforward. It results from (7.290) that if $\Delta\mathbf{l}<0$, $\delta^2 HS(\boldsymbol{\varepsilon}^*,\mathbf{p}^*)>0$ and therefore $HS(\boldsymbol{\varepsilon}^*,\mathbf{p}^*)$ has a minimum.

In conclusion, we can state the Hashin-Shtrikman's extremum principle as

$\delta^2 HS(\boldsymbol{\varepsilon}^*,\mathbf{p}^*)>0$, if $\Delta\mathbf{l}<0$, $HS(\boldsymbol{\varepsilon}^*,\mathbf{p}^*)$ has a minimum,

$\delta^2 HS(\boldsymbol{\varepsilon}^*,\mathbf{p}^*)<0$, if $\Delta\mathbf{l}<0$, $HS(\boldsymbol{\varepsilon}^*,\mathbf{p}^*)$ has a maximum.

Now to get bounds for the strain energy density difference $W(\mathbf{u}) - W(\bar{\mathbf{u}})$ between the real composite and a homogeneous comparison composite, we calculate successively the following integrals

$$\begin{aligned}\iiint_V \sigma^*_{ij}(\mathbf{r})\tilde{\varepsilon}(\mathbf{r})dV &= \iiint_V \sigma^*_{ij}(\mathbf{r})\tilde{u}_{i,j}(\mathbf{r})dV \\ &= \iint_{\partial V} \sigma^*_{ij}(\mathbf{r})n_j\tilde{u}_i(\mathbf{r})dS - \iiint_V \sigma^*_{ij,j}(\tilde{r})\tilde{u}_i(\mathbf{r})dV = 0\end{aligned} \tag{7.291}$$

Similarly,

$$\iiint_V \bar{\sigma}_{ij}\,\tilde{\varepsilon}_{ij}(\mathbf{r})dV = 0 \tag{7.292}$$

Therefore, the total potential energy of a kinematically admissible field $\mathbf{u}^*$ under prescribed displacement boundary conditions reads

$$\begin{aligned}\Pi(\mathbf{u}^*) &= \frac{1}{2}\iiint_V \sigma^*_{ij}(\mathbf{r})\varepsilon^*_{ij}(\mathbf{r})dV = \frac{1}{2}\iiint_V \left(\sigma^*_{ij}(\mathbf{r})\varepsilon^*_{ij}(\mathbf{r}) - \sigma^*_{ij}(\mathbf{r})\tilde{\varepsilon}_{ij}(\mathbf{r})\right)dV \\ &= \frac{1}{2}\iiint_V \sigma^*_{ij}(\mathbf{r})\left(\varepsilon^*_{ij}(\mathbf{r}) - \tilde{\varepsilon}_{ij}(\mathbf{r})\right)dV = \frac{1}{2}\iiint_V \sigma^*_{ij}(\mathbf{r})\bar{\varepsilon}_{ij}dV\end{aligned} \tag{7.293}$$

where

$$\begin{aligned}\boldsymbol{\sigma}^*_{ij}(\mathbf{r})\bar{\varepsilon}_{ij} &= \left(p^*_{ij}(\mathbf{r}) + L^{\mathrm{o}}_{ijkl}\,\varepsilon^*_{kl}(\mathbf{r})\right)\bar{\varepsilon}_{ij} \\ &= L^{\mathrm{o}}_{ijkl}\,\varepsilon^*_{kl}(\mathbf{r})\bar{\varepsilon}_{ij} + p^*_{ij}(\mathbf{r})\bar{\varepsilon}_{ij} + p^*_{ij}(\mathbf{r})\bar{\varepsilon}_{ij} - p^*_{ij}(\mathbf{r})\bar{\varepsilon}_{ij} \\ &= L^{\mathrm{o}}_{ijkl}\,\bar{\varepsilon}_{ij}\bar{\varepsilon}_{kl} + L^{\mathrm{o}}_{ijkl}\bar{\varepsilon}_{ij}\tilde{\varepsilon}_{kl}(\mathbf{r}) + 2p^*_{ij}(\mathbf{r})\bar{\varepsilon}_{ij} + p^*_{ij}(\mathbf{r})\varepsilon^*_{kl}(\mathbf{r}) - p^*_{ij}(\mathbf{r})\tilde{\varepsilon}_{kl}(\mathbf{r}) \\ &= L^{\mathrm{o}}_{ijkl}\,\bar{\varepsilon}_{ij}\bar{\varepsilon}_{kl} + \bar{\sigma}_{ij}\tilde{\varepsilon}_{ij}(\mathbf{r}) + 2p^*_{ij}(\mathbf{r})\bar{\varepsilon}_{ij} - p^*_{ij}(\mathbf{r})\varepsilon^*_{kl}(\mathbf{r}) + p^*_{ij}(\mathbf{r})\tilde{\varepsilon}_{kl}(\mathbf{r})\end{aligned} \tag{7.294}$$

Substituting (7.294) into (7.293) and by taking into account (7.292), one has

$$\begin{aligned}\Pi(\mathbf{u}) &= \frac{1}{2}\iiint_V \left(L^{\mathrm{o}}_{ijkl}\bar{\varepsilon}_{ij}\bar{\varepsilon}_{kl} - \Delta l^{-1}_{ijkl}(\mathbf{r})p^*_{ij}(\mathbf{r})p^*_{kl}(\mathbf{r}) + p^*_{ij}(\mathbf{r})\tilde{\varepsilon}_{kl}(\mathbf{r}) + 2p^*_{ij}(\mathbf{r})\right)dV \\ &= W(\mathbf{u})V = W(\mathbf{u})V + HS(\boldsymbol{\varepsilon}^*, \mathbf{p}^*)V\end{aligned} \tag{7.295}$$

where

$$W(\bar{\mathbf{u}}) = L^{\mathrm{o}}_{ijkl}\,\bar{\varepsilon}_{ij}\,\bar{\varepsilon}_{kl} \tag{7.296}$$

Hence

$$W(\mathbf{u}^*) - W(\bar{\mathbf{u}}) = HS(\boldsymbol{\varepsilon}^*, \mathbf{p}^*) \tag{7.297}$$

7.7.3 *Application: Hashin-Shtrikman Bounds for Linear Elastic Effective Properties*

The purpose of this section is to use the Hashin-Shtrikman extremum principle to get lower and upper bounds for effective properties of a linear elastic inclusion-matrix composite.

Consider an RVE with multiple phases, $\alpha = 0, 1, 2, \ldots, n$. The elastic tensors and compliance tensors in the phases are denotes by $\mathbf{L}^{\alpha}$ and $\mathbf{M}^{\alpha}$ where $\alpha = 0, 1, 2, \ldots, n$. V^{α} is the volumes of the inhomogeneity α, and V is the volume of the RVE. Further, we denote by $f^{\alpha} = V^{\alpha}/V$ the volume fractions of the α-phase.

Consider now the real and comparison composites (of course the associated RVE) subjected to displacement boundary conditions

$$u_i(\mathbf{r}) = \varepsilon^{o}_{ij} x_j \text{ when } \mathbf{r} \in \partial V \text{ and } \bar{u}_i = \varepsilon^{o}_{ij} x_j \text{ when } \mathbf{r} \in \partial V \tag{7.298}$$

The Hashin-Shtrikman extremum principle reads

$$HS^{-}(\boldsymbol{\varepsilon}^{*}, \mathbf{p}^{*}) \leq W(\mathbf{u}) - W(\bar{\mathbf{u}}) \leq HS^{+}(\boldsymbol{\varepsilon}^{*}, \mathbf{p}^{*}) \tag{7.299}$$

or

$$HS^{-}(\boldsymbol{\varepsilon}^{*}, \mathbf{p}^{*}) + W(\bar{\mathbf{u}}) \leq W(\mathbf{u}) \leq HS^{+}(\boldsymbol{\varepsilon}^{*}, \mathbf{p}^{*}) + W(\bar{\mathbf{u}}) \tag{7.300}$$

where the minimum $HS^{-}(\boldsymbol{\varepsilon}^{*}, \mathbf{p}^{*})$ corresponds to the case where $\Delta\mathbf{l}(\mathbf{r}) > 0$ and the maximum $HS^{+}(\boldsymbol{\varepsilon}^{*}, \mathbf{p}^{*})$ where $\Delta\mathbf{l}(\mathbf{r}) < 0$

Recall that

$$HS(\boldsymbol{\varepsilon}^{*}, \mathbf{p}^{*}) = -\frac{1}{2V} \iiint_V \left(\Delta l^{-1}_{ijkl}(\mathbf{r}) p^{*}_{ij}(\mathbf{r}) p^{*}_{kl}(\mathbf{r}) - p^{*}_{ij}(\mathbf{r}) \tilde{\varepsilon}_{kl}(\mathbf{r}) - 2 p^{*}_{ij}(\mathbf{r}) \bar{\varepsilon}_{ij} \right) dV \tag{7.301}$$

which can be written as

$$HS(\boldsymbol{\varepsilon}^{*}, \mathbf{p}^{*}) = I_1 + I_2 + I_3 \tag{7.302}$$

where

$$\begin{aligned} I_1 &= -\frac{1}{2V} \iiint_V \Delta l^{-1}_{ijkl}(\mathbf{r}) p^{*}_{ij}(\mathbf{r}) p^{*}_{kl}(\mathbf{r}) dV \\ I_2 &= \frac{1}{V} \iiint_V p^{*}_{ij}(\mathbf{r}) \bar{\varepsilon}_{ij} dV \\ I_3 &= \frac{1}{2V} \iiint_V p^{*}_{ij}(\mathbf{r}) \tilde{\varepsilon}_{kl}(\mathbf{r}) dV \end{aligned} \tag{7.303}$$

We assume that the polarization tensor is piecewise uniform so that one can write

$$\mathbf{p}^*(\mathbf{r}) = \sum_{\alpha=1}^{n} \mathbf{p}^{*\alpha}\theta^\alpha(\mathbf{r}) \tag{7.304}$$

We now determine successively the integrals I_1, I_2 and I_3. For simplicity the indices are omitted in few developments. With the property of piecewise uniform polarizations, I_1 is expressed by

$$\begin{aligned} I_1 &= -\frac{1}{2V}\iiint_V \Delta l^{-1}_{ijkl}(\mathbf{r})p^*_{ij}(\mathbf{r})p^*_{kl}(\mathbf{r})dV \\ &= -\frac{1}{2V}\iiint_V \mathbf{p}^*(\mathbf{r}) : \Delta\mathbf{l}^{-1}(\mathbf{r}) : \mathbf{p}^*(\mathbf{r})dV \\ &= -\frac{1}{2V}\sum_{\alpha=0}^{n}\iiint_V \mathbf{p}^{*\alpha} : \Delta\mathbf{L}^{\alpha^{-1}} : \mathbf{p}^{*\alpha}dV = \frac{1}{2}\sum_{\alpha=0}^{n} f^\alpha \mathbf{p}^{*\alpha} : \Delta\mathbf{L}^{\alpha^{-1}} : \mathbf{p}^{*\alpha} \end{aligned} \tag{7.305}$$

where

$$\Delta\mathbf{L}^{\alpha^{-1}} = (\mathbf{L}^\alpha - \mathbf{L}^o)^{-1}$$

Similarly, I_2 reads

$$I_2 = \frac{1}{V}\iiint_V p^*_{ij}(\mathbf{r})\bar{\varepsilon}_{ij}dV = \left(\frac{1}{V}\iiint_V \mathbf{p}^*(\mathbf{r})dV\right) : \bar{\varepsilon} = \sum_{\alpha=1}^{n} f^\alpha \mathbf{p}^{*\alpha} : \bar{\varepsilon} \tag{7.306}$$

while I_3 can be easily shown to be as

$$I_3 = \frac{1}{2V}\iiint_V p^*_{ij}(\mathbf{r})\tilde{\varepsilon}_{kl}(\mathbf{r})dV = -\frac{1}{2}\sum_{\alpha=1}^{n} f^\alpha \mathbf{p}^{*\alpha} : \mathbf{P}^\alpha : (\mathbf{p}^{*\alpha} - \bar{\mathbf{p}}^*) \tag{7.307}$$

where

$$\bar{\mathbf{p}}^* = \sum_{\alpha=0}^{n} f^\alpha \mathbf{p}^{*\alpha} \tag{7.308}$$

and $\mathbf{P}^\alpha$ the fourth-order polarization tensor initially introduced by Hill and discussed in the previous sections. Recall that $\mathbf{P}^\alpha$ is expressed in terms of the modified Green's function $\mathbf{\Gamma}$ of the infinite reference medium $\mathbf{L}^o$ as

$$\mathbf{P}^\alpha := \iiint_{V^\alpha} \Gamma(\mathbf{r} - \mathbf{r}')dV' \tag{7.309}$$

Note that $\mathbf{P}^{\alpha}$ depends on the shape of the α -th phase and the elastic constant $\mathbf{L}^{o}$ of the reference medium. Furthermore, Equation (7.309) expresses that $\mathbf{P}^{\alpha}$ is uniform, which is true in the considered case of ellipsoidal inclusions. As shown in the in Section 7.2, $\mathbf{P}^{\alpha}$ could be given as function of the Eshelby's tensor as

$$\mathbf{P}^{\alpha} = S^{\alpha} : \mathbf{L}^{o^{-1}} \tag{7.310}$$

To this end, we have all the ingredients to establish Hashin-Shtrikman bounds, which initially were developed in the case of two-phase isotropic composite materials and spherical inclusions. Under such conditions, the polarization tensor is deduced calculated from (7.310) as

$$\begin{aligned} S^{\alpha}_{ijkl} &= s^{o}_{1} E^{1}_{ijkl} + s^{o}_{2} E^{2}_{ijkl} \text{ with } s^{o}_{1} = \frac{1+\nu^{o}}{3(1-\nu^{o})} \text{ and } s^{o}_{2} = \frac{2(4-5\nu^{o})}{15(1-\nu^{o})} \\ L^{o^{-1}}_{ijkl} &= \frac{1}{3\,K^{o}} E^{1}_{ijkl} + \frac{1}{2\mu^{o}} E^{2}_{ijkl} \end{aligned} \tag{7.311}$$

Hence

$$P^{\alpha}_{ijkl} = \frac{1}{3\,K^{o} + 4\mu^{o}} E^{1}_{ijkl} + \frac{3(K^{o} + 2\mu^{o})}{5\mu^{o}(3\,K^{o} + 4\mu^{o})} E^{2}_{ijkl} \tag{7.312}$$

where ν^{o} is substituted by the relation

$$\nu^{o} = \frac{3\,K^{o} - 2\mu^{o}}{2\mu^{o}(3\,K^{o} + \mu^{o})}$$

Consider a two-phase material, which consists of a phase 1 with elastic properties (K^{1}, μ^{1}) and a phase 2 with elastic properties (K^{2}, μ^{2}). In addition we assume that $K^{2} > K^{1}$ and $\mu^{2} > \mu^{1}$.

As a first step we state that $K^{o} = K^{1}$ and $\mu^{o} = \mu^{1}$ so that $\Delta\mathbf{L}^{\alpha^{-1}}$ in Equation (7.305) reads

$$\Delta\mathbf{L}^{\alpha^{-1}} = 0 \text{ if } \alpha = 1 \tag{7.313}$$

and

$$\Delta\mathbf{L}^{\alpha^{-1}} = \left(\mathbf{L}^{2} - \mathbf{L}^{o}\right)^{-1} = 3\left(K^{2} - K^{1}\right)\mathbf{E}^{1} + 2\left(\mu^{2} - \mu^{1}\right)\mathbf{E}^{2} > 0 \text{ if } \alpha = 2 \tag{7.314}$$

Therefore, we are under the condition of the minimum of Hashin-Shtrikman functional described by $HS^{-}(\boldsymbol{\varepsilon}^{*}, \mathbf{p}^{*})$, which is calculated by choosing the following special cases of

- Stress polarization distribution in each phase such that

$$p_{ij}^{*^1} = 0 \text{ and } p_{ij}^{*^2} = p^* \delta_{ij} \tag{7.315}$$

- Displacement boundary conditions such that

$$\bar{u}_i = \bar{\varepsilon}\, \delta_{ij} x_j \text{ when } \mathbf{r} \in \partial V \tag{7.316}$$

Under the conditions (7.315) and (7.316), we have successively the following

$$I_1 = -\frac{1}{2}\sum_{\alpha=1}^{n} f^\alpha \mathbf{p}^{*^\alpha} : \Delta\mathbf{L}^{\alpha^{-1}} : \mathbf{p}^{*^\alpha} = -\frac{1}{2} f^2 \mathbf{p}^{*^2} : \Delta\mathbf{L}^{\alpha^{-1}} : \mathbf{p}^{*^2}$$

$$= -\frac{1}{2} f^2 \left(\frac{1}{3(K^2 - K^1)} E^1_{ijkl} + \frac{1}{2(\mu^2 - \mu^1)} E^2_{ijkl} \right) (p^*)^2 \delta_{ij}\, \delta_{kl}$$

$$= -\frac{f^2 (p^*)^2}{2(K^2 - K^1)}$$

$$I_2 = \sum_{\alpha=1}^{2} f^\alpha \mathbf{p}^{*\alpha} : \bar{\varepsilon} = \left(f^1 \mathbf{p}^{*^1} + f^2 \mathbf{p}^{*2} \right) : \bar{\varepsilon} = 3 f^2 p^* \bar{\varepsilon}$$

$$I_3 = -\frac{1}{2}\sum_{\alpha=1}^{2} f^\alpha \mathbf{p}^{*^\alpha} : \mathbf{P}^\alpha : (\mathbf{p}^{*^\alpha} - \bar{\mathbf{p}}^*) = -\frac{1}{2} f^2 \mathbf{p}^{*^2} : \mathbf{P}^2 : \left(\mathbf{p}^{*^2} - \bar{\mathbf{p}}^* \right)$$

$$\bar{p}^*_{ij} = f^2 p^* \delta_{ij}$$

$$I_3 = -\frac{1}{2} f^2 p^* \delta_{ij} P^2_{ijkl} (p^* \delta_{kl} - f^2 p^* \delta_{kl})$$

$$= -\frac{1}{2} f^2 (p^*)^2 (1 - f^2) P^2_{ijkl}\, \delta_{ij}\, \delta_{kl} = -\frac{1}{2} f^1 f^2 (p^*)^2 P^2_{ijkl}\, \delta_{ij}\, \delta_{kl}$$

$$= -\frac{1}{2}\frac{f^1 f^2 (p^*)^2}{K^{\mathrm{o}} + \frac{4}{3}\mu^{\mathrm{o}}}$$

and

$$W(\bar{\mathbf{u}}) = L^{\mathrm{o}}_{ijkl}\, \bar{\varepsilon}_{ij}\, \bar{\varepsilon}_{kl} = \frac{9}{2} K^1 \bar{\varepsilon}^2$$

Hence

$$\widetilde{HS}^-(p^*) = W(\bar{\mathbf{u}}) + HS^-(\boldsymbol{\varepsilon}^*, \mathbf{p}^*) = \frac{9}{2} K^1 \bar{\varepsilon}^2 - \frac{f^2 (p^*)^2}{2(K^2 - K^1)} + 3 f^2 p^* \bar{\varepsilon} - \frac{1}{2}\frac{f^1 f^2 (p^*)^2}{K^{\mathrm{o}} + \frac{4}{3}\mu^{\mathrm{o}}} \tag{7.317}$$

In Equation (7.317) the trial parameter p^* is calculated by the stationary condition

$$\frac{\partial\left(\widetilde{HS}^{-}(p^*)\right)}{\partial p^*}=0$$

which leads to

$$p^* = \frac{3\bar{\varepsilon}}{\dfrac{1}{(K^2-K^1)}+\dfrac{f^1}{K^1+\frac{4}{3}\mu^1}} \tag{7.318}$$

Substituting (7.318) into (7.317) and by taking into consideration the extremum condition (7.300), one obtains a lower bound of the bulk modulus. That is

$$\bar{K} \geq K^1 + \frac{f^2}{\frac{1}{(K^2-K^1)}+\frac{f^1}{K^1+\frac{4}{3}\mu^1}} \tag{7.319}$$

The second step supposes that $K^{\mathrm{o}} = K^2$ and $\mu^{\mathrm{o}} = \mu^2$. Therefore, $\Delta\mathbf{L}^{\alpha^{-1}}$ in equation (7.305) reads

$$\Delta\mathbf{L}^{\alpha^{-1}} = 0 \text{ if } \alpha = 2 \tag{7.320}$$

and

$$\Delta\mathbf{L}^{\alpha^{-1}} = \left(\mathbf{L}^1 - \mathbf{L}^o\right)^{-1} = 3\left(K^1 - K^2\right)\mathbf{E}^1 + 2\left(\mu^1 - \mu^2\right)\mathbf{E}^2 < 0 \text{ if } \alpha = 1 \tag{7.321}$$

Therefore, the definition (7.321) leads to the minimum of Hashin-Shtrikman functional described by $\widetilde{HS}^{+}(\boldsymbol{\varepsilon}^*, \mathbf{p}^*)$, which is calculated by choosing the following special cases of stress polarization distribution in each phase such that

$$p_{ij}^{*^1} = p^*\delta_{ij} \text{ and } p_{ij}^{*^2} = 0 \tag{7.322}$$

Hence

$$\widetilde{HS}^{+}(p^*) = \frac{9}{2}K^2\bar{\varepsilon}^2 - \frac{f^1(p^*)^2}{2(K^1-K^2)} + 3f^1p^*\bar{\varepsilon} - \frac{1}{2}\frac{f^1f^2(p^*)^2}{K^2+\frac{4}{3}\mu^2} \tag{7.323}$$

where the trial parameter is found by using the stationary condition

$$\frac{\partial\left(\widetilde{HS}^{+}(p^*)\right)}{\partial p^*}=0$$

which leads to

$$p^* = \frac{3\bar{\varepsilon}}{\dfrac{1}{(K^1-K^2)}+\dfrac{f^2}{K^2+\frac{4}{3}\mu^2}} \tag{7.324}$$

Substituting (7.324) into (7.323), equation (7.300) leads to an upper bound of the bulk modulus

$$\bar{K} \le K^2 + \frac{f^1}{\dfrac{1}{(K^1 - K^2)} + \dfrac{f^2}{K^2 + \frac{4}{3}\mu^2}} \tag{7.325}$$

Combining (7.319) with (7.325) leads to the Hashin-Shtrikman bounds of the bulk moduls

$$K^1 + \frac{f^2}{\dfrac{1}{(K^2 - K^1)} + \dfrac{f^1}{K^1 + \dfrac{4}{3}\mu^1}} \le \bar{K} \le K^2 + \frac{f^1}{\dfrac{1}{(K^1 - K^2)} + \dfrac{f^2}{K^2 + \dfrac{4}{3}\mu^2}} \tag{7.326}$$

Finally, by following similar procedures as the above and by choosing appropriate boundary conditions, one gets the Hashin-Shtrikman bounds for shear modulus as

$$\mu^1 + \frac{f^2}{\dfrac{1}{(\mu^2 - \mu^1)} + \dfrac{6f^1(K^1 + 2\mu^1)}{5(3K^1 + 4\mu^1)\mu^1}} \le \bar{\mu} \le \mu^2 + \frac{f^1}{\dfrac{1}{(\mu^1 - \mu^2)} + \dfrac{6f^2(K^2 + 2\mu^2)\mu^2}{5(3K^2 + 4\mu^2)}} \tag{7.327}$$

As an application of Equation (7.327), the generalized self-consistent model of Christensen and Lo (Equation (7.205)) is compared to Hashin-Shtrikman bounds in the case of a two-phase particulate composite material.

The material parameters used in this comparison are $\mu^1/\mu^M = 135.14$, $\nu^1 = 0.20$ and $\nu^M = 0.35$. It is seen from Fig. 7.6 that the effective shear modulus from the three-phase model Christensen and Lo is bounded by the Hashin-Shtrikman lower and upper bounds.

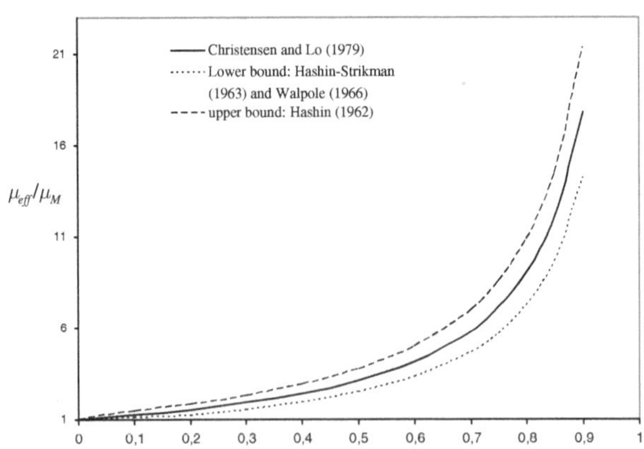

Fig. 7.6 Comparison between Hashin-Shtrikman bound and Christensen and Lo. Generalized self-consistent model

7.8 On Possible Extensions of Linear Micromechanics to Nonlinear Problems

Nonlinear problems in continuum micromechanics for inhomogeneous material with at least one nonlinear constituent (nonlinear elastic, viscoelastic, elastoplastic, elastoviscoplastic) consist in providing accurate estimates for the material response for any load state and load history at reasonable computational cost. The mean difficulty in attaining this goal results from the typical strong intraphase fluctuations of stress and strain fields in nonlinear inhomogeneous materials and in the hereditary nature of most inelastic behaviors. Therefore, the responses of the constituents can vary markedly at the microscale in comparison with the linear elastic behavior. For example, a two-phase elastoplastic composite material effectively behaves as multiphase materials and phase averages have less predictive capabilities than in the linear elastic case.

Descriptions for viscoelastic inhomogeneous materials are closely related to those for elastic composites. Relaxation moduli and creep compliances can be obtained by applying mean field theories in the Laplace Transform space, where the problem becomes equivalent to the elastic one for the same microgeometries. For correspondence principles between descriptions for elastic and viscoelastic inhomogeneous materials, the reader could refer to Hashin [20] for details. However, extensions of linear continuum micromechanics theories and their bounding methods to plastic time-dependent or time independent behaviors have proven to be challenging.

Historically, continuum nonlinear micromechanics for inelastic behavior devoted mainly to crystalline materials was the initial precursor in developing theoretical frameworks for nonlinear homogenization techniques of composite materials. In fact, crystalline materials were at the origin much more developed that "regular" inclusion-matrix" composites, and where the main interest was to relate the mechanical response of an aggregate of crystals (known as a polycrystal) to the fundamental mechanisms of single crystal deformation. These methodologies lead to the well known crystal plasticity frameworks, whose classification is made in terms of elastoplastic, viscoplastic and elastoviscoplastic behaviors. The last two categories are defined as to be time-dependent rigid plastic and time-dependent plastic behaviors, respectively.

It has been shown that approximated solutions for the local problem can be obtained in the linear case, leading to pertinent estimations of the macroscopic behavior, which are capable of accounting for the influence of morphological parameters and phase spatial distribution on the global behavior. Unfortunately, due to the nonapplicability of the superposition principle, on which most development are based in linear cases (e.g., use of elementary solutions such as Eshelby's one), the philosophy behind these approaches cannot be transported directly to nonlinear behaviors.

In order to take advantage of the knowledge acquired in linear cases, linearization of the local constitutive laws could be one of the interesting approaches. The procedure consists of replacing the nonlinear field equations in the RVE by linear equations in the same RVE, whose solution can be evaluated exactly or approximately via the use of the tools developed for linear materials. The linearization is completed with a set of complimentary equations, which characterizes the parameters defining the linear problem (e.g., elasticity moduli). Typically, these equations are nonlinear so that the character of the initial problem is conserved. However, in this case, the problem does not deal with field equations but with equations involving a finite set of variables which can be solved with the appropriate numerical tools. In most cases, simple algorithms lead to a solution.

The methodology presented above is referred to as the "nonlinear extension" of a linear model. The local linear problem resulting from the linearization procedure is identical to the homogenization problem for linear composites, referred to as linear comparison composite (LCC). This virtual LCC results solely from the linearization step and has no physical existence. Further, its moduli are distinct from the initial elasticity moduli of the real nonlinear composite. Although it is often the case, the LCC does not necessarily have the same spatial distribution of phases, or the same number of constituents, as the real nonlinear composite.

One of the difficulties of this method lies in the use of the linear model to obtain the nonlinear macroscopic behavior. Typically, linear models do not provide detailed description of the local fields in the LCC but only averaged strains or stresses in the phases. This information is sufficient in the case of linear problems since the macroscopic stress can be obtained from averaged strains in the phases via the following equation.

$$\bar{\boldsymbol{\sigma}} = \sum_r f^{(r)} \langle \boldsymbol{\sigma} \rangle_{V_r} = \sum_r f^{(r)} \mathbf{L}^{(r)} : \langle \boldsymbol{\varepsilon} \rangle_{V_r}.$$

Due to the nonlinear behavior of the constituents, this property does not hold for nonlinear composites where the macroscopic stress is given by:

$$\bar{\boldsymbol{\sigma}} = \sum_r f^{(r)} \langle \boldsymbol{\sigma} \rangle_{V_r} = \sum_r f^{(r)} \left\langle \frac{\partial \omega}{\partial \boldsymbol{\varepsilon}} (\boldsymbol{\varepsilon}) \right\rangle_{V_r} \neq \sum_r f^{(r)} \frac{\partial \omega}{\partial \boldsymbol{\varepsilon}} \left(\langle \boldsymbol{\varepsilon} \rangle_{V_r} \right)$$

where $\omega(\boldsymbol{\varepsilon})$ is the local free energy, $f^{(r)}$ the volume fraction of each phase and $\mathbf{L}^{(r)}$ their stiffness.

These problems have been brought to line since the pioneer work of Kröner [33]. Basically, the linearization procedure raised above was initially introduced in a context of a tangent formation leading to the Hill's self-consistent model in elastoplasticity [26] and to Huchinson approach in viscoplasticity [30]. It was early recognized that the incremental approaches based on the tangent stiffness tensors of the phases overestimated the flow stress of the material, and the origin of this error was traced to the anisotropic nature of the tangent stiffness tensor during plastic deformation. As will be discussed in this section, one of major difficulties of the tangent formulation lies in computing the Eshelby's tensor in each time increment by taking into account the anisotropy of the tangent moduli.

This limitation has motivated the development of the secant methods [5, 53], which deal with the elasto-plastic deformation within the framework of non-linear elasticity. However, the secant approaches have quickly shown their limitations, especially in the case of high fluctuating fields where stress and strain phase averages are not sufficient to capture correctly the nonlinear behavior. Several attempts were made to determine the correct this problem from energy considerations [44] or statistically based theories [7], which finally led to the so-called "modified" secant approximation [47] leading to the concept of second-order moment of strains. Other approaches attempt to introduce more rigorous bounds, using nonlinear extensions of the Hashin–Shtrikman variational principle , [42, 46, 52, 56, 57].

Indeed, secant approaches cannot simulate the mechanical behavior under non-proportional loading paths (e.g., cyclic deformation), and this renewed the interest in incremental approaches based on the tangent stiffness tensors. It was found that much better approximation of the flow stress was obtained when only the isotropic part of the tangent stiffness tensors was used in the analyses [17, 18]. More recently, Doghri and Ouaar [12] have obtained good predictions for the elasto-plastic response of sphere-reinforced composites by using the isotropic version of the stiffness tensor only to compute Eshelby's tensor, while the anisotropic version is used in all the other operations, allowing the study of nonproportional loading paths. The tangent formulation has also been adopted in a different manner by Molinari et al. [39, 40], and by Masson and Zaoui [26] and Masson et al. [37] leading to the well-known affine method.

Systematic comparisons between the different nonlinear methods generated by the concept of linearization of continuum micromechanics were carried out by taking as a reference numerical results of finite element homogenization schemes. This has been recently emphasized in various excellent papers [8, 9, 41].

The above general principles will be addressed in this section, in the particular case where the LCC is obtained from the secant moduli of the nonlinear constituents. Two approaches, the "classical" approach and the "modified" approach will be presented and compared. These two approaches are both relatively simple to use and differ only from the set of complimentary equations characterizing the LCC. The tangent formulation will be also addressed and compared with the secant formulation.

7.8.1 The Secant Formulation

For general purposes, the secant method solves the following field equations

$$\begin{aligned}
&\boldsymbol{\varepsilon} = \frac{1}{2}\left(\nabla \mathbf{u} + {}^{t}\nabla \mathbf{u}\right) \\
&\operatorname{div}\boldsymbol{\sigma} = 0 \\
&\langle \boldsymbol{\varepsilon} \rangle = \bar{\boldsymbol{\varepsilon}} \\
&\boldsymbol{\sigma}(\mathbf{r}) = \mathbf{l}^{\mathrm{sec}}(\mathbf{r}, \boldsymbol{\varepsilon}(\mathbf{r})) : \boldsymbol{\varepsilon}(\mathbf{r}),
\end{aligned} \tag{7.328}$$

where $\mathbf{l}^{\mathrm{sec}}$ is the local secant modulus tensor. It typically fluctuates within a phase due to the fluctuation of the local strain $\boldsymbol{\varepsilon}(\mathbf{r})$. Therefore, the secant modulus tensor is highly heterogeneous. Its fluctuation results from the nonlinearity of the problem associated with its dependency on the local strain.

In the case of isotropic materials, $\mathbf{l}^{\mathrm{sec}}$ is given by

$$\mathbf{l}^{\mathrm{sec}}(\varepsilon) = 3\,k\mathbf{E}^{1} + 2\mu^{\mathrm{sec}}\left(\varepsilon_{\mathrm{eq}}\right)\mathbf{E}^{2}. \tag{7.329}$$

in which one considers that most isotropic materials are linear under hydrostatic load and nonlinear under shear. Then their behavior can be written with the following expressions

$$\sigma_{\mathrm{m}} = 3\,k\varepsilon_{\mathrm{m}}, \quad \mathbf{s} = 2\mu^{\mathrm{sec}}\left(\varepsilon_{\mathrm{eq}}\right)\mathbf{e}, \ \mu^{\mathrm{sec}}\left(\varepsilon_{\mathrm{eq}}\right) = \frac{\sigma_{\mathrm{eq}}}{3\varepsilon_{\mathrm{eq}}},$$

where

$$\sigma_{\mathrm{m}} = \frac{\sigma_{kk}}{3}, \ \varepsilon_{\mathrm{m}} = \frac{\varepsilon_{kk}}{3}, \ \sigma_{\mathrm{eq}} = \left(\frac{3}{2}\mathbf{s} : \mathbf{s}\right)^{\frac{1}{2}}, \ s_{ij} = \sigma_{ij} - \sigma_{\mathrm{m}}\delta_{ij},$$

$$\varepsilon_{\mathrm{eq}} = \left(\frac{2}{3}\mathbf{e} : \mathbf{e}\right)^{\frac{1}{2}}, \ e_{ij} = \varepsilon_{ij} - \varepsilon_{\mathrm{m}}\delta_{ij}.$$

Indeed, the heterogeneity of the secant modulus tensor depends on which type of nonlinear behavior is displayed by the constituents and also on the amount of applied strain. To set up a direct homogenization procedure of such highly heterogeneous materials is very difficult unless systematic approximations are used, which, in general, relies on a linearization procedure with appropriate complementary laws.

As a first attempt and within a general procedure, the problem could be seen at a given strain state as a linear problem with the following local constitutive law

$$\boldsymbol{\sigma}(\mathbf{r}) = \mathbf{l}^{\mathrm{lin}}(\mathbf{r}) : \boldsymbol{\varepsilon}(\mathbf{r}), \tag{7.330}$$

with

$$\mathbf{l}^{\mathrm{lin}}(\mathbf{r}) = \mathbf{l}^{\mathrm{sec}}(\mathbf{r}, \boldsymbol{\varepsilon}(\mathbf{r})). \tag{7.331}$$

The definition (7.330) is the linear model required for the linearization procedure of the local behavior, whereas the definition (7.331) corresponds to additional or complementary relationships, so that the nonlinear behavior can be captured. When these two steps are accomplished, the problem becomes a classical one and an appropriate "classical" linear homogenization scheme can be chosen to obtain the nonlinear macroscopic behavior. However, Equations (7.330) and (7.331) are still not suitable for analytical calculations due to the infinite number of complementary equations required for the definition (7.331).

Therefore, approximations are needed, which, clearly, need to render a finite number of complementary equations with a certain accuracy in describing the heterogeneous nature of the nonlinear behavior. For such a purpose, approximations are introduced both in the step of linearization and complementary equations. The linear model in Equation (7.330) may be assumed piecewise uniform for the stiffness tensor $\mathbf{l}^{\mathrm{lin}}(\mathbf{r})$, so that, for a given phase $r(r = 1, \ldots, n)$ one has $\mathbf{l}^{\mathrm{lin}}(\mathbf{r}) = \mathbf{L}^{(r)}$. In addition, the complementary equations are reduced to a finite number corresponding to the identified number of constituents or phases, which lead to a definition of stiffness tensors $\mathbf{L}^{(r)}$ at some effective piecewise uniform strains $\tilde{\boldsymbol{\varepsilon}}^{(r)}$, representing the strain distribution in each phase, and therefore requiring an accurate model to be determined. The n complementary equations read

$$\mathbf{L}^{(r)} = \mathbf{L}^{(r)}_{\mathrm{sec}}\left(\tilde{\boldsymbol{\varepsilon}}^{(r)}\right), \tag{7.332}$$

where the nonlinearity of the problem lies in the dependency of each individual effective strain $\underline{\tilde{\boldsymbol{\varepsilon}}}^{(r)}$ on the stiffness $\mathbf{L}^{(r)}$ of the different phases, so that, n nonlinear problems have to be solved, requiring in general simple iterative procedures. Once the tensors $\mathbf{L}^{(r)}$ are determined the problem becomes a classical one by taking advantages of the homogenization approaches developed in linear elasticity.

The choice of the appropriate linear homogenization scheme to describe the microstructure of the real nonlinear composite material defines the so called linear comparison composite, for which the overall effective stiffness $\bar{\mathbf{L}}$ is expressed formally as

$$\bar{\mathbf{L}} = \bar{\mathbf{L}}\left(f^{(r)}, \mathbf{L}^{(r)}, \ldots\right) \equiv \bar{\mathbf{L}}(\bar{\varepsilon}), \tag{7.333}$$

which depends on the stiffness of each constituent and some morphological aspects related to the microstructure. The overall constitutive law is then nonlinear and formally given by

$$\bar{\boldsymbol{\sigma}} = \bar{\mathbf{L}}(\bar{\boldsymbol{\varepsilon}}) : \bar{\boldsymbol{\varepsilon}}. \tag{7.334}$$

In the following, two methods to define the effective strains $\tilde{\boldsymbol{\varepsilon}}^{(r)}$ are discussed and compared. The first approach is known as the classical secant method. It simply defines the effective strains as the average strain in each phase. The second method, called the "modified" secant method [47], could be seen as a refinement of the first method by introducing the second order moment of the strain field.

7.8.1.1 The Classical Method

The classical method has been extensively used to deal with the nonlinear behavior of composite materials. It consists of defining the effective strain $\tilde{\boldsymbol{\varepsilon}}^{(r)}$ as the mean value of the local strain field over the considered phase. That is

$$\tilde{\boldsymbol{\varepsilon}}^{(r)} = \frac{1}{V^r} \iiint\limits_{V^r} \boldsymbol{\varepsilon}(\mathbf{r}) dV = \langle \boldsymbol{\varepsilon}(\mathbf{r}) \rangle_{V^r}. \tag{7.335}$$

The main advantage behind the assumption lies in the expression of effective strains $\tilde{\boldsymbol{\varepsilon}}^{(r)}$ as a function of the applied strain $\boldsymbol{\varepsilon}$ by means of the average concentration tensors $\mathbf{A}^{(\mathbf{r})}$

$$\tilde{\boldsymbol{\varepsilon}}^{(r)} = \mathbf{A}^{(r)} : \bar{\boldsymbol{\varepsilon}}, r(r = 1, \ldots, n), \tag{7.336}$$

which are determined by appropriate explicit or implicit linear continuum mechanics theories that are extensively discussed in the previous sections. This provides the **n**

concentration tensors in terms of the stiffness tensors $\mathbf{L}^{(r)}$ of each phase for explicit schemes and the overall stiffness $\bar{\mathbf{L}}$ for implicit schemes. That is

$$\mathbf{A}^{(r)} = \mathbf{A}^{(r)}\left(\bar{\mathbf{L}}, \mathbf{L}^{(s)}, s = 1 \ldots, n\right) \tag{7.337}$$

Note that through the definition of a linear comparison composite, upper and lower bounds of the effective properties $\bar{\mathbf{L}}$, can be found using the linear variational principles presented in Section 7.7.

Finally, since $\mathbf{L}^{(r)}$ depends on the corresponding effective strain $\tilde{\boldsymbol{\varepsilon}}^{(r)}$ through Equation (7.332), Equation (7.336) together with expression (7.337) provide n nonlinear equations, whose solutions determine the overall nonlinear property $\bar{\mathbf{L}}$ by means of the constitutive equation (7.334). As noticed before, such a scheme requires in general iterative procedure and suitable convergence criteria.

The classical secant method can be illustrated in case of two-phase materials. In fact, combination of

$$\langle \boldsymbol{\varepsilon}(\mathbf{r}) \rangle_V = \bar{\boldsymbol{\varepsilon}} \text{ and } \begin{cases} \tilde{\boldsymbol{\varepsilon}}^{(1)} = \mathbf{A}^{(1)} : \bar{\boldsymbol{\varepsilon}} \\ \tilde{\boldsymbol{\varepsilon}}^{(2)} = \mathbf{A}^{(2)} : \bar{\boldsymbol{\varepsilon}} \end{cases}, \tag{7.338}$$

leads to

$$f^{(1)}\mathbf{A}^{(1)} + f^{(2)}\mathbf{A}^{(2)} = \mathbf{I}. \tag{7.339}$$

In addition, from the constitutive law of each constituent, we have

$$\bar{\boldsymbol{\sigma}}^{(1)} = \mathbf{L}^{(1)} : \tilde{\boldsymbol{\varepsilon}}^{(1)}, \ \bar{\boldsymbol{\sigma}}^{(2)} = \mathbf{L}^{(2)} : \tilde{\boldsymbol{\varepsilon}}^{(2)}, \tag{7.340}$$

and

$$\bar{\boldsymbol{\sigma}} = f^{(1)}\bar{\boldsymbol{\sigma}}^{(1)} + f^{(2)}\bar{\boldsymbol{\sigma}}^{(2)}. \tag{7.341}$$

Equation (7.338) gives

$$\bar{\mathbf{L}} = f^{(1)}\mathbf{L}^{(1)} : \mathbf{A}^{(1)} + f^{(2)}\mathbf{L}^{(2)} : \mathbf{A}^{(2)}. \tag{7.342}$$

Then, substituting (7.339) in (7.342) yields

$$\begin{cases} \mathbf{A}^{(1)} = \frac{1}{f^{(1)}}\left(\mathbf{L}^{(1)} - \mathbf{L}^{(2)}\right)^{-1}\left(\bar{\mathbf{L}} - \mathbf{L}^{(2)}\right) \\ \mathbf{A}^{(2)} = \frac{1}{f^{(2)}}\left(\mathbf{L}^{(2)} - \mathbf{L}^{(1)}\right)^{-1}\left(\bar{\mathbf{L}} - \mathbf{L}^{(1)}\right) \end{cases}. \tag{7.343}$$

Therefore, when the linear homogenization model is identified to obtain the effective stiffness as

$$\bar{\mathbf{L}} = \bar{\mathbf{L}}\left(f^{(1)}, \mathbf{L}^{(1)}, \mathbf{L}^{(2)}, \ldots\right) \tag{7.344}$$

the solution to the nonlinear problem is given by the following set of equations

$$\begin{cases} \tilde{\boldsymbol{\varepsilon}}^{(1)} = \frac{1}{f^{(1)}}\left(\mathbf{L}^{(1)} - \mathbf{L}^{(2)}\right)^{-1} : \left(\bar{\mathbf{L}} - \mathbf{L}^{(2)}\right) : \bar{\boldsymbol{\varepsilon}} \\ \tilde{\boldsymbol{\varepsilon}}^{(2)} = \frac{1}{f^{(2)}}\left(\mathbf{L}^{(2)} - \mathbf{L}^{(1)}\right)^{-1} : \left(\bar{\mathbf{L}} - \mathbf{L}^{(1)}\right) : \bar{\boldsymbol{\varepsilon}} \\ \mathbf{L}^{(1)} = \mathbf{L}^{(r)}_{\text{sec}}\tilde{\boldsymbol{\varepsilon}}^{(1)}\Big), \ \mathbf{L}^{(2)} = \mathbf{L}^{(2)}_{\text{sec}}\left(\tilde{\boldsymbol{\varepsilon}}^{(2)}\right) \end{cases}. \tag{7.345}$$

When the two phases are isotropic

$$\begin{aligned} \mathbf{L}^{(1)}\left(\tilde{\boldsymbol{\varepsilon}}^{(1)}\right) &= 3\,k^{(1)}\mathbf{E}^1 + 2\mu^{(1)}_{\text{sec}}\left(\tilde{\varepsilon}^{(1)}_{\text{eq}}\right)\mathbf{E}^2 \text{ and } \mathbf{L}^{(2)}\left(\boldsymbol{\varepsilon}^{(2)}\right) \\ &= 3\,k^{(2)}\mathbf{E}^1 + 2\mu^{(2)}_{\text{sec}}\left(\tilde{\varepsilon}^{(2)}_{\text{eq}}\right)\mathbf{E}^2 \end{aligned} \tag{7.346}$$

and the linear comparison composite displays an overall isotropy such that

$$\bar{\mathbf{L}}(\bar{\boldsymbol{\varepsilon}}) = 3\bar{k}\mathbf{E}^1 + 2\bar{\mu}\left(\bar{\varepsilon}_{\text{eq}}\right)\mathbf{E}^2 \tag{7.347}$$

where the linear homogenization scheme gives

$$\bar{k} = \bar{k}\left(k^{(1)}, k^{(2)}, \mu^{(1)}, \mu^{(2)}, f^{(1)}, \ldots\right), \ \bar{\mu} = \bar{\mu}\left(k^{(1)}, k^{(2)}, \mu^{(1)}, \mu^{(2)}, f^{(1)}, \ldots\right) \tag{7.348}$$

The set of nonlinear equations (7.345) is reduced to

$$\begin{cases} \tilde{\varepsilon}_{\mathrm{m}}^{(1)} = A_{\mathrm{m}}^{(1)} \bar{\varepsilon}_{\mathrm{m}}, \ \tilde{\varepsilon}_{\mathrm{m}}^{(2)} = A_{\mathrm{m}}^{(2)} \bar{\varepsilon}_{\mathrm{m}}, \ \tilde{\varepsilon}_{\mathrm{eq}}^{(1)} = A_{\mathrm{eq}}^{(1)} \bar{\varepsilon}_{\mathrm{eq}}, \ \tilde{\varepsilon}_{\mathrm{eq}}^{(2)} = A_{\mathrm{eq}}^{(2)} \bar{\varepsilon}_{\mathrm{eq}} \\ A_{\mathrm{m}}^{(1)} = \dfrac{1}{f^{(1)}} \dfrac{\bar{k} - k^{(2)}}{k^{(1)} - k^{(2)}}, \ A_{\mathrm{eq}}^{(1)} = \dfrac{1}{f^{(1)}} \dfrac{\bar{\mu} - \mu^{(2)}}{\mu^{(1)} - \mu^{(2)}} \\ A_{\mathrm{m}}^{(2)} = \dfrac{1}{f^{(2)}} \dfrac{\bar{k} - k^{(1)}}{k^{(2)} - k^{(1)}}, \ A_{\mathrm{eq}}^{(2)} = \dfrac{1}{f^{(2)}} \dfrac{\bar{\mu} - \mu^{(1)}}{\mu^{(2)} - \mu^{(1)}} \\ \mu^{(1)} = \mu_{\mathrm{sec}}^{(1)}\left(\tilde{\varepsilon}_{\mathrm{eq}}^{(1)}\right), \ \mu^{(2)} = \mu_{\mathrm{sec}}^{(2)}\left(\tilde{\varepsilon}_{\mathrm{eq}}^{(2)}\right). \end{cases} \tag{7.349}$$

Further, if the materials are incompressible, the linear homogenization model gives

$$\bar{\mu} = \bar{\mu}\left(\mu^{(1)}, \mu^{(2)}, f^{(1)}, \ldots\right), \tag{7.350}$$

and the nonlinear set of equations become

$$\begin{cases} \tilde{\varepsilon}_{\mathrm{eq}}^{(1)} = A_{\mathrm{eq}}^{(1)} \bar{\varepsilon}_{\mathrm{eq}}, \tilde{\varepsilon}_{\mathrm{eq}}^{(2)} = A_{\mathrm{eq}}^{(2)} \bar{\varepsilon}_{\mathrm{eq}} \\ A_{\mathrm{eq}}^{(1)} = \dfrac{1}{f^{(1)}} \dfrac{\bar{\mu} - \mu^{(2)}}{\mu^{(1)} - \mu^{(2)}}, A_{\mathrm{eq}}^{(2)} = \dfrac{1}{f^{(2)}} \dfrac{\bar{\mu} - \mu^{(1)}}{\mu^{(2)} - \mu^{(1)}} \\ \mu^{(1)} = \mu_{\mathrm{sec}}^{(1)}\left(\tilde{\varepsilon}_{\mathrm{eq}}^{(1)}\right), \mu(2) = \mu_{\mathrm{sec}}^{(2)}\left(\tilde{\varepsilon}_{\mathrm{eq}}^{(2)}\right). \end{cases} \tag{7.351}$$

As discussed above, when the appropriate linear homogenization scheme is chosen, the classical method becomes relatively easy to implement through an iterative algorithm.

7.8.1.2 Modified Secant Method

The classical secant method for describing the nonlinear behavior of composite materials assumes basically homogeneous strain field within each phase, and therefore neglects any intraphase fluctuations of local fields. This results in few discrepancies and limitations, which was behind the principal motivations in developing the modified secant approach.

Let us first recall the basis of the classical secant method, which leads to a certain number of inconsistencies. As shown above, the classical method derives the average stress $\bar{\boldsymbol{\sigma}}^{(r)}$ over a phase $r(r = 1, \ldots, n)$ as

$$\bar{\boldsymbol{\sigma}}^{(r)} = \mathbf{L}^{(r)} : \bar{\boldsymbol{\varepsilon}}^{(r)} = \mathbf{L}_{\mathrm{sec}}^{(r)}\left(\bar{\boldsymbol{\varepsilon}}^{(r)}\right) : \bar{\boldsymbol{\varepsilon}}^{(r)} \tag{7.352}$$

which implies the existence of a phase strain energy potential $\omega^{(r)}\left(\bar{\boldsymbol{\varepsilon}}^{(r)}\right)$ determined with respect to the average strain $\boldsymbol{\varepsilon}^{(r)}$, such that

$$\bar{\boldsymbol{\sigma}}^{(r)}=\frac{\partial\omega^{(r)}}{\partial\bar{\boldsymbol{\varepsilon}}^{(r)}}\left(\bar{\boldsymbol{\varepsilon}}^{(r)}\right). \tag{7.353}$$

In the case of incompressible materials, (7.352) reads

$$\bar{\sigma}_{\mathrm{eq}}^{(r)}=\frac{\partial\omega_{\mathrm{eq}}^{(r)}}{\partial\bar{\varepsilon}_{\mathrm{eq}}^{(\mathrm{r})}}\left(\bar{\varepsilon}_{\mathrm{eq}}^{(\mathrm{r})}\right), \tag{7.354}$$

where

$$\bar{\sigma}_{\mathrm{eq}}^{(r)}=\left(\frac{1}{V^{(r)}}\int\limits_{V^{(r)}}\boldsymbol{\sigma}(\mathbf{r})dV\right)_{\mathrm{eq}},\bar{\varepsilon}_{\mathrm{eq}}^{(r)}=\left(\frac{1}{V^{(r)}}\int\limits_{V^{(r)}}\boldsymbol{\varepsilon}(\mathbf{r})dV\right)_{\mathrm{eq}} \tag{7.355}$$

One can also define the following equivalent average strain as

$$\bar{\bar{\varepsilon}}_{\mathrm{eq}}^{(r)}=\frac{1}{V^{(r)}}\int\limits_{V^{(r)}}\varepsilon_{\mathrm{eq}}(\mathbf{r})dV \tag{7.356}$$

The first discrepancy of the classical method results from the equalities (7.353) – (7.354), which are satisfied only if the strain field, is homogeneous in each phase. Or, in general, the nonlinear behavior leads to highly intraphase fluctuations, and as a result one can show that

$$\bar{\boldsymbol{\sigma}}^{(r)}=\frac{1}{V^{(r)}}\int\limits_{V^{(r)}}\boldsymbol{\sigma}(\mathbf{r})dV=\frac{1}{V^{(r)}}\int\limits_{V^{(r)}}\frac{\partial\omega^{(r)}}{\partial\boldsymbol{\varepsilon}}(\boldsymbol{\varepsilon})dV\neq\frac{\partial\omega^{(r)}}{\partial\bar{\boldsymbol{\varepsilon}}^{(r)}}\left(\bar{\boldsymbol{\varepsilon}}^{(r)}\right) \tag{7.357}$$

or in the case of incompressible materials

$$\bar{\sigma}_{\mathrm{eq}}^{(r)}=\frac{1}{V^{(r)}}\int\limits_{V^{(r)}}\frac{\partial\omega_{\mathrm{eq}}^{(r)}}{\partial\varepsilon_{\mathrm{eq}}}\left(\varepsilon_{\mathrm{eq}}\right)dV\neq\frac{\partial\omega_{\mathrm{eq}}^{(r)}}{\partial\bar{\varepsilon}_{\mathrm{eq}}^{(\mathrm{r})}}\left(\bar{\varepsilon}_{\mathrm{eq}}^{(\mathrm{r})}\right) \tag{7.358}$$

The result (7.358) is shown in the following.

In fact, one has

$$\begin{aligned}\bar{\sigma}_{\mathrm{eq}}^{(r)}&=\left(\frac{1}{V^{(r)}}\int\limits_{V^{(r)}}\boldsymbol{\sigma}(\mathbf{r})dV\right)_{\mathrm{eq}}=\left(\frac{1}{V^{(r)}}\int\limits_{V^{(r)}}\frac{\partial\omega^{(r)}}{\partial\boldsymbol{\varepsilon}}(\boldsymbol{\varepsilon})dV\right)_{\mathrm{eq}}\\&=\left(\frac{1}{V^{(r)}}\int\limits_{V^{(r)}}\frac{\partial\omega_{\mathrm{eq}}^{(r)}}{\partial\varepsilon_{\mathrm{eq}}}\frac{\partial\varepsilon_{\mathrm{eq}}}{\partial\boldsymbol{\varepsilon}}(\boldsymbol{\varepsilon})dV\right)_{\mathrm{eq}}\end{aligned} \tag{7.359}$$

which leads to

$$\bar{\sigma}_{\mathrm{eq}}^{(r)} = \left(\frac{1}{V^{(r)}} \int\limits_{V^{(r)}} \frac{\partial \omega_{\mathrm{eq}}^{(r)}}{\partial \varepsilon_{\mathrm{eq}}} (\varepsilon_{\mathrm{eq}}) \frac{2\mathbf{e}}{3\varepsilon_{\mathrm{eq}}} (\boldsymbol{\varepsilon}) dV \right)_{\mathrm{eq}} \tag{7.360}$$

where the strain deviator $\mathbf{e}$ is defined in Equation (7.329).

On the other hand, the convexity of the function $\mathbf{e} : \mathbf{e}$ leads to the following inequality

$$\left(\frac{1}{V^{(r)}} \int\limits_{V^{(r)}} \frac{\partial \omega_{\mathrm{eq}}^{(r)}}{\partial \varepsilon_{\mathrm{eq}}} (\varepsilon_{\mathrm{eq}}) \frac{2\mathbf{e}}{3\varepsilon_{\mathrm{eq}}} (\boldsymbol{\varepsilon}) dV \right)_{\mathrm{eq}} \leq \frac{1}{V^{(r)}} \int\limits_{V^{(r)}} \frac{\partial \omega_{\mathrm{eq}}^{(r)}}{\partial \varepsilon_{\mathrm{eq}}} (\varepsilon_{\mathrm{eq}}) dV \tag{7.361}$$

Since for most nonlinear composite the function $\frac{\partial \omega_{\mathrm{eq}}^{(r)}}{\partial \varepsilon_{\mathrm{eq}}} (\varepsilon_{\mathrm{eq}})$ is concave, one can easily show that

$$\frac{1}{V^{(r)}} \int\limits_{V^{(r)}} \frac{\partial \omega_{\mathrm{eq}}^{(r)}}{\partial \varepsilon_{\mathrm{eq}}} (\varepsilon_{\mathrm{eq}}) dV < \frac{\partial \omega_{\mathrm{eq}}^{(r)}}{\partial \bar{\bar{\varepsilon}}_{\mathrm{eq}}^{(\mathrm{r})}} \left(\bar{\bar{\varepsilon}}_{\mathrm{eq}}^{(\mathrm{r})} \right) = \frac{\partial \omega_{\mathrm{eq}}^{(r)}}{\partial \bar{\varepsilon}_{\mathrm{eq}}^{(\mathrm{r})}} \left(\bar{\varepsilon}_{\mathrm{eq}}^{(\mathrm{r})} \right) \tag{7.362}$$

where we further assume that $\bar{\bar{\varepsilon}}_{\mathrm{eq}}^{(\mathrm{r})} = \bar{\varepsilon}_{\mathrm{eq}}^{(\mathrm{r})}$. With (7.362) and (7.361), the statement (7.358) is proved.

Another limitation of the classical method results from the fact that the definition of the macroscopic properties does not necessarily rely on the definition of a macroscopic potential $W(\bar{\boldsymbol{\varepsilon}})$, so that

$$\bar{\boldsymbol{\sigma}} = \frac{\partial W}{\partial \boldsymbol{\varepsilon}} (\bar{\boldsymbol{\varepsilon}}). \tag{7.363}$$

In fact, it turned out that in some cases of nonlinear composite materials, the following property of the macroscopic potential of isotropic materials

$$\frac{\partial^2 W}{\partial \bar{\varepsilon}_{\mathrm{eq}} \partial \bar{\varepsilon}_{\mathrm{m}}} (\bar{\varepsilon}_{\mathrm{eq}}, \bar{\varepsilon}_{\mathrm{m}}) = \frac{\partial \bar{\sigma}_{\mathrm{m}}}{\partial \bar{\varepsilon}_{\mathrm{eq}}} = \frac{\partial^2 W}{\partial \bar{\varepsilon}_{\mathrm{m}} \partial \bar{\varepsilon}_{\mathrm{eq}}} (\bar{\varepsilon}_{\mathrm{eq}}, \bar{\varepsilon}_{\mathrm{m}}) = \frac{\partial \bar{\sigma}_{\mathrm{eq}}}{\partial \bar{\varepsilon}_{\mathrm{m}}}, \tag{7.364}$$

is not fulfilled by the classical secant approach.

The modified secant method took its inspiration from the above statement. It was developed in accordance to the following. The first step is the definition of the macroscopic potential from the Hill lemma

$$W(\bar{\varepsilon}) = \frac{1}{V} \int\limits_{V} \varepsilon(\mathbf{r}) : \mathbf{L}(\mathbf{r}) : \varepsilon(\mathbf{r}), dV = \bar{\varepsilon} : \bar{\mathbf{L}} : \bar{\varepsilon}, \tag{7.365}$$

and its derivative with respect to the stiffness $\mathbf{L}^{(r)}$ of the r – phase in the linear comparison composite

$$\frac{\partial W}{\partial \mathbf{L}^{(r)}}(\bar{\boldsymbol{\varepsilon}}) = \bar{\boldsymbol{\varepsilon}} : \frac{\partial \bar{\mathbf{L}}}{\partial \mathbf{L}^{(r)}} : \bar{\boldsymbol{\varepsilon}} = \frac{1}{V}\int_V \boldsymbol{\varepsilon}(\mathbf{r}) : \frac{\partial \mathbf{L}(\mathbf{r})}{\partial \mathbf{L}^{(r)}} : \boldsymbol{\varepsilon}, (\mathbf{r})dV$$
$$+\frac{2}{V}\int_V \boldsymbol{\varepsilon}(\mathbf{r}) : \mathbf{L}(\mathbf{r}) : \frac{\partial \boldsymbol{\varepsilon}(\mathbf{r})}{\partial \mathbf{L}^{(r)}}\, dV \tag{7.366}$$

where the local stiffness $\mathbf{L}^{(r)}(\mathbf{r})$ is assumed to be piecewise uniform

$$\mathbf{L}^{(r)}(\mathbf{r}) = \sum_{r=1}^{n} \mathbf{L}^{(r)}\theta^{(r)}(\mathbf{r}), \tag{7.367}$$

With (7.367), the first term on the right-hand side of (7.366) yields

$$\frac{1}{V}\int_V \boldsymbol{\varepsilon}(\mathbf{r}) : \frac{\partial \mathbf{L}(\mathbf{r})}{\partial \mathbf{L}^{(r)}} : \boldsymbol{\varepsilon}(\mathbf{r})\, dV = f^{(r)}\frac{1}{V^{(r)}}\int_{V^{(r)}} \varepsilon_{ij}(\mathbf{r})\varepsilon_{kl}(\mathbf{r})\, dV$$
$$= f^{(r)}\left\langle \varepsilon_{ij}(\mathbf{r})\varepsilon_{kl}(\mathbf{r})\right\rangle_{V^{(r)}} \tag{7.368}$$

while the second term writes

$$\frac{1}{V}\int_V \boldsymbol{\varepsilon}(\mathbf{r}) : \mathbf{L}(\mathbf{r}) : \frac{\partial \boldsymbol{\varepsilon}(\mathbf{r})}{\partial \mathbf{L}^{(r)}}\, dV = \left\{\frac{1}{V}\int_V \boldsymbol{\sigma}(\mathbf{r})\, dV\right\} : \left\{\frac{1}{V}\int_V \frac{\partial \boldsymbol{\varepsilon}(\mathbf{r})}{\partial \mathbf{L}^{(r)}}\, dV\right\} = 0 \tag{7.369}$$

To establish (7.369) we used the Hill lemma in accordance to the fact that the strain field $\partial\varepsilon(r)/\partial\mathbf{L}^{(r)}$ is kinematically admissible, so that $\int_V \left(\partial\varepsilon(r)/\partial\mathbf{L}^{(r)}\right)dV = 0$.

According to (7.368) and (7.369), (7.366) is reduced to

$$\left\langle \varepsilon_{ij}(\mathbf{r})\varepsilon_{kl}(\mathbf{r})\right\rangle_{V^{(r)}} = \frac{1}{f^{(r)}}\bar{\boldsymbol{\varepsilon}} : \frac{\partial \bar{\mathbf{L}}}{\partial \mathbf{L}^{(r)}} : \bar{\boldsymbol{\varepsilon}}. \tag{7.370}$$

The fourth-order tensor $\left\langle \varepsilon_{ij}(\mathbf{r}) : \varepsilon_{kl}(\mathbf{r})\right\rangle_{V^{(r)}}$ corresponds to the second order moment of the strain field over the r – phase in the linear comparison composite material. It is calculated by (7.370) and therefore requires the definition of the linear homogenization scheme to express the macroscopic properties $\bar{\mathbf{L}}$ in terms of the local ones.

The diagonal term $\langle \boldsymbol{\varepsilon} : \boldsymbol{\varepsilon}\rangle_{V^{(r)}}$ of the second-order moment can be adopted in the modified second method as an alternative way to measure the intraphase fluctuation of the strain field better than the classical method. This comes from the convexity of the function $\boldsymbol{\varepsilon} : \boldsymbol{\varepsilon}$

$$\langle \boldsymbol{\varepsilon} : \boldsymbol{\varepsilon} \rangle_{V^{(r)}} \geq \langle \boldsymbol{\varepsilon} \rangle_{V^{(r)}} : \langle \boldsymbol{\varepsilon} \rangle_{V^{(r)}} \tag{7.371}$$

where the equal sign holds only if the strain field is homogeneous.

In the case of isotropic materials defined by (7.329), the second-order moment uses the equivalent strain such that

$$\frac{1}{f^{(r)}}\bar{\boldsymbol{\varepsilon}} : \frac{\partial \bar{\mathbf{L}}}{\partial \mu^{(r)}} : \bar{\boldsymbol{\varepsilon}} = \frac{1}{f^{(r)}}\bar{\boldsymbol{\varepsilon}} : \frac{\partial \bar{\mathbf{L}}}{\partial \mathbf{L}^{(r)}} :: \frac{\partial \mathbf{L}^{(r)}}{\partial \mu^{(r)}} : \bar{\boldsymbol{\varepsilon}} = 2\left\langle \varepsilon_{ij}\, \varepsilon_{kl} \right\rangle_{V^{(r)}} E^2_{ijkl} = 3\left(\bar{\bar{\varepsilon}}^{(r)}_{\text{eq}}\right)^2 \tag{7.372}$$

where $\bar{\bar{\varepsilon}}^{(r)}_{\text{eq}}$ is given by (7.356).

Finally, by adopting the second order moment of strain, the modified secant involves the following steps:

- The identification of the appropriate linear homogenization scheme, which give the overall stiffness $\bar{\mathbf{L}}$ as function of the phase stiffness $\mathbf{L}^{(r)}$ in the linear comparison composite. Then the derivatives in (7.373) can be accomplished.
- The resolution of the following n nonlinear set of equations

$$\mathbf{L}^{(r)} = \mathbf{L}^{(r)}\left(\bar{\bar{\varepsilon}}^{(r)}_{\text{eq}}\right),\ \bar{\bar{\varepsilon}}^{(r)}_{\text{eq}} = \left(\frac{1}{3f^{(r)}}\boldsymbol{\varepsilon} : \frac{\partial \bar{\mathbf{L}}}{\partial \mu^{(r)}} : \boldsymbol{\varepsilon}\right)^{\frac{1}{2}} \tag{7.373}$$

which gives the n unknown secant tensors $L^{(r)}_{ijkl}\left(\bar{\bar{\varepsilon}}^{(r)}_{\text{eq}}\right)$.

As in the case of the classical method, the modified method requires simple iterative algorithm to derive the overall properties of the nonlinear composite.

If the linear comparison composite has overall isotropy, one can easily show from (8.373) that

$$\bar{\bar{\varepsilon}}^{(r)}_{\text{eq}} = \left(\frac{1}{f^{(r)}}\left(\frac{1}{3}\frac{\partial \bar{k}}{\partial \mu^{(r)}}\bar{\varepsilon}^2_{\text{m}} + \frac{\partial \bar{\mu}}{\partial \mu^{(r)}}\bar{\varepsilon}^2_{\text{eq}}\right)\right)^{\frac{1}{2}} \tag{7.374}$$

where $\bar{k}$ and $\bar{\mu}$ are computed by a linear homogenization approach.

Let us illustrate the method in the case of a two-phase isotropic composite material, where phase (1) is softer and dispersed in phase (2). Suppose that Hashin-Shtrikman lower bounds are appropriate to derive the overall properties of the linear composite.

$$\begin{aligned}
\bar{K} &= K^{(2)} + \cfrac{f^{(1)}}{\cfrac{1}{\left(K^{(1)} - K^{(2)}\right)} + \cfrac{f^{(2)}}{K^{(2)} + \frac{4}{3}\,\mu^{(2)}}} \\
\bar{\mu} &= \mu^{(2)} + \cfrac{f^{(1)}}{\cfrac{1}{\left(\mu^{1} - \mu^{(2)}\right)} + \cfrac{6f^{(2)}\left(K^{(2)} + 2\mu^{(2)}\right)\mu^{(2)}}{5\left(3\,K^{(2)} + 4\mu^{(2)}\right)}},
\end{aligned} \tag{7.375}$$

from which one can derive explicit expressions for the second order moment of strain required to compute the tensor of secant moduli in each phase. The results are

$$\bar{\bar{\varepsilon}}^{(1)}_{\text{eq}} = \frac{1}{f^{(1)}} \frac{\bar{\mu} - \mu^{(2)}}{\mu^{(1)} - \mu^{(2)}} \bar{\varepsilon}_{\text{eq}}, \; \bar{\bar{\varepsilon}}^{(2)}_{\text{eq}} = \left(\left(N\bar{\varepsilon}^2_{\text{m}} + M\bar{\varepsilon}^2_{\text{eq}}\right)\right)^{\frac{1}{2}}, \tag{7.376}$$

with

$$N = \frac{1}{3f^{(2)}\mu^{(2)}} \left(\bar{k} - f^{(1)}k^{(1)} \left(\frac{1}{f^{(1)}} \frac{\bar{k} - k^{(2)}}{k^{(1)} - k^{(2)}} \right)^2 - f^{(2)}k^{(2)} \left(\frac{1}{f^{(2)}} \frac{\bar{k} - k^{(1)}}{k^{(2)} - k^{(1)}} \right)^2 \right),$$

$$M = \frac{1}{f^{(2)}\mu^{(2)}} \left(\bar{\mu} - f^{(1)}\mu^{(1)} \left(\frac{1}{f^{(1)}} \frac{\bar{\mu} - \mu^{(2)}}{\mu^{(1)} - \mu^{(2)}} \right)^2 \right.$$
$$\left. - \frac{12}{5} f^{(1)} f^{(2)} k^{(2)} \left(\frac{1}{f^{(1)}} \frac{\bar{\mu} - \mu^{(2)}}{\mu^{(1)} - \mu^{(2)}} \right)^2 \left(\frac{\mu^{(1)} - \mu^{(2)}}{3k^{(2)} + 4\mu^{(2)}} \right)^2 \right).$$

In the classical and modified secant nonlinear extensions presented above, the phase distribution is the same in the LCC and in the nonlinear composite. As explained in previous sections, this results from the choice of a particular linearization scheme. This option is pertinent and does not lead to any ambiguity in the choice of the linear homogenization model used to describe the morphology of the LCC.

However, another richer strategy can be used, in which the homogeneous domain for the secant moduli tensors does not correspond to the domain occupied by the constitutive phases. For example, one could define LCCs with more phases than the nonlinear composite. One can easily anticipate that this richer description of the local heterogeneity of the secant moduli will be closer to the real distribution of the moduli in the nonlinear composite. Hence, the prediction will be more suited. However, the evaluation of a large number of internal variables, the critical choice of a linear model and the difficulty related to the larger number of considered phases, complicate the use of this approach. There is a configuration where this description can be naturally called upon, at least theoretically; when the phase distribution of the nonlinear composite can be appropriately described with morphological patterns. Let us consider the simple case of Hashin's composite spheres assembly. In this case, the linear isotropic behavior of the microstructure can be well described with a three-phase self-consistent scheme based on the analytical solution of the problem of a composite inclusion embedded in an infinite medium.

7.8.2 The Tangent Formulation

The tangent formulation relies on an incremental form of the constitutive law

$$\dot{\boldsymbol{\sigma}}(\mathbf{r}) = \mathbf{l}^{\text{tg}}(\mathbf{r}, \boldsymbol{\varepsilon}(r)) : \dot{\boldsymbol{\varepsilon}}(\mathbf{r}), \tag{7.377}$$

where $\mathbf{l}^{\text{tg}}(\mathbf{r}, \boldsymbol{\varepsilon})$ is the tangent stiffness tensor which is typically anisotropic, even when the material is isotropic. Accordingly, in the case of an isotropic material described by Equations (7.328) and (7.329), the tangent tensor is given by

$$l^{\text{tg}}_{ijkl}(\mathbf{r}, \boldsymbol{\varepsilon}(\mathbf{r})) = 3kE^{1}_{ijkl} + 2\mu^{\text{sec}}(\varepsilon_{\text{eq}})E^{2}_{ijkl} + \frac{4}{3}\frac{d\mu^{\text{sec}}}{d\varepsilon_{\text{eq}}}(\varepsilon_{\text{eq}})\varepsilon_{\text{eq}}\tilde{e}_{ij}\tilde{e}_{kl} \tag{7.378}$$

$$\text{where } \tilde{e}_{ij} = \frac{e_{ij}}{\varepsilon_{\text{eq}}}.$$

The anisotropy of the local tangent modulus renders the development of nonlinear continuum micromechanics a challenging task.

7.8.2.1 The Kröner's Approach

The Kröner approach relies on the elastic Eshelby's solution and was initially motivated by the elastoplastic behavior of polycrystalline materials. This concept was first adopted by Budiansky and Mangasarian [6]. Their original idea was to model the first stage of the plastic deformation so that the elastic Eshelby's solution is applied without any major modifications. They argued that the favorable oriented grains which experience first a plastic deformation are represented by an ellipsoidal inclusion subject to stress-free plastic strain in interaction with an elastic infinite medium representing the other grains which still at the elastic regime. This approximation is also supported by the fact that the number of grain subjected to plastic deformation is low at the earlier stage of the plastic flow and hence the homogenization procedure can be performed by the dilute approximation. In other words, the interactions between grains can be neglected.

Subsequently, Kröner who initiated the self-consistent in elasticity propose similar description, which permits to describe the elastoplastic behavior beyond the earlier stages of the plastic flow. For such a purpose and contrary to Budiansky et al. analysis, Kröner considered the infinite medium in the Eshelby's scheme as the "unknown" homogeneous medium subjected to an average $\bar{\boldsymbol{\varepsilon}}^{\text{p}}$ at a certain stage of the plastic flow. To solve the interaction or localization problem for a given set of grains subjected to uniform plastic strain $\boldsymbol{\varepsilon}^{\text{p}}$ whereas the polycrystal in overall plastically deformed by $\bar{\boldsymbol{\varepsilon}}^{\text{p}}$, Kröner's approximation adopts the Eshelby's solution by describing the set of grains by an ellipsoidal inclusion subjected to an eigenstrain $\boldsymbol{\varepsilon}^{\text{p}}$ and surrounded by an infinite medium which in turn is subjected to a uniform deformation $\bar{\boldsymbol{\varepsilon}}^{\text{p}}$. In addition, the framework was developed in a particular case of homogeneous and isotropic elasticity as well as incompressible plasticity and spherical inclusions.

An intermediary step is required before to apply the Eshelby's solution. It consists of describing the topology of the present inclusion problem by an equivalent one where the infinite medium is purely elastic and the inclusion experiences an eigenstrain $\tilde{\boldsymbol{\varepsilon}} = \boldsymbol{\varepsilon}^{\mathrm{p}} - \bar{\boldsymbol{\varepsilon}}^{\mathrm{p}}$, then the Eshelby's solution can be applied directly as

$$\boldsymbol{\varepsilon} = \bar{\boldsymbol{\varepsilon}} + \mathbf{S} : \tilde{\boldsymbol{\varepsilon}} \tag{7.379}$$

where $\boldsymbol{\varepsilon}$ is the total strain in the inclusion, $\bar{\boldsymbol{\varepsilon}}$ the macroscopic applied strain, and $\mathbf{S}$ the Eshelbys' tensor . If we denote by $\boldsymbol{\sigma}$ the stress in the ellipsoidal inclusion and by $\mathbf{L}^{\mathrm{e}}$ the homogeneous elastic constants (which means they are the same for the ellipsoid and the infinite medium), the linear elastic constitutive law reads

$$\boldsymbol{\sigma} = \mathbf{L}^{\mathrm{e}} : (\boldsymbol{\varepsilon} - \tilde{\boldsymbol{\varepsilon}}) \tag{7.380}$$

Substituting (7.380) into (7.379) gives

$$\boldsymbol{\sigma} = \mathbf{L}^{\mathrm{e}} : \bar{\boldsymbol{\varepsilon}} + \mathbf{L}^{\mathrm{e}} : (\mathbf{S} - \mathbf{I}) : \tilde{\boldsymbol{\varepsilon}} \tag{7.381}$$

Taking into account the homogenized constitutive law $\bar{\boldsymbol{\sigma}} = \mathbf{L}^{\mathrm{e}} : \bar{\boldsymbol{\varepsilon}}$, one obtains the following interaction law

$$\boldsymbol{\sigma} = \bar{\boldsymbol{\sigma}} + \mathbf{L}^{\mathrm{e}} : (\mathbf{I} - \mathbf{S}) : (\bar{\boldsymbol{\varepsilon}}^{\mathrm{p}} - \boldsymbol{\varepsilon}^{\mathrm{p}}) \tag{7.382}$$

Recall that the Eshelby tensor $\mathbf{S}$ depends on the elastic constant $\mathbf{L}^{\mathrm{e}}$ and the shape of the inclusion. Therefore, by its general form, Equation (7.382) can capture the plastic anisotropy resulting from morphological aspects related to the irregular shape of grains.

As mentioned above, the Kröner approach was initially performed in the case of isotropic elastic materials and spherical inclusions. Under such conditions, one has

$$L^{\mathrm{e}}_{ijkl} = 3KE^{1}_{ijkl} + 2\mu E^{2}_{ijkl} \tag{7.383}$$

and

$$S_{ijkl} = s_1 E^{1}_{ijkl} + s_2 E^{2}_{ijkl} \text{ with } s_1 = \frac{1+\nu}{3(1-\nu)} \text{ and } s_2 = \frac{2(4-5\nu)}{15(1-\nu)} \tag{7.384}$$

Hence

$$\mathbf{L}^{\mathrm{e}} : (\mathbf{I} - \mathbf{S}) = 3\,K(1 - s_1)\mathbf{E}^{1} + 2\mu(1 - s_2)\mathbf{E}^{2} \tag{7.385}$$

Substituting (7.385) into (7.382), the incompressible plasticity yields

$$\boldsymbol{\sigma} = \bar{\boldsymbol{\sigma}} + 2\mu(1 - s_2)(\bar{\boldsymbol{\varepsilon}}^{\mathrm{p}} - \boldsymbol{\varepsilon}^{\mathrm{p}}) \tag{7.386}$$

To express the plastic flow, Equation (7.386) should be expressed in a rate or incremental form such that

$$\dot{\boldsymbol{\sigma}} = \dot{\bar{\boldsymbol{\sigma}}} + 2\mu(1 - s_2)(\dot{\bar{\boldsymbol{\varepsilon}}}^{\mathrm{p}} - \dot{\boldsymbol{\varepsilon}}^{\mathrm{p}}) \tag{7.387}$$

which can be also rewritten in terms of total strains as

$$\dot{\boldsymbol{\sigma}} = \dot{\bar{\boldsymbol{\sigma}}} + \frac{2\mu(1 - s_2)}{s_2}(\dot{\bar{\boldsymbol{\varepsilon}}} - \dot{\boldsymbol{\varepsilon}}) \tag{7.388}$$

Equation (7.388) relates local quantities to macroscopic ones, it constitutes the first step for a homogenization scheme, and it is crucial for an accurate prediction of the macroscopic behavior.

Clearly, Equation (7.388) shows that the interaction between the different quantities is purely elastic. This results from the description of the plastic strain as an eigenstrain leading to a purely inhomogeneous thermoelastic problem. In fact, during the plastic flow constraint exerted by the aggregate on a single grain become softer than in the elastic regime and change with the plastic deformation, however, the Kröner's model is governed by an elastic constraint, which also still elastic during the plastic flow. Therefore, this will result in stiff predictions of the overall behavior. The limitations of Kröner's model can be explicitly shown as follows by adopting the tangent formulation.

Let us denote by $\bar{\mathbf{L}}$ the tangent stiffness tensor the polycrystalline aggregate and by $\mathbf{l}$ the one of a single crystal. That is

$$\dot{\bar{\boldsymbol{\sigma}}} = \bar{\mathbf{L}} : \dot{\bar{\boldsymbol{\varepsilon}}}, \dot{\boldsymbol{\sigma}} = \mathbf{l} : \dot{\boldsymbol{\varepsilon}} \tag{7.389}$$

Substituting (7.389) into (7.388) leads to

$$\left(\bar{\mathbf{L}} + \frac{2\mu(1 - s_2)}{s_2}\mathbf{I}\right) : \dot{\bar{\boldsymbol{\varepsilon}}} = \left(\mathbf{l} + \frac{2\mu(1 - s_2)}{s_2}\mathbf{I}\right) : \dot{\boldsymbol{\varepsilon}} \tag{7.390}$$

or

$$\dot{\boldsymbol{\varepsilon}} = \left(\mathbf{l} + \frac{2\mu(1 - s_2)}{s_2}\mathbf{I}\right)^{-1} : \left(\bar{\mathbf{L}} + \frac{2\mu(1 - s_2)}{s_2}\mathbf{I}\right) : \dot{\bar{\boldsymbol{\varepsilon}}} \tag{7.391}$$

which expresses the strain concentration tensor $\mathbf{A}$ as

$$\mathbf{A} = \left(\mathbf{l} + \frac{2\mu(1 - s_2)}{s_2}\mathbf{I}\right)^{-1} : \left(\bar{\mathbf{L}} + \frac{2\mu(1 - s_2)}{s_2}\mathbf{I}\right) \tag{7.392}$$

Furthermore, we can readily show from $\dot{\bar{\boldsymbol{\varepsilon}}} = \langle \dot{\boldsymbol{\varepsilon}} \rangle$ that

$$\bar{\mathbf{L}} = \langle \mathbf{A} : \mathbf{l} \rangle \tag{7.393}$$

and therefore

$$\bar{\mathbf{L}} = \left\langle \mathbf{l} : \left(\mathbf{l} + \frac{2\mu(1 - s_2)}{s_2} \mathbf{I} \right)^{-1} : \left(\bar{\mathbf{L}} + \frac{2\mu(1 - s_2)}{s_2} \mathbf{I} \right) \right\rangle \tag{7.394}$$

Expression (7.394) reproduces the implicit character of the self-consistent scheme as already shown in elasticity. Clearly the nonlinearity is captured in (7.394) since the local tangent modulus depends on the plastic strain $\boldsymbol{\varepsilon}^{\mathrm{p}}$. However, it could easily be proved that the assumption of plastic eigenstrain leads to the Lin-Taylor bound which is equivalent to Voigt model in linear elasticity.

In fact, Taylor [51] and Lin [35] approaches rely on the assumption of homogeneous strain in the polycrystalline aggregate (the single grains experience the same strain) so that $\boldsymbol{\varepsilon} = \bar{\boldsymbol{\varepsilon}}$ and therefore the homogenized tangent modulus $\bar{\mathbf{L}}^{\mathrm{TL}}$ predicted by Taylor-Lin model simply reads $\bar{\mathbf{L}}^{\mathrm{TL}} = \langle \mathbf{l} \rangle$.

On the other hand, one can approximately state in (7.394) that

$$\frac{2\mu(1 - s_2)}{s_2} \approx \mu \tag{7.395}$$

Hence

$$\bar{\mathbf{L}} \approx \left\langle \mathbf{l} : (\mathbf{l} + \mu \mathbf{I})^{-1} . (\bar{\mathbf{L}} + \mu \mathbf{I}) \right\rangle \tag{7.396}$$

According to $\mathbf{l} << \mu$ and $\bar{\mathbf{L}} << \mu$, (7.396) is approximately equivalent to

$$\bar{\mathbf{L}} \approx \langle \mathbf{l} \rangle \tag{7.397}$$

which corresponds to Taylor-Lin solution. Again, such a treatment of the interactions between grains is the subject of criticisms of being purely elastic instead of elastoplastic. In addition, the consistency attributed to Kröner approach is not really true since in his procedure the equivalent homogeneous medium is taken as to be elastic even with assigned plastic deformation. This was fundamentally taken into consideration in the Hill's self-consistent model.

7.8.2.2 Hill's Self-Consistent Model

Hill was inspired by the self-consistent approach developed for inhomogeneous linear elasticity where the methodology consists in introducing the stress polarization tensor with the constraint Hill tensor (Section 7.3.2.) depending on the shape of the inclusion and the elastic constant of the equivalent homogeneous medium. For the nonlinear behavior, Hill adopted the same philosophy by solving successive linear problems at each loading increment that rely on the tangent formulation (7.377).

At this stage, Hill faced the problem of highly fluctuating field in nonlinear behavior. He proposed then systematically an inclusion approach relying on piecewise uniform tangent modulus associated with an average strain rate, so that one has for a single inclusion and the equivalent homogeneous medium the following constitutive relations

$$\dot{\bar{\boldsymbol{\sigma}}} = \bar{\mathbf{L}} : \dot{\bar{\boldsymbol{\varepsilon}}}, \dot{\boldsymbol{\sigma}} = \mathbf{l} : \dot{\boldsymbol{\varepsilon}}, \text{ or } \dot{\bar{\boldsymbol{\varepsilon}}} = \bar{\mathbf{M}} : \dot{\bar{\boldsymbol{\sigma}}}, \dot{\boldsymbol{\varepsilon}} = \mathbf{m} : \dot{\boldsymbol{\sigma}} \tag{7.398}$$

where $\bar{\mathbf{M}}$ and $\mathbf{m}$ denote the global and local tangent compliance tensors, respectively.

Clearly, since the homogenized nonlinear behavior is approached by a linear behavior as proposed by Hill in (7.398), the description does not define the tangent modulus uniquely. For example, any reference moduli $\mathbf{L}^\circ$ for which $\mathbf{L}^\circ : \dot{\boldsymbol{\varepsilon}} = 0$ for all $\dot{\bar{\boldsymbol{\varepsilon}}}$ can be added to $\bar{\mathbf{L}}$ and still yield the same relation between $\dot{\bar{\boldsymbol{\sigma}}}$ and $\dot{\bar{\boldsymbol{\varepsilon}}}$. However, the nature of Hill's model is such that it does select a particular characterization for $\bar{\mathbf{L}}$ among all the possibilities. In Hill's methodology, the shape and orientation of a particular grain is approximated by a similarly aligned ellipsoidal single crystal, which is taken to be embedded in an infinite homogeneous matrix whose moduli $\bar{\mathbf{L}}$ are the overall tangent moduli of the polycrystals to be determined. In this approximate way, the interaction between the grain under consideration and plastically deforming neighbors is taken into account.

Based on Eshelby's solution of an ellipsoidal inclusion having a tangent modulus $\mathbf{l}$ and embedded in an infinite medium homogeneous medium with properties $\bar{\mathbf{L}}$, one can write

$$\dot{\boldsymbol{\sigma}} = \dot{\bar{\boldsymbol{\sigma}}} + \mathbf{H} : \left(\dot{\bar{\boldsymbol{\varepsilon}}} - \dot{\boldsymbol{\varepsilon}}\right) \tag{7.399}$$

Note that equation (7.399) is obtained by following exactly the different steps leading to (7.60). One only needs to substitute the elastic moduli by the tangent ones.

Equation (7.399) involves the constraint Hill's tensor extensively discussed in elasticity and it is given by

$$\mathbf{H} = \bar{\mathbf{L}} : \left(\mathbf{S}^{-1} - \mathbf{I}\right) \tag{7.400}$$

where $\mathbf{S}$ is the Eshelby tenor depending on the shape of the inclusion and on the overall tangent stiffness $\bar{\mathbf{L}}$. At this stage, it should be noticed that a fundamental difference with the Eshelby's solution in linear elasticity is that the Hill's framework relies on the determination of Eshelby's tensor with respect to an anisotropic tangent modulus. This constitutes one of major difficulties in implementing the Hill's model.

Substituting Equation (7.398) into (7.399), one has

$$(\mathbf{l}+\mathbf{H}):\dot{\boldsymbol{\varepsilon}}=(\bar{\mathbf{L}}+\mathbf{H}):\dot{\bar{\boldsymbol{\varepsilon}}} \tag{7.401}$$

from which the strain rate in a single crystal is expressed in terms of the macroscopic strain rate as

$$\dot{\boldsymbol{\varepsilon}}=(\mathbf{l}+\mathbf{H})^{-1}:(\bar{\mathbf{L}}+\mathbf{H}):\dot{\bar{\boldsymbol{\varepsilon}}} \tag{7.402}$$

Equation (7.402) enables us to define the strain concentration tensor in elastoplastic behavior as

$$\dot{\boldsymbol{\varepsilon}}=\mathbf{A}:\dot{\bar{\boldsymbol{\varepsilon}}} \quad \text{with} \quad \mathbf{A}=(\mathbf{l}+\mathbf{H})^{-1}:(\bar{\mathbf{L}}+\mathbf{H}) \tag{7.403}$$

which depends on the overall tangent modulus and geometrical aspects of the inclusion.

Similarly, Equations (7.398) and (7.399) give the stress increment in the inclusion as

$$\dot{\boldsymbol{\sigma}}=\left(\mathbf{m}+\tilde{\mathbf{H}}\right)^{-1}:\left(\bar{\mathbf{M}}+\tilde{\mathbf{H}}\right):\dot{\bar{\boldsymbol{\sigma}}} \tag{7.404}$$

or by introducing the stress concentration tensor as

$$\dot{\boldsymbol{\sigma}}=\mathbf{B}:\dot{\bar{\boldsymbol{\sigma}}} \quad \text{with} \quad \mathbf{B}=\left(\mathbf{m}+\tilde{\mathbf{H}}\right)^{-1}:\left(\bar{\mathbf{M}}+\tilde{\mathbf{H}}\right) \tag{7.405}$$

where $\tilde{\mathbf{H}}$ is the inverse of the Hill's tensor $\mathbf{H}$

We can also readily get from (7.403) and (7.405) the following relationship between strain and stress concentration tensors

$$\mathbf{l}:\mathbf{A}=\mathbf{B}:\bar{\mathbf{L}} \quad \text{and} \quad \mathbf{m}:\mathbf{B}=\mathbf{A}:\bar{\mathbf{M}} \tag{7.406}$$

Finally, the homogenization procedure, which relies on $\dot{\bar{\boldsymbol{\varepsilon}}}=\langle\dot{\boldsymbol{\varepsilon}}\rangle, \bar{\mathbf{L}}=\langle\mathbf{l}:\mathbf{A}\rangle$, $\dot{\bar{\boldsymbol{\sigma}}}=\langle\dot{\boldsymbol{\sigma}}\rangle$ and $\bar{\mathbf{M}}=\langle\mathbf{m}:\mathbf{B}\rangle$ leads to

$$\bar{\mathbf{L}}=\left\langle\mathbf{l}:(\mathbf{l}+\mathbf{H})^{-1}:(\bar{\mathbf{L}}+\mathbf{H})\right\rangle \tag{7.407}$$

$$\bar{\mathbf{M}}=\left\langle\mathbf{m}:\left(\mathbf{m}+\tilde{\mathbf{H}}\right)^{-1}:\left(\bar{\mathbf{M}}+\tilde{\mathbf{H}}\right)\right\rangle \tag{7.408}$$

7.8.2.3 Illustrations in the Case of Conventional Polycrystalline Materials

The main purpose of this section is to explore the feasibility of the self-consistent method developed by Hill to predict stress-strain behavior of conventional polycrystalline materials from the elastoplastic properties of single crystal constituents. We will focus on FCC metallic materials by presenting briefly the main features of plastic deformation at the continuum level of single crystals under conventional loading conditions of strain rates and temperature. Our attention is not to describe exhaustively the various mechanisms of plastic deformation from physical metallurgy point of view, for such a purpose the reader could refer to more specialized text book and technical papers which are recommended at the end of this chapter. Instead, we will follow a mechanistic procedure to bridge the scales between the single crystal level to the polycrystalline one.

The physical aspects of single crystal plasticity were established during the earlier part of the last century, in 1900 to 1938, with the contribution of Ewing and Rosenhain [14], Bragg [4], Taylor and co-workers [48, 49, 50, 51], Polanyi [43], Schmid [45], and others. Their experimental measurements established that at room temperature the major source of plastic deformation is the dislocation movements through the crystal lattices. These motions occur on certain crystal planes in certain crystallographic directions, and the crystal structure of metals is not altered by the plastic flow. The mathematical presentation of these physical phenomena of plastic deformation in single crystals was pioneered by Taylor [51] when he investigated the plastic deformation of polycrystalline materials in terms of single crystal deformation. More rigorous and rational formulations have been provided by Hill [27], Hill and Rice [28], Asaro and Rice [1], and by Hill and Havner [29]. A comprehensive review of this subject can be found in Asaro [2].

The kinematics of single crystal deformation and resulting elastoplastic constitutive laws are based on an idealization of dislocation movement by a collective one leading to slips in certain directions on specific crystallographic planes. This process occurs when the resolved shear stress on one or more of these slip systems reaches a critical values. As plastic deformation proceeds, the critical yield stresses associated with the slip systems increases. This contributes to the strain hardening of the polycrystalline aggregate.

Consider a single crystal with N possible slip systems. Each system g is characterized by the unit normal $\mathbf{n}^g$ to the plane along which the collective movement of dislocations occurs, and by the direction $\mathbf{m}^g$ of dislocation gliding, which is co-linear to the Burgers vector $\mathbf{b}^g$ of gliding dislocations on the system g, so that $\mathbf{b}^g = b\mathbf{m}^g$, where b is the magnitude of the Burgers vector. The mathematical tool treating the collective movement of dislocations consider each dislocation line as the boundary of a cutting surface S^g with a unit normal $\mathbf{n}^g$, across which the discontinuities of the displacement vector are uniform and characterized by the Burgers vector $\mathbf{b}^g$ so that $b_i^g n_i^g = 0$. This transformation can be described at each material point by a second-order tensor $\boldsymbol{\beta}^{\mathrm{p}}(\mathbf{r})$ as

$$\beta_{ij}^{\mathrm{p}}(\mathbf{r}) = b_i^g n_j^g \delta(S^g) \tag{7.409}$$

where $\delta(S^g)$ is the Dirac function given by

$$\delta(S^g) = \iint_{S^g} \delta(\mathbf{r} - \mathbf{r}')dS' \tag{7.410}$$

If many dislocations with the same Burgers vector $\mathbf{b}^g$ and same cutting surfaces are present in the single crystal volume V, one can define and average transformation $\boldsymbol{\beta}^{\mathrm{p}}(\mathbf{r})$ expressed by

$$\beta_{ij}^{p} = b m_i^g n_j^g \frac{1}{V} \iiint\limits_V \delta(S^g) dV \tag{7.411}$$

Introducing the average plastic shear γ^g by

$$\gamma^g = b \frac{1}{V} \iiint\limits_V \delta(S^g) dV \tag{7.412}$$

leads to

$$\beta_{ij}^{\mathrm{p}} = \gamma^g m_i^g n_j^g \tag{7.413}$$

If we account for all dislocations present at the slip systems, (7.413) can be extended to

$$\beta_{ij}^{\mathrm{p}} = \sum_g \gamma^g m_i^g n_j^g \tag{7.414}$$

The shear rate $\dot{\gamma}^g$ is calculated from (7.412) as

$$\dot{\gamma}^g = b \frac{\partial}{\partial t} \left\{ \frac{1}{V} \iiint\limits_V \delta(S^g) dV \right\} \tag{7.415}$$

Equation (7.415) describing the creation and movement of dislocations at the continuum level corresponds to the Orowan relation.

The rate of plastic distortion $\dot{\boldsymbol{\beta}}^{\mathrm{p}}$ is the sum of the contributions of the shear rates $\dot{\gamma}^g$ from all the active slip systems. That is

$$\dot{\beta}_{ij}^{\mathrm{p}} = \sum_g \dot{\gamma}^g m_i^g n_j^g \tag{7.416}$$

where the symmetric part give the plastic strain rate

$$\dot{\varepsilon}_{ij}^{p} = \frac{1}{2} \left(\dot{\beta}_{ij}^{\mathrm{p}} + \dot{\beta}_{ji}^{\mathrm{p}} \right) = \sum_g \dot{\gamma}^g R_{ij}^g \tag{7.417}$$

where

$$R_{ij}^g = \frac{1}{2} \left(m_i^g n_j^g + m_j^g n_i^g \right) \tag{7.418}$$

and the antisymmetric part the plastic spin

$$\dot{w}_{ij}^{p} = \frac{1}{2}\left(\dot{\beta}_{ij}^{\mathrm{p}} - \dot{\beta}_{ji}^{\mathrm{p}}\right) = \sum_{g} \dot{\gamma}^{g} \tilde{R}_{ij}^{g} \tag{7.419}$$

where

$$\tilde{R}_{ij}^{g} = \frac{1}{2}\left(m_i^g n_j^g - m_j^g n_i^g\right) \tag{7.420}$$

The second-order tensors $\mathbf{R}^g$ and $\tilde{\mathbf{R}}^g$ are also called the orientation tensors, they only depend on the orientation of the considered single crystal.

Let $\boldsymbol{\sigma}$ denote the stress in the single crystal. The so-called resolved shear stress on a slip system g is given by

$$\tau^{g} = \sigma_{ij} R_{ij}^{g} \tag{7.421}$$

Within the framework of time independent plasticity (any viscous effect is neglected), a slip system g is considered to be active if the resolved shear stress τ^g reaches a critical value τ_{c}^{g}, which depends on the previous deformation history of the single crystal leading to notion of strain hardening state. It is generally assumed that the deformation history of a given slip system g only depends on the amplitude of shear strain associated to n active systems, so that one can write

$$\tau_{\mathrm{c}}^{g} = \tilde{F}^{g}\left(\gamma^{1}, \gamma^{1}, \ldots, \gamma^{n}\right) \tag{7.422}$$

When the amount of shear is small enough, we can use a linear approximation of (7.422) as

$$\tau_{\mathrm{c}}^{g} \approx \tilde{F}^{g}(0,0,\ldots,0) + \sum_{h} \frac{\partial \tilde{F}^{g}}{\partial \gamma^{h}}(0,0,\ldots,0)\gamma^{h} \tag{7.423}$$

where $\tilde{F}^{g}(0,0,\ldots,0)$ can be seen as the initial critical shear stress of the slip system g, it is generally assumed to be the same for all slip systems. Therefore, (7.423) can be expressed by

$$\tau_{\mathrm{c}}^{g} = \tau^{\mathrm{o}} + \sum_{h} H^{gh}\gamma^{h} \tag{7.424}$$

where $\tau^{\mathrm{o}} = \tilde{F}^{g}(0,0,\ldots,0)$ and $H^{gh} = \frac{\partial \tilde{F}^{g}}{\partial \gamma^{h}}(0,0,\ldots,0)$ is the strain hardening matrix, which describes the hardening interactions between the different slip systems. Note that the diagonal terms of the matrix H^{gh} define the self-hardening due to the plastic shear in the same system, whereas the nondiagonal components correspond to the latent-hardening due to shear slip on the other systems. The matrix coefficients can be evaluated by experimental characterization performed on single crystals.

At any stage of the deformation process the rates of changes of critical shear stress are deduced from (7.424) as

$$\dot{\tau}_{\mathrm{c}}^{g} = \sum_{h} H^{gh} \dot{\gamma}^{h} \tag{7.425}$$

It follows from the above definitions that a slip system is potentially active if $\tau^g = \tau_{\mathrm{c}}^g$ and load or unload, respectively, depends on whether

$$\dot{\tau}^{g} = \dot{\tau}_{\mathrm{c}}^{g} \qquad \text{with} \qquad \dot{\gamma}^{g} \geq 0 \tag{7.426}$$

or

$$\dot{\tau}^{g} < \dot{\tau}_{\mathrm{c}}^{g} \qquad \text{with} \qquad \dot{\gamma}^{g} = 0 \tag{7.427}$$

A system is inactive if $\tau^g < \tau_{\mathrm{c}}^g$ and then $\dot{\gamma}^g = 0$

Relation (7.426) is known as the consistency condition whose resolution for each active system g determine the shear rate $\dot{\gamma}^g$ on this system. Taking into account Equations (7.421) and (7.425), the consistency condition writes

$$\dot{\sigma}_{ij} R_{ij}^{g} = \sum_{h} H^{gh} \dot{\gamma}^{h} \tag{7.428}$$

From the definition (7.417) of the plastic strain rate, the total strain rate, which is the sum of the elastic and plastic parts is given by

$$\dot{\boldsymbol{\varepsilon}} = \dot{\boldsymbol{\varepsilon}}^{\mathrm{e}} + \dot{\boldsymbol{\varepsilon}}^{p} = \mathbf{L}^{\mathrm{e}-1} : \dot{\boldsymbol{\sigma}} \sum_{g} \dot{\gamma}^{g} \mathbf{R}^{g} \tag{7.429}$$

or

$$\dot{\boldsymbol{\sigma}} = \mathbf{L}^{\mathrm{e}} : \left(\dot{\boldsymbol{\varepsilon}} - \sum_{g} \dot{\gamma}^{g} \mathbf{R}^{g} \right) \tag{7.430}$$

Note that, for a given state of stress $\boldsymbol{\sigma}$, $\dot{\boldsymbol{\sigma}}$ is uniquely related to $\dot{\boldsymbol{\varepsilon}}$ if the hardening matrix H^{gh}, governing the determination of shear rate in different slip systems, is positive semi-definite [27], while only for certain hardening laws, the shear rates $\dot{\gamma}^g$ are always unique. At least one set of shear rates exists which satisfies the constitutive relations (7.415) and (7.426) thru (7.428) for a prescribed strain rate $\dot{\boldsymbol{\varepsilon}}$ (or prescribed stress $\dot{\boldsymbol{\sigma}}$). If there are N non-zero $\dot{\gamma}^g$, they satisfy N equations resulting from the combination of the consistency condition $\dot{\tau}^g = \dot{\tau}_{\mathrm{c}}^g$ and the constitutive relations (7.430).

In fact, substituting (7.430) into (7.428) yields to the following set of equations

$$\sum_h \mathbf{Q}^{gh} \dot{\gamma}^h = \mathbf{R}^g : \mathbf{L}^e : \dot{\boldsymbol{\varepsilon}}$$

where

$$\mathbf{Q}^{gh} = H^{gh} + \mathbf{R}^g : \mathbf{L}^e : \mathbf{R}^h \tag{7.432}$$

These equations are associated with the loading system together with the constraints $\dot{\gamma}^h \geq 0$

Only for certain hardening laws will the N × N matrix $\mathbf{Q}^{gh}$ be necessarily nonsingular. But for perfect plasticity ($H^{gh} = 0$), for example, it is always possible to choose at least a set of linearly independent slip systems among the potentially active such that this matrix is nonsingular and the auxiliary equations (7.432) are satisfied. Thus, for perfect plasticity $\mathbf{Q}^{gh}$ is never greater than 5×5 matrix. If its inverse is denoted by $\tilde{\mathbf{Q}}^{gh}$, the N nonzero shear rates for this choice of active slip systems are expressed by

$$\dot{\gamma}^g = \mathbf{d}^g : \dot{\boldsymbol{\varepsilon}} \quad \text{where} \quad \mathbf{d}^g = \sum_h \tilde{\mathbf{Q}}^{gh} \mathbf{L}^e : \mathbf{R}^h \tag{7.433}$$

Recall that the tangent moduli and compliances of the considered single crystal are, respectively, defined by

$$\dot{\boldsymbol{\sigma}} = \mathbf{l} : \dot{\boldsymbol{\varepsilon}} \quad \text{and} \quad \dot{\boldsymbol{\varepsilon}} = \mathbf{m} : \dot{\boldsymbol{\sigma}} \tag{7.434}$$

From the foregoing kinematics of single crystal plastic deformation, the main feature of the tangent moduli and compliances is that they depend on the set of active slip systems which in turn depends on the prescribed strain rate $\dot{\boldsymbol{\varepsilon}}$ (or stress $\dot{\boldsymbol{\sigma}}$). Well-known in crystal plasticity framework, the definition of tangent moduli and compliances leads to a multi-branches description. It also should be noticed that regarding Equation (7.434), the inverse of tangent moduli does not exist in all situations, in other words $\mathbf{l}$ may present singularities. Such a case is typically the one of a perfect plastic behavior where the problem of homogenization is treated by using directly the tangent moduli rather than its inverse since there is no restrictions on the strain rate, whereas the stress rate is subjected to certain conditions regarding the regions in stress rate space.

Substituting (7.433) into (7.430) and in comparison to (7.434), one obtains the following expression for single crystal tangent moduli

$$l_{ijmn} = L^e_{ijkl} : \left(I_{klmn} - \sum_g R^g_{kl} d^g_{mn} \right) \tag{7.435}$$

Using (7.432) and (7.433), (7.435) can be explicitly rewritten as

$$\mathbf{l} = \mathbf{L}^{\mathrm{e}} : \left(I - \sum_{g,h} \mathbf{R}^{g} \left(H^{gh} + \mathbf{R}^{g} : \mathbf{L}^{\mathrm{e}} : \mathbf{R}^{h} \right)^{-1} L^{\mathrm{e}} : \mathbf{R}^{h} \right) \tag{7.436}$$

It can readily shown that tangent moduli $\mathbf{l}$ as given by (7.436) satisfies the symmetries

$$l_{ijkl} = l_{klij} \text{ if } H^{gh} = H^{hg}.$$

In the case of isotropic elastic constant of single crystals where

$$L^{\mathrm{e}}_{ijkl} = 3KE^{1}_{ijkl} + 2\mu E^{2}_{ijkl}$$

the tangent modulus writes

$$l_{ijkl} = 3KE^{1}_{ijkl} + 2\mu E^{2}_{ijkl} - 4\mu^{2} \sum_{g,h} R^{g}_{ij} \left(H^{gh} + 2\mu R^{g}_{pq} R^{h}_{pq} \right)^{-1} R^{h}_{kl} \tag{7.437}$$

where the plastic incompressibility is used.

In summary, the single crystal tangent modulus described by (7.436) is unique for a given strain rate $\dot{\boldsymbol{\varepsilon}}$ even if the shear rates $\dot{\gamma}^{g}$ are not. Referring to (7.437), one can remark that even though the elasticity is approximated as to be isotropic, the tangent moduli are anisotropic in nature. This results from the typical process of the plastic flow, which relies on the number of active self-systems and their interactions governing the strain hardening behavior.

Then equations (7.407) and (7.408) generated by Hill's self-consistent tangent formaltion can be used to estimate the overall tagent stiffness or compliance of a conventional polycrystalline aggregate. Contrary to (7.407), (7.408) requires the inverse of $\mathbf{l}$ for each grain to exist. In practical situations of polycrystalline materials, equation (7.407) is widely used to avoid the difficulties associated with (7.408) when any of the single crystal tangent modulus $\mathbf{l}$ presents a singularity.

Note that in (7.407) by substituting the elastoplastic tangent modulus by the elastic ones lead exactly to the description of overall elastic moduli in linear inhomogeneous elasticity. This simply results from the Hill's linearization procedure of the elastoplastic behavior. Particularly, equation (7.407) gives accurate results in comparison with Kröner's model. In fact, simplifications made in expression (7.396) are not allowed here since $\mathbf{l}$ and $\mathbf{H}$ are of the same order of magnitude and then contrary to Taylor-Lin model, Hill's method captures at least partially the fluctuations of strains between each grain. However, the Hill's model still an approximation by taking piecewise uniform mechanical properties and therefore any intraphase fluctuation, which naturally results from the elastoplastic behavior is disregarded is this approach.

It can also be noticed from expressions (7.436) and (7.407) that the determination of the homogenized properties described by the macroscopic tangent modulus $\bar{\mathbf{L}}$ is not an easy task. This is due to the implicit nature of (7.407) and also to the anisotropy of the tangent modulus, which makes the calculation of the constraint Hill tensor (7.407) a complicated procedure. In general, the determination of $\bar{\mathbf{L}}$ through the Hill's approach requires an iterative procedure in which at a given state of deformation an initial guess of $\bar{\mathbf{L}}$ has to be made, then equation (7.407) can used to obtain an improved value for $\bar{\mathbf{L}}$. This procedure is repeated sufficiently until a convergent value is obtained.

The major difficulty behind the Hill's model is discussed in the following. The average tangent modulus $\bar{\mathbf{L}}$ depends on the prescribed values of strain rate $\dot{\boldsymbol{\varepsilon}}$; but, unlike the corresponding single crystal tangent moduli, it will not have only a finite number of branches. Instead, $\bar{\mathbf{L}}$ will, in general, varies continuously as the direction of prescribed strain rate varies in strain rate space. That is, as pointed out by Hutchinson [30], $\bar{\mathbf{L}}$ is a homogeneous function of degree zero of $\dot{\boldsymbol{\varepsilon}}$. For practical situations, the Hill's model is appropriate for monotonic radial loading conditions but by adding further assumptions regarding the anisotropy of the tangent modulus to make easier the calculation of the Hill's fourth-order tensor. Comprehensive discussions about feasible ways in implementing the Hill's approach to describe practical situations can be found in the excellent paper of Doghri et Ouaar (2003) or recently in the contribution by Pierard et al. (2007), who succeeded in carrying out a systematic comprison between the classical secant method, the modified one, and the tangent formulation in the case of a two-phase nonlinear composite material.

In general, the treatment of the difficulties generated by the Hill's approach was the center of different investigations leading to the emergence of nonlinear mean field theories with different varieties of linearization sequences of the nonlinear behavior. Within these procedures, one can distinguish between the secant, the tangent and affine approaches. The classical secant formulation was developed by Berveiller and Zaoui [5] for crystalline materials and adapted later by Tandon and Weng [53] to the case of two-phase elastoplastic composite materials. The secant formulation which could be seen an intermediate method between the Kröner one and the incremental method of Hill, reduces deeply the complexity of Hill's model by assuming isotropic homogeneous plastic flow in each phase. The first variety of tangent mean field theory was introduced by Hutchinson [31] to to describe steady-state creep behavior of crystalline materials. It turned out through the Hutchinson's analysis that the use of a power-law creep leads also to another variant of the secant description (see below). More recently, Molinari et al. [40] derived a tangent formulation for viscoplastic power-law by adopting a sequence of linearization similar to the linear thermoelasticity.

7.8.2.4 On Time-Dependent Behavior of Polycrystalline Materials

Time-dependent behavior of polycrystalline materials has been first formulated by Hutchinson by assuming a power law describing the shear rate on slip systems for a given single crystal. Hutchinson [31] assumed a steady creep behavior for single crystal for which the shear rate induced in a slip system g by a given resolved shear stress τ^g is described by a rate sensitive criterion leading to a nonlinear viscous behavior. That is

$$\dot{\gamma}^g = \dot{\gamma}^o \left(\frac{\tau^g}{\tau_c^g} \right)^n \tag{7.438}$$

where $\dot{\gamma}^o$ is a reference rate and n the inverse of rate sensitivity. When $n >> 1$, the shear increase of the considered slip system is negligible unless τ^g is very close to τ_c^g. This statement is equivalent to conditions (7.426) for slip systems activation in time-independent plasticity. The critical shear stress τ_c^g, often called reference stress in time-dependent plasticity, is strongly dependent on temperature. The exponent n depends also on temperature, although somewhat less strongly, and usually falls between 3 and 8 for metals. A survey on the temperature ranges where the steady creep of polycrystal and single crystal can be approximated by a power law was given by Ashby and Frost [3].

If N is the total number of all slip systems, the total strain rate is the sum of the contributions of all these systems and it is given by

$$\dot{\varepsilon}_{ij} = \sum_{g=1}^{N} \dot{\gamma}^g R_{ij}^g = m_{ijpq}^{\text{sec}} \sigma_{pq} \tag{7.439}$$

where

$$m_{ijpq}^{\text{sec}} = \sum_{g=1}^{N} \left(\frac{\dot{\gamma}^o}{\tau_c^g} \right) \left(\frac{\tau^g}{\tau_c^g} \right)^{n-1} R_{ij}^g R_{pq}^g \tag{7.440}$$

where $\mathbf{m}^{\text{sec}}$ is the so-called secant viscoplastic compliance moduli of the considered single crystal. As reported by Hutchinson, the compliances are homogeneous of degree $n - 1$ in the stress, so that

$$\mathbf{m}^{\text{sec}}(\lambda \boldsymbol{\sigma}) = \lambda^{n-1} \mathbf{m}^{\text{sec}}(\boldsymbol{\sigma}) \tag{7.441}$$

Let now define the stress potential $\psi(\boldsymbol{\sigma})$ and strain rate potential $\phi(\dot{\boldsymbol{\varepsilon}})$ such that

$$\dot{\boldsymbol{\varepsilon}} = \frac{\partial \psi}{\partial \boldsymbol{\sigma}} \quad \text{and} \quad \boldsymbol{\sigma} = \frac{\partial \phi}{\partial \dot{\boldsymbol{\varepsilon}}} \tag{7.442}$$

The viscoplastic constitutive law (7.439) leads to following typical relationships between the dissipation $\boldsymbol{\sigma} : \dot{\boldsymbol{\varepsilon}}, \psi(\boldsymbol{\sigma})$ and $\phi(\dot{\boldsymbol{\varepsilon}})$

$$\boldsymbol{\sigma} : \dot{\boldsymbol{\varepsilon}} = (n+1)\psi(\boldsymbol{\sigma}) = \frac{n+1}{n}\phi(\dot{\boldsymbol{\varepsilon}}) = \sum_{g=1}^{n} \tau^g \dot{\gamma}^g \tag{7.443}$$

Substituting (7.442) into (7.443) and by the derivative of (7.443) with respect to stresses, one obtains

$$\frac{\partial}{\partial \sigma_{ij}}(\sigma_{kl} : \dot{\varepsilon}_{kl}) = (n+1)\frac{\partial \psi}{\partial \sigma_{ij}} = \frac{\partial \psi}{\partial \sigma_{ij}} + \sigma_{kl} : \frac{\partial^2 \psi}{\partial \sigma_{ij} \partial \sigma_{kl}} \tag{7.444}$$

which leads to

$$n\frac{\partial \psi}{\partial \sigma_{ij}} = \sigma_{kl} : \frac{\partial^2 \psi}{\partial \sigma_{ij} \partial \sigma_{kl}} \tag{7.445}$$

Combining (7.444) with (7.439), we can readily show that

$$m_{ijkl}^{\text{sec}} = \frac{1}{n}\frac{\partial^2 \psi}{\partial \sigma_{ij} \partial \sigma_{kl}} \tag{7.446}$$

where

$$\frac{\partial^2 \psi}{\partial \sigma_{ij} \partial \sigma_{kl}} = \frac{\partial \dot{\varepsilon}_{kl}}{\partial \sigma_{ij}} = m_{ijkl}^{\text{tg}} \tag{7.447}$$

In Equation (7.447) $\mathbf{m}^{\text{tg}}$ correspond to the tangent compliance moduli.

On the other hand, a Taylor expansion of equation (7.439) at the vicinity of a point $\tilde{\boldsymbol{\sigma}} - \mathbf{d}$ can be written as

$$\dot{\varepsilon}_{ij} = \left.\frac{\partial \dot{\varepsilon}_{ij}}{\partial \sigma_{kl}}\right|_{\sigma=\tilde{\sigma}} \sigma_{kl} + \tilde{\dot{\varepsilon}}_{ij} = m_{ijkl}^{\text{tg}}(\tilde{\boldsymbol{\sigma}}) : \sigma_{kl} + \tilde{\dot{\varepsilon}}_{ij} \tag{7.448}$$

where $\tilde{\dot{\varepsilon}}_{ij}$ is called the back extrapolated strain rate.

From (7.448) and (7.446), the relation between the grain's secant and tangent moduli are

$$\mathbf{m}^{\text{tg}} = n\,\mathbf{m}^{\text{sec}} \tag{7.449}$$

At the polycrystal level, the macroscopic constitutive law is assumed to be similar to the one of single crystals, so that one can write

$$\bar{\dot{\varepsilon}}_{ij} = \bar{M}_{ijpq}^{\text{sec}} \bar{\sigma}_{pq} \tag{7.450}$$

where $\bar{\mathbf{M}}^{sec}$ is the macroscopic secant compliance moduli. In the same way as the single crystal level, Taylor development of (7.450) at the vicinity of the macroscopic stress leads to the definition of a macroscopic tangent modulus $\bar{M}^{tg}$ as

$$\dot{\bar{\varepsilon}}_{ij} = \bar{M}^{tg}_{ijpq}(\bar{\boldsymbol{\sigma}})\bar{\sigma}_{pq} + \dot{\varepsilon}^{o}_{ij} \tag{7.451}$$

where

$$\bar{M}^{tg}_{ijpq}(\bar{\boldsymbol{\sigma}}) = \left.\frac{\partial \dot{\bar{\varepsilon}}_{ij}}{\partial \bar{\sigma}_{kl}}\right|_{\bar{\sigma}=\bar{\sigma}}$$

and $\dot{\varepsilon}^{o}_{ij}$ a macroscopic extrapolated strain rate.

Hutchinson [31] has shown that the macroscopic tangent and secant moduli are linked by a similar relation as for single crystals, i.e., $\bar{\mathbf{M}}^{tg} = n\bar{\mathbf{M}}^{sec}$. This is straightforward derived by defining the macroscopic dissipation $\bar{\boldsymbol{\sigma}} : \dot{\bar{\boldsymbol{\varepsilon}}}$ and macroscopic strain rate potential $\boldsymbol{\Phi}(\dot{\bar{\boldsymbol{\varepsilon}}})$ and stress potential $\boldsymbol{\Psi}(\bar{\boldsymbol{\sigma}})$

Note that Equations (7.448) and (7.451) are exact only when they describe the strain rate associated with the stress used as a reference for the expansion, otherwise they are only approximate. This will not present a limitation for treatment of the grain, since the stress and the strain rate are assumed to be uniform within the framework of the self-consistent scheme. As a result, the actual value of stress in the considered grain can be selected to perform the expansion.

Starting from the linearized equation (7.451), the macroscopic tangent compliance moduli $\bar{\mathbf{M}}^{tg}$ can be estimated by adopting a Hill's type self-consistent method, which consists in considering each grain with tangent compliance moduli $\mathbf{m}^{tg}$ and prescribed reference strain rate $\dot{\bar{\boldsymbol{\varepsilon}}}$ embedded in an infinite homogenized medium having the properties $\bar{\mathbf{M}}^{tg}$ and prescribed reference strain rate $\dot{\boldsymbol{\varepsilon}}^{o}$

Following the same procedure as for the Hill's formulation, the Eshelby's solutions extended for a tangent formulation give the interaction relation linking the local to the macroscopic quantities

$$\dot{\boldsymbol{\varepsilon}} = \dot{\bar{\boldsymbol{\varepsilon}}} + \tilde{\mathbf{H}} : (\bar{\boldsymbol{\sigma}} - \boldsymbol{\sigma}) \tag{7.452}$$

where $\tilde{\mathbf{H}}$ is the inverse of the constraint Hill's tensor expressed by

$$\tilde{\mathbf{H}} = \left(\mathbf{S}^{-1} - \mathbf{I}\right)^{-1} : \bar{\mathbf{M}}^{tg} \tag{7.453}$$

Note that the Eshelby tensor in (7.453) depends on the tangent compliance moduli together with the shape of the considered grain. As reported by Lebensohn and Tomé [34], the relation $\bar{\mathbf{M}}^{tg} = n\bar{\mathbf{M}}^{sec}$ enables to express the equations in terms of the secant compliance moduli as

$$\tilde{\mathbf{H}} = n\left(\mathbf{S}^{-1} - \mathbf{I}\right)^{-1} : \bar{\mathbf{M}}^{\text{sec}} \tag{7.454}$$

Substituting (7.439) and (7.450) into (7.452) yields

$$\boldsymbol{\sigma} = \mathbf{B} : \bar{\boldsymbol{\sigma}} \qquad \text{with} \qquad \mathbf{B} = \left(\mathbf{m}^{\text{sec}} + \tilde{\mathbf{H}}\right)^{-1} : \left(\bar{\mathbf{M}}^{\text{sec}} + \tilde{\mathbf{H}}\right) \tag{7.455}$$

Finally, the homogenization procedure using $\bar{\boldsymbol{\sigma}} = \langle \boldsymbol{\sigma} \rangle$ and $\bar{\mathbf{M}}^{\text{sec}} = \langle \mathbf{m}^{\text{sec}} : \mathbf{B} \rangle$ leads to

$$\bar{\mathbf{M}}^{\text{sec}} = \left\langle \mathbf{m}^{\text{sec}} : \left(\mathbf{m}^{\text{sec}} + \tilde{\mathbf{H}}\right)^{-1} : \left(\bar{\mathbf{M}}^{\text{sec}} + \tilde{\mathbf{H}}\right) \right\rangle \tag{7.456}$$

As a conclusion, the viscoplastic self-consistent model initially developed by Hutchinson [31] to deal with steady creep of polycrystalline materials has shown how to identify a secant formulation to a tangent one. The application of the Eshelby's solution required for the self-consistent scheme has also shown the utility of combining both the secant and tangent moduli to solve interaction problem. Since the developments of Hutchinson, the viscoplastic self-consistent model was adopted by many authors as an alternative strategy for tackling the problem of large plastic deformations by simply neglecting the elastic deformation. This way of thinking was successively adopted by Molinari et al. [40] to describe the texture development in cubic polycrystals. For more information regarding the numerical implementation and limitations of the method, the reader may refer to the work of Lebensohn and Tomé [34].

7.9 Illustrations in the Case of Nanocrystalline Materials

As discussed above, continuum micromechanics principles can be adapted to capture an intrinstic size effect required to describe the deformation responses of NC materials. This will be illustrated by the contribution of Jiang and Weng [32] that invokes the concept of two-phase composite with grain interiors and grain boundaries playing the role of constitutive phases.

Jiang and Weng's framework relies on the generalized self-consistent model of Christensen and Lo to account, within a phenomenological manner, for the plastic anisotropy of variously oriented grains, and the stress heterogeneity of the grains and grain-boundary phases. The polycrystalline material is replaced by a micro-continuum domain constituted of equiaxed grains exhibiting distinct crystallographic orientations embedded in a matrix, as depicted in (Figs. 7.7a, b). The composite inclusion problem used to determine the stress-strain state over a grain is presented in (Fig. 7.7c). It considers a spherical grain surrounded by a grain boundary phase of finite thickness, the system is surrounded by an infinite medium representing the unknown effective properties of the polycrystal.

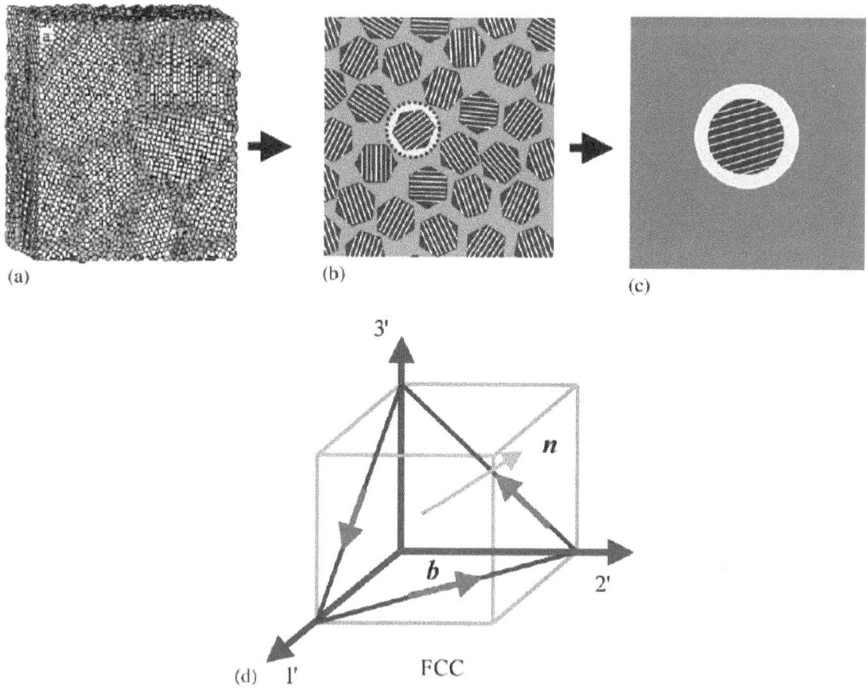

Fig. 7.7 Rationale for the generalized self-consistent polycrystal model (Jiang and Weng, 2004)

Both grain and grain boudary are modeled as ductile phases capable of undergoing plastic deformation at room-temperature. In a grain the process is governed by crystallographic slips. A slip direction and slip-plane normal of a faced-centered cubic crystal, such as copper, are schematically shown in (Fig. 7.7d).

7.9.1 Volume Fractions of Grain and Grain-Boundary Phases

In NC materials, the grain size (typically below 100 nm) is such that the grain boundary volume is no more negligible. In terms of the grain size (diameter) d and grain-boundary thickness δ, the volume fraction of the grains can be approximated by

$$c_{\mathrm{g}} = \left(\frac{d}{d + \delta} \right)^3 \tag{7.457}$$

7.9.2 Linear Comparison Composite Material Model

Within the framework of Jiang and Weng, the overall elastoplastic response of the NC polycrystal is calculated through a linear comparison composite model,

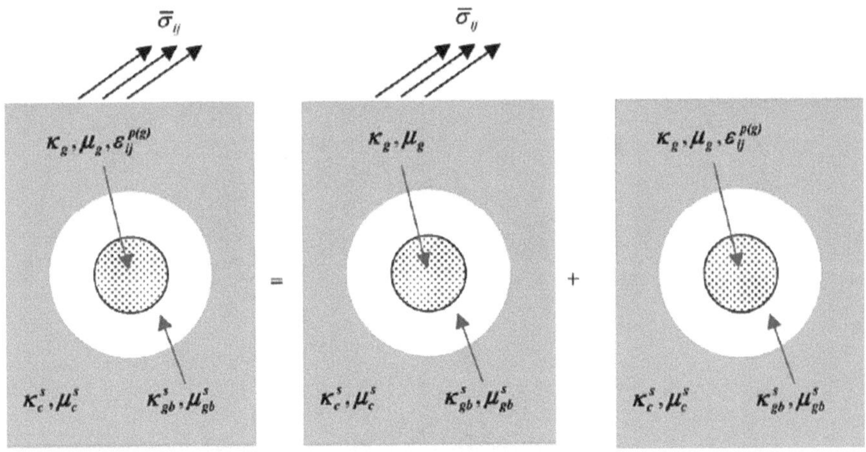

Fig. 7.8 Superposition of two linear problems (Jiang and Weng, 2004)

using the secant moduli of the grain-boundary phase to represent its elasto-plastic state and the eigenstrain in the inclusion to represent the plastic strain of the crystallite. This was accomplished by superposing Christensen and Lo's [11] generalized self-consistent scheme and Luo and Weng's [36] three-phase concentrated eigenstrain problem. Such a superposition is schematically shown in (Fig. 7.8). Both solutions were given for elastically isotropic constituents, and thus for simplicity the crystallites were also taken to be elastically isotropic while retaining there plastic anisotropy. At a given stage of external loading, the secant bulk and shear moduli of the nanocrystalline polycrystal (composite) and the grain-boundary phase are denoted by (κ_c^s, μ_c^s) and $(\kappa_{gb}^s, \mu_{gb}^s)$, respectively, and the elastic moduli of the grains by (κ_g^s, μ_g^s). The plastic strain of the grain is represented by $\varepsilon_{ij}^{p(g)}$. In this approach, the secant moduli are taken as linear elastic moduli at a given level of the applied stress, and thus the said superposition principle can be applied. Such secant moduli of course need to be adjusted as the applied stress increases.

7.9.2.1 Christensen and Lo's Solutions

The generalized self-consistent scheme presented in Section 7.6. was adopted by Jiang and Weng to solve the localization problem that relate the external applied stress $\bar{\sigma}_{ij}$ to the mean stresses of the grain (inclusion) for a given orientation by taking into account the mechanical properties of grain boundary phase (matrix) as

$$\bar{\sigma}_{ij}^{(g)}(\mathrm{CL}) = \frac{1}{3}\bar{\alpha}_g\bar{\sigma}_{kk}\delta_{ij} + \bar{\beta}_g\bar{\sigma}'_{ij}, \quad \bar{\sigma}_{ij}^{(g)}(\mathrm{CL}) = \frac{1}{3}\bar{\alpha}_{gb}\bar{\sigma}_{kk}\delta_{ij} + \bar{\beta}_{gb}\bar{\sigma}'_{ij}, \tag{7.458}$$

where in (7.458) the applied stress is decomposed into hydrostatic and deviatoric components as $\bar{\sigma}_{ij} = (1/3)\delta_{ij}\bar{\sigma}_{kk} + \bar{\sigma}'_{ij}$, and the different constants are given below. The subscript (CL) refers to Christensen–Lo solution.

$$
\begin{aligned}
\bar{\alpha}_{\mathrm{g}} &= \frac{1}{p}(3\kappa_{\mathrm{c}}^{\mathrm{s}} + 4\mu_{\mathrm{c}}^{\mathrm{s}})(3\kappa_{\mathrm{gb}}^{\mathrm{s}} + 4\mu_{\mathrm{gb}}^{\mathrm{s}})\frac{\kappa_{\mathrm{g}}}{\kappa_{\mathrm{c}}^{\mathrm{s}}},\\
\bar{\beta}_{\mathrm{g}} &= 2\mu_{\mathrm{g}}\left[\bar{a}_1 - \frac{21}{5(1-2\nu_{\mathrm{g}})}\bar{a}_2\right],\\
\bar{\alpha}_{\mathrm{gb}} &= \frac{1}{p}(3\kappa_{\mathrm{c}}^{\mathrm{s}} + 4\mu_{\mathrm{c}}^{\mathrm{s}})(3\kappa_{\mathrm{g}} + 4\mu_{\mathrm{gb}}^{\mathrm{s}})\frac{\kappa_{\mathrm{gb}}^{\mathrm{s}}}{\kappa_{\mathrm{c}}^{\mathrm{s}}},\\
\bar{\beta}_{\mathrm{gb}} &= 2\mu_{\mathrm{gb}}\left[\bar{b}_1 - \frac{21}{5(1-2\nu_{\mathrm{gb}}^{\mathrm{s}})}\frac{1-c_{\mathrm{g}}^{5/3}}{1-c_{\mathrm{g}}}\bar{b}_2\right],
\end{aligned}
\tag{7.459}
$$

and

$$
p = (3\kappa_{\mathrm{g}} + 4\mu_{\mathrm{gb}}^{\mathrm{s}})(3\kappa_{\mathrm{gb}}^{\mathrm{s}} + 4\mu_{\mathrm{c}}^{\mathrm{s}}) - 12c_{\mathrm{g}}(\kappa_{\mathrm{g}} - \kappa_{\mathrm{gb}}^{\mathrm{s}})(\mu_{\mathrm{c}}^{\mathrm{s}} - \mu_{\mathrm{gb}}^{\mathrm{s}}) \tag{7.460}
$$

The constants $\bar{a}_1$, $\bar{a}_2$, $\bar{b}_1$, and $\bar{b}_2$ are given in Jiang and Weng's paper.

7.9.2.2 Luo and Weng's Eigenstrain Problem

In this consideration an eigenstrain such as the plastic strain $\varepsilon_{ij}^{\mathrm{p(g)}}$ exists in the inclusion but no external stress is applied. Luo and Weng's solution derives average stresses in the grain and grain-boundary phases due to prescribed dilatational eigenstrain $\varepsilon_{mm}^{\mathrm{p(g)}}$ and a deviatoric eigenstrain $\varepsilon_{ij}^{\prime\mathrm{p(g)}}$. That is

$$
\begin{aligned}
\bar{\sigma}_{ij}^{(\mathrm{g})}(\mathrm{LW}) &= \kappa_{\mathrm{g}}(\tilde{\alpha}_{\mathrm{g}} - 1)\varepsilon_{mm}^{\mathrm{p(g)}}\delta_{ij} + 2\mu_{\mathrm{g}}(\tilde{\beta}_{\mathrm{g}} - 1)\varepsilon_{ij}^{\prime\mathrm{p(g)}},\\
\bar{\sigma}_{ij}^{(\mathrm{gb})}(\mathrm{LW}) &= \kappa_{\mathrm{gb}}^{\mathrm{s}}\tilde{\alpha}_{\mathrm{gb}}\varepsilon_{mm}^{\mathrm{p(g)}}\delta_{ij} + 2\mu_{\mathrm{gb}}^{\mathrm{s}}\tilde{\beta}_{\mathrm{gb}}\varepsilon_{ij}^{\prime\mathrm{p(g)}},
\end{aligned}
\tag{7.461}
$$

where

$$
\begin{aligned}
\tilde{\alpha}_{\mathrm{g}} &= \frac{3\kappa_{\mathrm{g}}}{p}[(3\kappa_{\mathrm{gb}}^{\mathrm{s}} + 4\mu_{\mathrm{c}}^{\mathrm{s}}) - 4c_{\mathrm{g}}(\mu_{\mathrm{c}}^{\mathrm{s}} - \mu_{\mathrm{gb}}^{\mathrm{s}})],\\
\tilde{\beta}_{\mathrm{g}} &= \tilde{a}_1 - \frac{21}{5(1-2\nu_{\mathrm{g}})}\tilde{a}_2,\\
\tilde{\alpha}_{\mathrm{gb}} &= -\frac{12c_{\mathrm{g}}\kappa_{\mathrm{g}}}{p}(\mu_{\mathrm{c}}^{\mathrm{s}} - \mu_{\mathrm{gb}}^{\mathrm{s}}),\\
\tilde{\beta}_{\mathrm{gb}} &= \tilde{b}_1 - \frac{21}{5(1-2\nu_{\mathrm{gb}}^{\mathrm{s}})}\frac{1-c_{\mathrm{g}}^{5/3}}{1-c_{\mathrm{g}}}\tilde{b}_2.
\end{aligned}
\tag{7.462}
$$

The constants $\tilde{a}_1$, $\tilde{a}_2$, $\tilde{b}_1$, and $\tilde{b}_2$ are given in Jiang and Weng's paper. The subscript (LW) stands for Luo–Weng solution.

7.9.2.3 The Superposed Solution of Jiang and Weng

Under the simultaneous influence of an external stress and eigenstrain, the total mean stresses in the grain for a given orientation and its surrounding grain boundary are the sum of the two solutions

$$\begin{aligned}\bar{\sigma}_{ij}^{(g)} &= \bar{\sigma}_{ij}^{(g)}(\mathrm{CL}) + \bar{\sigma}_{ij}^{(g)}(\mathrm{LW}),\\ \bar{\sigma}_{ij}^{(gb)} &= \bar{\sigma}_{ij}^{(gb)}(\mathrm{CL}) + \bar{\sigma}_{ij}^{(gb)}(\mathrm{LW}),\end{aligned} \tag{7.463}$$

The corresponding total mean strain components are

$$\begin{aligned}\bar{\varepsilon}_{ij}^{(g)} &= \bar{\varepsilon}_{ij}^{(g)}(\mathrm{CL}) + \bar{\varepsilon}_{ij}^{(g)}(\mathrm{LW}),\\ \bar{\varepsilon}_{ij}^{(gb)} &= \bar{\varepsilon}_{ij}^{(gb)}(\mathrm{CL}) + \bar{\varepsilon}_{ij}^{(gb)}(\mathrm{LW}),\end{aligned} \tag{7.464}$$

where

$$\begin{aligned}&\bar{\varepsilon}_{ij}^{(g)}(\mathrm{CL}) = \frac{\bar{\alpha}_g}{9\kappa_g}\bar{\sigma}_{kk}\delta_{ij} + \frac{\bar{\beta}_g}{2\mu_g}\bar{\sigma}'_{ij}, \quad \bar{\varepsilon}_{ij}^{(g)}(\mathrm{LW}) = \frac{1}{3}\tilde{\alpha}_g\varepsilon_{mm}^{p(g)}\delta_{ij} + \tilde{\beta}_g\varepsilon_{ij}^{\prime p(g)};\\ &\bar{\varepsilon}_{ij}^{(gb)}(\mathrm{CL}) = \frac{\bar{\alpha}_{gb}}{9\kappa_{gb}^s}\bar{\sigma}_{kk}\delta_{ij} + \frac{\bar{\beta}_{gb}}{2\mu_{gb}^s}\bar{\sigma}'_{ij}, \quad \bar{\varepsilon}_{ij}^{(gb)}(\mathrm{LW}) = \frac{1}{3}\tilde{\alpha}_{gb}\varepsilon_{mm}^{p(g)}\delta_{ij} + \tilde{\beta}_{gb}\varepsilon_{ij}^{\prime p(g)}.\end{aligned} \tag{7.465}$$

The overall strains of the NC material under a given level of external stress $\bar{\sigma}_{ij}$ then follow from the orientational average over all grain orientations and their respective grain boundaries, as

$$\bar{\varepsilon}_{ij} = \langle \bar{\varepsilon}_{ij} \rangle = c_g \left\langle \bar{\varepsilon}_{ij}^{(g)}(\theta, \varphi, \psi) \right\rangle + c_{gb} \left\langle \bar{\varepsilon}_{ij}^{(gb)}(\theta, \varphi, \psi) \right\rangle, \tag{7.466}$$

where (θ, φ, ψ) represent the Euler angles of the rotation (or orientation) of a grain with respect to a base lattice that are aligned along the external loading coordinates, as indicated in (Fig. 7.8). The overbar on the strain signifies that it was calculated from the mean stress of the oriented grain and grain-boundary phase in the CL and LW models, whereas the brackets $\langle\ \rangle$ represent the orientational average taken over all grain orientations. The above grain and grain-boundary stresses are all (θ, φ, ψ)-dependent. The transformation matrix connecting the global {1; 2; 3} and the local {1'; 2'; 3'} coordinates carries the components $a_{ij} = \cos(i', j)$

$$[a_{ij}] = \begin{bmatrix} \cos\theta\cos\varphi\cos\psi - \sin\varphi\sin\psi & \cos\theta\sin\varphi\cos\psi + \cos\varphi\sin\psi & -\sin\theta\cos\psi \\ -\cos\theta\cos\varphi\sin\psi - \sin\varphi\cos\psi & -\cos\theta\sin\varphi\sin\psi + \cos\varphi\cos\psi & \sin\theta\cos\psi \\ \sin\theta\cos\varphi & \sin\theta\sin\varphi & \cos\theta \end{bmatrix} \quad (7.467)$$

7.9.3 Constitutive Equations of the Grains and Grain Boundary Phase

Jiang and Weng adopted the following deformation mechanisms in grain and grain boundary phases.

Plastic deformation in the grains is taken to be caused by crystallographic slip. The shear stress τ and shear strain γ^{p} are simply related by a Ludwick type equation as

$$\tau = \tau_0 + h(\gamma^{\mathrm{p}})^n, \quad (7.468)$$

where τ_0 is the initial flow stress, and h and n are, respectively, the strength coeficient and work-hardening exponent. For coarse-grained materials both τ_0 and h increase with $d^{-1/2}$ [56]

$$\tau = \tau_0^{\infty} + kd^{-1/2}, \quad h = h^{\infty} + ad^{-1/2}, \quad (7.469)$$

where the superscript ∞ signifies the value of a grain with an infinite grain size (i.e., free crystal), and k and a are material constants. Multiple slips in the constituent grains will introduce latent hardening. This is described in Jiang and Weng's paper by assuming that the flow stress of a slip system, say system i, due to the strain hardening of a latent system j, can be written as

$$\overset{(i)}{\tau}(d, \gamma^{\mathrm{p}}) = (\tau_0^{\infty} + k_0 d^{-1/2}) + (h^{\infty} + ad^{-1/2}) \times \sum_j [\alpha + (1-\alpha)\cos\overset{(i,j)}{\theta}\cos\overset{(i,j)}{\phi}](\overset{(j)}{\gamma^{\mathrm{p}}})^n, \quad (7.470)$$

where angles $\overset{(i,j)}{\theta}$ and $\overset{(i,j)}{\phi}$ define the angles between the slip directions and slip-plane normals of systems i and j, and the summation over j extends to all active slip systems in the considered grain. In particular, the condition $\alpha = 1$ evidently results in the isotropic hardening whereas $\alpha = 0$ corresponds to the kinematic hardening [55].

The increase of flow stress with $d^{-1/2}$ in Equation (7.470) cannot continue to hold as the grain size decreases to the nanometer range, due to the fact that dislocation activities would become increasingly restricted by the grain boundary. Consequently in Jiang and Weng's calculations, the constitutive Equation (7.470) is used up to a critical grain size, and below that the flow stress will no longer increase and stay constant. For copper the cut-off value is taken at 7.2 nm, as determined by Wang et al. [54].

For a slip system of a given grain to be in the plastic state its flow stress in equation (7.470) must be equal to its resolved shear stress, given by

$$\tau_{\mathrm{s}}^{(\mathrm{g})} = \nu_{ij}\bar{\sigma}_{ij}^{(\mathrm{g})}, \tag{7.471a}$$

where the grain stress $\bar{\sigma}_{ij}^{(\mathrm{g})}$ varies from grain to grain, and ν_{ij} is the Schmid tensor of a slip system, defined as $\nu_{ij} = (b_i n_j + b_j n_i)/2$, in which b_i and n_i are the unit slip direction and slip plane normal, respectively, of the considered slip system (see Fig. 7.7d).

The stress–strain relation of each oriented grain is simply given by

$$\bar{\sigma}_{ij}^{(\mathrm{g})} = L_{ijkl}^{(\mathrm{g})}\left(\bar{\varepsilon}_{kl}^{(\mathrm{g})} - \varepsilon_{kl}^{\mathrm{p(g)}}\right), \tag{7.471b}$$

Where the stiffness tensor $\mathbf{L}^{(g)}$ has the bulk and shear moduli $(3\kappa_g, 2\mu_g)$, and

$$\varepsilon_{ij}^{\mathrm{p(g)}} = \sum_k \overset{(k)}{\nu}_{ij} \overset{(k)}{\gamma^{\mathrm{p}}} \tag{7.472}$$

summing over all active slip systems in the considered grain. As for stresses, the plastic strain of each grain $\varepsilon_{ij}^{\mathrm{p(g)}}$ also varies from one grain-orientation to the other. Owing to plastic incompressibility we further have $\varepsilon_{mm}^{\mathrm{p(g)}} = 0$ and $\varepsilon_{ij}^{'\mathrm{p(g)}} = \varepsilon_{ij}^{\mathrm{p(g)}}$ in Equation (7.472).

Concerning grain-boundary phase, Jiang and Weng adopted a Drucker's [13] type yield function to model its constitutive relation. That is

$$\sigma_{\mathrm{e}} = \sigma_y^{(\mathrm{gb})} + mp + h_{\mathrm{gb}}(\varepsilon_{\mathrm{e}}^{\mathrm{p}})^{n_{\mathrm{gb}}}, \tag{7.473}$$

where von Mises' effective stress and effective plastic strain are defined as

$$\sigma_{\mathrm{e}}^{(\mathrm{gb})} = \left(\frac{3}{2}\sigma_{ij}^{'(\mathrm{gb})}\sigma_{ij}^{'(\mathrm{gb})}\right)^{1/2}, \quad \varepsilon_{\mathrm{e}}^{\mathrm{p(gb)}} = \left(\frac{2}{3}\varepsilon_{ij}^{\mathrm{p(gb)}}\varepsilon_{ij}^{\mathrm{p(gb)}}\right)^{1/2} \tag{7.474}$$

in terms of the deviatoric stress σ'_{ij} and plastic strain $\varepsilon_{ij}^{\mathrm{p}}$, and $p = -(1/3)\sigma_{kk}$ is the hydrostatic pressure. Constants $\sigma_y^{(\mathrm{gb})}$ and h_{gb} are not grain-size dependent; together with m and n_{gb} they form the material constants of the grain-boundary phase. The plastic strain was taken to be incompressible, that is, the uncorrelated motion of atoms inside the grain boundary would not result in any significant amount of volume change.

7.9.4 Application to a Nanocystalline Copper

The developped theory is applied to evaluate the stress–strain relation and yield strength of copper during the coarse to nano grain transition, and the results are

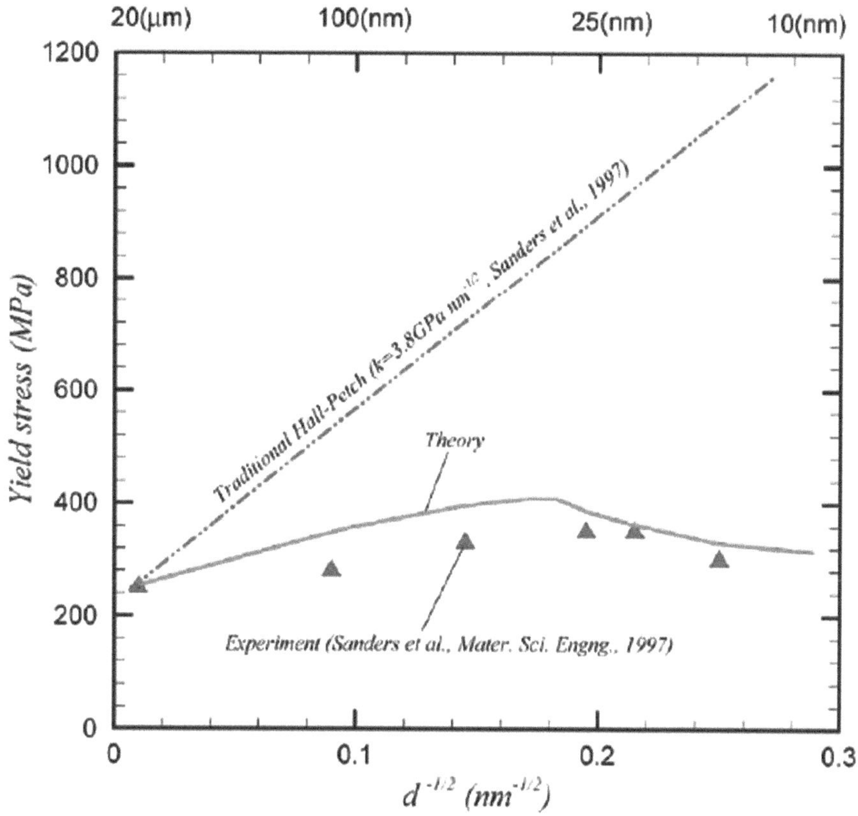

Fig. 7.9 Transition from a positive to a negative slope in the Hall-Petch plot of yield strength of Cu [32]

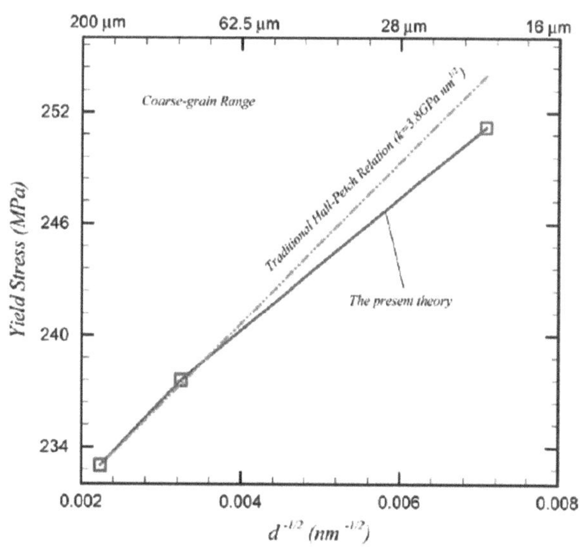

Fig. 7.10 Departure from the Hall-Petch relation as the grain size decreases [32]

compared with experimental tensile data on Cu and the classical Hall-Petch law. The calculated results suggest that plastic deformation of the grain-boundary phase plays a very significant role in changing the nature of plastic behavior of nanocrystalline materials. The yield strength of a coarse-grained material basically follows the Hall-Petch relation, but as the grain size decreases it gradually deviates from it (Fig. 7.10), and eventually decreases after attaining a maximum at a critical grain size (Fig. 7.9). Thus, the slope of the Hall-Petch plot is negative in the very fine grain-size region and, as the grain size approaches zero, its yield strength also asymptotically approaches that of the grain-boundary phase. When the yield strength follows the Hall-Petch relation, plastic deformation of the polycrystal is contributed solely by the constituent

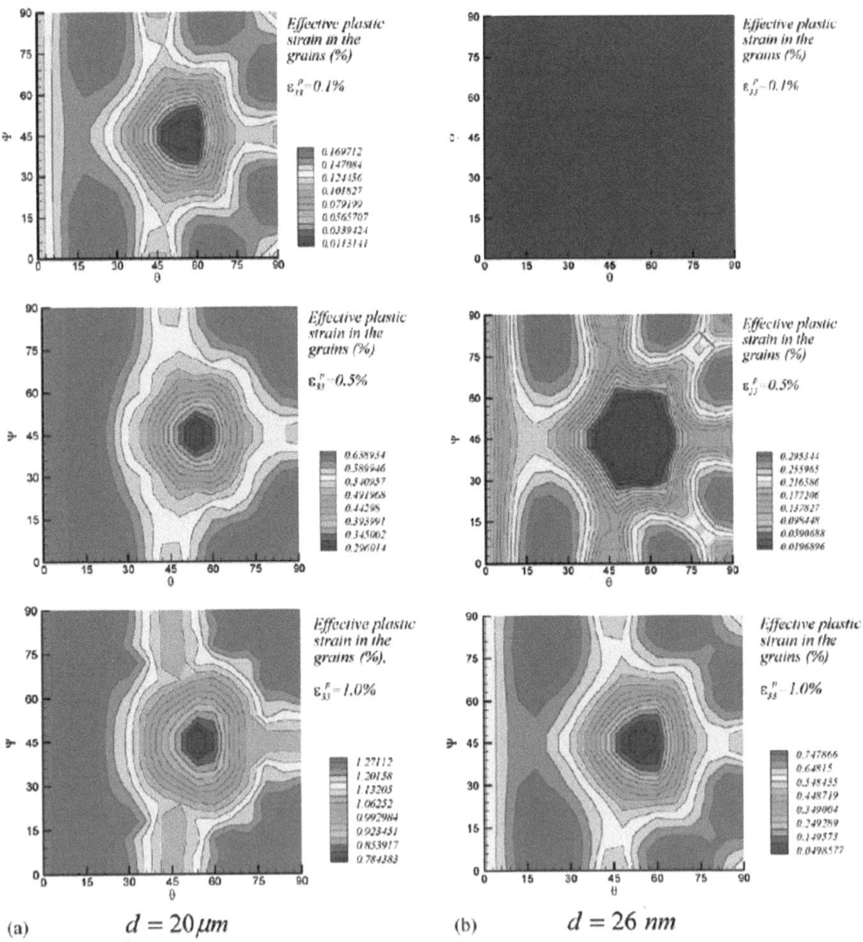

Fig. 7.11 Map for the evolution of the effective plastic strain in the constituent grains [32]

grains, but when the Hall-Petch plot shows a negative slope its plastic behavior is dominated by the grain boundary. During the transition from the Hall-Petch relation to one with a negative slope, both grains and grain boundaries contribute competitively to the overall plastic deformation of the material. It is also concluded from maps for the evolution of the effective plastic strain in the constituent grains (Fig. 7.11), and of the evolution of the overall effective stress of the grain-boundary phase (Fig. 7.12) in terms of the orientation of the grain, that plastic deformation in the grain would relieve the overall effective stress of its surrounding grain boundary.

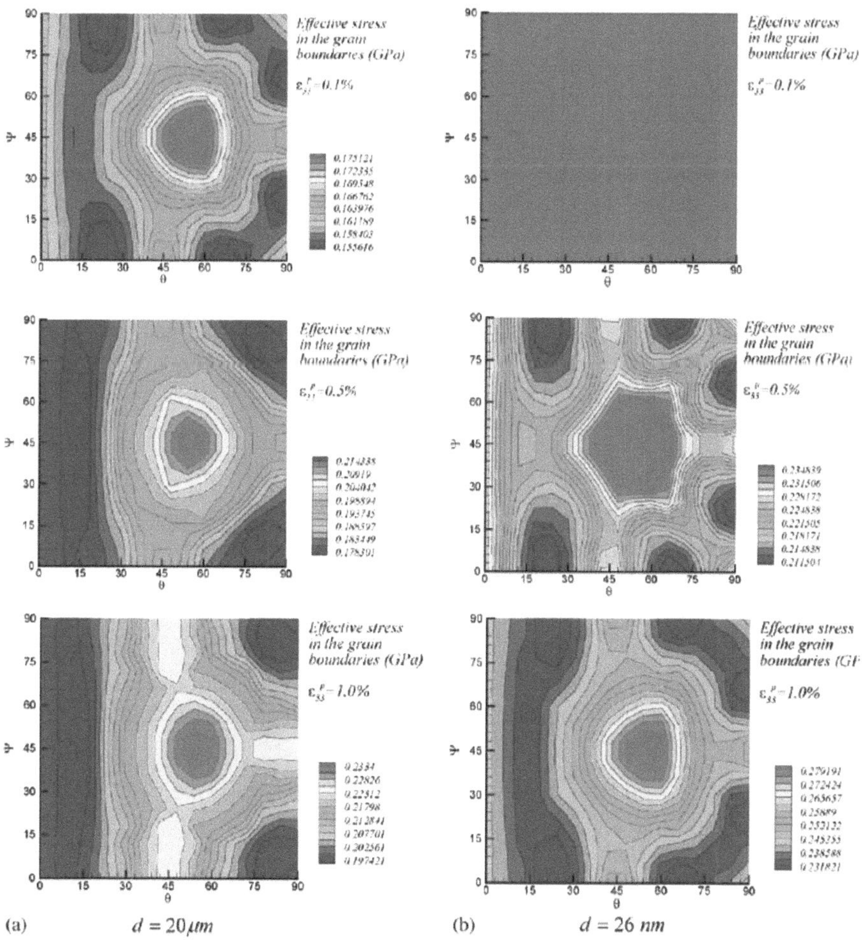

Fig. 7.12 Map for the evolution of the overall effective stress of the grain-boundary phase in terms of the orientation of the grain it encloses [32]

References

1. Asaro, R.J. and J.R. Rice, Strain localization in ductile single crystals. Journal of Mechanics and Physics of Solids **25**, 309–338, (1977)
2. Asaro, R.J. Crystal plasticity. Journal of Applied Mechanics., **50**, 921–934, (1983)
3. Ashby, M.F. and H.J. Frost, The kinematics of inelastic deformation above 0 K. In: Argon, A.S. (ed.), *Constitutive equations in plasticity*. MIT Press, Boston, 116, (1975)
4. Bragg, W.H. and W.L. Bragg, *The crystalline state*. Bell, London, (1933)
5. Berveiller, M. and A. Zaoui, An extension of the self-consistent scheme to plastically-flowing polycrystals. Journal of Mechanics and Physics of Solids, **26**, 325–344, (1979)
6. Budiansky, B. and O.L. Mangasarian, Plastic stress concentration at circular hole in infinite sheet subjected to equal biaxial tension. Journal of Applied **27**, 59–64, (1960)
7. Buryachenko, V., The overall elastoplastic behavior of multiphase materials with isotropic components. Acta Mechanica **119**, 93–117, (1996)
8. Capolungo, L., C. Jochum, M. Cherkaoui, and J. Qu, Homogenization method for strength and Inelastic Behavior of Nanocrystalline Materials. International Journal of Plasticity, **21**(1), 67–82, (2005a).
9. Capolungo, L., M. Cherkaoui, J. Qu, A self consistent model for the inelastic deformation of nanocrystalline materials. Journal of Engineering Materials and Technology, **127**(4), 400–407, (2005b).
10. Chaboche, J.L., P. Kanouté, and A. Roos, On the capabilities of mean-field approaches for the description of plasticity in metal matrix composites. International Journal of Plasticity, **21**(7), 1409–1434, (2005, July)
11. Christensen, R.M. and K.H. Lo, Solutions for effective shear properties in three phase sphere and cylinder models. Journal of the Mechanics and Physics of Solids, **27**(4), 315–330, (1979 August)
12. Doghri, I. and A. Ouaar, Homogenization of two-phase elasto-plastic composite materials and structures: Study of tangent operators, cyclic plasticity and numerical algorithms. International Journal of Solids and Structures, **40**(7), 1681?1712, (2003, April)
13. Drucker, D.C., Some implications of work hardening and ideal plasticity. Quarterly of Applied Mathematics, **7**(4), 411?418, (1950)
14. Ewing, J.A., and W. Rosenhain, Experiments in micro-metallurgy: Effects of strain. Preliminary notice. Philosophical Transactions of the Royal Society of London. Series **A199**, 85–90, (1900)
15. Eshelby, J.D., The determination of the elastic field of an ellipsoidal inclusion and related problems. Proceedings of the Royal Society of London, Series A **241**, 376–396, (1957)
16. Garboczi, E.J. and D.P. Bentz, Analytical formulas for interfacial transition zone properties. Advanced Cement Based Materials, **6**(3–4), 99–108 (1997, October-November)
17. Gilormini, P., A shortcoming of the classical nonlinear extension of the self consistent model. C. R. Acad. Sci. Paris, Série IIb **320**, 115–122, (1995)
18. González, C. and J. LLorca, A self-consistent approach to the elasto-plastic behaviour of two-phase materials including damage. Journal of the Mechanics and Physics of Solids, **48**(4), 675–692, (2000, April)
19. Hashin, Z., Strength of materials: By Peter Black, published by Pergamon Press, Oxford, 1966; 454 pp., price: 45s Materials science and engineering, **1**(3), 198–199, (1966, September)
20. Hashin, Z., Thermal expansion of polycrystalline aggregates: II. Self consistent approximation. Journal of the Mechanics and Physics of Solids, **32**(2), 159–165, (1984)
21. Hashin, Z. and P.J.M. Monteiro, An inverse method to determine the elastic properties of the interphase between the aggregate and the cement paste. Cement and Concrete Research, **32**(8), 1291–1300, (2002, August)
22. Hashin, Z. and B.W. Rosen, The elastic moduli of fiber-reinforced materials. Journal of Applied Mechanics-Transactions of the ASME, **31**, 223–232, (1964).

23. Hashin, Z. and S. Shtrikman, On some variational principles in anisotropic and nonhomogeneous elasticity. Journal of the Mechanics and Physics of Solids, **10**(4), 335–342, (1962, October–December)
24. Hervé, E. and O. Pellegrini, The elastic constants of a material containing spherical coated holes. Archives Mechanics, **47**(2), 223–246, (1995)
25. Herve, E. and A. Zaoui, n-Layered inclusion-based micromechanical modelling. International Journal of Engineering Science, **31**(1), 1–10, (1993, January)
26. Hill, R., Continuum micro-mechanics of elastoplastic polycrystals. Journal of Mechanics and Physics of Solids **13**, 89–101, (1965)
27. Hill, R., Generalized constitutive relations for incremental deformation of metals by multislip Journal of Mechanics and Physics of Solids **14**, 99, (1966)
28. Hill, R. and J.R. Rice, Constitutive analysis of elasto-plastic crystals at arbitrary strains. Journal of Mechanics and Physics of Solids **20**, 401–413, (1972)
29. Hill, R., and K.S. Havner, Perspectives in the mechanics of elastoplastic crystals, Journal of Mechanics and Physics of Solids **30**, 5–22, (1982)
30. Hutchinson, J.W., Elastic-plastic behaviour of polycrystalline metals and composites Proceedings of the Royal Society of London **A319**, 247–272, (1970)
31. Hutchinson, J.W., Bounds of self-consistent estimates for creep of polycrystalline materials. Proceedings of the Royal Society of London **A348**, 101–127, (1976)
32. Jiang, B. and G.J. Weng, A generalized self-consistent polycrystal model for the yield strength of nanocrystalline materials. Journal of the Mechanics and Physics of Solids, **52**(5), 1125–1149, (2004 May)
33. Kröner, E., Zur plastischen Verformung des Vielkristalls. Acta Metall **9**, 155–161, (1961)
34. Lebensohn, R. and C.N. Tomé, A self-consistent anisotropic approach for the simulation of plastic deformation and texture development of polycrystals applications to zirconium alloys Acta Metall Mater **41**, 2611–2624, (1993)
35. Lin, T.H., Analysis of elastic and plastic strains of a face-centred cubic crystal. Journal of the Mechanics and Physics of Solids, **5**(2), 143–149, (1957, March)
36. Luo, H.A. and G.J. Weng, On Eshelby's inclusion problem in a three-phase spherically concentric solid, and a modification of Mori–Tanaka's method. Mechanics of Materials, **6**(4), 347–361, (1987, December)
37. Masson, R., M. Bornert, P. Suquet, and A. Zaoui, An affine formulation for the prediction of the effective properties of nonlinear composites and polycrystals. Journal of the Mechanics and Physics of Solids, **48**(6–7), 1203–1227, (2000, June)
38. Masson, R. and A. Zaoui Self-consistent estimates for the rate-dependent elastoplastic behaviour of polycrystalline materials Journal of Mechanics and Physics of Solids **47**(7), 1543–1568, (1999)
39. Molinari, A., S. Ahzi, and R. Kouddane, On the self-consistent modeling of elastic-plastic behavior of polycrystals. Mechanics of Materials, **26**(1), 43–62, (1997, July–August)
40. Molinari, A., G.R. Canova, and S. Ahzi, A self-consistent approach of the large deformation polycrystal viscoplasticity. Acta Metalurgica **35**(12), 2983–2994, (1987)
41. Pierard, O., C. Gonźlez, J. Segurado, J. LLorca, and I. Doghri, Micromechanics of elasto-plastic materials reinforced with ellipsoidal inclusions. International Journal of Solids and Structures, **44**(21), 6945–6962, (2007, October)
42. Ponte Castañeda, P., The effective mechanical properties of nonlinear isotropic composites. Journal of Mechanics and Physics of Solids **39**(1), 45–71, (1991)
43. Polanyi, von M., Rontgenographische Bestimmung von Ksistallanordnunge. Naturwissenschaften **10**, 411, (1922)
44. Qiu, Y.P. and G.J. Weng, A theory of plasticity for porous materials and particle-reinforced composites. Journal of Applied Mechanics **59**, 261–268, (1992)
45. Schmid, E., Remarks on the vivid deformation of crystals. Zeitschrift fur Physik, **22**, 328–333, (1924, Feb–Mar)
46. Suquet, P., Overall potentials and extremal surfaces of power law or ideally plastic materials Journal of Mechanics and Physics of Solids **41**, 981–1002, (1993)

47. Suquet, P., Overall properties of nonlinear composites: A modified secant moduli theory and its link with Ponte Castañeda's nonlinear variational procedure. C.R. Academiae Scientiarum Paris **320** (Série IIb), 563–571, (1995)
48. Taylor, G.I., and C.F. Elam, The distortion of an aluminum crystal during a tensile test. Proceedings of the Royal Society **A102**, 647, (1923)
49. Taylor, G.I., and C.F. Elam, The plastic extension and fracture of aluminum crystals. Proceedings of the Royal Society, **A108**, 28–51, (1925)
50. Taylor, G.I., Plastic deformation of crystal. Proceedings of the Royal Society **A148**, 362–404, (1934)
51. Taylor, G.I., Plastic strain in metals. Journal of Institute Metals **62**, 307, (1938)
52. Talbot, D., and J. Willis, Variational principles for inhomogeneous nonlinear media. IMA Journal of Applied Mathematics **35**, 39–54, (1985)
53. Tandon, G.P., and G.J. Weng, A theory of particle-reinforced plasticity. Journal of Applied Mechanics **55**, 126–135, (1988)
54. Wang, N., Wang, Z., Aust, K.T., and U. Erb, Effect of grain size on mechanical properties of nanocrystalline materials. Acta Metall. Mater **43**, 519–528, (1995)
55. Weng, G.J., Kinematic hardening rule in single crystals. International Journal of Solids Structure, **15**, 861–870, (1979)
56. Weng, G.J., A micromechanical theory of grain-size dependence in metal plasticity. Journal of the Mechanics and Physics of Solids, **31**, 193–203, (1983)
57. Willis, J.R., Variational and related methods for the overall properties of composites. Advances in Applied Mechanics, **21**, 1–78, (1981)
58. Willis, J.R., The overall response of composite materials. Journal of Applied Mechanics **50**, 1202–1209, (1983)

Chapter 8
Innovative Combinations of Atomistic and Continuum: Mechanical Properties of Nanostructured Materials

8.1 Introduction

Currently, due to advances in nanotechnology, many investigations are devoted to nanoscale science and developments of nanocomposites. Nanomaterials in general can be roughly classified into two categories. On one hand, if the characteristic length of the microstructure, such as the grain size of a polycrystal material, is in the nanometer range, it is called a nanostructured material. On the other hand, if at least one of the overall dimensions of a structural element is in the nanometer range, it may be called a nano-sized structural element. Thus, this may include nanoparticles, nanofilms, and nanowires [2, 10, 47].

Why so much interest in nanomaterials or nanocomposites? Nanocomposites/nanomaterials are of interest because of their unusual mechanical, thermo-mechanical, electrical, optical, and magnetic properties as compared to composites of similar constituents, volume proportion, and shape/orientation of reinforcements. Here are some examples to name a few:

- Nanophase ceramics are of particular interest because they are more ductile at elevated temperatures as compared to the coarse-grained ceramics.
- Nanostructured semiconductors are known to show various nonlinear optical properties. Semiconductor Q-particles also show quantum confinement effects which may lead to special properties, like luminescence in silicon powders and silicon germanium quantum dots as infrared optoelectronic devices. Nanostructured semiconductors are used as window layers in solar cells.
- Nanosized metallic powders have been used for the production of gas tight materials, dense parts, and porous coatings. Cold welding properties combined with the ductility make them suitable for metal-metal bonding, especially in the electronic industry.
- Single nanosized magnetic particles are mono-domains and one expects that also in magnetic nanophase materials the grains correspond with domains, while boundaries on the contrary to disordered walls. Very small particles have special atomic structures with discrete electronic states, which give rise to

M. Cherkaoui, L. Capolungo, *Atomistic and Continuum Modeling of Nanocrystalline Materials*, Springer Series in Materials Science 112,
DOI 10.1007/978-0-387-46771-9_8,

special properties in addition to the super-paramagnetism behavior. Magnetic nanocomposites have been used for mechanical force transfer (ferrofluids), for high-density information storage and magnetic refrigeration.

- Nanostructured metal clusters and colloids of mono- or plurimetallic composition have a special impact in catalytic applications. They may serve as precursors for new type of heterogeneous catalysts (Cortex-catalysts) and have been shown to offer substantial advantages concerning activity, selectivity, and lifetime in chemical transformations and electrocatalysis (fuel cells). Enantioselective catalysis was also achieved using chiral modifiers on the surface of nanoscale metal particles.
- Nanostructured metal-oxide thin films are receiving a growing attention for the realization of gas sensors (NOx, CO, CO_2, CH_4 and aromatic hydrocarbons) with enhanced sensitivity and selectivity. Nanostructured metal-oxide (MnO_2) find application for rechargeable batteries for cars or consumer goods. Nanocrystalline silicon films for highly transparent contacts in thin film solar cell and nanostructured titanium oxide porous films for its high transmission and significant surface area enhancement leading to strong absorption in dye-sensitized solar cells.
- Polymer-based composites with a high content of inorganic particles leading to a high dielectric constant are interesting materials for photonic band gap structure produced by the LIGA.

However, nanocomposites of SiC-reinforced Al_2O_3 matrices were reported to display no size dependency of the nano-inclusion, decreased fracture toughness with reduction of inclusion size, or even increased mechanical properties with reduction of inclusion size for fixed inclusion volume ratio [58]. These contradictory size dependencies (or size nondependencies) on nanoscale particulates could possibly point to the quality of the interfacial bonding between nano-inclusions and whether the matrix material is superior, inferior, or similar as a result of processing techniques.

The size dependency in the area of nanotechnology is well known and has been investigated in terms of surface/interface energies, stresses, and strains [8, 9, 10, 12, 53]. The classical Eshelby's solution [15] of an embedded inclusion neglects the presence of surface or interface energies (stresses, strains) and indeed, the effects of those are negligible except in the size range of tens of nanometers, where one contends with a significant surface-to-volume ratio. Thus, due to the large ratio of surface area to volume in nanosized objects, the behavior of surfaces and interfaces becomes a prominent factor controlling the nanomechanical properties of nanostructured materials.

The reduced coordination of atoms near a free surface induces a corresponding redistribution of electronic charge, which alters the binding situation [51]. As a result, the energy of these atoms will, in general, be different from that of the atoms in the bulk. In a similar vein, atoms at an interface of two materials experience a different local environment than atoms in the bulk of the materials, and the equilibrium position and energy of these atoms will, in general, be different from those of the atoms in the bulk. Therefore, in the case of nanocomposites the elastic properties of the interface should be given due consideration. There are different ways in which the properties of the surface can be defined and introduced. For example, if one considers an "interface" separating two otherwise homogeneous phases, the interfacial property may be defined either in terms of an *interphase*, or by introducing the concept of a *dividing surface*. While "interface" refers to the surface area between two phases, "interphase" corresponds to the volume defined by the narrow region sandwiched between the two phases.

In the approach of interface where a single dividing surface is used to separate the two homogeneous phases, the interface contribution to the thermodynamic properties is defined as the excess over the values that would obtain if the bulk phases retained their properties constant up to an imaginary surface (of zero thickness) separating the two phases [9, 10]. As pointed out by Dingreville (2007), for realistic bimaterials, there typically exist two distinctive length parameters, namely, the atomic spacing (lattice parameter) d, and the radius of curvature of the interface D, where D is generally several order of magnitude greater than d for most of the problems of engineering interest. Thus, if one measures the characteristic length of these inhomogeneities by D, the radius of curvature of the interface between an inhomogeneity and its surrounding medium, the discrete atomic structure of the material is smeared (homogenized) into a continuum. This is like observing the interface from a far distance so that one cannot see the atomic structure, nor the thickness of the interphase. All one sees is that the properties jump from one bulk value to the other across the interface. Consequently, one may perceive that field quantities (stress, displacement, etc.) are discontinuous at the interface when measured by the mesoscopic length scale D [7]. Several attempts [11–13, 18, 24, 25, 35, 36, 41, 52–55, 57, 63] which have been made in analyzing the nanocomposites by considering interfacial effect are based on this viewpoint. [7] develops the interfacial conditions for the displacement, strain and stress fields across the interface of bimaterials and shows that none of the above works has taken the interface effects fully into account. The various solutions for the Eshelby's nano-inclusion problems that have appeared in the literature recently assume an elastically isotropic surface/interface and are concerned with the case of spherical inhomogeneities problems. Generally, the problem is solved using the generalized Young-Laplace equations for solids [48] and the general expressions for the displacements in an infinite region containing a spherical inhomogeneity from [39] in terms of Legendre polynomial of order two. Although [7] establishes the relationship between microscopic properties (measured by d) and mesoscopic jumps of these properties across the interface measured by D by taking into account the "3-D nature" of the surface/interface [7, 8], the solution of the full boundary value problem remains very complex to solve.

The concept of surface/interface stress in solids was first introduced by Gibbs [19] as part of his treatment of the thermodynamics of surface and interfaces. Qualitatively speaking, the surface free energy is defined as a reversible work per unit area to create a surface. The surface stress is a reversible work per area to stretch a surface elastically. The surface tension is defined as the excess of the appropriate thermodynamic potential of the system with an interface, per unit area of the interface, compared to that of the homogeneous bulk phase occupying the same volume. During the last decade, the importance of stress and strain effects on surface/interface physics has been extensively recognized. It has provoked a great theoretical, computational, and experimental activity that has allowed a better understanding of the stress effects on surface physics. Among them we can quote:

- From a thermodynamic viewpoint, proper definitions of surface stress and surface strain have been introduced. The thermodynamic properties of stressed surfaces have been rationalized and great progress in the numerical calculations of surface stress and strain based on atomistic models has been made.
- Comparison between results obtained by atomistic calculations and results obtained by usual theory of elasticity have been extensively studied and the limit of validity of this classical theory thus discussed.
- Stress-induced surface instabilities have been extensively studied. It is, for example, the case for the well-known Asaro-Tiller-Grienfeld instability without external flux. It is also the ca se of step bunching mediated by elastic step-step interactions or even the case of strain-driven surface diffusion instability in presence of impinging flux.
- Surface elasticity has been recognized as an important quantity for a better understanding of some surface two-dimensional phase transitions. We can quote, in particular, surface stress effects on surface melting. A possible role of the surface stress on surface reconstructions has been also mentioned.
- Important improvements have been obtained to understand stress release in complex materials at the atomic scale. We can mention, for example, the interplay between surface relaxation and surface segregation or between surface relaxation and chemical ordering of alloy surfaces. It is also the case for the notion of local pressure maps which has been used as a tool to predict the stress release upon atomic rearrangements.
- The surface/interface effect on effective properties of particulate composite containing nano-inhomogeneities has been investigated.

The purpose of this chapter is to review the important developments in the understanding of interface/surface effect on nanomaterials. The discussion in the above – presenting an overview of the challenges and recent advances related to the fundamental understanding of interfacial effects – is particularly relevant to nanocomposite (NC) materials in which the interfaces/interphases of interest are grain boundaries and twin boundaries. Indeed, their plastic response is largely influenced by energy relaxation processes – such as grain boundary sliding and dislocation emission – occurring at the grain boundaries. The activation of such plastic mechanisms will necessarily affect the local stress state at grain boundaries. Given the limited thickness of grain boundaries (e.g., 1–2 nm) – which are typically modeled as an interphase – the use of interfacial approaches appear more appropriate for the following two reasons: (1) the use of stress, which is a continuum variable is less ambiguous, and (2) simulations would be less computationally expensive provided sufficiently accurate models can be developed. To illustrate this second point let us consider the case of grain growth via grain boundary coalescence. This case study was shown as an example of application of molecular dynamic simulations in Chapter 4. Clearly, these simulations are computationally intensive. Moreover, while the grain growth mechanism can be depictured, the fundamental driving force activating the motion of grain boundaries is not revealed with such simulations. More focused studies on interfacial effects are thus necessary to answer this question. Clearly, a continuum-based interpretation of atomic scale processes would be less intensive the molecular dynamic simulations. While current understanding on interfacial behavior has not yet allowed reaching this objective, critical advances have been made in the field and it is likely that future continuum models will be based on these approaches. This chapter will briefly present a review on interfacial effects prior to discussing recent advances allowing to account for local atomic scale processes – we limit ourselves to elasticity here – within a continuum mechanics framework.

8.2 Surface/Interface Structures

8.2.1 What Is a Surface?

Using the common sense, a **surface** can be can defined as the shell of a macroscopic object (the inside) in contact with its environment (the outside world). The surface of an object determines its optical appearance, stickiness, wetting behavior, frictional behavior, and chemical reactivity, e.g.,

- in **large** objects with small surface area A to volume V ratio (A/V) the physical and chemical **properties** are primarily defined by the **bulk** (inside)
- in **small** objects with a large A/V-ratio the **properties** are strongly influenced by the **surface**

In a solid the density of atoms is on the order of 10^{23} atoms/cm^3, so only a few number of surface atoms compared to the number of bulk atoms.

8.2.2 Dispersion, the Other A/V Relation

The dispersion is the ratio of the number of surface atoms to the total number of the atoms in a particle (Fig. 8.1).

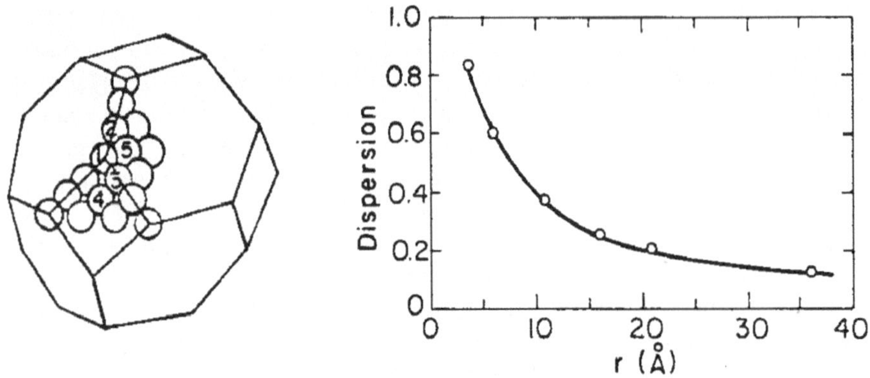

Fig. 8.1 Variation of the dispersion with particle size for close-packed cubic

8.2.3 *What Is an Interface?*

An **interface** is the separating layer between two condensed phases (usually molecular dimensions). At the border of a solid or liquid in contact with vapor there is usually no abrupt change in density, but a more or less continuous transition from high density to low density. The interface consists either of evaporating bulk material or condensing material from the gas phase (Fig. 8.2).

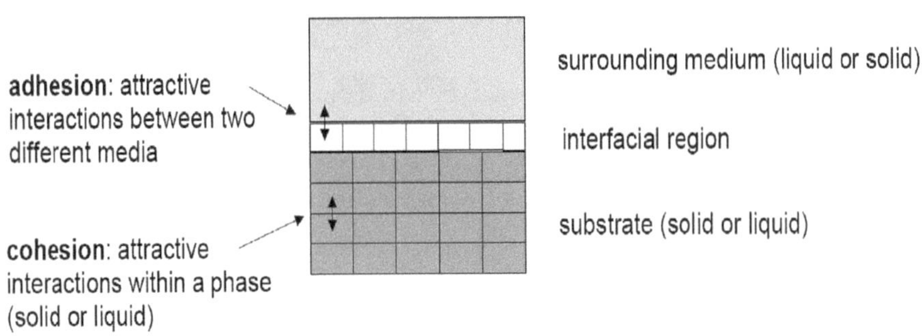

Fig. 8.2 Illustration of Interface

8.2.4 *Different Surface and Interface Scenarios*

8.2.4.1 Liquid/Vapor Interface (Fig. 8.3)

- Liquids are highly mobile and disordered
- Constant evaporation and recondensation at surface

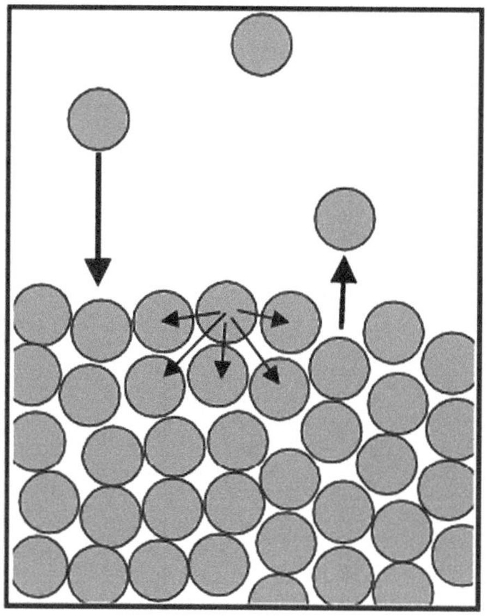

Fig. 8.3 Liquid/vapor interface

8.2.4.2 Solid/Vapor Interface (Fig. 8.4)

- Solids are highly immobile
- Crystalline solids are highly ordered/structured

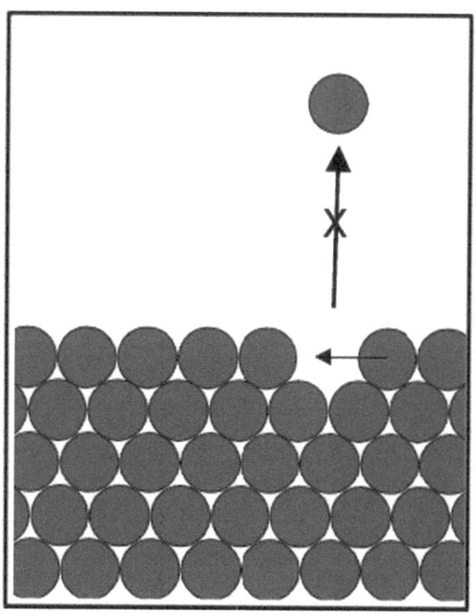

Fig. 8.4 Solid/vapor Interface

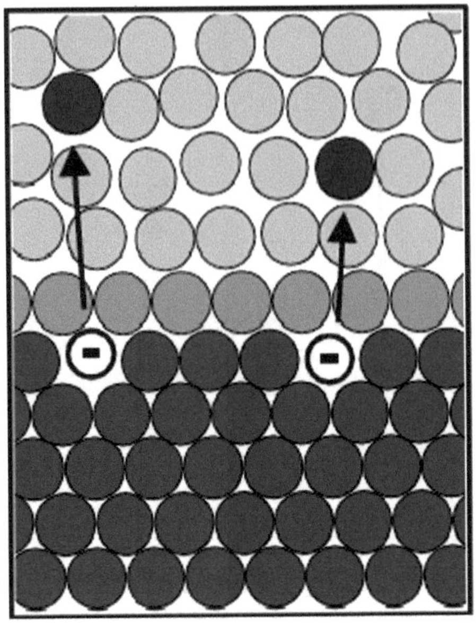

Fig. 8.5 Solid/liquid interface

- Usually there is no evaporation of surface atoms and molecules but only lateral diffusion (depends on the temperature)

8.2.4.3 Solid/Liquid Interface (Fig. 8.5)

- Liquid can dissolve surface atoms therefore this may lead to surface charges
- Liquid molecules at the interface can be much higher ordered than in the bulk

8.2.4.4 Liquid/Liquid Interface (Fig. 8.6)

- Both phases are highly mobile so the shape of interface is controlled by surface tension
- Depending on solubility molecules will migrate from one phase to the other so the shape of interface is controlled by chemical potential (partition cœfficient.)

8.2.4.5 Solid/Solid Interface (Fig. 8.7)

- If two crystalline solids are in atomic contact the different lattice constants will generate strain at interface
- If both materials react together new compound will be formed in contact region (**interphase**)
- At high temperature, interdiffusion is possible (e.g., Cr and Au)

Fig. 8.6 Liquid/liquid interface

Fig. 8.7 Solid/solid interface

8.3 Surface/Interface Physics

In the past decades, the science of solid surfaces has developed largely with the emphasis to gain insight into the microscopic structure of surfaces on an atomic scale. The importance of stress and strain effects on surface physics are reviewed

[26, 44]. The elastic, thermodynamic, and atomistic definitions of surface stress and surface strain are presented in a complementary way so that the surface stress and surface strain concepts based on a proper definition of surface elastic energy in terms of excess quantities are presented in depth. This leads to a natural link between surface stress and surface energy known as Shuttleworth's relation [56]. With an ever-increasing knowledge about the crystallographic structure, the electronic, magnetic, and dynamical properties of surfaces, and with the ability to engineer surface and interface systems with particular properties, experimental and theoretical studies on macroscopic aspects of surfaces fell out of fashion.

Surface and interface are characterized by some quantities which need to be well defined and understood.

8.3.1 Surface Energy

Surface energy quantifies the disruption of intermolecular bonds that occurs when a surface is created. Qualitatively speaking, the surface free energy is defined as a reversible work per unit area to create a surface. The specific free energy of a surface must be positive, since otherwise the solid would gain energy upon fragmentation and, therefore, would not be stable. Cutting a solid body into pieces disrupts its bonds, and therefore consumes energy. If the cutting is done reversibly, then conservation of energy means that the energy consumed by the cutting process will be equal to the energy inherent in the two new surfaces created. The unit surface energy of a material would therefore be half of its energy of cohesion, all other things being equal; in practice, this is true only for a surface freshly prepared in vacuum. Surfaces often change their form away from the simple "cleaved bond" model just implied above. They are found to be highly dynamic regions, which readily rearrange or react, so that energy is often reduced by such processes as passivation or adsorption.

As first described by Thomas Young in 1805 in the Philosophical Transactions of the Royal Society of London, it is the interaction between the forces of cohesion and the forces of adhesion which determines whether or not wetting, the spreading of a liquid over a surface, occurs. If complete wetting does not occur, then a bead of liquid will form, with a contact angle which is a function of the surface energies of the system. Surface energy is most commonly quantified using a contact angle goniometer and a number of different methods. Thomas Young described surface energy as the interaction between the forces of cohesion and the forces of adhesion which, in turn, dictate if wetting occurs. If wetting occurs, the drop will spread out flat. In most cases, however, the drop will bead to some extent and by measuring the contact angle formed where the drop makes contact with the solid the surface energies of the system can be measured.

Surface energy derives from the unsatisfied bonding potential of molecules at a surface, giving rise to "free energy." This is in contrast to molecules within a material which have less energy because they are subject to interactions with like molecules in all directions. Molecules at the surface will try to reduce this free energy by interacting with molecules in an adjacent phase. When one of the bulk phases is a gas, the free energy per unit area is termed the surface energy for solids, and the **surface tension** in liquids. One manifestation of surface energy is a state of tension at the surface of a liquid, which is why work is required to increase the surface area of a liquid, hence the above physical definition. However, when both phases are condensed (i.e., solid-solid, solid-liquid, and immiscible liquid-liquid interfaces) the free energy per unit area of the interface is called the **interfacial energy**.

The term surface energy is also closely linked with **surface hydrophobicity**. Whereas surface energy describes interactions with a range of materials, surface hydrophobicity describes these interactions with water only. Because water has a huge capacity for bonding, a material of high surface energy (i.e., high bonding potential) can enter into more interactions with water and consequently will be more hydrophilic. Therefore hydrophobicity generally decreases as surface energy increases. Hydrophilic surfaces such as glass therefore have high surface energies, whereas hydrophobic surfaces such as PTFE or polystyrene have low surface energies.

Precise characterization of solid material surfaces and fluid interfaces plays a vital role in research, innovation, and product development in many industrial and academic areas. Measurement of contact angles and surface/interfacial tensions provides a better understanding of the interactions between phases, regardless of whether they are gas, liquid, or solid. The surface/interfacial tension of multiphase liquid systems provides essential information about the stability of foams, emulsions, dispersions, gels, aerosols etc. The wettability and surface energy of solid surfaces plays an important role in many processes, such as controlled capillary action, spreading of coatings, adhesion, and absorption into porous solids to name just a few. Contact angle and surface/interfacial tension measurement is a rapid and accurate characterization tool for emerging state-of-the-art surface engineering techniques.

8.3.2 Surface Tension and Liquids

The **surface tension** is a property of the surface of a liquid that causes it to behave as an elastic sheet. It allows insects, such as the water strider, to walk on water. It allows small objects, even metal ones such as needles, razor blades, or foil fragments, to float on the surface of water, and it is the cause of capillary action. The physical and chemical behavior of liquids cannot be understood without taking surface tension into account. It governs the shape that small masses of liquid can assume and the degree of contact a liquid can make with another substance. Applying Newtonian physics to the forces that arise due to surface tension accurately predicts many liquid behaviors that are so commonplace that most people take them for granted. Applying thermodynamics to those same forces further predicts other more subtle liquid behaviors. Information source http://en.wikipedia.org/wiki/Surface_tension.

8.3.2.1 Physical Cause

Surface tension is caused by the attraction between the molecules of the liquid by various intermolecular forces. In the bulk of the liquid each molecule is pulled equally in all directions by neighboring liquid molecules, resulting in a net force of zero. At the surface of the liquid, the molecules are pulled inwards by other molecules deeper inside the liquid and are not attracted as intensely by the molecules in the neighboring medium (be it vacuum, air, or another liquid). Therefore all of the molecules at the surface are subject to an inward force of molecular attraction which can be balanced only by the resistance of the liquid to compression. This inward pull tends to diminish the surface area, and in this respect a liquid surface resembles a stretched elastic membrane. Thus the liquid squeezes itself together until it has the locally lowest surface area possible. Another way to view it is that a molecule in contact with a neighbor is in a lower state of energy than if it were not in contact with a neighbor. The interior molecules all have as many neighbors as they can possibly have. But the boundary molecules have fewer neighbors than interior molecules and are therefore in a higher state of energy. For the liquid to minimize its energy state, it must minimize its number of boundary molecules and must therefore minimize its surface area.

8.3.2.2 Surface Tension in Everyday Life

Some examples of the effects of surface tension seen with ordinary water are

- Beading of rain water on the surface of a waxed automobile. Water adheres weakly to wax and strongly to itself, so water clusters into drops. Surface tension gives them their near-spherical shape, because a sphere has the smallest possible surface area to volume ratio.
- Formation of drops occurs when a mass of liquid is stretched. The animation shows water adhering to the faucet gaining mass until it is stretched to a point where the surface tension can no longer bind it to the faucet. It then separates and surface tension forms the drop into a sphere. If a stream of water were running from the faucet, the stream would break up into drops during its fall. Gravity stretches the stream, then surface tension pinches it into spheres.
- Flotation of objects denser than water occurs when the object is non-wettable and its weight is small enough to be born by the forces arising from surface tension.

- Separation of oil and water is caused by a tension in the surface between dissimilar liquids. This type of surface tension goes by the name "interface tension," but its physics are the same.
- Tears of wine is the formation of drops and rivulets on the side of a glass containing an alcoholic beverage. Its cause is a complex interaction between the differing surface tensions of water and ethanol.

Figure 8.8 shows water striders standing on the surface of a pond. It is clearly visible that their feet cause indentations in the water's surface and it is intuitively evident that the surface with indentations has more surface area than a flat surface. If surface tension tends to minimize surface area, how is it that the water striders are increasing the surface area? Recall that what nature really tries to minimize is potential energy. By increasing the surface area of the water, the water striders have increased the potential energy of that surface. But note also that the water striders' center of mass is lower than it would be if they were standing on a flat surface. So their potential energy is decreased. Indeed when you combine the two effects, the net potential energy is minimized. If the water striders depressed the surface any more, the increased surface energy would more than cancel the decreased energy of lowering the insects' center of mass. If they depressed the surface any less, their higher center of mass would more than cancel the reduction in surface energy. The photo of the water striders also illustrates the notion of surface tension being like having an elastic film over the surface of the liquid. In the surface depressions at their feet it is easy to see that the reaction of that imagined elastic film is exactly countering the weight of the insects.

Surface tension is responsible for the shape of liquid droplets. Although easily deformed, droplets of water tend to be pulled into a spherical shape by the cohesive forces of the surface layer. The spherical shape minimizes then necessary "wall tension" of the surface layer according to Laplace's law. At left is a single early morning dewdrop in an emerging dogwood blossom. Surface tension and adhesion determine the shape of this drop on a twig. It dropped a short time later, and took a more nearly spherical shape as it fell. Falling drops take a variety of shapes due to oscillation and the effects of air friction. The relatively high surface tension of water accounts

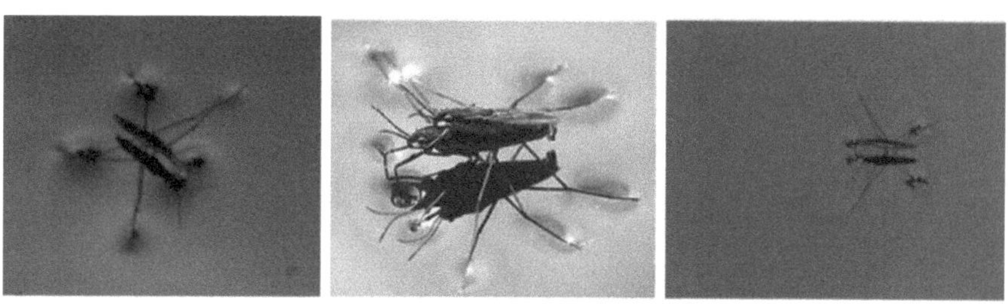

Fig. 8.8 Surface tension and the water strider. http://en.wikipedia.org/wiki/Surface_tension Source: Wikipidia

for the ease with which it can be nebulized, or placed into aerosol form. Low surface tension liquids tend to evaporate quickly and are difficult to keep in an aerosol form. All liquids display surface tension to some degree. The surface tension of liquid lead is utilized to advantage in the manufacture of various sizes of lead shot. Molten lead is poured through a screen of the desired mesh size at the top of a tower. The surface tension pulls the lead into spherical balls, and it solidifies in that form before it reaches the bottom of the tower (Fig. 8.9).

8.3.2.3 Basic Physics Definitions

Surface tension, represented by the symbol σ, γ, or T, is defined as the force along a line of unit length, where the force is parallel to the surface but perpendicular to the line. One way to picture this is to imagine a flat soap film bounded on one side by a taut thread of length, L. The thread will be pulled toward the interior of the film by a force equal to $2\gamma L$ (the factor of 2 is because

Fig. 8.9 Surface tension and droplets

the soap film has two sides hence two surfaces). Surface tension is therefore measured in forces per unit length. Its SI unit is newton per meter (N/m).

An equivalent definition, one that is useful in thermodynamics, is work done per unit area. As such, in order to increase the surface area of a mass of liquid by an amount, δA, a quantity of work, $\gamma\delta A$, is needed. This work is stored as potential energy. Consequently surface tension can be also measured in SI system as joules per meter^2 (J/m^2).

8.3.3 Surface Tension and Solids

In his seminal work, Shuttleworth [56] made a distinction between the surface Helmholtz free energy F, and the surface tension γ. In the paper, the surface tension and the surface Helmholtz free energy are defined, and a thermodynamic relation between them is derived. Shuttleworth pointed out that the surface tension of a crystal face is related to the surface free energy by the relation

$$\gamma = F + A\frac{\mathrm{d}F}{\mathrm{d}A}, \tag{3.1}$$

where A is the area of the surface. For a one-component liquid, surface free energy and tension are equal. For crystals the surface tension is not equal to the surface energy. The standard thermodynamic formula of surface physics are reviewed, and it is found that the surface free energy appears in the expression for the equilibrium contact angle, and in the Kelvin expression for the excess vapor pressure of small drops, but that the surface tension appears in the expression for the difference in pressure between the two sides of a curved surface. The surface tensions of inert-gas and alkali-halide crystals are calculated from expressions for their surface energies and are found to be negative. The surface tensions of homopolar crystals are zero if it is possible to neglect the interaction between atoms that are not nearest neighbors.

8.3.3.1 Origin of Surface Tension for a Crystal

For simplicity a crystal at 0 K is considered, and the forces between any two atoms are supposed to depend only on their separation. If it is not possible to neglect the interaction between atoms that are not nearest neighbors, then the equilibrium separation of atoms in an isolated plane will be different from that in a three-dimensional lattice, since the number of non-nearest-neighbor atoms will be different in the two cases. The lattice constant of an isolated (100) plane of atoms of an inert-gas crystal is 0.643% greater than that of the three-dimensional crystal. Lennard-Jones and Dent [34] have shown that the lattice constant of an isolated (100) plane of ions of an alkali-halide crystal is about 5% less than that of the three-dimensional crystal. In order that an isolated plane should have the same spacing as that of the crystal it is necessary to apply external forces to the edges of the plane and tangential to it: the forces are compression for inert-gas crystals and tension for alkali-halide crystals. If the stressed plane is now moved towards the crystal, until it becomes the surface plane, the external forces needed to keep it with the three-dimensional lattice constant will be reduced. When all the atoms are on the positions they would occupy if no surface existed and they were in the center of the crystal, then the tangential force it is necessary to apply to the surface plane is reduced to half of that which must be applied to an isolated plane. This state is not stable, for in equilibrium the distance between the outermost plane of atoms and the next is different from that in the center of the crystal; when the surface plane takes up its equilibrium position this movement causes a further change in the tangential force which must be applied. Similar, but smaller, tangential forces must be applied to successive planes in the crystal surface. *The surface tension is the total force per unit length that must be applied tangentially to the surface in order that the surface planes have the same lattice spacing as the underlying crystal.*

8.4 Elastic Description of Free Surfaces and Interfaces

Dingreville [7] discusses essential concepts and definitions relative to the elastic description of surfaces and interfaces. The concept of surface/interfacial excess energy is first reformulated from the continuum mechanics point of view by considering a single dividing surface separating the two homogeneous phases (as opposed to the interface considered as an interphase). It is shown that the well-known Shuttleworth relationship between the interfacial excess energy and interfacial excess stress is valid only when the interface is free of transverse stresses. To account for the transverse stress, a new relationship is derived between the interfacial excess energy and interfacial excess stress. At the same time, the concept of transverse interfacial excess strain is also introduced, and a complementary Shuttleworth equation is derived that relates the interfacial excess energy to the newly introduced transverse interfacial excess strain. This new formulation of interfacial excess stress and excess strain naturally leads to the definition of an in-plane interfacial stiffness tensor, a transverse interfacial compliance tensor, and a coupling tensor that accounts for the Poisson's effect of the interface. These tensors fully describe the elastic behavior of a coherent interface upon deformation. A semi-analytical method is subsequently presented to calculate the interfacial elastic properties. The cases of free surfaces and interfaces are distinguished. As an illustration, he presents numerical examples for low-index surfaces (111), (100), and (110) of face-centered cubic transition metals.

8.4.1 Definition of Interfacial Excess Energy

The surface free (excess) energy, γ_n, of a near surface atom is defined by the difference between its total energy and that of an atom deep in the interior of a large bicrystal. Clearly, γ_n depends on the location of the atom. In addition, γ_n is a function of the intrinsic bicrystal interface properties, as well as a function of the relative surface deformation. If there are N atoms surrounding an area A in the deformed configuration, then the total surface free energy associated with area A is given by $\sum_{n=1}^{N} \gamma_n$ and the Gibbs surface free energy density is defined by

$$\Gamma = \frac{1}{A}\sum_{n=1}^{N} \gamma_n. \tag{8.1}$$

Note that the above definition is in the deformed configuration. It can be viewed as the Eulerian description of the surface free energy density. For solid crystal surfaces, the Lagrange description of the surface free energy density can be defined by

$$\Gamma = \frac{1}{A_0}\sum_{n=1}^{\infty} \gamma_n = \frac{1}{A_0}\sum_{n=1}^{\infty}\left(E^{(n)} - E^{(0)}\right), \tag{8.2}$$

where $E^{(n)}$ is the total energy of the atom n surrounding the area A_0, and $E^{(0)}$ is the total energy of an atom in a perfect lattice far away from the free surface. A_0 is the area originally occupied in the undeformed configuration by the same atoms that occupy the area A in the deformed configuration. It can be easily shown that the two areas are related through

$$A = A_0\left(1 + \varepsilon^s_{\eta\eta}\right), \tag{8.3}$$

where $\varepsilon^s_{\alpha\beta}$ is the Lagrange surface strain relative to the undeformed crystal lattice. Although the sum in Equation (8.2) involves an infinite number of atoms, the difference $E^{(n)} - E^{(0)}$ is non-zero only for atoms within a few atomic layers near the interface. So, in practice, the sum in Equation (8.2) only involves a very limited number of terms. It should also be pointed out that the surface energy density calculated from Equation (8.2) contains contributions not only from atoms on the surface, but from all atoms near the interface.

8.4.2 Surface Elasticity

Dingreville [7] shows that the elastic behavior of the interface is fully characterized by five tensors, namely $\mathbf{\Gamma}^{(1)}$, $\mathbf{\Gamma}^{(2)}$, $\mathbf{H}$, $\mathbf{\Lambda}^{(1)}$, and $\mathbf{\Lambda}^{(2)}$. The first term $\mathbf{\Gamma}^{(1)}$ is a two-dimensional, second-order tensor representing the internal excess stress of the interface. It is the part of interfacial stress that exists when the surface strain and transverse stress are absent. The second term $\mathbf{\Gamma}^{(2)}$ is a the two-dimensional, fourth-order tensor that represents the interface's in-plane elasticity, while the third term $\mathbf{H}$ is a third-order tensor that measures the Poisson's effect of the interface. $\mathbf{\Lambda}^{(1)}$ represents the part of transverse interfacial deformation that exists even when the remote traction at the in-plane strain vanishes. This is the reason why $\mathbf{\Lambda}^{(1)}$ is called the interfacial "relaxation" tensor. The fourth-order tensor $\mathbf{\Lambda}^{(2)}$ representing the transverse compliance of the interface is called the interfacial transverse compliant tensor. It has been pointed out that although $\mathbf{\Lambda}^{(1)}$ and $\mathbf{H}$ affect the in-plane interfacial excess stress and transverse interfacial excess strain, they do not explicitly appear in the interfacial excess energy.

$\mathbf{\Gamma}^{(1)}, \mathbf{\Gamma}^{(2)}, \mathbf{H}, \mathbf{\Lambda}^{(1)}$, and $\mathbf{\Lambda}^{(2)}$ can be calculated analytically for a given bimaterial with known interatomic potentials as shown later on in this chapter. Once these tensors are known, the elastic behavior of the interface is fully characterized.

8.4.3 Surface Stress and Surface Strain

The interfacial excess in-plane stress $\Sigma^s_{\alpha\beta}$ is determined by

$$\Sigma^s_{\alpha\beta} = \Gamma^{(1)}_{\alpha\beta} + \Gamma^{(2)}_{\alpha\beta\kappa\lambda}\varepsilon^s_{\kappa\lambda} + H_{j\alpha\beta}\sigma^t_j, \tag{8.4}$$

and the interfacial excess transverse strain Δ^t_k is given by

$$\Delta^t_k = \Lambda^{(1)}_k + \Lambda^{(2)}_{kj}\sigma^t_j - H_{k\alpha\beta}\varepsilon^s_{\alpha\beta}. \tag{8.5}$$

8.5 Surface/Interfacial Excess Quantities Computation

Dingreville [7] exposed an approach combining continuum mechanics and atomistic simulations to develop a nanomechanics theory for modeling and predicting the macroscopic behavior of nanomaterials. This nanomechanics theory exhibits the simplicity of the continuum formulation while taking into account the discrete atomic structure and interaction near surfaces/interfaces.

First, Dingreville [7] revisited the theory of interfaces to better understand its behavior and effects on the overall behavior of nanostructures. Second, atomistic tools are provided in order to efficiently determine the properties of free surfaces and interfaces. Third, he proposes a continuum framework that casts the atomic level information into continuum quantities that can be used to analyze, model, and simulate macroscopic behavior of nanostructured materials. In particular, he studies the effects of surface free energy on the effective modulus of nanoparticles, nanowires, and nanofilms as well as nanostructured crystalline materials and proposes a general framework valid for any shape of nanostructural elements/nano-inclusions (integral forms) that characterize the size-dependency of the elastic properties. This approach bridges the gap between discrete systems (atomic-level interactions) and continuum mechanics. Finally this continuum outline is used to understand the effects of surfaces on the overall behavior of nanosize structural elements (particles, films, fibers, etc.) and nanostructured materials. In terms of engineering applications, this approach proves to be a useful tool for multi-scale modeling of heterogeneous materials with nanometer-scale microstructures and provides insights on surface properties for several material systems; these will be very useful in many fields including surface science, tribology, fracture mechanics, adhesion science and engineering, and more. It will accelerate the insertion of nanosize structural elements, nanocomposite, and nanocrystalline materials into engineering applications. The related papers are Dingreville et al. [10]; Dingreville and Qu [8, 9].

8.6 On Eshelby's Nano-Inhomogeneities Problems

Homogenization methods have been recognized as a rapid developing scheme in the past decades due to a strong desire for tailoring material microstructures. There are several techniques to establish the relationship between the effective properties and the microstructure of a heterogeneous material [15, 16, 23, 64]. Eshelby [15] was the first to address rigorously the problem of determination of elastic states of an embedded inclusion in the context of classical elasticity. This seminal work of Eshelby [15], both with and without modifications, has been employed to tackle a diverse set of problems: Localized thermal heating, residual strains, dislocation induced plastic strains, phase transformations, overall or effective elastic, plastic and viscoplastic properties of composites, viscoelastic properties of composites, damage in heterogeneous materials, quantum dots, interconnect reliability, microstructural evolution, to name a few. The micromechanical modeling approach initiated by Eshelby [15] consists of two fundamental operations [45]:

- localization, which determines the relationship between the microscopic (local) fields and the macroscopic (global) loading,
- homogenization, which employs averaging techniques to approximate macroscopic behavior.

In Eshelby's work, inhomogeneities are defined as embedded particles with material properties differing from the surrounding host material or matrix while eigenstrains are stress-free strains such as lattice parameter mismatch, thermal expansion, inelastic strains, etc. In its present form, Eshelby's formalism does not include the effects of the elastic surface properties (residual surface tension, surface moduli) of inhomogeneities and their elastic state is entirely based on bulk properties [7]. Thus, the classical solution of an embedded inclusion neglects the presence of surface or interface energies and, therefore, the effects of those are negligible except in the size range of tens of nanometers, where one contends with a significant surface-to-volume ratio. For most technological problems (until recently where nanomaterials have been growing explosively) inclusions were of the order of microns and rarely were one concerned with nano-inclusions or related size effects. At the micron and higher length scales, the surface-to-volume ratios are negligible and indeed Eshelby's original assumptions hold true and so does his solution. In other words, each particle in composite materials can be treated as a continuous medium and, therefore, continuum mechanics equations can be used to describe the deformation of conventional composite materials. There are many approaches that attempt to combine continuum mechanics and surface/interface properties to develop a nanomechanics theory for modeling and predicting the macroscopic behavior of nanomaterials. This nanomechanics theory exhibits the simplicity of the continuum formulation while taking into account the discrete atomic structure and interaction near surfaces/interfaces. The purpose of this report is to summarize these several attempt to incorporate surface/interface energy in continuum mechanics-based micromechanics theories.

8.7 Background in Nano-Inclusion Problem

8.7.1 The Work of Sharma et al.

The work by Sharma et al. [54] is one of the pioneering works to address the problem of combining surface elasticity with Eshelby's formalism to analyze inhomogeneities with size-dependent surface effects. They reformulate the inhomogeneity problem in terms of generalized energy functionals (rather than the stress-based approach of Eshelby), permitting a simple way to include surface/interface effects. In their study, the surface stress tensor, $\boldsymbol{\sigma}^s$, is related to the deformation dependent surface energy $\boldsymbol{\Gamma}(\boldsymbol{\varepsilon}^s)$ by:

$$\sigma^s_{\alpha\beta} = \tau_0 \delta_{\alpha\beta} + \frac{\partial \boldsymbol{\Gamma}}{\partial \varepsilon^s_{\alpha\beta}}, \tag{8.6}$$

where, $\varepsilon^s_{\alpha\beta}$ is the 2×2 strain tensor for surfaces, $\delta_{\alpha\beta}$ represents the Kronecker delta for surfaces while τ_0 is the residual surface tension. By making the assumption that the surface adheres to the bulk without slipping, and in the absence of body forces, they summarize equilibrium and constitutive equations for isotropic case as:

In the bulk:

$$\begin{cases} \operatorname{div}(\boldsymbol{\sigma}^b) = 0, \\ \boldsymbol{\sigma}^b = \mathrm{C} : \boldsymbol{\varepsilon}^b. \end{cases} \tag{8.7}$$

On the surface/interface:

$$\begin{cases} \boldsymbol{\sigma}^b \cdot \mathbf{n} + \operatorname{div}_s(\boldsymbol{\sigma}^s) = 0, \\ \mathbf{n} \cdot \boldsymbol{\sigma}^b \cdot \mathbf{n} = \sigma^s : \boldsymbol{\kappa}, \\ \boldsymbol{\sigma}^s = \tau_0 \mathbf{I}_2 + 2(\mu^s - \tau_0)\boldsymbol{\varepsilon}^s + (\lambda^s + \tau_0)\mathbf{Tr}(\boldsymbol{\varepsilon}^s)\mathbf{I}_2, \end{cases} \tag{8.8}$$

where, $\mathbf{C}$ is the stiffness tensor of the isotropic bulk, λ^s and μ^s characterizes Lamé constants (which render the surface energy deformation dependent) for isotropic interface. $\boldsymbol{\kappa}$ represents the curvature tensor of the surface or interface, $\mathbf{n}$ is the normal vector on the interface or surface, $\mathbf{I}_2$ represents the 2×2 identity tensor. Then, Sharma et al. [54] consider a spherical inhomogeneity, of radius R_0, located in an infinite matrix, and undergoing a dilatation eigenstrain (generally, but not necessarily, nonzero), $\varepsilon_{11}^* = \varepsilon_{22}^* = \varepsilon_{33}^* = \varepsilon^*$, and subjected to far-field triaxial stress, $\boldsymbol{\sigma}^\infty$. The free energy of the spherically symmetric system, in the presence of surface effects, is then written as:

$$\Pi = 4\pi \int_0^{R_0} r^2 \Psi_I \mathrm{d}r + 4\pi R_0^2 \int_0^{\varepsilon_{ij}^s} \sigma_{ij}^s \mathrm{d}\varepsilon_{ij}^s + 4\pi \int_{R_0}^{R_\infty} r^2 \Psi_M \mathrm{d}r. \tag{8.9}$$

In Equation (8.9), Ψ_I and Ψ_M are the bulk elastic energy densities of the inhomogeneity and the matrix, respectively. By setting the variation of the free energy to be zero, i.e., $\delta\Pi = 0$, Sharma et al. [54] derive analytical solution of the radially symmetric (due to the spherically symmetric nature of the problem) displacement field, $u(r)$, from the Euler-Lagrange equations and the appropriate boundary conditions. After, they present an application of their work to the classical problem of stress concentration at a void.

8.7.2 *The Work by Lim et al.*

Lim et al. [36] analyze the influence of interface stress on the elastic field within a nanoscale inclusion by focusing special attention on the case of nonhydrostatic eigenstrain. From the viewpoint of practicality, they assume that the inclusion (of radius R) is spherically shaped and embedded into an infinite solid, within which an axisymmetric eigenstrain is prescribed

$$\boldsymbol{\varepsilon}^* = \varepsilon_{11}^* \mathbf{e}_1 \otimes \mathbf{e}_1 + \varepsilon_{11}^* \mathbf{e}_2 \otimes \mathbf{e}_2 + \varepsilon_{33}^* \mathbf{e}_3 \otimes \mathbf{e}_3,$$

where $\mathbf{e}_1, \mathbf{e}_2$ and $\mathbf{e}_3$ are, respectively, the base vectors along the x_1, x_2, and x_3 directions. For simplicity, both the matrix and inclusion are assumed elastically isotropic with the same elastic modulus in their work. since the deformation is axisymmetric about the x_3-axis, the displacements will be confined to meridian planes, having a component, u_r along the radius, r, and a component, u_θ, in the direction of increasing θ. For convenience, the analysis has been carried out in spherical coordinates (r, θ, φ) with the origin at the center of the inclusion. Within and outside the sphere, the displacement,

$$\mathbf{u} = u_r \mathbf{e}_r + u_\theta \mathbf{e}_\theta,$$

satisfies the following Navier's equation (with no body forces):

$$(1+\Lambda)\nabla(\nabla\cdot\mathbf{u})+\nabla^2\mathbf{u}=\mathbf{0}, \tag{8.10}$$

where $\Lambda=\lambda/\mu$, $\nabla=\mathbf{e}_r(\partial/\partial r)+\mathbf{e}_\theta(\partial/r\partial\theta)$. The strain tensor, ε, and stress tensor, $\boldsymbol{\sigma}$, are defined as:

$$\begin{cases}\boldsymbol{\varepsilon}=\frac{1}{2}\left[\nabla\mathbf{u}+(\nabla\mathbf{u})^{\mathbf{T}}\right],\\ \boldsymbol{\sigma}=2\mu(\boldsymbol{\varepsilon}-\boldsymbol{\varepsilon}^*)+\lambda\mathrm{Tr}(\boldsymbol{\varepsilon}-\boldsymbol{\varepsilon}^*)\mathbf{I}.\end{cases} \tag{8.11}$$

This stress field must fulfill the stress jump condition at the interface ($r=R$):

$$[\boldsymbol{\sigma}\cdot\mathbf{n}]=-\mathrm{div}_s(\boldsymbol{\sigma}^s), \tag{8.12}$$

where:

$$\boldsymbol{\sigma}^s=\tau_0\mathbf{I}_2+2(\mu^s-\tau_0)\boldsymbol{\varepsilon}^s+(\lambda^s+\tau_0)\mathrm{Tr}(\varepsilon^s)\mathbf{I}_2+\underbrace{\tau_0\nabla_s\mathbf{u}}.$$

The underlined term, as pointed out by Lim, is often omitted in some studies such Sharma et al. [54]; Sharma and Ganti [53]; Duan et al. [12]. Following the works by Goodier [20] and Love [38], Lim et al. express the solution to Equation (8.10) in terms of two types of spherical solid harmonical functions, ϕ and ω_n, as:

$$\mathbf{u}=\left[\frac{\partial\phi}{\partial r}+\left(r^2\frac{\partial\omega_n}{\partial r}+\alpha_n r\omega_n\right)\right]\mathbf{e}_r+\left(\frac{\partial\phi}{r\partial\theta}+r\frac{\partial\omega_n}{\partial\theta}\right)\mathbf{e}_\theta, \tag{8.13}$$

with

$$\nabla^2\phi=0,\qquad \nabla^2\omega_n=0, \tag{8.14}$$

and

$$\alpha_n=-2\frac{3n+1+n\Lambda}{n+5+(n+3)\Lambda}.$$

The general axisymmetric solution of Equation (8.14) is of the following form

$$\sum_{n=0}^{\infty}\left(b_n r^n+\frac{c_n}{r^{n+1}}\right)P_n(\cos\theta), \tag{8.15}$$

where P_n is the $n-$order Legendre polynomial.

Solution outside the inclusion:

$$\phi = \frac{c_0}{r} + \frac{c_2}{r^3} P_2(\cos\theta), \qquad \omega_{-3} = \frac{c_2'}{r^3} P_2(\cos\theta). \tag{8.16}$$

Solution within the inclusion:

$$\phi = b_2 r^2 P_2(\cos\theta), \qquad \omega_2 = b_2' r^2 P_2(\cos\theta), \qquad \omega_0 = b_0. \tag{8.17}$$

The continuity condition for displacements together with the equilibrium condition (8.12) at $r = R$ yield six independent equations to solve for b_0, b_2, b_2', c_0, c_2, c_2'. Therefore the displacement, strain, and stress fields within and outside of the inclusion are determined in closed-form. Then, Lim et al. [36] have carried out numerical simulations to investigate the sensitivity of the elastic field to the surface/interfacial excess energy. They concludes that the strain state of the elastic system is size dependent (in the sense that it is dependent on $(r/R)^2$), differing significantly from the classic result obtained from the classical linear elasticity. Numerical computation indicates that such a size dependence is quite remarkable when the radius of the inclusion is below tens of nanometer. Different elastic constants of the interface may cause the interface to either shrink or dilate, implying that there exists local softening or hardening at the interface of the inclusion and the matrix. Another important conclusion is that interface stress results in nonuniform elastic field inside the spherical inclusion when the eigenstrain is nonhydrostatic even if uniform. These results indicate that interface stress plays a significant role in the elastic behavior of embedded inclusions of nanoscale size.

8.7.3 The Work by Yang

Yang [63] analyzes the effective bulk modulus of a composite material consisting of spherical inclusions at dilute concentrations. The consider an infinite elastic matrix containing a spherical inclusion of radius a and a spherical coordinate system (r, θ, φ) is also used such that the origin coincides with the center of the inclusion. Yang provides a set of five basis equations for determining the stress state in a composite material containing spherical inclusions at dilute concentrations:

$$\begin{cases} \mu^t \nabla^2 u_i^t + (\lambda^t + \mu^t)\nabla \cdot \mathbf{u}^t = 0, \\ \varepsilon_{ij}^t = \frac{1}{2}\left(u_{i,j}^t + u_{j,i}^t\right), \\ \sigma^t = \lambda^t \varepsilon_{kk}^t \delta_{ij} + 2\mu^t \varepsilon_{ij}^t, \\ \sigma_{ij}^s = \gamma \delta_{ij} + \frac{\partial \gamma}{\partial \varepsilon_{ij}^s}, \\ \left(\sigma_{ij}^M - \sigma_{ij}^I\right) n_i n_j = \sigma_{ij}^s \kappa_{ij} \quad \text{at} \quad r = a, \end{cases} \tag{8.18}$$

where M denotes the matrix and I denotes the inclusion. The superscript $t(=M, I)$ represents the field in the matrix $t = M$ and the inclusion $t = I$. κ_{ij} represents the curvature tensor of the interface. To obtain a closed-form solution, Yang consider the case in which the interface is isotropic and $\sigma_{\theta\theta}^s = \sigma_{\varphi\varphi}^s = \sigma^s$. Making use of these five basic equations, Yang determines the nonzero components of the displacement vectors and stress tensors within and outside the inclusion with some algebraic manipulations.

First case

Yang considers a stress-free spherical shell with initial outer radius b and initial inner radius a and a stress-free spherical inclusion with initial radius a. The spherical inclusion is embedded into the spherical shell to form a composite, in which the center of the spherical inclusion is the same as that of the spherical shell. The interfacial stresses between the spherical shell and the inclusion then create internal stresses in the composite. The matrix (spherical shell) is under tension, while the inclusion is under compression. From the viewpoint of the theory of linear elasticity, the reference state of the composite is stress-free at this stage. The nonzero components of the displacement vectors is

$$\begin{cases} u_r^M(r) = -\frac{2\sigma^s r}{a\Delta}\left(\frac{a^3}{r^3} + \frac{2a^3}{b^3}\right), & \text{for} \quad a \le r \le b, \\ u_r^I(r) = -\frac{2\sigma^s r}{a\Delta}\left(1 + \frac{2a^3}{b^3}\right), & \text{for} \quad 0 \le r \le a, \end{cases} \tag{8.19}$$

where

$$\Delta = 4\mu^M + 3K^I - \frac{2a^3}{b^3}\left[2\mu^M + 3\left(\lambda^M - K^I\right)\right].$$

It is followed from Equation (8.19) that the interface between the inclusion and the matrix moves toward the center of the inclusion under the action of the interfacial stress.

Second case

Then, Yang considers that a sphere of radius b having a spherical inclusion of initial radius a at its center is subjected to a radial strain ε_0 on the external surface. The reference state of the composite is different from the stress-free configuration assumed in the theory of linear elasticity. It should be the geometrical configuration involving the deformation created by the interfacial stresses. Using Equation (8.19), one obtains the reference radius of the particle as

$$\tilde{a} = a + u_r^I(a) = a - \frac{2\sigma^s}{\Delta}\left(1 + \frac{2a^3}{b^3}\right). \tag{8.20}$$

Following the same procedure as in the above section, Yang derives the nonzero components of the resultant displacement vectors as

$$\begin{cases} u_r^M(r) = \frac{r}{\tilde{\Delta}}\left[\varepsilon_0\Lambda - \frac{2\sigma^s}{\tilde{a}}\left(\frac{\tilde{a}^3}{r^3} + \frac{2\tilde{a}^3}{b^3}\right)\right], & \text{for} \quad \tilde{a} \le r \le b, \\ u_r^I(r) = \frac{r}{\tilde{\Delta}}\left[3\varepsilon_0(2\mu^M + \lambda^M) - \frac{2\sigma^s}{\tilde{a}}\left(1 + \frac{2\tilde{a}^3}{b^3}\right)\right], & \text{for} \quad 0 \le r \le \tilde{a}, \end{cases} \tag{8.21}$$

where:

$$\tilde{\Delta} = 4\mu^M + 3K^I - \frac{2\tilde{a}^3}{b^3}\left[2\mu^M + 3(\lambda^M - K^I)\right],$$

$$\Lambda = 4\mu^M + 3K^I + \frac{\tilde{a}^3}{b^3}\left[2\mu^M + 3(\lambda^M - K^I)\right].$$

The effective bulk modulus of a composite material is then derived in terms of total elastic energy, in the sense that if the composite material is replaced by an equivalent linearly elastic and homogeneous material, it must store the same amount of elastic energy as the actual composite material for the same applied stress or applied strain. Considering only the dilute condition $f = (a/b)^3 << 1$ Yang obtains

$$K_{eff} = K^M\left[1 + \Pi f\left(1 - \frac{2\sigma^s}{a\Delta}\right)\right], \tag{8.22}$$

where

$$\Pi = -\frac{1 - K^I/K^M}{1 + (K^I - K^M)/(K^M + 4\mu^M/3)}.$$

Yang concludes that unlike the classical result, in the theory of linear elasticity, the effective bulk modulus is a function of the interfacial stress and the size of the inclusion. The interfacial stress enhances the effective bulk modulus of composite materials having inclusions softer than the matrix, while it reduces the effective bulk modulus of composites having inclusions stiffer than the matrix. The effect of the interfacial stress is negligible for large inclusions in which case the effective bulk modulus reduces to the classical result obtained from the theory of linear elasticity.

8.7.4 The Work by Sharma and Ganti

Sharma and Ganti [53] have revisited and modified the classical formulation of Eshelby for embedded inclusions by incorporating surface/interface stresses, tension, and energies. The latter effects, as it is stated in the previous sections, come into prominence at inclusion sizes in the nanometer range. Sharma and Ganti consider an arbitrary shaped inclusion Ω embedded in an infinite amount of material. By definition of an inclusion, they suppose a prescribed stress-free transformation strain within the domain of the inclusion as shown by Fig. 8.10. The eigenstrain is considered to be uniform. Equation (8.6) defines the relationship between the surface stress tensor, $\boldsymbol{\sigma}^s$, and the deformation dependent surface energy $\Gamma(\boldsymbol{\varepsilon}^s)$. They summarize equilibrium and constitutive equations for isotropic case as:

In the bulk:

$$\begin{cases} \operatorname{div}(\boldsymbol{\sigma}^b) = 0, \\ \boldsymbol{\sigma}^b = \lambda \mathbf{I}_3 \operatorname{Tr}(\varepsilon) + 2\mu\varepsilon. \end{cases} \tag{8.23}$$

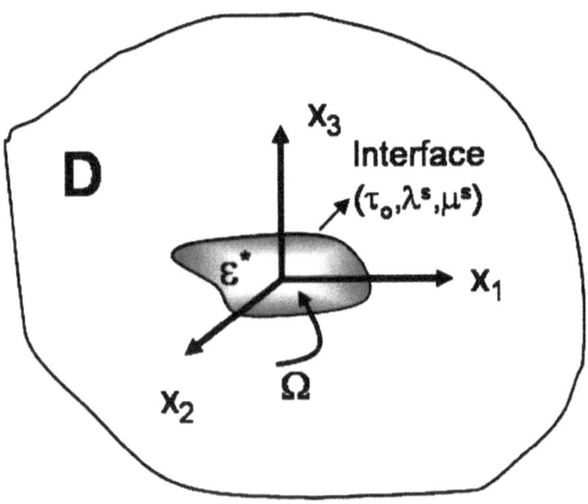

Fig. 8.10 Schematic of the problem

On the surface/interface:

$$\begin{cases} [\boldsymbol{\sigma}^b \cdot \mathbf{n}] + \mathrm{div}_s(\boldsymbol{\sigma}^s) = 0, \\ \boldsymbol{\sigma}^s = \tau_0 \mathbf{I}_2 + 2(\mu^s - \tau_0)\varepsilon^s + (\lambda^s + \tau_0)\mathrm{Tr}(\varepsilon^s)\mathbf{I}_2, \end{cases} \tag{8.24}$$

where $\mathbf{I}_3$ represents the 3×3 identity tensor. Noting that the transformation strain is only nonzero within the inclusion domain ($\mathbf{x} \varepsilon \Omega$), they write the bulk-constitutive law for the inclusion-matrix as follows:

$$\boldsymbol{\sigma}^b = \mathbf{C} : \{\boldsymbol{\varepsilon} - \boldsymbol{\varepsilon}^* H(z(\mathbf{x}))\}, \tag{8.25}$$

where H is the Heaviside function and $z(\mathbf{x})$ is defined as:

$$\{z(\mathbf{x}) > 0 | \mathbf{x} \in \Omega\}, \quad \{z(\mathbf{x}) < 0 | \mathbf{x} \notin \Omega\}. \tag{8.26}$$

Taking the divergence of Equation (8.25) and making use of the stress jump condition Eq. (7.19) they obtain

$$\nabla \cdot \boldsymbol{\sigma}^b = \nabla \cdot (\mathbf{C} : \boldsymbol{\varepsilon}) - \underbrace{\nabla \cdot \{\mathbf{C} : \varepsilon^* H(z(\mathbf{x}))\} + \delta(\bar{z}(\mathbf{x}))\mathrm{div}_s \sigma^s} = \mathbf{0}. \tag{8.27}$$

$\delta(\cdot)$ is the Dirac delta function while $\bar{z}(\mathbf{x})$ defines the interface. Using the underlined term as representing a body force in conjunction with the elastic Greens function, they write the displacement field due to both the eigenstrain and the surface effect as

$$\mathbf{u} = \int_V \mathbf{G}^T(\mathbf{y} - \mathbf{x}) \cdot (\nabla \cdot \{\mathbf{C} : \boldsymbol{\varepsilon}^* H(\mathbf{y})\}) \mathrm{d}V_y + \underbrace{\int_S \mathbf{G}^T(\mathbf{y} - \mathbf{x}) \cdot \mathrm{div}_s \boldsymbol{\sigma}(\mathbf{y}) \mathrm{d}S_y}. \tag{8.28}$$

Making use of Gauss theorem to cast Equation (8.28) and invoking the linearized strain-displacement law:

$$\boldsymbol{\varepsilon} = \mathrm{sym}(\nabla \otimes \mathbf{u}),$$

one obtains

$$\boldsymbol{\varepsilon} = \mathbf{S} : \boldsymbol{\varepsilon}^* + \mathrm{sym}\left\{ \nabla_x \otimes \underbrace{\int_S \mathbf{G}^T(\mathbf{y} - \mathbf{x}) \cdot \mathrm{div}_s \boldsymbol{\sigma}(\mathbf{y}) \mathrm{d}S_y} \right\}, \tag{8.29}$$

where $\mathbf{S}$ is the classical size independent Eshelby tensor. Further simplification does not appear feasible without additional assumptions regarding inclusion shape. One notes that Equation (8.29) implicitly gives the modified Eshelby's tensor for inclusions incorporating surface energies. This relation is implicit since the surface stress depends on the surface strain, which in turn is the projection of the conventional strain ($\boldsymbol{\varepsilon}$) on the tangent plane of the inclusion-matrix interface. In terms of the surface projection tensor, $\mathbf{P} = \mathbf{I}_3 - \mathbf{n} \otimes \mathbf{n}$, the surface divergence of the surface stress tensor can be written as

$$\mathrm{div}_s(\boldsymbol{\sigma}^s) = \mathrm{div}_s(\mathbf{C}^s : \mathbf{P} : \boldsymbol{\varepsilon} : \mathbf{P} + \tau_0 \mathbf{P}). \tag{8.30}$$

From Equation (8.30), one notices that the surface divergence of surface stress tensor can only be uniform if the classical "bulk" strain as well as the projection tensor is uniform over the inclusion surface. Gurtin et al. [21] consider that:

$$\mathrm{div}_s \mathbf{P}^s = 2\kappa \mathbf{n}, \tag{8.31}$$

here κ is the mean curvature of the inclusion. For a general ellipsoid the curvature is nonuniform and varies depending upon the location at the surface. Only for the special cases of spherical and cylindrical shape is the mean curvature uniform hence leading them to conclude the following:

Proposition: *Eshelby's original conjecture that only inclusions of the ellipsoid family admit uniform elastic state under uniform eigenstrains must be modified in the context of coupled surface/interface-bulk elasticity. Only inclusions that are of a constant curvature admit a uniform elastic state, thus restricting this remarkable property to spherical and cylindrical inclusions.*

Spherical and cylindrical inclusions are endowed with a constant curvature and thus according to the previous section must admit a uniform elastic state in coupled bulk-surface elasticity. The new Eshelbys tensor will, of course, be size-dependent because of the presence of curvature terms. Due to the constant curvature, Equation (8.29) can be simplified considerably. The surface divergence of the surface stress can be simply taken out of the differential and integral operators. The surface integral is converted into a volume integral and we can then write:

$$\boldsymbol{\varepsilon} = \mathbf{S} : \boldsymbol{\varepsilon}^* - (2\kappa s)\mathbf{C}^{-1} : (\mathbf{S} : \mathbf{I}_3), \tag{8.32}$$

where the scalar s is defined from the relation:

$$\boldsymbol{\sigma}^s = s\mathbf{P}.$$

Sharma and Ganti make then three applications of their work: *Size-Dependent Stress Concentration at a Spherical Void, Size-Dependent Overall Properties of Composites* and *Size-Dependent Strain and Emission Wavelength in Quantum Dots*. They point out several limitations of their work :

- Isotropic behavior was assumed throughout. This is a rather dubious assumption when one is concerned with surfaces and interfaces. Unfortunately, matters are unlikely to be analytically tractable once the assumption of isotropy is abandoned. Numerical formulation of the coupled-surface bulk elasticity may be necessary to remove this restriction.

- Analytical formulas were restricted to the spherical and cylindrical shape. This limits their ability to study the effect of shape on the size-dependent elastic state of nano-inclusions. Derivation of the modified Eshelby tensor for the general ellipsoid (which surely must proceed numerically) would be a useful extension of the present work.
- It would be also of interest to see the behavior of nonsmooth inclusion shapes, e.g., parallelepipeds. Polyhedral inclusions with vertices essentially possess zero curvature everywhere except at the corners where singularities exist.
- Slip, twist, and wrinkling of surfaces/interfaces were ignored. One can expect some interesting physics to emerge from inclusion of such effects. Slip and twist of elastic interfaces were recently included by Gurtin et al. [21] to supplement the original formulation, [22]. These notions are closely linked to the concept of coherency-incoherency and their discussion in relation to Eshelby's problems is relegated to a future work.

8.7.5 The Work of Sharma and Wheeler

In their work, Sharma and Wheeler [55] use a tensor virial method of moments [4], to derive an approximate solution to the relaxed elastic state of embedded ellipsoidal inclusions that incorporates surface/interface energies since the direct use of the integral equation (8.29) is not very convenient for their purposes. This is the first extension of the previous work [53] on incorporation of surface/interface energies in the elastic state of inclusions to the ellipsoidal shape. They only consider the effect of surface tension (i.e., τ_0) and ignore deformation dependent surface elasticity. They state that, this assumption is reasonable for small strains and indeed, as has been found in some technological applications, the deformation dependent surface elasticity effects can often be small compared to surface tension effect. Of course, in certain classes of problems, essential physics is lost by abandoning the deformation dependent surface elasticity (e.g., effective properties of nanocomposites, dislocation nucleation in flat nanosized thin films). For the authors viewpoint, Since the effect of surface tension manifests itself as a residual type effect (i.e., independent of external loading), they employ Eshelby's classic *gendanken* of cutting and welding operations [15] to put a physical perspective on the problem. They state also that, taking the inclusion (containing a prescribed physical eigenstrain, say, a thermal expansion mismatch strain or that due to lattice mismatch) out of the matrix but with a surface tension equivalent to the interfacial tension of inclusion-matrix. Then, from a classical perspective the inclusion should relax to a strain equal to the physical eigenstrain. However, in the context of coupled surface-bulk elasticity, an additional strain ensues due to the presence of interfacial tension. Thus the total effective eigenstrain is equal to the superposition of the initial prescribed eigenstrain (due to a physical mechanism) and the strain state of an *isolated unembedded* inclusion under the action of a surface tension.

Sharma and Wheeler [55] consider an isolated (i.e., unembedded) triaxial ellipsoid made from the same material as the inclusion (Fig. 8.11). Mathematically

$$\boldsymbol{\varepsilon}^{*T}(\mathbf{x}) = \boldsymbol{\varepsilon}^{*P}(\mathbf{x}, \text{physicalcause}) + \boldsymbol{\varepsilon}^{I}(\mathbf{x}, \tau_0, \kappa), \tag{8.33}$$

where the superscripts T, P, and I stand for "total," "physical," and "isolated," respectively.

Therefore, if one is able to evaluate $\boldsymbol{\varepsilon}^{I}$, Eshelby's classical tensor type concept can be employed to determine the elastic state of the inclusion incorporating surface energy such as:

$$\boldsymbol{\varepsilon}(\mathbf{x}) = \hat{\mathbf{S}} : \boldsymbol{\varepsilon}^{*T}(\mathbf{x}), \tag{8.34}$$

where $\hat{\mathbf{S}}$ is a modified Eshelby's tensor.using the tensor virial method developed by Chandrasekhar [4] and the first-order moment approximation (the total eigenstrain is uniform), Sharma and Wheeler write the surface contributed eigenstrain of the ellipsoidal inclusion in the following simple manner:

$$\boldsymbol{\varepsilon}^{I} = -\frac{2}{V}\mathbf{C}^{-1} : \mathbf{M}, \tag{8.35}$$

where

$$\mathbf{M}_{ij} = \tau_0 \int_S (\delta_{ij} - n_i n_j)\mathrm{d}s.$$

The final (interior) strains and stresses of the embedded ellipsoidal inclusion are expressed as

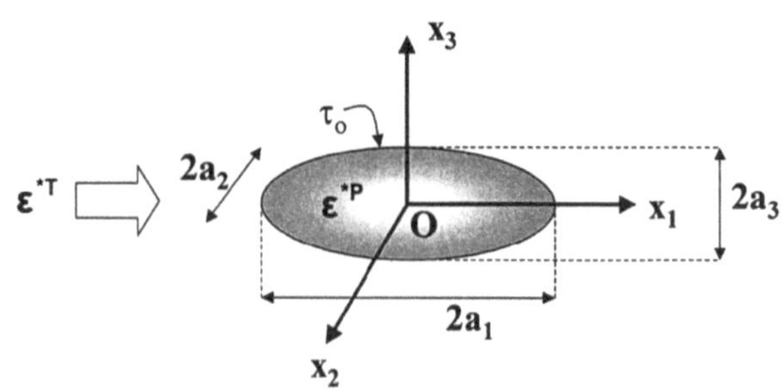

Fig. 8.11 Schematic of the problem for the isolated ellipsoidal particle under a surface tension

$$
\begin{cases} \boldsymbol{\varepsilon} = \mathbf{S} : \left\{ \boldsymbol{\varepsilon}^{*P} - \frac{2}{V}\mathbf{C}^{-1} : \mathbf{M} \right\}, \\ \boldsymbol{\sigma} = \mathbf{C} : (\mathbf{S} - \mathbf{I}) : \left\{ \boldsymbol{\varepsilon}^{*P} - \frac{2}{V}\mathbf{C}^{-1} : \mathbf{M} \right\}. \end{cases} \tag{8.36}
$$

Sharma and Wheeler derive then the basic expressions for higher-order virial moments but they point out the evident fact that due to the lengthy and tedious expressions involved, implementation is somewhat inconvenient beyond the first-order approximation.

When discussing the applications to quantum dots, the differing properties of the inclusion and the matrix are taken into account using Eshelby's equivalent condition [15, 45].

To conclude, Sharma and Wheeler state that the discarding of deformation dependent surface elasticity prohibits use of their results for calculations of effective properties of composites. Also their work shares with its preceding companion article [53] much of the same limitations. For example, they have presented a completely isotropic formulation while interfacial/surface properties can be fairly anisotropic. In addition, they have assumed a perfectly coherent interface. In dealing with nano-inclusions it is important also to consider the degree of coherency.

8.7.6 The Work by Duan et al.

Duan et al. [12] investigate the effective moduli of solids containing nano-inhomogeneities in conjunction with the composite spheres assemblage model (CSA), the Mori-Tanaka method (MTM), and the generalized self-consistent method (GSCM).

The basic set of equations for solving elastostatic problems of heterogeneous solids within the framework of linear and infinitesimal elasticity consists of

$$
\begin{cases} \nabla \cdot \boldsymbol{\sigma}^b = 0, \\ \boldsymbol{\varepsilon}^b = \frac{1}{2}\left[\nabla \mathbf{u} + (\nabla \mathbf{u})^{\mathrm{T}} \right], \\ \boldsymbol{\sigma}^b = \mathbf{C}^b : \boldsymbol{\varepsilon}^b, \end{cases} \tag{8.37}
$$

for the bulk materials and

$$
\begin{cases} \boldsymbol{\sigma}^s = \mathbf{C}^s : \boldsymbol{\varepsilon}^s = 2\mu_s \boldsymbol{\varepsilon}^s + \lambda_s (Tr \boldsymbol{\varepsilon}^s) \mathbf{I}_2, \\ \mathbf{n} \cdot \left[\boldsymbol{\sigma}^b \right] \cdot \mathbf{n} = -\boldsymbol{\sigma}^s : \boldsymbol{\kappa}, \\ \mathbf{P} \cdot \left[\boldsymbol{\sigma}^b \right] \cdot \mathbf{n} = -\nabla_s \cdot \boldsymbol{\sigma}^s. \end{cases} \tag{8.38}
$$

for the analysis of the mechanical equilibrium of the interface between the two different media.

To define the effective elastic moduli of a composite Duan use the usual concept of homogeneous boundary conditions imposed on a representative volume element (RVE). In the presence of interface effect (stress discontinuity), the average strain and average stress are

$$\begin{cases} \bar{\boldsymbol{\varepsilon}} = (1-f)\bar{\boldsymbol{\varepsilon}}^{(2)} + f\bar{\boldsymbol{\varepsilon}}^{(1)}, \\ \bar{\boldsymbol{\sigma}} = (1-f)\bar{\boldsymbol{\sigma}}^{(2)} + f\bar{\boldsymbol{\sigma}}^{(1)} + \frac{f}{V_1}\int_{\Gamma}([\bar{\boldsymbol{\sigma}}]\cdot\mathbf{n})\otimes\mathbf{x}\mathrm{d}\Gamma, \end{cases} \tag{8.39}$$

where where $\bar{\boldsymbol{\varepsilon}}^{(k)}$ and $\bar{\boldsymbol{\sigma}}^{(k)}(k=1,2)$ denote volume averages of the strain and stress over the respective phases in the RVE, f and $1-f$ denote the volume fractions of the inhomogeneity and matrix, respectively. As usual, the effective elastic moduli of the composite can be determined by subjecting the external surface S to homogeneous displacement or traction boundary conditions. They first derive formulas relating the average stress (strain) in the inhomogeneities and at the interface to the applied stress (strain) under both types of boundary condition since these formulas are needed to calculate the effective moduli of the composite according to the dilute concentration approximation and GSCM schemes. Then, they derive formulas relating the average stress (strain) in the inhomogeneities and at the interface to the average stress (strain) in the matrix, again under both types of boundary condition: this is required in MTM.

In this first case, assuming that, **R** and **T** define a strain concentration tensor in the inhomogeneity and a strain concentration tensor at the interface, respectively (regarding the applied strain), the effective stiffness tensor, $\mathbf{C}^{(3)}$, of the composite is given by

$$\mathbf{C}^{(3)} = \mathbf{C}^{(2)} + f\left[\mathbf{C}^{(1)} - \mathbf{C}^{(2)}\right] : \mathbf{R} + f\mathbf{C}^{(2)} : \mathbf{T}. \tag{8.40}$$

For the MTM, they assume that **M** and **H** define a strain concentration tensor in the inhomogeneity and a strain concentration tensor at the interface respectively (regarding the average strain in the matrix). The effective stiffness tensor, $\mathbf{C}^{(3)}$, of the composite is then given by

$$\mathbf{C}^{(3)} = \mathbf{C}^{(2)} + f\left\{\left[\mathbf{C}^{(1)} - \mathbf{C}^{(2)}\right] : \mathbf{M} + f\mathbf{C}^{(2)} : \mathbf{H}\right\} : [\mathbf{I} + f(\mathbf{M}-\mathbf{I})]^{-1}. \tag{8.41}$$

Then, Duan et al. obtain the strain concentration tensors in the three schemes by solving the corresponding boundary-value problems for predicting the effective moduli of a composite containing spherical nano-inhomogeneities with the interface effect.

For a composite with spherical inhomogeneities, the configuration of MTM is a spherical inhomogeneity with radius $r = R_0$ embedded in an infinite matrix subjected to an imposed remote field equal to the as-yet-unknown average stress (strain) field in the matrix of the composite.

The configuration of the CSA consists of two concentric spheres with radii $r = R_0$ and R_1, which correspond to the radius of inhomogeneity and the outer radius of matrix, respectively. The boundary conditions are imposed at the outer boundary of the matrix $r = R_1$.

The configuration of the GSCM is a three-phase model, i.e., a spherical inhomogeneity ($r = R_0$) with a matrix shell (outer radius $r = R_1$) embedded in an infinite effective medium (i.e., the composite material) and boundary conditions are specified at infinity. In the GSCM scheme, conventional stress and displacement continuity conditions are assumed to prevail at the interface between the matrix shell and the effective medium $r = R_1$.

The solutions for finding the effective moduli of the composite with spherical inhomogeneities are given in the spherical coordinate system (r, θ, φ). Because of the different configurations of MTM, CSA, and GSCM, the solutions for the three schemes should satisfy different interface and boundary conditions. The interface conditions at the interface of the inhomogeneity and matrix ($r = R_0$) consist of displacement continuity conditions and Equation (8.38). For the GSCM the interface between the matrix/effective medium ($r = R_1$) is perfectly bonded.

8.7.6.1 Bulk Modulus

To predict the effective bulk modulus of the composite with spherical inhomogeneities, Duan et al. assume that the configurations for CSA, MTM, and GSCM undergo a hydrostatic deformation. For CSA, MTM, and GSCM, the displacement solutions for finding the bulk modulus of the composite are

$$u_r^{(k)} = F_k r + \frac{G_k}{r^2}, \quad u_\theta^{(k)} = 0, \quad u_\varphi^{(k)} = 0, \tag{8.42}$$

where F_k and G_k ($k = 1, 2$ for CSA and MTM, $k = 1, 2, 3$ for GSCM) are constants to be determined from the boundary conditions and the interface conditions.

After some tedious algebra, the bulk modulus using the CSA, MTM, and GSCM schemes are obtained from the respective formulas. Like the classical case without the interface effect, Duan realize that the three schemes give the same result for the effective bulk modulus of the composite with the interface effect.

8.7.6.2 Shear Modulus

In order to obtain closed-form expressions for the effective shear modulus the authors employ only the MTM and GSCM, as only bounds can be obtained for the shear modulus of CSA. To this end, they impose deviatoric strain. After some straightforward but tedious algebra, the effective shear modulus using the MTM and GSCM are obtained.

In the subsequent numerical calculations Duan et al. [12] consider a heterogeneous solid containing spherical voids. The numerical results are presented for aluminum.

8.7.7 The Work by Huang and Sun

Huang and Sun [25] consider the change of the elastic fields induced by the interface energies and the interface stresses from the reference configuration to the current configuration. Until now, two kinds of fundamental equation are necessary in the solution of boundary-value problems for stress fields with surface/interface effect. The first is the surface/interface constitutive relations, and the second is the discontinuity conditions of the stress across the interface, namely, the Young-Laplace equations [48]. For Huang and Sun, even if an infinitesimal analysis is employed, these equations should be established within the framework of finite deformation in the first place. In the authors viewpoint, the reasons for this are:

- In the study of the mechanical behavior of a composite material or a structure, what one is concerned with is the mechanical response from the reference configuration to the current configuration. During the deformation process, the size and the shape of the interface will change, hence the curvature tensor in the governing equations will change too. This means that the deformation will change the residual elastic field induced by the interface energy, and the effect of the interface energy manifests itself precisely through the change of the residual elastic field due to the change of configuration. Therefore, this is essentially a finite deformation problem.
- For the interface energy model, there should be a residual elastic field due to the presence of the interface energy (and the interface stress) in the material, even though there is no external loading. Thus, by taking into account the change of the residual elastic field due to the change of configuration, the influence of the liquid-like surface tension on the effective properties of a composite material can also be included. Therefore, in their paper, they focus on the discussions of the interface energy model.
- Recently, Huang and Wang [24] derived the constitutive relations for hyperelastic solids with the surface/interface energy effect at finite deformation. These constitutive relations are expressed in terms of the free energy of the interface per unit area at the current configuration, denoted by γ.

For an isotropic interface, they show that, even if the infinitesimal deformation approximation is used, the interface Piola-Kirchhoff stresses of the first and second kinds denoted respectively, by $\mathbf{\Pi}^s$ and $\mathbf{T}^s$ and the Cauchy stress of the interface $\boldsymbol{\sigma}^s$ are not the same. They conclude that in the study of the interface energy effect on the mechanical properties of a heterogeneous material, only starting from a finite deformation theory can one correctly choose an appropriate infinitesimal interface stress to be used in the governing equations.

Then, Huang and Sun derive the approximate expressions of the changes of the interface stress and the Young-Laplace equation due to the change of configuration under infinitesimal deformation. As an application of their theory, the authors also give the analytical expressions for the effective moduli of a composite reinforced by spherical particles. It is shown that a liquid-like surface/interface tension also affects the effective moduli, which has not been discussed in the literature.

The difference between this work and those of Sharma and Ganti [53], and Duan et al. [12] is that here, starting from the finite deformation theory proposed by Huang and Sun [25], they have derived the infinitesimal deformation approximations of the interface constitutive relation and the Lagrangian description of the Young-Laplace equation by considering the change of configuration. Hence one can explicitly demonstrate the necessity of using the asymmetric interface stress in the Young-Laplace equation and show the influence of the residual surface/interface tension γ_0^* on the effective elastic moduli.

8.7.8 Other Works

It is worth pointing out some works related to the surface/interface energy, stress and tension. Among them, one can note the work by Mi and Kouris [41] who investigate the effect of surface/interface elasticity in the presence of nanoparticles, embedded in a semi-infinite elastic medium. The work is motivated by the technological significance of self-organization of strained islands in multilayered systems. Islands, adatom-clusters, or quantum dots are modeled as inhomogeneities, with properties that differ from the ones of the surrounding material.

Then follows the work of Ferrari [18] who derives the solution of the problem of a large elastostatic matrix, with an embedded eigenstraining inclusion. The inclusion is modeled as an array of discrete points, in accordance with the theory of doublet mechanics (DM), while the matrix is viewed as a conventional continuum. The integration of the two representations affords the simultaneous access to atomic-scale stress and deformation analysis, and retention of the modeling benefits associated with the macroscopic continuum treatment of non-critical material regions. Thus, the theory presented appears suitable for the analysis of the mechanical states in nanotechnological devices, embedded within constraining matrices, biological and otherwise.

Chen et al. [5] have formulated a theoretical framework to examine the size effect due to both nonlocal effect and interface effect for a composite material. The nonlocal effect is considered by idealizing the matrix material as a micropolar material model. The interface constitutive relations and the generalized Young-Laplace equations for a micropolar material with interface effect are presented. A micropolar micromechanics with interface effect is employed to predict the effective moduli of a fiber-reinforced composite material. The effective bulk modulus is found to be the same as that predicted by the classical micromechanics with interface effect.

8.8 General Solution of Eshelby's Nano-Inhomogeneities Problem

This part is devoted to our work on nano-inhomogeneities problem [29].

8.8.1 Atomistic and Continuum Description of the Interphase

8.8.1.1 Atomic Level Caracterization

To evaluate the elastic properties of a given interfacial region from a discrete medium viewpoint, consider a given interface between two materials A and B. Figure 8.12 illustrates schematically the two different views based on two different length scales of the nano-inhomogeneities problem. Consider then a bimaterial system containing N equivalent atoms. The total energy $E^{(n)}$ of atom n is given by

$$\begin{aligned} E^{(n)} = E_0 + \sum_{m \neq n} E(r^{nm}) + \frac{1}{2!}\sum_{m \neq n}\sum_{p \neq n} E(r^{nm}, r^{np}) + \cdots \\ + \frac{1}{N!}\sum_{m \neq n}\sum_{p \neq n}\cdots\sum_{q \neq n} E(r^{nm}, r^{np}, \cdots, r^{nq}), \end{aligned} \tag{8.43}$$

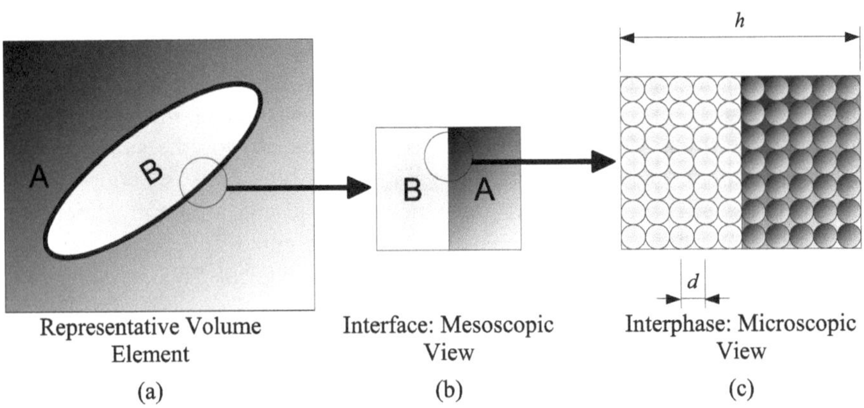

Fig. 8.12 Concept of interface-interphase for nanocomposites: different views based on different length scales

where,

$$r^{nm} = \sqrt{\left(r_1^{nm}\right)^2 + \left(r_2^{nm}\right)^2 + \left(r_3^{nm}\right)^2}$$

is the scalar distance between atom m and atom n and E is the interatomic potentials function which may include pair potentials such as the Lennard-Jones potential as well as multi-body potentials such as the Embedded Atom Method (EAM) potentials. Thus, the total energy of this ensemble containing N such atoms is $E = \sum_{n=1}^{N} E^{(n)}$. If one considers a single solid of infinite extent subjected to a macroscopically uniform strain field ε_{ij}, Johnson [28] demonstrates that the elastic stiffness tensor, C_{ijkl} of the bulk crystal is given by

$$C_{ijkl} = \frac{1}{N} \sum_{n=1}^{N} \sum_{p \neq n} \sum_{q \neq n} \frac{1}{\Omega_n} \frac{r_j^{pn} r_l^{qn} \partial^2 E^{(n)}}{\partial r_i^{pn} \partial r_k^{qn}} \Bigg|_{r^{mn}}, \tag{8.44}$$

where Ω_n is the atomic volume of atom n. However, when considering an atomic ensemble containing nonequivalent atoms (which is the case for systems containing grain boundaries and interfaces) subjected to a macroscopically uniform deformation, internal relaxations occur [40] and Equation (8.44) can be interpreted as a description of the homogeneous elastic response of the ensemble [7]. To take into account the inner displacements, an atomic level mapping between the undeformed, $\hat{r}_i^n$, and deformed, r_i^n, configurations is defined by

$$r_i^n - \hat{r}_i^n = \left(\bar{\varepsilon}_{ij}^{\pm} + \tilde{\varepsilon}_{ij}^{n}\right) \hat{r}_j^n, \tag{8.45}$$

where $\bar{\varepsilon}_{ij}^{\pm}$ corresponds to a homogeneous deformation of atom n and $\tilde{\varepsilon}_{ij}^{n}$ describes the "inner" relaxation (or additional "nonhomogeneous" deformation) of atom n with respect to a homogeneous deformation. Note that the positive (or negative) sign should be selected if atom n is in the phase A (or B). The "T" stress decomposition [49] can then be used to describe the homogeneous deformation of the bimaterial assembly by an in-plane deformation $\varepsilon_{\alpha\beta}^{s}$ and a transverse loading σ_i^t, (see Appendix 1). Thus, following Appendix 1, one gets

$$\bar{\varepsilon}_{ij}^{\pm} = A_{ij\alpha\beta}^{\pm} \varepsilon_{\alpha\beta}^{s} + B_{ijk}^{\pm} \sigma_k^t, \tag{8.46}$$

with,

$$\begin{cases} A_{ij\alpha\beta}^{\pm} = \delta_{i\alpha}\delta_{j\beta} - \frac{1}{2}\left(\gamma_{j\alpha\beta}^{\pm}\delta_{3i} + \gamma_{i\alpha\beta}^{\pm}\delta_{3j}\right), \\ B_{ijk}^{\pm} = \frac{1}{2}\left(M_{jk}^{\pm}\delta_{3i} + M_{ik}^{\pm}\delta_{3j}\right), \end{cases} \tag{8.47}$$

where $M^{\pm}_{jk}$ and $\gamma^{\pm}_{i\alpha\beta}$ are given in Appendix 1. The tensors $A^{\pm}_{ij\alpha\beta}$ and $B^{\pm}_{ijk}$ characterize the homogeneous behavior of the bimaterial. At this point, one can get the difference in position of two atoms, m and n, near their relaxed state as

$$r_i^{mn} - \hat{r}_i^{mn} = A^{mn}_{i\alpha\beta}\varepsilon_{\alpha\beta} + B^{mn}_{ik}\sigma^t_k + \left(\tilde{\varepsilon}^m_{ij}\hat{r}^m_j - \tilde{\varepsilon}^n_{ij}\hat{r}^n_j\right), \tag{8.48}$$

where $A^{mn}_{i\alpha\beta}$ and B^{mn}_{ik} are defined in Appendix 2. The energy density of an atom n about its equilibrium configuration is given as [7, 28]

$$\begin{aligned} w^n = \frac{1}{\Omega_n}\sum_{\substack{m=1\\ m\neq n}}^{N-1}\Bigg[& E^{(n)}(r^{nm})\Big|_{r^{nm}=\hat{r}^{nm}} + \frac{\partial E^{(n)}}{\partial r_i^{nm}}\Big|_{r^{nm}=\hat{r}^{nm}}\left(r_i^{nm} - \hat{r}_i^{nm}\right) \\ & + \frac{1}{2}\sum_{\substack{p=1\\ p\neq n}}^{N-1}\frac{\partial^2 E^{(n)}}{\partial r_i^{nm}\partial r_k^{np}}\Big|_{r^{nm}=\hat{r}^{nm}}\left(r_i^{nm} - \hat{r}_i^{nm}\right)\left(r_k^{np} - \hat{r}_k^{np}\right) + \cdots\Bigg]. \end{aligned} \tag{8.49}$$

The total strain energy density of the interphase containing N atoms is $\Sigma^N_{n=1} w^n$. Making use of Equation (8.48) in Equation (8.49) yields for the total strain energy of the atomic assembly,

$$\begin{aligned} E = E_0 & + \overline{\mathbf{A}}^{(1)} : \hat{\varepsilon}^s + \overline{\mathbf{B}}^{(1)} \cdot \overline{\boldsymbol{\sigma}}^t + \frac{1}{2}\hat{\varepsilon}^s : \overline{\mathbf{A}}^{(2)} : \hat{\varepsilon}^s + \frac{1}{2}\overline{\boldsymbol{\sigma}}^t \cdot \overline{\mathbf{B}}^{(2)} \cdot \overline{\boldsymbol{\sigma}}^t \\ & + \overline{\boldsymbol{\sigma}}^t \cdot \overline{\mathbf{Q}} : \hat{\varepsilon}^s + \sum_{n=1}^{N-1}\left(\mathbf{K}^n + \mathbf{D}^n : \hat{\varepsilon}^s + \mathbf{G}^n \cdot \overline{\boldsymbol{\sigma}}^t\right) : \tilde{\varepsilon}^n \\ & + \frac{1}{2}\sum_{n=1}^{N-1}\sum_{m=1}^{N-1}\tilde{\varepsilon}^n : \mathbf{L}^{mn} : \tilde{\varepsilon}^m. \end{aligned} \tag{8.50}$$

Equation (8.50) shows that the tensors $\overline{\mathbf{A}}^{(1)}$, $\overline{\mathbf{B}}^{(1)}$, $\overline{\mathbf{A}}^{(2)}$, $\overline{\mathbf{B}}^{(2)}$, and $\overline{\mathbf{Q}}$ describe the homogeneous behavior of the assembly upon a deformation configuration $\left(\hat{\varepsilon}^s, \overline{\boldsymbol{\sigma}}^t\right)$ while the tensors $\overline{\mathbf{K}}^n$, $\overline{\mathbf{D}}^n$, $\overline{\mathbf{G}}^n$, and $\overline{\mathbf{L}}^{mn}$ represent the components of perturbation response of the system introduced by the nonequivalency of the atomic ensemble such as in grain boundaries or interface and account for the accommodation of internal relaxations upon a deformation configuration $\left(\hat{\varepsilon}^s, \overline{\boldsymbol{\sigma}}^t\right)$. Their expressions are derived by Dingreville [7] and are reported herein in Appendix 2. Now, one can determine the atomic level stress associated within an atom n. The virial stress on atom n is given by

$$\overline{\boldsymbol{\sigma}}^n_{ij} = \frac{1}{2\Omega_n}\sum_{m\neq n}\frac{\partial E}{\partial r_i^{nm}} r_j^{nm}. \tag{8.51}$$

Expanding this atomic level stress, $\overline{\boldsymbol{\sigma}}_{ij}^{n}$, with respect to r_i^{nm} near the equilibrium configuration, $\hat{r}_i^{nm}$, of the bimaterial, gives

$$\overline{\boldsymbol{\sigma}}_{ij}^{n} = \overline{\boldsymbol{\sigma}}_{ij}^{n}\Big|_{r^{nm}=\hat{r}^{nm}} + \sum_{m\neq n} \frac{\partial \overline{\boldsymbol{\sigma}}_{ij}^{n}}{\partial r_k^{nm}}\Bigg|_{r^{nm}=\hat{r}^{nm}} \left(r_k^{nm} - \hat{r}_k^{nm}\right). \tag{8.52}$$

Making use of Equation (8.48), the atomic level stress, $\overline{\boldsymbol{\sigma}}_{ij}^{n}$, takes the following form

$$\overline{\boldsymbol{\sigma}}_{ij}^{n} = \tau_{ij}^{n} + \overline{C}_{ij\alpha\beta}^{s,n}\varepsilon_{\alpha\beta} + \overline{M}_{kij}^{t,n}\sigma_k^{t} + \sum_{m\neq n} T_{ijkl}^{mn}\tilde{\varepsilon}_{kl}^{m}, \tag{8.53}$$

where, τ_{ij}^{n}, $\overline{C}_{ij\alpha\beta}^{s,n}$, $\overline{M}_{kij}^{t,n}$ and T_{ijkl}^{mn} are known constants given in terms of the interatomic potential E and its partial derivative with respect to the interatomic distance r. Derivations and expressions of these tensors are given in Dingreville [7]. In Equation (8.53), there are $6N$ unknowns, $\tilde{\varepsilon}_{kl}^{m}$, which describe the internal relaxations. The conditions of mechanical equilibrium and traction continuity across the interface yield

$$\overline{\boldsymbol{\sigma}}_{j}^{t,n} = \sigma_j^{t}. \tag{8.54}$$

Using Equation (8.54) in Equation (8.53) and some algebra manipulations, the expressions of the $6N$ unknowns are derived by Dingreville [7] as

$$\tilde{\varepsilon}_{\alpha\beta}^{s,n} = \eta_{\alpha\beta}^{s,n} - M_{i\alpha\beta}^{s,n}\sigma_i^{t} + Q_{\alpha\beta\hat{\kappa}\lambda}^{s,n}\varepsilon_{\hat{\kappa}\lambda}, \tag{8.55}$$

$$\tilde{\varepsilon}_{i}^{t,n} = \eta_{i}^{t,n} + M_{ij}^{t,n}\sigma_j^{t} - Q_{i\alpha\beta}^{t,n}\varepsilon_{\alpha\beta}. \tag{8.56}$$

Now, the atomic level stress, $\overline{\boldsymbol{\sigma}}_{ij}^{n}$ (see Equation (8.53)) can be fully determined. Therefore, from some algebra manipulations, the atomic level in-plane stress, $\overline{\boldsymbol{\sigma}}_{\alpha\beta}^{s,n}$ is given as

$$\overline{\boldsymbol{\sigma}}_{\alpha\beta}^{s,n} = \pi_{\alpha\beta}^{n} + C_{\alpha\beta\hat{\kappa}\lambda}^{s,n}\varepsilon_{\hat{\kappa}\lambda} + Q_{i\alpha\beta}^{n}\sigma_i^{t}, \tag{8.57}$$

with,

$$\begin{cases} \pi_{\alpha\beta}^{n} &= \tau_{\alpha\beta}^{n} + \Sigma_{m=1}^{N-1} T_{\alpha\beta 3k}^{nm}\eta_k^{t,m} + \Sigma_{m=1}^{N-1} T_{\alpha\beta\hat{\kappa}\lambda}^{nm}\eta_{\hat{\kappa}\lambda}^{s,m}, \\ C_{\alpha\beta\hat{\kappa}\lambda}^{s,n} &= \overline{C}_{\alpha\beta\hat{\kappa}\lambda}^{s,n} - \Sigma_{m=1}^{N-1} T_{\alpha\beta 3i}^{nm} Q_{i\hat{\kappa}\lambda}^{t,m} + \Sigma_{m=1}^{N-1} T_{\alpha\beta\mu\nu}^{nm} Q_{\mu\nu\hat{\kappa}\lambda}^{s,m}, \\ Q_{i\alpha\beta}^{n} &= \overline{M}_{i\alpha\beta}^{t,n} + \Sigma_{m=1}^{N-1} T_{\alpha\beta 3j}^{nm} M_{ji}^{t,m} - \Sigma_{m=1}^{N-1} T_{\alpha\beta\hat{\kappa}\lambda}^{nm} M_{i\hat{\kappa}\lambda}^{s,m}. \end{cases} \tag{8.58}$$

Similarly, far away from the interface region, the bulk in-plane stress, $\overline{\boldsymbol{\sigma}}_{\alpha\beta}^{\pm,s}$, is determined by (see Appendix 1),

$$\overline{\boldsymbol{\sigma}}_{\alpha\beta}^{\pm,s} = \left(C_{\alpha\beta\hat{\kappa}\lambda}^{\pm} - C_{\alpha\beta 3j}^{\pm}\gamma_{j\hat{\kappa}\lambda}^{\pm}\right)\varepsilon_{\hat{\kappa}\lambda} + \gamma_{i\alpha\beta}^{\pm}\sigma_i^t. \tag{8.59}$$

With Equations (8.57) and (8.59), the interfacial region (interphase in the present work) excess in-plane stress is thus determined by

$$\Sigma_{\alpha\beta}^s = \frac{1}{V_0}\sum_{n=1}^{N}\Omega_n\left(\overline{\boldsymbol{\sigma}}_{\alpha\beta}^{s,n} - \overline{\boldsymbol{\sigma}}_{\alpha\beta}^{\pm,s}\right) = \frac{A_0}{V_0}\left(\Gamma_{\alpha\beta}^{(1)} + \Gamma_{\alpha\beta\hat{\kappa}\lambda}^{(2)}\varepsilon_{\hat{\kappa}\lambda}^s + H_{j\alpha\beta}\sigma_j^t\right), \tag{8.60}$$

where A_0 is the area of the interface concerned, V_0 is the volume of the associated interphase (interfacial region), and

$$\begin{cases} \Gamma_{\alpha\beta}^{(1)} = \frac{1}{A_0}\sum\limits_{n=1}^{N}\Omega_n\pi_{\alpha\beta}^n, \\ \Gamma_{\alpha\beta\hat{\kappa}\lambda}^{(2)} = \frac{1}{A_0}\sum\limits_{n=1}^{N}\Omega_n\left[C_{\alpha\beta\hat{\kappa}\lambda}^{s,n} - C_{\alpha\beta\hat{\kappa}\lambda}^{\pm} - C_{\alpha\beta 3j}^{\pm}\gamma_{j\hat{\kappa}\lambda}^{\pm}\right], \\ H_{i\alpha\beta} = \frac{1}{A_0}\sum\limits_{n=1}^{N}\Omega_n\left[Q_{i\alpha\beta}^n - \gamma_{i\hat{\kappa}\lambda}^{\pm}\right]. \end{cases} \tag{8.61}$$

Similarly the transverse excess strain given by Equation (8.56), is determined as

$$\Delta_k^t = \frac{1}{V_0}\sum_{n=1}^{N}\Omega_n\tilde{\varepsilon}_k^{t,n} = \frac{A_0}{V_0}\left(\Lambda_k^{(1)} + \Lambda_{kj}^{(2)}\sigma_j^t - H_{k\alpha\beta}\varepsilon_{\alpha\beta}\right), \tag{8.62}$$

where

$$\begin{cases} \Lambda_k^{(1)} = -\frac{1}{A_0}\sum\limits_{n=1}^{N}\Omega_n\eta_k^{t,n}, \\ \Lambda_{kj}^{(2)} = \frac{1}{A_0}\sum\limits_{n=1}^{N}\Omega_n M_{jk}^{t,n}. \end{cases} \tag{8.63}$$

The tensors, $\boldsymbol{\Gamma}^{(1)}$, $\boldsymbol{\Gamma}^{(2)}$, $\boldsymbol{\Lambda}^{(1)}$, $\boldsymbol{\Lambda}^{(2)}$, and $\mathbf{H}$, are the so-called interfacial elastic properties. For a given interatomic potential function, $E^{(n)}$, numerical evaluation of the analytical expressions of these tensors requires knowledge of the relaxed state, $\hat{r}^{mn}$, of the interface. To obtain $\hat{r}^{mn}$, a preliminary molecular static (MS) simulation may be conducted. This is why the method is called semi-analytical [7, 9]. At this point one can get the elastic properties, C_{ijkl}^c, of the interphase associated to this interface. This is done in the following section.

8.8.1.2 Interphase Stiffness Tensor

Instead of reporting the excess stress and strain to the interface area, A_0, this work attributes then to the volume of the interfacial region named interphase. Thus Equations (8.60) and (8.62) describe, respectively, the interphase excess in-plane stress and its transverse excess strain. It is therefore conceivable to attribute to this interfacial region effective elastic properties. To this end, one can make a comparison between Equations (8.60) and (8.126) on one hand and Equations (8.62) and (8.125) on this other hand, that is,

$$\begin{cases} \frac{A_0}{V_0}\left(\boldsymbol{\Gamma}^{(1)} + \boldsymbol{\Gamma}^{(2)} : \hat{\boldsymbol{\varepsilon}}^s + \mathbf{H} \cdot \bar{\boldsymbol{\sigma}}^t\right) = \hat{\tau}^s + \mathbf{C}^s : \boldsymbol{\varepsilon}^s + \boldsymbol{\gamma} \cdot \boldsymbol{\sigma}^t, \\ \frac{A_0}{V_0}\left(\boldsymbol{\Lambda}^{(1)} + \boldsymbol{\Lambda}^{(2)} \cdot \bar{\boldsymbol{\sigma}}^t - \mathbf{H} : \hat{\boldsymbol{\varepsilon}}^s\right) = -\mathbf{M} \cdot \tau^t + \mathbf{M} \cdot \boldsymbol{\sigma}^t - \boldsymbol{\gamma} : \boldsymbol{\varepsilon}^s. \end{cases} \tag{8.64}$$

It follows that

$$\begin{cases} \mathbf{C}^s = \frac{A_0}{V_0}\boldsymbol{\Gamma}^{(2)}, \\ \boldsymbol{\gamma} = \frac{A_0}{V_0}\mathbf{H}, \\ \mathbf{M} = \frac{A_0}{V_0}\boldsymbol{\Lambda}^{(2)}. \end{cases} \tag{8.65}$$

The 21 components of the interphase stiffness tensor, C^c_{ijkl}, are completely determined from Equation (8.65). Thus, one gets from the last equation of Equation (8.65),

$$C^c_{3j3k} = M^{-1}_{jk} = \left[\frac{A_0}{V_0}\Lambda^{(2)}\right]^{-1}_{jk}, \tag{8.66}$$

which gives the six components C^c_{3k3j} of C^c_{ijkl}. Next, using the second equation of Equation (8.65), one gets a linear system of nine equations to solve for the nine components $C^c_{3k\alpha\beta}$ of C^c_{ijkl}

$$\Lambda^{(2)}_{jk} C^c_{3k\alpha\beta} = H_{j\alpha\beta}, \tag{8.67}$$

Finally, the first equation of Equation (8.65) gives the six components $C^c_{\alpha\beta\hat{\kappa}\lambda}$ of C^c_{ijkl} by

$$C^c_{\alpha\beta\hat{\kappa}\lambda} = \frac{A_0}{V_0}\left(\Gamma^{(2)}_{\alpha\beta\hat{\kappa}\lambda} + H_{j\hat{\kappa}\lambda} C^c_{3j\alpha\beta}\right). \tag{8.68}$$

The interphase elastic properties are therefore completely determined using Equations (8.66), (8.67) and (8.68) and the tensors $\boldsymbol{\Gamma}^{(2)}$, $\mathbf{H}$, and $\boldsymbol{\Lambda}^{(2)}$ obtained from MS simulations and the analytical expressions Equations (8.61) and (8.63).

8.8.1.3 Particular Case of Isotropic Interface

In the case of isotropic interface, $\mathbf{\Gamma}^{(2)}$ is defined by 2 parameters K_s and μ_s by

$$\begin{aligned}\Gamma^{(2)}_{\alpha\beta\hat{\kappa}\lambda} &= (K_s - \mu_s)\delta_{\alpha\beta}\delta_{\hat{\kappa}\lambda} + \mu_s(\delta_{\alpha\hat{\kappa}}\delta_{\beta\lambda} + \delta_{\alpha\lambda}\delta_{\beta\hat{\kappa}}) \\ &= \lambda_s\delta_{\alpha\beta}\delta_{\hat{\kappa}\lambda} + \mu_s(\delta_{\alpha\hat{\kappa}}\delta_{\beta\lambda} + \delta_{\alpha\lambda}\delta_{\beta\hat{\kappa}}),\end{aligned} \tag{8.69}$$

where, λ_s and μ_s can be seen as the Lamé constants of the interface. The Lamé constants of the interphase, λ_c and μ_c, in this case are as follows

$$\begin{cases} \mu_c = \frac{A_0}{V_0}\mu_s, \\ \lambda_c = \frac{2A_0}{V_0}\left(\frac{\mu_s\lambda_s}{2\mu_s - \lambda_s}\right). \end{cases} \tag{8.70}$$

For interphases such $V_0/A_0 = t$, the thickness of the interphase (this is true for rectangular interface or even spherical interface since t is very small), Equation (8.70) leads to

$$\begin{cases} \mu_s = \mu_c t, \\ \lambda_s = \frac{2\mu_c\nu_c t}{1-\nu_c}, \end{cases} \tag{8.71}$$

where $\nu_c = 1/2(1 + \mu_c/\lambda_c)$ is Poisson coefficient of the interphase. It worthy pointing that, Equation (8.71) is similar to Equation (69) in the work by Wang et al. [61] or Equation (7) in the work of Duan et al. [14] for interface representation of thin and stiff interphase for spherical particles. Note that the result of Wang et al. [61] is based on the interface stress model in Duan et al. [11, 12, 13] which assumes displacement continuity and stress jump across the interface and isotropic interface. The stress discontinuities across an interface are equilibrated by the interface stress through the so-called generalized Young-Laplace equations. The identification of the parameters μ_s and λ_s with respect to the interphase parameters μ_c and λ_c is related to these features of the interface model by Duan et al. [13]. The connection between interphase and interface models is then done since, in the case of spherical concentric coating inhomogeneity, the same features (displacement continuity and stress jump across the interface) are observed for thin and stiff interphase. In the other hand, the fully interface approach of Dingreville [7] assumes the displacement discontinuity and stress discontinuity across the interface. The displacement discontinuity is related to the tensors $\mathbf{\Lambda}^{(1)}$, $\mathbf{\Lambda}^{(2)}$, and $\mathbf{H}$ and the stress jump is the same as in the interface model [12, 13, 53]. Implicitly the results of Wang et al. [61] assume that the inhomogeneity and the interphase display positive stiffness behavior and thus μ_s and λ_s should be always positive whereas the present result suggests the possible presence of negative stiffness [32] region around the nano-inhomogeneity depending on the nature of the interface (μ_s and λ_s may be positive or negative). The correspondence between the two results in the case of isotropic (spherical) interfaces leads to very small values of the components of the tensors $\mathbf{\Lambda}^{(1)}$, $\mathbf{\Lambda}^{(2)}$ and

H so that the displacement continuity across the interface can be assumed in the interface approach by Dingreville [7]. Therefore, it is obvious from Equation (8.66) or Equation (8.67) that the interphase is stiff. The thin assumption is also verified due to the small nature of the interfacial region.

8.8.1.4 Nano-Particles and Negative Stiffness Behavior

Equation (8.70) shows that the interphase properties, μ_c and λ_c, can take positive or negative value depending on the interface elastic properties. This observation is very interesting since the works of Lakes and Drugan [32]; Lakes et al. [33]; Lakes [30, 31] have shown that included materials possessing negative stiffness behavior can lead to extremely high macroscopic damping properties and high stiffness. Negative stiffness is one way to state that portions of the stress-strain curve of a material have negative values. The existence of such behavior is suggested by the existence of multiple local minimums, or energy wells, predicted by Landau theory for ferroelastic materials [17]. Indeed, extreme damping behavior has been experimentally observed in bi-phase materials containing trace elements of single domain crystals undergoing phase transformation [27, 33]. Negative bulk modulus behavior has also been observed in single cells of polymer foams. Negative stiffness behavior is qualitatively well understood to be the material stiffness analogue of the bi-stable force versus displacement curves characteristic of beam buckling or the snap through behavior observed when a lateral force is applied to a post-buckled beam. It is imperative to state that negative stiffness material behavior cannot exist alone in nature as it is inherently instable as it implies that the stiffness tensor of the material is not positive definite. Naturally occurring negative stiffness materials are therefore transitory occurrences at best. However, negative stiffness is not excluded by any physical law. Objects with negative stiffness are unstable if they have free surfaces but can be stabilized when constrained by rigid boundaries as in the case of the buckled tubes studied by Lakes [31]. The sole requirement is that the macroscopic behavior of a heterogeneous system containing negative stiffness elements be described by a positive definite stiffness tensor. Further work has also shown that included phases with negative stiffness may also lead to extreme thermal expansion, and piezoelectricity [62], thereby giving further impetus to research the creation of such materials. The ability to create composites containing such phases for practical application is an open, and very active, area of research. Thus from the present modeling schemes one realizes that a nanoparticle embedding leads to local domains of negative stiffness. Therefore this is a very promising area of research in material design strategies.

To end this section recall that the main objective of this work is to solve the Eshelby's problem of nano-inhomogeneities in continuum viewpoint for ellipsoidal shape of the nanoparticles and general materials and interfaces anisotropies. The atomistic description and information have been put in continuum framework and thus the initial problem is transformed to a three-phase composite problem which can be efficiently solved by the well-developed theories of micromechanics. The current problem consists of a nanoparticle surrounded by an interphase with a stiffness tensor previously determined by Equations (8.66), (8.67) and (8.68). The coated nano-inhomogeneity is then embedded in the host material. In the following section, a recent multi-phase micromechanical scheme of Lipinski et al. [37] is presented to solve this Eshelby's nano-inhomogeneities problem.

8.8.2 Micromechanical Framework for Coating-Inhomogeneity Problem

Many micromechanical schemes have been successfully used to obtain effective elastic constants of heterogeneous solids. For a comprehensive exposition, one can refer to the monographs of Aboudi [1], Nemat-Nasser and Hori [46], Milton [43], Torquato [59], and Qu and Cherkaoui [50]. In the present paper, the coated inhomogeneities micromechanical scheme first developed by Cherkaoui et al. [6] and extended by Lipinski et al. [37] is used to compute the effective properties of the nanocomposite. Micromechanical schemes are based on two distinct steps: (i) localization, which determines the relationship between the microscopic (local) fields and the macroscopic (global) loading, and (ii) homogenization, which employs averaging techniques to approximate macroscopic behavior. The topology of the multi-coated inhomogeneity problem (see Fig. 8.13(a)) by Lipinski et al. [37] consists of an inhomogeneity phase occupying a volume, V_1, whose mechanical behavior is described by the elastic stiffness tensor, $\mathbf{C}^1$. Surrounding this inhomogeneity phase are $(n-1)$ layers of coatings of another materials whose elastic behaviors are described by their respective stiffness tensors, $\mathbf{C}^i$ and that occupies a volume, V_i, $i \in \{2, 3, \ldots, n\}$. The multi-coated inhomogeneity is embedded in a host material described by the elastic stiffness tensor, $\mathbf{C}^0$. It is important to note that this derivation is limited to the case of small perturbation theory and the interfaces matrix-coating, coating-coating, and coating-inhomogeneity are assumed to be perfect, thus ensuring continuity of traction and displacements across these boundaries. In the special of nano-inhomogeneity problem considers herein, the topology consists of nano-inhomogeneity of elastic tensor, $\mathbf{C}^1$, surrounded by an interphase (characterized in Section 8.1) of elastic tensor, $\mathbf{C}^2 = \mathbf{C}^c$. Surrounding this coated-inhomogeneity is a shell of the matrix material of elastic tensor, $\mathbf{C}^3$. This composite inhomogeneity is then embedded in the effective nanocomposite material described by the elastic stiffness tensor, $\mathbf{C}^{eff}$ (see Fig. 8.13(b)). It is further assumed that the layers of this composite inhomogeneity are concentric and homothetic. In the following, the two general steps of this micromechanical-based homogenization scheme are outlined.

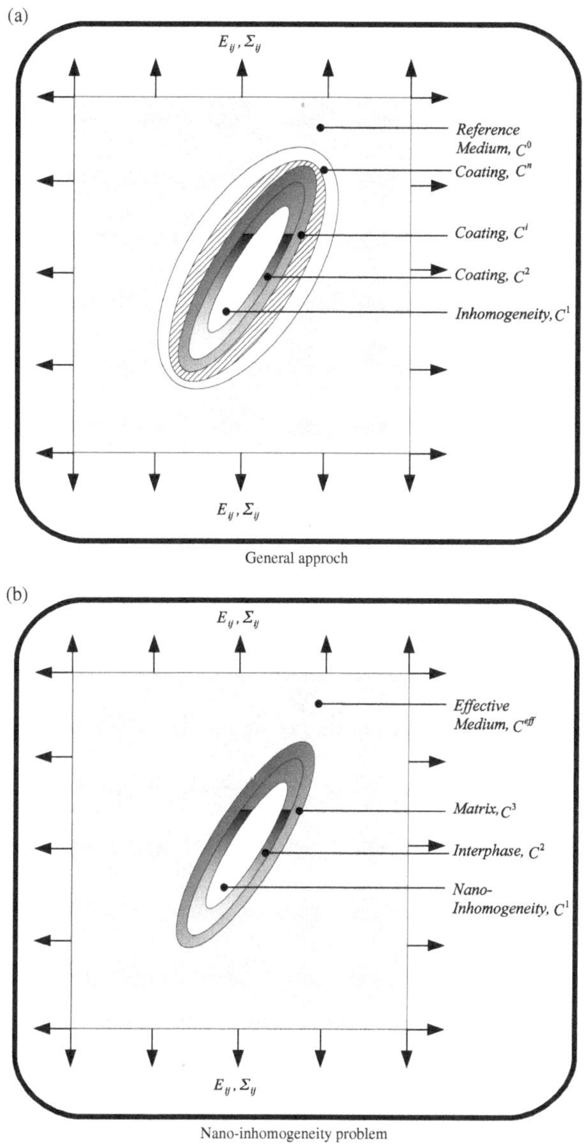

Fig. 8.13 Topology of a multi-coated inhomogeneity embedded in a limitless matrix. Σ_{ij} and E_{ij} represent the macroscopically applied stresses and strains, respectively

8.8.2.1 Integral Equation and Localization

The beginning point of this homogenization scheme is based on the integral equation of Zeller and Dederichs [64] who have proposed to model the composite material shown in Fig. 8.13 as a homogeneous material whose elastic behavior varies spatially, that is

$$\mathbf{C}(\mathbf{r}) = \mathbf{C}^0 + \delta\mathbf{C}(\mathbf{r}), \tag{8.72}$$

where $\mathbf{r} \in V$, V is the volume of the homogeneous medium, $\delta\mathbf{C}(\mathbf{r})$ is the spatially dependent elastic stiffness tensor variation and $\mathbf{C}^0$ represents the elastic stiffness tensor of the reference material which is constant for all $\mathbf{r}$. Based on the local equilibrium equation

$$\mathrm{div}(\bar{\boldsymbol{\sigma}}) = 0, \tag{8.73}$$

where, $\bar{\boldsymbol{\sigma}}$, is the stress tensor and by employing Green's formalism, one gets the simplified equation for the strain field, $\hat{\boldsymbol{\varepsilon}}$, at any point in the medium as [16, 64]

$$\hat{\boldsymbol{\varepsilon}}(\mathbf{r}) = \mathbf{E} - \int_V \boldsymbol{\Gamma}^0(\mathbf{r} - \mathbf{r}') : \delta\mathbf{C}(\mathbf{r}') : \boldsymbol{\varepsilon}(\mathbf{r}')\mathrm{d}\mathbf{r}'. \tag{8.74}$$

In Equation (8.74), $\mathbf{E}$ represents the uniform strain field of the medium (macroscopic strain field that has no spatial dependence), and $\Gamma^0(\mathbf{r} - \mathbf{r}')$ is the modified Green's tensor which is related to the second order Green's tensor, $\mathbf{G}^0(\mathbf{r} - \mathbf{r}')$, by

$$\Gamma^0_{ijkl} = -\frac{1}{2}\left(\frac{\partial^2 G^0_{ki}}{\partial r_j r_l} + \frac{\partial^2 G^0_{kj}}{\partial r_i r_l}\right). \tag{8.75}$$

Here the superscript 0 denotes that the Green's tensors are computed using the elastic properties, $\mathbf{C}^0$, of the reference medium. The fluctuation part of the elastic constants with respect to the reference medium is given by the relation

$$\delta\mathbf{C}(\mathbf{r}) = \sum_{k=0}^{n} \Delta\mathbf{C}^{(k/0)}\theta_k(\mathbf{r}), \qquad \text{with} \qquad \Delta\mathbf{C}^{(k/0)} = \left(\mathbf{C}^k - \mathbf{C}^0\right). \tag{8.76}$$

The characteristic function $\theta_k(\mathbf{r})$ of phase k, occupying the volume V_k, is defined as:

$$\theta_k(\mathbf{r}) = \begin{cases} 1\,\forall\,\mathbf{r} \in V_k \\ \\ 0\,\forall\,\mathbf{r} \notin V_k \end{cases}, \quad \text{with } k \in \{0, 1, 2, \ldots, n\}. \tag{8.77}$$

For the following, certain notation conventions need to be mentioned. The volume V_I of the composite inhomogeneity, I, consists of the inhomogeneity and $(n-1)$ coatings and the volume fraction, φ_k, of phase k are such that

$$V_I = \sum_{k=1}^{n} V_k \qquad \text{and} \qquad \varphi_k = \frac{V_k}{V_I}, \qquad k \in \{1, 2, \ldots, n\}. \tag{8.78}$$

The average strain, $\bar{\hat{\boldsymbol{\varepsilon}}}^I$, in the composite inhomogeneity, I, is defined as

$$\bar{\hat{\boldsymbol{\varepsilon}}}^I = \frac{1}{V_I}\int_{V_I} \hat{\boldsymbol{\varepsilon}}(\mathbf{r})\mathrm{d}\mathbf{r} = \mathbf{E} - \mathbf{T}^I(\mathbf{C}^0) : \boldsymbol{\tau}^I, \tag{8.79}$$

where

$$\begin{cases} \mathbf{T}^I(\mathbf{C}^0) = \frac{1}{V_I}\int_{V_I}\int_{V_I}\boldsymbol{\Gamma}^0(\mathbf{r}-\mathbf{r}')\mathrm{d}\mathbf{r}\mathrm{d}\mathbf{r}', \\ \boldsymbol{\tau}^I = \sum_{k=1}^{n} \varphi_k \Delta\mathbf{C}^{(k/0)} : \bar{\hat{\boldsymbol{\varepsilon}}}^k, \\ \bar{\hat{\boldsymbol{\varepsilon}}}^k = \frac{1}{V_k}\int_{V_k} \hat{\boldsymbol{\varepsilon}}(\mathbf{r})\mathrm{d}\mathbf{r}. \end{cases} \tag{8.80}$$

From Equation (8.79) it is obvious that if one can find a local strain concentration tensors, $\mathbf{a}^k$, such as

$$\bar{\hat{\varepsilon}}^k = \mathbf{a}^k : \bar{\hat{\varepsilon}}^I, \tag{8.81}$$

then, the strain localization tensor, $\mathbf{A}^I$, in the composite inhomogeneity, I, can be valued such us

$$\begin{cases} \bar{\hat{\varepsilon}}^I = \mathbf{A}^I : \mathbf{E}, \\ \mathbf{A}^I = \left[\mathbf{I}_4 + \mathbf{T}^I(\mathbf{C}^0) : \left(\sum_{k-1}^{n} \varphi_k \Delta\mathbf{C}^{(k/0)} : \mathbf{a}^k\right)\right]^{-1}, \end{cases} \tag{8.82}$$

where $\mathbf{I}_4$ is the fourth-order identity tensor. Next, to complete the localization step, the local strain localization tensors, $\mathbf{a}^k$, must be found. If one introduces a strain localization tensor, $\mathbf{A}^k$, in each phase, k, such as

$$\bar{\hat{\boldsymbol{\varepsilon}}}^I = \mathbf{A}^k : \mathbf{E}, \tag{8.83}$$

one can verify the following relationships using Equation (8.81)

$$\begin{cases} \mathbf{A}^k = \mathbf{a}^k : \mathbf{A}^I, \\ \mathbf{A}^I = \langle \mathbf{A}^k \rangle = \sum_{k=1}^{n} \left(\varphi_k \mathbf{A}^k\right), \\ \mathbf{I}_4 = \langle \mathbf{a}^k \rangle = \sum_{k=1}^{n} \left(\varphi_k \mathbf{a}^k\right). \end{cases} \tag{8.84}$$

Here, the notation, $\langle \boldsymbol{\zeta} \rangle$, denotes the average value of the quantity, $\boldsymbol{\zeta}$, over the whole volume of the composite inhomogeneity, I. Equation (8.84) constitutes the solution of the posed problem given as function of the unknown n local localization tensors, $\mathbf{a}^k$, which can be determined if one takes into account the boundary conditions through the different interfaces in the composite inhomogeneity. Interfacial operators [60] are a very convenient mathematical tool that efficiently calculates the stress or strain jump across a material interface (an interface separating two dissimilar materials). These operators are derived by writing the equations for the continuity of displacement and traction across the material interface (hypothesis of perfect interface). The derivation begins with the general case of two solid phases k and $(k+1)$, with oelastic constants $\mathbf{C}^k$ and $\mathbf{C}^{k+1}$ separated by a surface with unit normal, $\mathbf{N}$, directed from phase k to phase $(k+1)$. Using the elastic constants of these two phases, the strain jump across the material interface is given as follows [60]

$$\varepsilon_{ij}^{k+1}(\mathbf{r}) - \varepsilon_{ij}^{k}(\mathbf{r}) = P_{ijmn}^{k+1}\left(C_{mnpq}^{k} - C_{mnpq}^{k+1}\right)\varepsilon_{pq}^{k}(\mathbf{r}). \tag{8.85}$$

The interfacial operator, P_{ijmn}^{k+1}, dependent only on the constituent material properties and the unit normal of the interface, is defined as

$$P_{ijmn}^{k+1} = \frac{1}{2}\left[\left(h^{k+1}\right)_{im}^{-1} N_j N_n + \left(h^{k+1}\right)_{jm}^{-1} N_i N_n\right], \tag{8.86}$$

where, $h_{ip}^{k+1} = C_{ijpq}^{k+1} N_j N_q$, is Christoffel's matrix. This leads to the following general expressions that relate the strain field in phase k to that in phase $(k+1)$, in tensorial form as:

$$\begin{cases} \boldsymbol{\varepsilon}^{k+1}(\mathbf{r}) = \left[\mathbf{I}_4 + \mathbf{P}^{k+1} : (\mathbf{C}^k - \mathbf{C}^{k+1})\right] : \boldsymbol{\varepsilon}^k(\mathbf{r}), \\ \boldsymbol{\varepsilon}^{k}(\mathbf{r}) = \left[\mathbf{I}_4 + \mathbf{P}^{k} : (\mathbf{C}^{k+1} - \mathbf{C}^{k})\right] : \boldsymbol{\varepsilon}^{k+1}(\mathbf{r}). \end{cases} \tag{8.87}$$

In the following, some notation conventions need to be defined:

$$\Omega_j = \bigcup_{k=1}^{j} V_k \quad \text{and} \quad \Delta\mathbf{C}^{(p/q)} = \mathbf{C}^p - \mathbf{C}^q. \tag{8.88}$$

Next, as a first approximation, if one applies Equation (8.87) to the inhomogeneity, phase $k = 1$, and the first ellipsoidal coating, phase $k = 2$ and also taking the average value of strain in the coating and substituting $\boldsymbol{\varepsilon}^1(\mathbf{r})$ by its average value $\bar{\boldsymbol{\varepsilon}}^1$, one gets:

$$\bar{\varepsilon}^2 = \left[\mathbf{I}_4 + \mathbf{T}^2(\mathbf{C}^2) : \Delta\mathbf{C}^{(1/2)}\right] : \bar{\boldsymbol{\varepsilon}}^1, \tag{8.89}$$

where,

$$\mathbf{T}^2(\mathbf{C}^2) = \frac{1}{V_2}\int_{V_2} \mathbf{P}^2 d\mathbf{r}.$$

Cherkaoui et al. [6] have shown that:

$$\mathbf{T}^2(\mathbf{C}^2) = \mathbf{T}^{\Omega_1}(\mathbf{C}^2) - \frac{\varphi_1}{\varphi_2}\left[\mathbf{T}^{\Omega_2}(\mathbf{C}^2) - \mathbf{T}^{\Omega_1}(\mathbf{C}^2)\right], \tag{8.90}$$

where the interaction tensors are defined by

$$\mathbf{T}^{\Omega_j}(\mathbf{C}^2) = \frac{1}{V_{\Omega_j}} \int_{V_{\Omega_j}} \int_{V_{\Omega_j}} \mathbf{\Gamma}^2(\mathbf{r} - \mathbf{r}')\mathrm{d}\mathbf{r}\mathrm{d}\mathbf{r}', \qquad j = 1, 2.$$

Because these tensors are not size-dependent but shape dependent, it is obvious that in the specific case of homothetic inhomogeneities, one can verify the following relations

$$\mathbf{T}^2(\mathbf{C}^2) = \mathbf{T}^{\Omega_1}(\mathbf{C}^2) = \mathbf{T}^{\Omega_2}(\mathbf{C}^2). \tag{8.91}$$

Next, Equation (8.89) can be rewritten as

$$\bar{\boldsymbol{\varepsilon}}^2 = \boldsymbol{\vartheta}^{(1/2)} : \bar{\boldsymbol{\varepsilon}}^1, \tag{8.92}$$

where the fourth-order localization tensor, $\boldsymbol{\vartheta}^{(1/2)}$, is defined by

$$\boldsymbol{\vartheta}^{(1/2)} = \mathbf{I}_4 + \mathbf{\Lambda}_1 : \Delta\mathbf{C}^{(1/2)},$$

$$\mathbf{\Lambda}_1 = \left(\mathbf{T}^{\Omega_1}(\mathbf{C}^2) - \frac{\varphi_1}{\varphi_2}\left[\mathbf{T}^{\Omega_2}(\mathbf{C}^2) - \mathbf{T}^{\Omega_1}(\mathbf{C}^2)\right]\right).$$

One can verify from Equation (8.81) that

$$\boldsymbol{\varepsilon}^2 = \boldsymbol{\vartheta}^{(1/2)} : \mathbf{a}^1 : \boldsymbol{\varepsilon}^I, \qquad \mathbf{a}^2 = \boldsymbol{\vartheta}^{(1/2)} : \mathbf{a}^1. \tag{8.93}$$

With some algebra manipulations, one can get a recurrent relationship between the strain local localization tensor, $\mathbf{a}^k$, in coating k and $\mathbf{a}^1$ as

$$\left\{\begin{aligned} \mathbf{\Pi}^1 &= \mathbf{I}_4, \\ \mathbf{\Pi}^2 &= \boldsymbol{\vartheta}^{(1/2)}, \\ \forall k, \mathbf{a}^k &= \mathbf{\Pi}^k : \mathbf{a}^1, \end{aligned}\right. \tag{8.94}$$

where

$$\begin{cases} \Pi^k = \dfrac{\sum_{j=1}^{k-1}\left(\varphi_j \boldsymbol{\vartheta}^{(j/k)} : \Pi^j\right)}{\sum_{j=1}^{k-1}\varphi_j}, \\ \boldsymbol{\vartheta}^{(j/k)} = \mathbf{I}_4 + \left(\mathbf{T}^{\Omega_{k-1}}(\mathbf{C}^k) - \dfrac{\sum_{n=1}^{k-1}\varphi_n}{\varphi_k}\Delta\mathbf{T}^k\right) : \Delta\mathbf{C}^{(j/k)}, \\ \Delta\mathbf{T}^k = \left[\mathbf{T}^{\Omega_k}(\mathbf{C}^k) - \mathbf{T}^{\Omega_{k-1}}(\mathbf{C}^k)\right], \\ \mathbf{T}^{\Omega_p}(\mathbf{C}^k) = \frac{1}{V_{\Omega_p}}\int_{V_{\Omega_p}}\int_{V_{\Omega_p}}\Gamma^k(\mathbf{r}-\mathbf{r}')\mathrm{d}\mathbf{r}\mathrm{d}\mathbf{r}'. \end{cases} \tag{8.95}$$

Recall that, in the specific case of homothetic inhomogeneities, $\Delta\mathbf{T}^k = 0$. Furthermore, from the third expression of Equation (8.84), one can determine $\mathbf{a}^1$, and thus definitively complete the localization step of the micromechanical model

$$\mathbf{a}^1 = \left(\sum_{k=1}^{n}\varphi_k\Pi^k\right)^{-1}. \tag{8.96}$$

8.8.2.2 Homogenization

The homogenization step starts by relating the macroscopic stress and strain to each other through Hooke's law for elastic solids.

$$\Sigma_{ij} = C_{ijkl}^{eff}E_{kl}. \tag{8.97}$$

In Equation (8.97), $\mathbf{C}^{eff}$ is the effective elastic stiffness tensor of the composite material. $\boldsymbol{\Sigma}$ and $\mathbf{E}$ are the volume average, over the whole material, of the stress and the strain, respectively:

$$\begin{cases} \boldsymbol{\Sigma} = \sum_{k=0}^{n}\varphi_k\overline{\boldsymbol{\sigma}}^k, \\ \mathbf{E} = \sum_{k=0}^{n}\varphi_k\overline{\boldsymbol{\varepsilon}}^k. \end{cases} \tag{8.98}$$

The constitutive laws for each material phase are given below

$$\overline{\boldsymbol{\sigma}}^k = \mathbf{C}^k : \overline{\boldsymbol{\varepsilon}}^k. \tag{8.99}$$

Making use of Equations (8.81), (8.83) and (8.99) in Equation (8.98), one gets from Equation (8.97)

$$\mathbf{C}^{eff} = \mathbf{C}^0 + \sum_{k=1}^{n}\varphi_k\left(\mathbf{C}^k - \mathbf{C}^0\right) : \mathbf{A}^k. \tag{8.100}$$

8.8.2.3 Application to the Present Nano-Inhomogeneities Problem

The previous multi-coating micromechanics-based scheme is applied to the nanocomposites. The nano-inhomogeneity of stiffness tensor, $\mathbf{C}^1$, is surrounded by an interphase of stiffness tensor, $\mathbf{C}^2$, characterized in Section 8.8.1. The generalized self-consistence scheme (GSCS) is used herein to determine the effective stiffness tensor of the nanocomposite. The GSCS supposes that the composite nano-inhomegeneity (nano-inhomegeneity + interphase) is surrounded by a shell of the matrix material of stiffness tensor, $\mathbf{C}^3$, and embedded in the effective medium (see Fig. 8.13(b)). The effective elastic stiffness tensor, $\mathbf{C}^{eff}$, of the nanocomposite is defined by

$$\mathbf{C}^{eff} = \mathbf{C}^3 + \sum_{k=1}^{2} \varphi_k \left(\mathbf{C}^k - \mathbf{C}^3\right) : \mathbf{A}^k. \tag{8.101}$$

The strain localization tensors, $\mathbf{A}^k$, are defined by $\mathbf{A}^k = \mathbf{a}^k : \mathbf{A}^I$ such as

$$\mathbf{A}^I = \left[\mathbf{I}_4 + \mathbf{T}^I(\mathbf{C}^{eff}) : \left(\sum_{k=1}^{3} \varphi_k \Delta\mathbf{C}^{(k/eff)} : \hat{\mathbf{a}}^k\right)\right]^{-1}, \tag{8.102}$$

where: $\Delta\mathbf{C}^{(k/eff)} = \left(\mathbf{C}^k - \mathbf{C}^{eff}\right)$. Recall that, the strain localization tensors, $\mathbf{a}^k$, are already evaluated in the localization step, Equations (8.94), (8.95) and (8.96).

8.8.2.4 Analitycal Solution for Spherical Isotropic Nano-Inhomogeneity

In the case of spherical isotropic configuration, all the above tensors are also isotropic. If $\mathbf{X}$ is one of these tensors then it can be written as

$$\hat{\mathbf{X}} = \hat{\mathbf{S}}_X \mathbf{J} + \hat{\mathbf{D}}_X \mathbf{K}, \tag{8.103}$$

where $\hat{\mathbf{S}}_X$ and $\hat{\mathbf{D}}_X$ are, respectively, the spherical and deviatoric parts of $\hat{\mathbf{X}}$, and the fourth-order tensors, $\mathbf{J}$ and $\mathbf{K}$, are defined as function of Kronecker symbol, δ, by

$$I^h_{ijkl} = \frac{1}{3}\delta_{ij}\delta_{kl},$$
$$I^d_{ijkl} = \frac{1}{2}\left(\delta_{ik}\delta_{jl} + \delta_{il}\delta_{jk} - \frac{2}{3}\delta_{ij}\delta_{kl}\right),$$

and have the following properties:

$$\mathbf{K} : \mathbf{K} = \mathbf{K}, \qquad \mathbf{J} : \mathbf{J} = \mathbf{J},$$
$$\mathbf{J} : \mathbf{K} = \mathbf{K} : \mathbf{J} = 0.$$

All the interaction tensors, $\mathbf{T}^{\Omega_p}(\mathbf{C}^i), p = \{1, 2, 3, \ldots, n\}$ in Equation 8.95 are such as $\mathbf{T}^{\Omega_p}(\mathbf{C}^i) = \mathbf{T}^I(\mathbf{C}^i)$ where $\mathbf{T}^I(\mathbf{C}^i)$ is defined as follows [3]

$$\begin{cases} \mathbf{T}^I(\mathbf{C}^i) = \dfrac{1}{3\hat{\kappa}_i + 4\hat{\mu}_i}\mathbf{J} + \dfrac{3(\hat{\kappa}_i + 2\hat{\mu}_i)}{5\hat{\mu}_i(3\hat{\kappa}_i + 4\hat{\mu}_i)}\mathbf{K}, \\ \mathbf{C}^i = 3\hat{\kappa}_i\mathbf{J} + 2\hat{\mu}_i\mathbf{K}, \end{cases} \tag{8.104}$$

where $\hat{\mu}_i$ and $\hat{\kappa}_i$ are the shear and bulk moduli of the ith elastic isotropic phase. The expressions of the various strain concentration tensors are listed in Appendix 3. The effective properties of the present nano-composite are obtained from Equation 8.101 as follows

$$\begin{cases} \hat{\mu}^{eff} = \hat{\mu}_3 + \sum_{i=1}^{2} \varphi_i(\hat{\mu}_i - \hat{\mu}_3)\hat{\mathbf{D}}^i_A, \\ \hat{\kappa}^{eff} = \hat{\kappa}_3 + \sum_{i=1}^{2} \varphi_i(\hat{\kappa}_i - \hat{\kappa}_3)\hat{\mathbf{S}}^i_A, \end{cases} \tag{8.105}$$

where, $\hat{\mathbf{D}}^i_A$ and $\hat{\mathbf{S}}^i_A$, are defined in Appendix 3. Equation (8.105) is two nonlinear equations which must be solved for $\hat{\mu}^{eff}$ and $\hat{\kappa}^{eff}$.

8.8.3 *Numerical Simulations and Discussions*

8.8.3.1 Spherical Inhomogeneities and Isotropic Material

All the theoretical aspects exposed up to here, are hereafter applied to predict effective properties of isotropic elastic composite containing spherical nano-voids. The numerical results are presented for aluminum with bulk modulus and Poisson ratio are respectively $k_3 = 75.2\,\text{GPa}$ and $\nu_3 = 0.3$. In order to show the effectiveness of the models derived herein, the two sets of surface moduli used in Duan et al. [12] are considered. As it has been done by Duan et al. [12], the free-surface properties are taken from the papers of Miller and Shenoy [42] and are set to equal to the interfacial properties. These free-surface properties are obtained from molecular dynamic (MD) simulations [42]. The elastic properties of the two surfaces named A and B are given in Table 8.1. With these surface properties, the associated interphase properties, μ_c and λ_c (Lamé constants), are determined using Equation (8.70) which is recalled here in the case of spherical nanoparticles

Table 8.1 Elastic properties of surfaces A and B

Surface	$k_s\ (\text{J} \cdot \text{m}^{-2})$	$\hat{\mu}_s\ (\text{J} \cdot \text{m}^{-2})$
A [1 0 0]	−5.457	−6.2178
B [1 1 1]	12.9327	−0.3755

$$\begin{cases} \mu_c = \dfrac{1}{t}\mu_s, \\ \lambda_c = \dfrac{2}{t}\left(\dfrac{\mu_s\lambda_s}{2\mu_s - \lambda_s}\right). \end{cases} \quad (8.106)$$

It is assumed a thickness, t, for the interphase in the subsequent numerical calculations. Next, Equation (8.105) is used to calculate the effective properties of the nanocomposite. In what follows, κ_C and μ_C represent the classical results without the interfacial effect. The normalized bulk modulus κ^{eff}/κ_C for both surface properties as a function of the void radius is plotted in Fig. 8.14. Figure 8.14 shows that κ^{eff}/κ_C decreases (increases) with an increase of void size due to the surface effect. The variation of the bulk modulus κ^{eff}/κ_C with void volume fraction, φ_1, for two different void radii is shown in Fig. 8.15. The normalized shear modulus κ^{eff}/μ_C calculated for both surface properties as a function of the void radius is shown in Fig. 8.16. The variation of the normalized shear modulus with void volume fraction is shown in Fig. 8.17. Conclusions in Duan et al. [12], that is the surface effect is much more pronounced for surface A than for surface B, are verified. All the figures presented with the models derived herein are similar to those in Duan et al. [12]. Thus the results from these first numerical simulations are very encouraging since they show that the present modeling schemes are able to reproduce the results in the work of Duan et al. [12]. The case of spherical isotropic nanoparticle with isotropic interface elastic properties is a particular case of the more general framework in this paper. So in order to show the capability of the present models to deal with various particles shape and interfaces/materials anisotropy, some other numerical simulations are performed in the following sections.

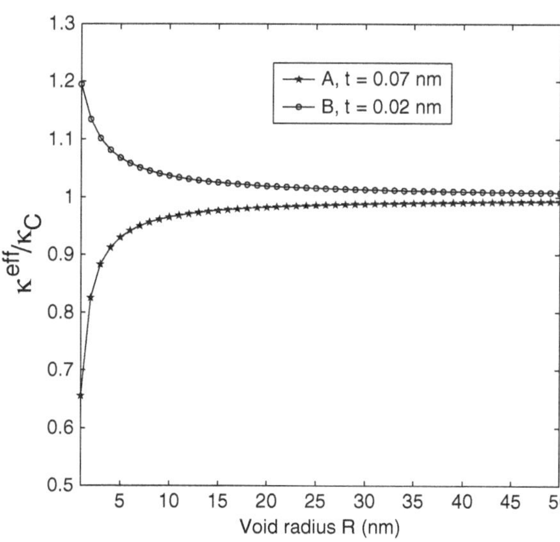

Fig. 8.14 Effective bulk modulus as a function of void radius, $\varphi_1 = 0.3$

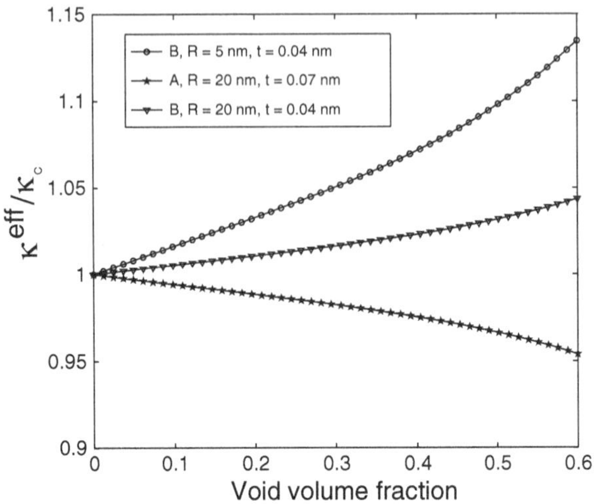

Fig. 8.15 Effective bulk modulus as a function of void volume fraction

8.8.3.2 Ellipsoidal Inhomogeneities and Isotropic Material

Consider now the same material and surface properties as in Section 8.3.1. Then consider three different shapes of nanovoids: an oblate spheroid void ($a = b > c$), a prolate spheroid void ($a = b < c$), and a more general ellipsoid void ($a < b < c$) where a, b and c are the semiaxes of the ellipsoid (see Fig. 8.18). The surface elastic properties in Table 8.1 are set to equal to the interfacial properties of the different shapes of the nanovoids in what follows. The stiffness tensor of the aluminum matrix, $\mathbf{C}^3$, is isotropic and it is given by

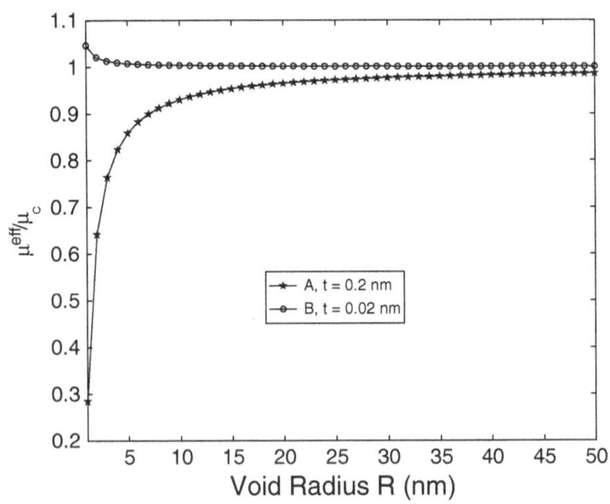

Fig. 8.16 Effective shear modulus as a function of void radius. $\varphi_1 = 0.3$

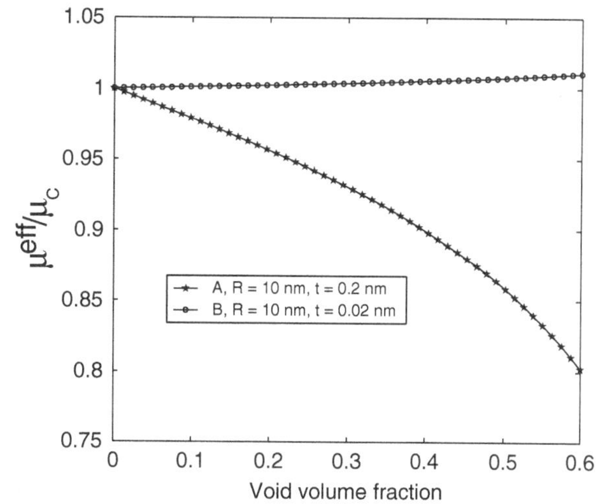

Fig. 8.17 Effective shear modulus as a function of volume fraction

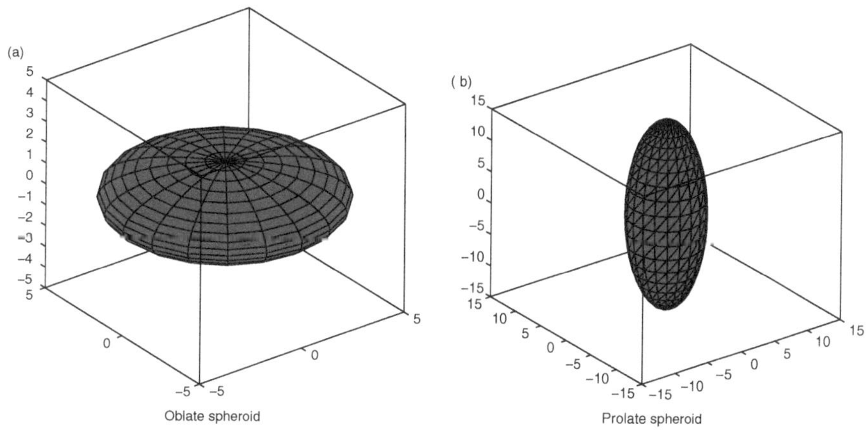

Fig. 8.18 Nanovoid Shapes

$$\mathbf{C}^3 = 3\hat{\kappa}_3\mathbf{J} + 2\hat{\mu}_3\mathbf{K},$$

where $\hat{\kappa}_3 = 75.2\,\text{GPa}$ and $\hat{\mu}_3 = 34.71\,\text{GPa}$.

Oblate Spheroid Nano-Voids

It is first considered an oblate spheroid nanovoid with semiaxes $a = 5\,\text{nm}$, $b = a$ and $c = a/3$. The interphase's thickness is assumed to be $t = 0.02\,\text{nm}$. Then using Equation (8.70) one gets the isotropic stiffness tensors of the interphases associated to surfaces A, $\mathbf{C}_A^2$, and B, $\mathbf{C}_B^2$, respectively, as

$$\mathbf{C}_A^2 = 3\hat{\kappa}_A^2\mathbf{J} + 2\hat{\mu}_A^2\mathbf{K}, \qquad \mathbf{C}_B^2 = 3\hat{\kappa}_B^2\mathbf{J} + 2\hat{\mu}_B^2\mathbf{K}$$

Table 8.2 Elastic properties of the interphases associated to surfaces A and B (oblate spheroid)

Surface i	$\hat{\kappa}_i^2$ (GPa)	$\hat{\mu}_i^2$ (GPa)
A [1 0 0]	−310.5667	−563.2
B [1 1 1]	41.7169	−34.0131

where the interphases bulk and shear moduli are defined in Table 8.2. Next the effective stiffness matrix of the nanocomposite containing $\varphi_1 = 30\%$ of oblate spheroid nanovoids is determined from Equation (8.101) for each surface. For purpose of comparison, the effective stiffness matrix, $\mathbf{C}_C^{eff}$, of the same voided (oblate spheroid shape) composite without surface effect is also computed. The results are as follows

Effective stiffness matrix for surface A for an oblate spheroid shape (GPa)

$$\mathbf{C}_A^{eff} = \begin{bmatrix} 62.6621 & 23.9149 & 15.7073 & 0 & 0 & 0 \\ 23.9149 & 62.6621 & 15.7073 & 0 & 0 & 0 \\ 15.7073 & 15.7073 & 32.4368 & 0 & 0 & 0 \\ 0 & 0 & 0 & 13.6668 & 0 & 0 \\ 0 & 0 & 0 & 0 & 13.6668 & 0 \\ 0 & 0 & 0 & 0 & 0 & 19.3736 \end{bmatrix}, \quad (8.107)$$

Effective stiffness matrix for surface B for an oblate spheroid shape (GPa)

$$\mathbf{C}_B^{eff} = \begin{bmatrix} 73.2006 & 29.8136 & 20.2262 & 0 & 0 & 0 \\ 29.8136 & 73.2006 & 20.2262 & 0 & 0 & 0 \\ 20.2262 & 20.2262 & 41.9145 & 0 & 0 & 0 \\ 0 & 0 & 0 & 16.5411 & 0 & 0 \\ 0 & 0 & 0 & 0 & 16.5411 & 0 \\ 0 & 0 & 0 & 0 & 0 & 21.6935 \end{bmatrix}, \quad (8.108)$$

Effective stiffness matrix without interface effect (oblate spheroid shape) (GPa)

$$\mathbf{C}_C^{eff} = \begin{bmatrix} 68.1533 & 24.6258 & 17.2370 & 0 & 0 & 0 \\ 24.6258 & 68.1533 & 17.2370 & 0 & 0 & 0 \\ 17.2370 & 17.2370 & 37.8677 & 0 & 0 & 0 \\ 0 & 0 & 0 & 15.9087 & 0 & 0 \\ 0 & 0 & 0 & 0 & 15.9087 & 0 \\ 0 & 0 & 0 & 0 & 0 & 21.7638 \end{bmatrix}. \quad (8.109)$$

For the above three effective stiffnesses matrices, the effective material is transversely isotropic (five independent constants). This anisotropic material behavior is due to the shape of the nanovoids. One can also verify the conclusion that the surface effect is much more pronounced for surface A than for surface B for the shear moduli when comparing $\mathbf{C}_A^{eff}$, $\mathbf{C}_B^{eff}$, and $\mathbf{C}_C^{eff}$.

Prolate Spheroid Nano-Void

Consider now a prolate spheroid nanovoid with semiaxes $a = 5\,\text{nm}$, $b = a$, and $c = 3a$. The effective stiffness matrices corresponding to surface A and surface B are determined for $\varphi_1 = 30\%$ of prolate spheroid nanovoids as

Effective stiffness matrix for surface A for a prolate spheroid shape (GPa)

$$\mathbf{C}_A^{eff} = \begin{bmatrix} 50.6710 & 18.9927 & 21.5411 & 0 & 0 & 0 \\ 18.9927 & 50.6710 & 21.5411 & 0 & 0 & 0 \\ 21.5411 & 21.5411 & 69.4285 & 0 & 0 & 0 \\ 0 & 0 & 0 & 18.1765 & 0 & 0 \\ 0 & 0 & 0 & 0 & 18.1765 & 0 \\ 0 & 0 & 0 & 0 & 0 & 15.8392 \end{bmatrix}. \quad (8.110)$$

Effective stiffness matrix for surface B for a prolate spheroid shape (GPa)

$$\mathbf{C}_B^{eff} = \begin{bmatrix} 55.6094 & 21.1245 & 24.1950 & 0 & 0 & 0 \\ 21.1245 & 55.6094 & 24.1950 & 0 & 0 & 0 \\ 24.1950 & 24.1950 & 74.2538 & 0 & 0 & 0 \\ 0 & 0 & 0 & 19.2974 & 0 & 0 \\ 0 & 0 & 0 & 0 & 19.2974 & 0 \\ 0 & 0 & 0 & 0 & 0 & 17.2425 \end{bmatrix}. \quad (8.111)$$

Both stiffness matrices show that the effective material is transversely isotropic. When one compares the effective stiffness matrices (8.110) and (8.111) (prolate spheroid shape) to the effective stiffness matrices (8.107) and (8.108) (oblate spheroid shape), one notices the effect of inhomogeneity shape on the effective behavior of the nanocomposite.

Ellipsoidal Nano-inhomogeneity

To close this section, consider an ellipsoid inhomogeneity shape such as $a = 5\,\text{nm}$, $b = 3a$, and $c = 5a$. In this case, the effective stiffness matrices corresponding to surfaces A and B are computed for $\varphi_1 = 30\%$ of ellipsoidal nanovoids. The effective stiffness matrix is also determined for the same nanovoided composite without interfacial effect. The results are presented below

Effective stiffness matrix for surface A for an ellipsoidal shape (GPa)

$$\mathbf{C}_A^{eff} = \begin{bmatrix} 32.2775 & 14.8422 & 14.9696 & 0 & 0 & 0 \\ 14.8422 & 64.5813 & 24.0633 & 0 & 0 & 0 \\ 14.9696 & 24.0633 & 69.1320 & 0 & 0 & 0 \\ 0 & 0 & 0 & 21.2488 & 0 & 0 \\ 0 & 0 & 0 & 0 & 14.0894 & 0 \\ 0 & 0 & 0 & 0 & 0 & 13.6601 \end{bmatrix}, \quad (8.112)$$

Effective stiffness matrix for surface B for an ellipsoidal shape (GPa)

$$\mathbf{C}_B^{eff} = \begin{bmatrix} 35.1351 & 16.1343 & 16.3993 & 0 & 0 & 0 \\ 16.1343 & 67.9275 & 25.9629 & 0 & 0 & 0 \\ 16.3993 & 25.9629 & 72.4041 & 0 & 0 & 0 \\ 0 & 0 & 0 & 21.9527 & 0 & 0 \\ 0 & 0 & 0 & 0 & 14.9226 & 0 \\ 0 & 0 & 0 & 0 & 0 & 14.5873 \end{bmatrix}. \quad (8.113)$$

Effective stiffness matrix without interface effect (ellipsoidal shape) (GPa)

$$\mathbf{C}_C^{eff} = \begin{bmatrix} 34.1511 & 15.5049 & 15.5097 & 0 & 0 & 0 \\ 15.5049 & 66.6753 & 24.4152 & 0 & 0 & 0 \\ 15.5097 & 24.4152 & 71.0479 & 0 & 0 & 0 \\ 0 & 0 & 0 & 22.0867 & 0 & 0 \\ 0 & 0 & 0 & 0 & 14.9111 & 0 \\ 0 & 0 & 0 & 0 & 0 & 14.3916 \end{bmatrix}. \quad (8.114)$$

With this ellipsoid nanovoids shape, the effective material is orthotropic (nine independent constants).

8.8.3.3 Ellispsoidal inhomogeneities and anisotropic material

In order to go far in the applications of the present modeling schemes, consider now a nanocomposite with ellipsoidal inhomogeneities and anisotropic interface elastic properties. The interfacial excess anisotropic elastic properties are taken from Dingreville and Qu [8]

$$\boldsymbol{\Gamma}^{(2)}(\mathrm{J/m^2}) = \begin{bmatrix} \Gamma_{1111} & \Gamma_{1122} & \Gamma_{1112} \\ \Gamma_{2211} & \Gamma_{2222} & \Gamma_{2212} \\ \Gamma_{1211} & \Gamma_{1222} & \Gamma_{1212} \end{bmatrix} = \begin{bmatrix} -10.679 & -14.908 & 0 \\ -14.908 & -10.510 & 0 \\ 0 & 0 & -2.489 \end{bmatrix},$$

$$\mathbf{H}(\mathrm{nm}) = \begin{bmatrix} H_{111} & H_{122} & H_{112} \\ H_{211} & H_{222} & H_{212} \\ H_{311} & H_{322} & H_{312} \end{bmatrix} = \begin{bmatrix} -0.0003 & 0.0002 & 0 \\ 0.0460 & 0.0570 & 0 \\ 0.3500 & 0.6180 & 0 \end{bmatrix},$$

$$\boldsymbol{\Lambda}^{(2)}(10^{-11}\mathrm{nm/Pa}) = \begin{bmatrix} \Lambda_{11}^{(2)} & \Lambda_{12}^{(2)} & \Lambda_{13}^{(2)} \\ \Lambda_{21}^{(2)} & \Lambda_{22}^{(2)} & \Lambda_{23}^{(2)} \\ \Lambda_{31}^{(2)} & \Lambda_{32}^{(2)} & \Lambda_{33}^{(2)} \end{bmatrix} = \begin{bmatrix} 0.494 & 0 & 0 \\ 0 & -0.185 & 0 \\ 0 & 0 & 0.121 \end{bmatrix}.$$

The stiffness matrix, $\mathbf{C}^2$ (100GPa), of the interphase associated to these interface elastic properties is computed using Equation (8.66), (8.67) and (8.68) and it is given as follows

$$\mathbf{C}^2 = \begin{bmatrix} 25.7648 & 46.8044 & 2.8926 & 0 & 0 & 0 \\ 46.8044 & 87.4148 & 5.1074 & 0 & 0 & 0 \\ 2.8926 & 5.1074 & 0.2868 & 0 & 0 & 0 \\ 0 & 0 & 0 & -0.1876 & 0 & 0 \\ 0 & 0 & 0 & 0 & 0.0703 & 0 \\ 0 & 0 & 0 & 0 & 0 & -0.7172 \\ . & & & & & \end{bmatrix}.$$

In the present case, the matrix is copper (Cu) which stiffness matrix, $\mathbf{C}^3$ (GPa) is given by

$$\mathbf{C}^3 = \begin{bmatrix} 167.3900 & 124.1000 & 124.1000 & 0 & 0 & 0 \\ 124.1000 & 167.3900 & 124.1000 & 0 & 0 & 0 \\ 124.1000 & 124.1000 & 167.3900 & 0 & 0 & 0 \\ 0 & 0 & 0 & 21.6450 & 0 & 0 \\ 0 & 0 & 0 & 0 & 21.6450 & 0 \\ 0 & 0 & 0 & 0 & 0 & 21.6450 \end{bmatrix}.$$

The effective stiffness matrix, $\mathbf{C}^{eff}$ (GPa), of this ellipsoidal nanovoided composite with $a = 5\,\mathrm{nm}$, $b = 3a$, $c = 5a$ and $\varphi_1 = 30\%$ is computed as

$$\mathbf{C}^{eff} = \begin{bmatrix} 29.1712 & 19.3567 & 20.2603 & 0 & 0 & 0 \\ 19.3567 & 58.7770 & 31.8219 & 0 & 0 & 0 \\ 20.2603 & 31.8219 & 61.6544 & 0 & 0 & 0 \\ 0 & 0 & 0 & 13.7425 & 0 & 0 \\ 0 & 0 & 0 & 0 & 9.6451 & 0 \\ 0 & 0 & 0 & 0 & 0 & 9.5319 \end{bmatrix}. \tag{8.115}$$

As it is shown by the matrix (8.115), the effective material displays orthotropic behavior.

The numerical results presented above show the capacity of the present models to efficiently handle the nano-inhomogeneity Eshelby's problem by taking into account the atomistic level informations. In contrast to the previous models which have been devoted to this problem, the present modeling approach is able to tackle any material/interface anisotropy and a general ellipsoidal inhomogeneity shape. It is shown from this modeling approach that a nanoparticles-reinforced composite can exhibit locally negative stiffness behavior. This observation is very interesting since it may lead to new avenues in materials design strategies. Many potential applications can be made: from damping to piezoelectricity, from low-k materials to magnetostriction of nanocristalline magnetic materials.

Appendix 1: "T" Stress Decomposition

Consider an inhomogeneous, linearly elastic solid with strain energy density per unit undeformed volume defined by

$$w = w_0 + \tau_{ij}\,\varepsilon_{ij} + \frac{1}{2}C_{ijkl}\,\varepsilon_{ij}\,\varepsilon_{kl}, \tag{8.116}$$

where ε_{ij} is the Lagrangian strain tensor. The corresponding second Piola-Kirchhoff stress tensor is thus given by

$$\sigma_{ij} = \frac{\partial w}{\partial \varepsilon_{ij}} = \tau_{ij} + C_{ijkl}\,\varepsilon_{ij}. \tag{8.117}$$

Equivalently, (8.117) can be written as

$$\sigma^s_{\alpha\beta} = \tau^s_{\alpha\beta} + C_{\alpha\beta\hat{\kappa}\lambda}\varepsilon_{\hat{\kappa}\lambda} + C_{\alpha\beta 3k}\varepsilon^t_k, \qquad \sigma^t_j = \tau^t_j + C_{3j\hat{\kappa}\lambda}\varepsilon_{\hat{\kappa}\lambda} + C_{3j3k}\varepsilon^t_k, \tag{8.118}$$

where the summation convention is implied, and the lowercase Roman subscripts go from 1 to 3 and the lowercase Greek subscripts go from 1 to 2, and

$$\sigma^s_{\alpha\beta} = \bar{\boldsymbol{\sigma}}_{\alpha\beta},\ \varepsilon_{\alpha\beta} = \hat{\varepsilon}_{\alpha\beta},\ \sigma^t_j = \bar{\boldsymbol{\sigma}}_{3j},\ \varepsilon^t_\alpha = 2\hat{\varepsilon}_{\alpha 3},\ \varepsilon^t_3 = \hat{\varepsilon}_{33},\ \tau^s_{\alpha\beta} = \tau_{\alpha\beta},\ \tau^t_j = \tau_{3j}. \tag{8.119}$$

Assuming that the second-order, C_{3k3j}, is invertible, the second part of Equation (8.118) can be rewritten as

$$\varepsilon_k^t = -M_{kj}\tau_j^t + M_{jk}\sigma_j^t - \gamma_{k\alpha\beta}\varepsilon_{\alpha\beta}, \tag{8.120}$$

where

$$M_{kj} = C_{3k3j}^{-1}, \qquad \gamma_{k\alpha\beta} = M_{kj}C_{3k\alpha\beta}. \tag{8.121}$$

Substituting (8.120) into the first of (8.118) yields

$$\sigma_{\alpha\beta}^s = \hat{\tau}_{\alpha\beta}^s + C_{\alpha\beta\hat{\kappa}\lambda}^s\varepsilon_{\hat{\kappa}\lambda} + \gamma_{j\alpha\beta}\sigma_j^t, \tag{8.122}$$

where

$$\hat{\tau}_{\alpha\beta}^s = \tau_{\alpha\beta}^s - \tau_j^t\gamma_{j\alpha\beta}, \qquad C_{\alpha\beta\hat{\kappa}\lambda}^s = C_{\alpha\beta\hat{\kappa}\lambda} - C_{\alpha\beta3j}\gamma_{j\hat{\kappa}\lambda}. \tag{8.123}$$

Using tensorial notation, Equation (8.116), (8.120) and (8.122) can be written, respectively, as

$$w = w_0 - \frac{1}{2}\boldsymbol{\tau}^t \cdot \mathbf{M} \cdot \boldsymbol{\tau}^t + \hat{\boldsymbol{\tau}}^s : \boldsymbol{\varepsilon}^s + \frac{1}{2}\boldsymbol{\varepsilon}^s : \mathbf{C}^s : \boldsymbol{\varepsilon}^s + \frac{1}{2}\boldsymbol{\sigma}^t \cdot \mathbf{M} \cdot \boldsymbol{\sigma}^t, \tag{8.124}$$

$$\boldsymbol{\varepsilon}^t = -\mathbf{M} \cdot \boldsymbol{\tau}^t + \mathbf{M} \cdot \boldsymbol{\sigma}^t - \boldsymbol{\gamma} : \boldsymbol{\varepsilon}^s, \tag{8.125}$$

$$\boldsymbol{\sigma}^s = \hat{\boldsymbol{\tau}}^s + \mathbf{C}^s : \boldsymbol{\varepsilon}^s + \boldsymbol{\gamma} \cdot \boldsymbol{\sigma}^t. \tag{8.126}$$

In addition, if the material is isotropic, that is

$$C_{ijkl} = \lambda\delta_{ij}\delta_{kl} + \hat{\mu}\left(\delta_{ik}\delta_{jl} + \delta_{il}\delta_{jk}\right), \tag{8.127}$$

where λ and μ are the Lamé constants. The other quantities, in this special case, are such as

$$\begin{cases} C_{3k3j} &= (\lambda + \hat{\mu})\delta_{3k}\delta_{3j} + \hat{\mu}\delta_{kj}, \\ M_{kj} &= -\frac{\lambda+\hat{\mu}}{\hat{\mu}(\lambda+2\hat{\mu})}\delta_{3k}\delta_{3j} + \frac{1}{\hat{\mu}}\delta_{kj}, \\ \gamma_{i\alpha\beta} &= \frac{\lambda}{\lambda+2\hat{\mu}}\delta_{3i}\delta_{\alpha\beta}, \\ C_{\alpha\beta\hat{\kappa}\lambda}^s &= \frac{2\lambda\hat{\mu}}{\lambda+2\hat{\mu}}\delta_{\alpha\beta}\delta_{\hat{\kappa}\lambda} + \hat{\mu}\left(\delta_{\alpha\hat{\kappa}}\delta_{\beta\lambda} + \delta_{\alpha\lambda}\delta_{\beta\hat{\kappa}}\right). \end{cases} \tag{8.128}$$

Appendix 2: Atomic Level Description

The difference in position of two atoms, m and n, near their relaxed state as

$$r_i^{mn} - \hat{r}_i^{mn} = A_{i\alpha\beta}^{mn}\varepsilon_{\alpha\beta} + B_{ik}^{mn}\sigma_k^t + \left(\tilde{\varepsilon}_{ij}^{m}\hat{r}_j^{m} - \tilde{\varepsilon}_{ij}^{n}\hat{r}_j^{n}\right), \tag{8.129}$$

where,

$$\begin{cases} A_{i\alpha\beta}^{mn} = \left(A_{ij\alpha\beta}^{\pm,m} + A_{ij\alpha\beta}^{\pm,n}\right)\hat{r}_j^{mn} - \left(A_{ij\alpha\beta}^{\pm,n}\hat{r}_j^{m} - A_{ij\alpha\beta}^{\pm,m}\hat{r}_j^{n}\right), \\ B_{ik}^{mn} = \left(B_{ijk}^{\pm,m} + B_{ijk}^{\pm,n}\right)\hat{r}_j^{mn} - \left(B_{ijk}^{\pm,n}\hat{r}_j^{m} - B_{ijk}^{\pm,m}\hat{r}_j^{n}\right). \end{cases} \tag{8.130}$$

The total strain energy of the atomic assembly (see Section 8.8.1.1),

$$\begin{aligned} E =& E_0 + \overline{\mathbf{A}}^{(1)} : \hat{\boldsymbol{\varepsilon}}^s + \overline{\mathbf{B}}^{(1)} \cdot \overline{\boldsymbol{\sigma}}^t + \frac{1}{2}\hat{\boldsymbol{\varepsilon}}^s : \overline{\mathbf{A}}^{(2)} : \hat{\boldsymbol{\varepsilon}}^s + \frac{1}{2}\overline{\boldsymbol{\sigma}}^t \cdot \overline{\mathbf{B}}^{(2)} \cdot \overline{\boldsymbol{\sigma}}^t \\ & + \overline{\boldsymbol{\sigma}}^t \cdot \overline{\mathbf{Q}} : \hat{\boldsymbol{\varepsilon}}^s + \sum_{n=1}^{N-1}\left(\mathbf{K}^n + \mathbf{D}^n : \hat{\boldsymbol{\varepsilon}}^s + \mathbf{G}^n \cdot \overline{\boldsymbol{\sigma}}^t\right) : \tilde{\boldsymbol{\varepsilon}}^n \\ & + \frac{1}{2}\sum_{n=1}^{N-1}\sum_{m=1}^{N-1}\tilde{\boldsymbol{\varepsilon}}^n : \mathbf{L}^{mn} : \tilde{\boldsymbol{\varepsilon}}^m. \end{aligned} \tag{8.131}$$

with

$$E_0 = \sum_n \frac{1}{\Omega_n}\sum_{m\neq n} E^{(n)}\Big|_{r^{mn}=\hat{r}^{mn}}, \tag{8.132}$$

$$\overline{A}_{\alpha\beta}^{(1)} = \sum_n \frac{1}{\Omega_n}\sum_{m\neq n}\frac{\partial E^{(n)}}{\partial r_i^{mn}}\Bigg|_{r^{mn}=\hat{r}^{mn}} A_{i\alpha\beta}^{mn}, \tag{8.133}$$

$$\overline{B}_k^{(1)} = \sum_n \frac{1}{\Omega_n}\sum_{m\neq n}\frac{\partial E^{(n)}}{\partial r_i^{mn}}\Bigg|_{r^{mn}=\hat{r}^{mn}} B_{ik}^{mn}, \tag{8.134}$$

$$\overline{A}_{\alpha\beta\kappa\lambda}^{(2)} = \sum_n \frac{1}{\Omega_n}\sum_{m\neq n}\frac{\partial^2 E^{(n)}}{\partial r_i^{mn}\partial r_k^{mn}}\Bigg|_{r^{mn}=\hat{r}^{mn}} A_{i\alpha\beta}^{mn}\,A_{k\kappa\lambda}^{mn}, \tag{8.135}$$

$$\overline{B}_{jl}^{(2)} = \sum_n \frac{1}{\Omega_n}\sum_{m\neq n}\frac{\partial^2 E^{(n)}}{\partial r_i^{mn}\partial r_k^{mn}}\Bigg|_{r^{mn}=\hat{r}^{mn}} B_{ij}^{mn}\,B_{kl}^{mn}, \tag{8.136}$$

$$\overline{Q}_{\alpha\beta j} = \sum_n \frac{1}{\Omega_n} \sum_{m \neq n} \left. \frac{\partial^2 E^{(n)}}{\partial r_i^{mn} \partial r_k^{mn}} \right|_{r^{mn} = \hat{r}^{mn}} A_{i\alpha\beta}^{mn} B_{kj}^{mn}, \tag{8.137}$$

$$\begin{aligned} K_{ij}^n &= \frac{1}{2\Omega_n} \hat{r}_j^n \left(\sum_{p \neq n} \left. \frac{\partial E^{(p)}}{\partial r_i^{pn}} \right|_{r^{mn} = \hat{r}^{mn}} - \left. \frac{\partial E^{(n)}}{\partial r_i^{pn}} \right|_{r^{mn} = \hat{r}^{mn}} \right) \\ &+ \frac{1}{2\Omega_n} \hat{r}_i^n \left(\sum_{p \neq n} \left. \frac{\partial E^{(p)}}{\partial r_j^{pn}} \right|_{r^{mn} = \hat{r}^{mn}} - \left. \frac{\partial E^{(n)}}{\partial r_j^{pn}} \right|_{r^{mn} = \hat{r}^{mn}} \right), \end{aligned} \tag{8.138}$$

$$\begin{aligned} D_{ij\alpha\beta}^n &= \frac{1}{2\Omega_n} \hat{r}_j^n \sum_{p \neq n} \sum_{q \neq n} \left[\left(\left. \frac{\partial^2 E^{(p)}}{\partial r_i^{pn} \partial r_k^{qn}} \right|_{r^{mn} = \hat{r}^{mn}} - \left. \frac{\partial^2 E^{(n)}}{\partial r_i^{pn} \partial r_k^{qn}} \right|_{r^{mn} = \hat{r}^{mn}} \right) A_{k\alpha\beta}^{pn} \right] \\ &+ \frac{1}{2\Omega_n} \hat{r}_i^n \sum_{p \neq n} \sum_{q \neq n} \left[\left(\left. \frac{\partial^2 E^{(p)}}{\partial r_j^{pn} \partial r_l^{qn}} \right|_{r^{mn} = \hat{r}^{mn}} - \left. \frac{\partial^2 E^{(n)}}{\partial r_j^{pn} \partial r_l^{qn}} \right|_{r^{mn} = \hat{r}^{mn}} \right) A_{l\alpha\beta}^{pn} \right], \end{aligned} \tag{8.139}$$

$$\begin{aligned} G_{ijv}^n &= \frac{1}{2\Omega_n} \hat{r}_j^n \sum_{p \neq n} \sum_{q \neq n} \left[\left(\left. \frac{\partial^2 E^{(p)}}{\partial r_i^{pn} \partial r_k^{qn}} \right|_{r^{mn} = \hat{r}^{mn}} - \left. \frac{\partial^2 E^{(n)}}{\partial r_i^{pn} \partial r_k^{qn}} \right|_{r^{mn} = \hat{r}^{mn}} \right) B_{kv}^{pn} \right] \\ &+ \frac{1}{2\Omega_n} \hat{r}_i^n \sum_{p \neq n} \sum_{q \neq n} \left[\left(\left. \frac{\partial^2 E^{(p)}}{\partial r_j^{pn} \partial r_l^{qn}} \right|_{r^{mn} = \hat{r}^{mn}} - \left. \frac{\partial^2 E^{(n)}}{\partial r_j^{pn} \partial r_l^{qn}} \right|_{r^{mn} = \hat{r}^{mn}} \right) B_{lv}^{pn} \right], \end{aligned} \tag{8.140}$$

$$\begin{aligned} L_{ijkl}^{mn} &= \frac{1}{2\Omega_n} \left(\sum_{p \neq n} \left. \frac{\partial^2 E^{(p)}}{\partial r_i^{pn} \partial r_k^{pn}} \right|_{r^{mn} = \hat{r}^{mn}} + \left. \frac{\partial^2 E^{(n)}}{\partial r_i^{pn} \partial r_k^{pn}} \right|_{r^{mn} = \hat{r}^{mn}} \right) \hat{r}_j^n \hat{r}_l^n \delta_{mn} \\ &+ \frac{1}{2\Omega_n} \left(\sum_{p \neq n} \left. \frac{\partial^2 E^{(p)}}{\partial r_j^{pn} \partial r_l^{pn}} \right|_{r^{mn} = \hat{r}^{mn}} + \left. \frac{\partial^2 E^{(n)}}{\partial r_j^{pn} \partial r_l^{pn}} \right|_{r^{mn} = \hat{r}^{mn}} \right) \hat{r}_i^n \hat{r}_k^n \delta_{mn} \\ &- \frac{1}{4\Omega_n} \left. \frac{\partial^2 E^{(n)}}{\partial r_i^{mn} \partial r_k^{mn}} \right|_{r^{mn} = \hat{r}^{mn}} \left(\hat{r}_j^n \hat{r}_l^m + \hat{r}_j^m \hat{r}^n \right) (1 - \delta_{mn}) \\ &- \frac{1}{4\Omega_n} \left. \frac{\partial^2 E^{(n)}}{\partial r_j^{mn} \partial r_l^{mn}} \right|_{r^{mn} = \hat{r}^{mn}} \left(\hat{r}_i^n \hat{r}_k^m + \hat{r}_i^m \hat{r}_k^n \right) (1 - \delta_{mn}). \end{aligned} \tag{8.141}$$

Appendix 3: Strain Concentration Tensors: Spherical Isotropic Configuration

The deviatoric and spherical parts of all concentration tensors needed to solve Equation (8.105) for $\hat{\mu}^{eff}$ and $\hat{\kappa}^{eff}$ are defined below.

Parts of $\boldsymbol{\vartheta}^{(i/j)}$

From Equation (8.95), one gets:

$$\begin{cases} \hat{\mathbf{S}}_{\vartheta}^{(i/j)} = \dfrac{3\hat{\kappa}_i + 4\hat{\mu}_j}{3\hat{\kappa}_j + 4\hat{\mu}_j}, \\ \hat{\mathbf{D}}_{\vartheta}^{(i/j)} = \dfrac{3\hat{\kappa}_j(2\hat{\mu}_i + 3\hat{\mu}_j) + 4\hat{\mu}_j(4\hat{\mu}_i + 2\hat{\mu}_j)}{5\hat{\mu}_j(3\hat{\kappa}_j + 4\hat{\mu}_j)}. \end{cases} \tag{8.142}$$

Parts of $\boldsymbol{\Pi}^j$

Note that:

$$\boldsymbol{\Pi}^1 = \mathbf{J} + \mathbf{K}, \qquad \text{and} \qquad \boldsymbol{\Pi}^2 = \boldsymbol{\vartheta}^{(1/2)}.$$

$$\begin{cases} \hat{\mathbf{S}}_{\Pi}^{j} = \dfrac{\sum_{k=1}^{j-1} \varphi_k \hat{\mathbf{S}}_{\Pi}^{k} \hat{\mathbf{S}}_{\vartheta}^{(k/j)}}{\sum_{k=1}^{j-1} \varphi_k}, \\ \hat{\mathbf{D}}_{\Pi}^{j} = \dfrac{\sum_{k=1}^{j-1} \varphi_k \hat{\mathbf{D}}_{\Pi}^{k} \hat{\mathbf{D}}_{\vartheta}^{(k/j)}}{\sum_{k=1}^{j-1} \varphi_k}, \end{cases} \tag{8.143}$$

for $j = 3$.

Parts of $\mathbf{a}^1$

From Equation (8.96), one gets:

$$\begin{cases} \hat{\mathbf{S}}_{a}^{1} = \left(\sum_{k=1}^{3} \varphi_k \hat{\mathbf{S}}_{\Pi}^{k}\right)^{-1}, \\ \hat{\mathbf{D}}_{a}^{1} = \left(\sum_{k=1}^{3} \varphi_k \hat{\mathbf{D}}_{\Pi}^{k}\right)^{-1}. \end{cases} \tag{8.144}$$

Parts of $\mathbf{a}^k$

From Equation (8.94), one gets:

$$\begin{cases} \hat{\mathbf{S}}_{a}^{k} = \hat{\mathbf{S}}_{\Pi}^{k} \hat{\mathbf{S}}_{a}^{1}, \\ \hat{\mathbf{D}}_{a}^{k} = \hat{\mathbf{D}}_{\Pi}^{k} \hat{\mathbf{D}}_{a}^{1}. \end{cases} \tag{8.145}$$

Parts of $\mathbf{A}^I$

From Equation (8.102), one gets:

$$\begin{cases} \hat{\mathbf{S}}_A^I = \left[1 + \Lambda_3 \sum_{i=1}^3 \varphi_i \left(\hat{\kappa}_i - \hat{\kappa}^{eff}\right) \hat{\mathbf{S}}_a^k\right]^{-1}, \\ \hat{\mathbf{D}}_A^I = \left[1 + \Lambda_4 \sum_{i=1}^3 \varphi_i \left(\hat{\mu}_i - \hat{\mu}^{eff}\right) \hat{\mathbf{D}}_a^k\right]^{-1}, \end{cases} \tag{8.146}$$

where: $\Lambda_3 = \dfrac{3}{4\hat{\mu}^{eff} + 3\hat{\kappa}^{eff}}$, $\Lambda_4 = \dfrac{6\left(\hat{\kappa}^{eff} + 2\hat{\mu}^{eff}\right)}{5\hat{\mu}^{eff}\left(4\hat{\mu}^{eff} + 3\hat{\kappa}^{eff}\right)}$.

Parts of $\mathbf{A}^k$

From Eq. (8.42), one gets:

$$\begin{cases} \hat{\mathbf{S}}_A^k = \hat{\mathbf{S}}_a^k \hat{\mathbf{S}}_A^I, \\ \hat{\mathbf{D}}_A^k = \hat{\mathbf{D}}_a^k \hat{\mathbf{D}}_A^I. \end{cases} \tag{8.147}$$

References

1. Aboudi, J., *Mechanics of composite materials: A unified micromechanical approach.* Elsevier, Amsterdam, (1991)
2. Alymov, M.I. and M.K. Shorshorov, Surface tension of ultrafine particles. Nanostructured Materials **12**, 365–368, (1999)
3. Barhdadi, E.H., P. Lipinski and M. Cherkaoui, Four phase model: A new formulation to predict the effective elastic moduli of composites. Journal of Engineering Materials and Technology **129**, 313–320, (2007)
4. Chandrasekhar, S., *Ellipsoidal figures of equilibrium.* Yale University Press, New Haven, CT, (1969)
5. Chen, H., G. Hu and Z. Huang, Effective moduli for micropolar composite with interface effect. International Journal of Solids and Structures **44**, 8106–8118, (2007)
6. Cherkaoui, M., H. Sabar and M. Berveiller, Micromechanical approach of the coated inclusion problem and applications to composite materials. Journal of Engineering Materials and Technology **116**, 274–278, (1994)
7. Dingreville, R., Modeling and characterization of the elastic behavior of interfaces in nanostructured materials: from an atomistic description to a continuum approach. Ph.D. thesis, George W. Woodruff School of Mechanical Engineering, Atlanta, GA, USA, (2007)
8. Dingreville, R. and J. Qu, Interfacial excess energy, excess stress and excess strain in elastic solids-planar interfaces. Journal of the Mechanics and Physics of Solids **56**(5), 1944–1954 (2007a)
9. Dingreville, R. and J. Qu, A semi-analytical method to compute surface elastic properties. Acta Materialia **55**, 141–147, (2007b)

10. Dingreville, R., J. Qu and M. Cherkaoui, Surface free energy and its effect on the elastic behavior of nano-sized particules, wires and films. Journal of the Mechanics and Physics of Solids **53**, 1827–1854, (2005)
11. Duan, H.L., J. Wang, Z.P. Huang and B.L. Karihaloo, Eshelby formalism for nano-inhomogeneities. Proceedings of the Royal Society A **461**, 3335–3353, (2005a)
12. Duan, H.L., J. Wang, Z.P. Huang and B.L. Karihaloo, Size-dependent effective elastic constants of solids containing nano-inhomogeneities with interface stress. Journal of the Mechanics and Physics of Solids **53**, 1574–1596, (2005b)
13. Duan, H.L., J. Wang, Z.P. Huang and Z.Y. Luo, Stress concentration tensors of inhomogeneities with interface effects. Mechanics of Materials **37**, 723–736, (2005c)
14. Duan, H.L., X. Yi, Z.P. Huang and J. Wang, A unified scheme for prediction of effective moduli of multiphase composites with interface effects. Part I: Theoretical framework. Mechanics of Materials **39**, 81–93, (2007)
15. Eshelby, J.D., The determination of the elastic field of an ellipsoidal inclusion and related problems. Proceedings of the Royal Society A **241**, 376–396, (1957)
16. Eshelby, J.D., *Elastic inclusions and inhomogeneities*. Vol. 2 of Progress in Solid Mechanics. Amsterdam: North-Holland, (1961)
17. Faulk, F., Ginzburg-landau theory of static domain walls in shape memory alloys. Zeitschrift für Physik **51**(B), 177–185, (1983)
18. Ferrari, M., Nanomechanics, and biomedical nanomechanics: Eshelby's inclusion and inhomogeneity problems at the discrete/continuum interface. Biomedical Microdevices **2**(4), 273–281, (2000)
19. Gibbs, J.W., The Scientific Papers of J. Willard Gibbs. Vol. 1. Longmans-Green, London, (1906)
20. Goodier, J.N., Concentration of stress around spherical and cylindrical inclusions and flaws. Journal of Applied Mechanics **55**, 39–44, (1933)
21. Gurtin, M.E., J. Weissmuller and F. Larche, The general theory of curved deformable interfaces in solids at equilibrium. Philosophical Magazine A **78**, 1093, (1998)
22. Gurtin, M.E. and A.I. Murdoch, A continuum theory of elastic material surfaces. Archive for Rational Mechanics and Analysis **59**, 389, (1975)
23. Hashin, Z., The elastic moduli of heterogeneous materials. Journal of Applied Mechanics **29**, 143–150, (1962)
24. Huang, Z.P. and J. Wang, A theory of hyperelasticity of multi-phase media with surface/interface energy effect. Acta Mechanica 182, 195–210, (2006)
25. Huang, Z.P. and L. Sun, Size-dependent effective properties of a heterogeneous material with interface energy effect: From finite deformation theory to infinitesimal strain analysis. Acta Mechanica **190**, 151–163, (2007)
26. Ibach, H., The role of surface stress in reconstruction, epitaxial growth and stabilization of mesoscopic structures. Surface Science Reports **29**, 193–263, (1997)
27. Jaglinski, T. and R. Lakes, Anelastic instability in composites with negative stiffness inclusions. Philosophical Magazine Letters **84**(12), 803–810, (2004)
28. Johnson, R.A., Relationship between two-body interatomic potentials in a lattice model and elastic constants. Physical Review B **6**(6), 2094–2100, (1972)
29. Koutsawa, Y., M. Cherkaoui, J. Qu and E.M. Daya, Atomistic-continuum interphase model for effective properties of composite materials containing ellipsoidal nano-inhomogeneities. Journal of the Mechanics and Physics of Solids Under Review. February (2008)
30. Lakes, R.S., Extreme damping in compliant composites with a negative stiffness phase. Philosphical Magazine Letters **81**, 95–100, (2001a)
31. Lakes, R.S., Extreme damping in composite materials with a negative stiffness phase. Physical Review Letters **86**, 2897–2900, (2001b)
32. Lakes, R.S. and W.J., Drugan, Dramatically stiffer elastic composite materials due to a negative stiffness phase? Journal of the Mechanics and Physics of Solids **50**, 979–1009, (2002)

33. Lakes, R.S., T. Lee, A. Bersie and Y.C. Wang, Extreme damping in composite materials with negative stiffness inclusions. Nature **410**, 565–567, (2001)
34. Lennard-Jones, J.E. and B.M. Dent, Cohesion at a crystal surface. Transactions of the Farady Society **24**, 0092–0107, (1928)
35. Le Quang, H. and Q.-C. He, Size-dependent effective thermoelastic properties of nanocomposites with spherically anisotropic phases. Journal of the Mechanics and Physics of Solids **55**, 1889–1921, (2007)
36. Lim, C.W., Z.R. Li and L.H. He, Size dependent, non-uniform elastic field inside a nano-scale spherical inclusion due to interface stress. International Journal of Solids and Structures **43**, 5055–5065, (2006)
37. Lipinski, P., E.H. Barhdadi and M. Cherkaoui, Micromechanical modeling of an arbitrary ellipsoidal multi-coated inclusion. Philosophical Magazine **86**(10), 1305–1326, (2006)
38. Love, A.E.H., *Mathematical theory of elasticity*. Dover Publications, New York, Netherlands, (1944)
39. Lur'e, A.I., Three-dimensional Problems of Theory of Elasticity. Interscience, New York, (1964)
40. Martin, J.W., Many-body forces in metals and the brugger elastic constants. Journal of Physics C **8**(18), 2837–2857, (1975)
41. Mi, C. and D.A. Kouris, Nanoparticles under the influence of surface/interface elasticity. Journal of Mechanics of Materials and Structures **1**(4), 763–791, (2006)
42. Miller, R.E. and V.B. Shenoy, Size-dependent elastic properties of nanosized structural elements. Nanotechnology **11**, 139–147, (2000)
43. Milton, G.W., *The theory of composites*. Cambridge University Press, Cambridge, (2002)
44. Müller, P. and A. Saül, Elastic effects on surface physics. Surface Science Reports **54**, 157–258, (2004)
45. Mura, T., *Micromechanics of defects in solids*. Martinus-Nijhoff, Netherlands, (1987)
46. Nemat-Nasser, S. and M. Hori, *Micromechanics: overall properties of heterogeneous materials*, second ed. Edition. Elsevier, Amsterdam, (1999)
47. Pei, Z.W. and H.L. Hwang, Formation of silicon nano-dots in luminescent silicon nitride. Applied Surface Science **212**, 760–764, (2003)
48. Povstenko, Y.Z., Theoretical investigation of phenomena caused by heterogeneous surface tension in solids. Journal of the Mechanics and Physics of Solids **41**, 1499–1514, (1993)
49. Qu, J. and J.L. Bassani, Interfacial fracture-mechanics for anisotropic bimaterials. Journal of Applied Mechanics **60**(2), 422–431, (1993)
50. Qu, J. and M. Cherkaoui, *Fundamentals of micromechanics of solids*, Wiley Edition. John Wiley & Sons, Inc., Hoboken, NJ, (2006)
51. Sander, D., Surface stress: implications and measurements. Current Opinion in Solid State and Materials Science **7**, 51–57, (2003)
52. Sharma, P. and S. Ganti, Interfacial elasticity corrections to size-dependent strain-state of embedded quantum dots. Physique Status Solida B **234**, R10–R12, (2002)
53. Sharma, P. and S. Ganti, Size-dependent eshelby's tensor for embedded nano-inclusions incorporating surface/interface energies. Journal of Applied Mechanics **71**, 663–671, (2004)
54. Sharma, P., S. Ganti and N. Bhate, Effect of surfaces on the size-dependent elastic state of nano-inhomogeneities. Applied Physics Letters **82**(4), 535–537, (2003)
55. Sharma, P. and L.T. Wheeler, Size-dependent elastic state of ellipsoidal nano-inclusions incorporating surface/interface tension. Journal of Applied Mechanics **74**, 447–454, (2007)
56. Shuttleworth, R., The surface tension of solids. Proc. Phys. Soc. A **63**, 444–457, (1950)
57. Sun, L., Y.M. Wu, Z.P. Huang and J.X. Wang, Interface effect on the effective bulk modulus of a particle-reinforced composite. Acta Mechanica Sinica **20**, 676–679, (2004)

58. Teik-Cheng, L., Size-dependency of nano-scale inclusions. Journal of Materials Science Letters **40**, 3841–3842, (2005)
59. Torquato, S., *Random heterogeneous materials: Microstructure and macroscopic properties*. Springer, New York, (2002)
60. Walpole, L.J., Elastic behavior of composite materials: theoretical foundations. Advances in Applied Mechanics **21**, 169–242, (1981)
61. Wang, J., H.L. Duan, Z. Zhang and Z.P. Huang, An anti-interpenetration model and connections between interphase and interface models in particle-reinforced composites. International Journal of Mechanical Sciences **47**, 701–718, (2005)
62. Wang, Y.C. and R.S. Lakes, Extreme thermal expansion, piezoelectricity, and other coupled field properties in composites with a negative stiffness phase. Journal of Applied Physics **90**(12), 6458–6465, (2001)
63. Yang, F.Q., Effect of interfacial stresses on the elastic behavior of nanocomposite materials. Journal of Applied Physics **99**(5), 054306, (2006)
64. Zeller, R. and P.H. Dederichs, Elastic constants of polycrystals. Physica Status Solidi B **55**, 831–842, (1973)

Chapter 9
Innovative Combinations of Atomistic and Continuum: Plastic Deformation of Nanocrystalline Materials

In this last chapter, novel techniques allowing us to face the challenges presented in Chapter 3 (e.g., how to perform the scale transition from the atomistic scale to a higher scale) will be introduced. Recall that the activity of several mechanisms operating in NC materials (e.g., grain boundary dislocation emission, grain boundary sliding/migration) was revealed by atomistic simulations. Unfortunately, due to the limitations inherent in atomistic modeling, presented in detail in Chapter 4, and mainly arising from the computational expanse of atomistic simulations, most simulations are performed at either strain rates, temperatures, or stress states several orders of magnitude larger than that relevant to both quasi-static and shock loading applications. With these considerations, the critical issue arising from atomistic simulations consists of predicting the overall effect of each mechanism. For example, in the case of the emission of dislocation from grain boundaries, it is critical to predict the frequency at which a dislocation is emitted when a nanocrystalline (NC) sample is subjected to monotonic loading. Additionally, it is also necessary to know the effect of each emission and penetration event.

In order to address these issues several strategies can be employed. First, one must recognize that grain boundary motion and dislocation emission from grain boundaries are thermally activated mechanisms. Phenomenological approaches based on statistical mechanics (e.g., using, either implicitly or explicitly, Boltzman distributions to "sample" all acceptable microstrates) appear well suited to face the aforementioned challenges. Models based on thermal activation will be discussed in this chapter. Typically, such models introduce a constitutive relation describing the response of either a grain interior or a grain boundary segment (represented either as a new phase or as an interface). A prediction of the overall material's response is then obtained by introducing the newly developed constitutive model into either micromechanical schemes or finite element codes. Depending on the desired model output (e.g., local stress/strain states, overall state of a statistically representative sample), both continuum micromechanics and finite element methods may be more appropriate.

M. Cherkaoui, L. Capolungo, *Atomistic and Continuum Modeling of Nanocrystalline Materials*, Springer Series in Materials Science 112,
DOI 10.1007/978-0-387-46771-9_9,

In the case of NC material, using the finite element method, grain boundary sliding can be modeled via the use of interface elements. However, the emission of dislocations from grain boundaries can typically not be treated rigorously via the use of dislocation density approaches. Indeed, the transfer of dislocations from a given element to its neighbors cannot be accounted for in finite element simulations. To overcome this challenge, higher-order schemes need to be developed. Some examples of such schemes will be given in this chapter.

While continuum micromechanics–based models can account for dislocation emission in a rather direct fashion, the treatment of imperfect interfaces (e.g., grain boundary sliding) is usually not accounted for (the problem of moving interfaces is addressed in [1]). Similarly to the case of finite elements, novel micromechanical schemes – to be presented in this chapter – need to be developed to overcome this challenge.

There is an alternative to approaches based on phenomenological representations of thermally activated mechanisms combined with scale transition techniques. A novel numerical method, which in essence aims at reducing the number of degrees of freedom associated with atomistic simulations by combining the finite element method and molecular statics simulations, referred to as the quasi-continuum (QC) method, will be reviewed prior to presenting applications to the case of bicrystal modeling.

9.1 Quasi-continuum Methods

The idea behind the quasi-continuum method (QC), introduced by Tadmor et al. [2], is relatively simple. It consists of a framework combining *finite element methods* with atomistic static simulations such that the number of degrees of freedom of the system can be substantially reduced compared to a purely atomistic simulation on the same system. Such an approach will allow a gain in computational time. From the point of view of physics, within a physical system subjected to exterior constraints, some regions may be areas of local effects of interests while other regions will behave as a continuum. Therefore, it is advantageous to introduce a framework essentially capable of treating regions either as a continuum or as a discrete system.

Consider a physical system composed of N atoms subjected to external loads. The potential energy of the system can be written as the sum of each atom's energy – obtained via the use of an interatomic potential – to which the work done, because of applied forces, is removed. Denoting E_i and f_i the energy of atom i and its applied load, respectively, one can write the system's potential energy, Φ, as follows:

$$\Phi = \sum_{i=1}^{N} (E_i(\boldsymbol{u}) - \boldsymbol{f}_i \cdot \boldsymbol{u}_i) \tag{9.1}$$

In the equation above, displacements are denoted with vector $\boldsymbol{u}_i$ and $\boldsymbol{u} = \{\boldsymbol{u}_1, \boldsymbol{u}_2..., \boldsymbol{u}_N\}$. The practical problem to be solved is to find the local displacements such that the potential energy (9.1) is at a minimum. So far, the problem presented above is that of molecular statics simulations (e.g., at 0 K).

The QC method minimizes (9.1) by approximating the total energy of all atoms without requiring an explicit calculation of $E_i \forall i \in [1, N]$. Also, fully atomistic regions can evolve during the deformation. In order to achieve this objective, the physical system is represented by repatoms (*atoms* representing a group of atoms).

Let us introduce the deformation gradient $\boldsymbol{F} = \boldsymbol{I} + \frac{\partial \boldsymbol{u}}{\partial \boldsymbol{X}}$ where $\boldsymbol{I}$ denotes the identity matrix and $\boldsymbol{X}$ denotes the reference configuration of a given atom. In regions of the system where the deformation gradient evolves gradually, it is not necessary to calculate explicitly (e.g., via molecular statics) the position of all atoms. Instead, the position of a given number of atoms, the repatoms, is calculated explicitly while the positions of all atoms within a volume defined by their repatoms (e.g., an element) are calculated by interpolation. This is similar to the finite element method. For the sake of clarity, consider an element of the physical system, delineated by repatoms A, B, and C (see Fig. 9.1):

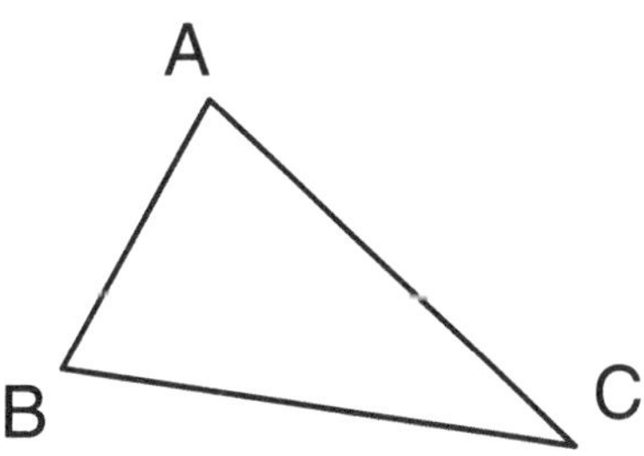

Fig. 9.1 System element defined by repatoms A, B, and C

The displacement of any atom with ABC can thus be written as:

$$u(X_i) = N^A(X_i)u_A + N^B(X_i)u_B + N^C(X_i)u_C \tag{9.2}$$

Here, $N^{A,B,C}$ are linear interpolation functions. Clearly, the density of repatoms shall be increased in regions where local effects (e.g., dislocation cores, stacking faults) are expected to occur. Additionally, "remeshing" must be performed during each step to accurately treat local effects. For detailed discussion on the matter, the reader is referred to [2, 3]. With the above discretization technique the contribution of the potential energy arising from each atom can already be estimated more rapidly.

The computational efficiency can be furthermore improved by use of the Cauchy-Born rule. If, as given by Equation (9.2), linear functions are used to interpolate the displacement fields within a given element, then the deformation gradient will be uniform within this element. The Cauchy-Born rule suggests that in this case the deformation gradient at the microscale is the same as that at the macroscale. As a result, all atoms within the element will be in the same energy state. Therefore, the energy of a given element can be written as the product of the energy density within an element by the element volume. This can be done by calculating the energy of a periodically repeated cell in which all atoms' displacements are imposed by the deformation gradient $\boldsymbol{F}$. Therefore, the contribution of the potential energy arising from each atom's energy contribution can be written as:

$$E^{atom} = \sum_{i=1}^{N} E_i(\boldsymbol{u}) \approx \sum_{i=1}^{N_{element}} \Omega_i E_i^{element}(\boldsymbol{u}) \tag{9.3}$$

Consider now a physical system in which some elements will be bounded by free surfaces, or which will contain any defect (e.g., dislocation core) or interface. Then, the use of the Cauchy-Born rule will not allow quantifying the energy contributions arising from interfaces and defects. In this case, energy calculation shall be conducted via use of a nonlocal formulation. There are two types of nonlocal formulations: (1) energy based and (2) force based. In the case of the energy-based nonlocal formulation, the total energy of the system from each is obtained via explicit calculation of each repatom's energy such that one has:

$$E^{atom} = \sum_{i=1}^{N} E_i(\boldsymbol{u}) \approx \sum_{i=1}^{N_{rep}} n_i E_i^{rep}(\boldsymbol{u}) \tag{9.4}$$

Here, n_i is the weight associated with repatom i. Note that in the expression in the above, the summation is conducted over all repatoms. Also, with this method, the QC method will lead to the same solution as molecular statics in the regions of atomic scale geometrical representation. The nonlocal formulation thus will be more computationally costly than the local one.

To optimize the QC method, coupled local-nonlocal formulations have been introduced in which the atom's total energy is given by:

$$E^{atom} \approx \sum_{i=1}^{N_{rep,loc}} n_i E_i^{rep,loc}(\boldsymbol{u}) + \sum_{i=1}^{N_{rep,non-loc}} n_i E_i^{rep,non-loc}(\boldsymbol{u}) \tag{9.5}$$

The weight of each repatom is calculated from tessellation based on the element's mesh. Calculation of the energy of local repatoms is obtained via the same procedure as that employed in the purely local QC formulation. If M denotes the number of elements surrounding repatom α then one has:

$$n_i E_i^{rep,loc} = \sum_{i=1}^{M} n_\alpha^i \Omega_0 E_i^{element} \tag{9.6}$$

Here, Ω_0 defines the atomic volume and n^i_α the number of atoms in element i. In addition, it is now necessary to define a criterion allowing to one identify local repatoms from nonlocal ones. The criterion used is typically based on the eigenvalues, λ, of the stretch tensor $\boldsymbol{U} = \sqrt{\boldsymbol{F}^T\boldsymbol{F}}$:

$$\max\left|\lambda_i^a - \lambda_j^b\right| < \varepsilon \qquad \forall i,j \in [1,3]; \tag{9.7}$$

Here, λ_i^a denotes the *ith* eigenvalues of element "a". This criterion is used on all elements a,b within a cutoff distance of the repatom of interest.

The QC method was applied to the case of pure Cu bicrystals. In particular, several symmetric and asymmetric tilt grain boundaries were subjected to pure shear strain. Each bicrystal interface was constructed from the CSL model (see Chapter 5 for more detail on the CSL notation). It was shown that three possible mechanisms can be activated: (1) grain boundary sliding, (2) partial dislocation emission, or (3) grain boundary migration. Figure 9.2, shows a partial dislocation emitted from a <110>Σ9(221) 38.94° copper grain boundary with thickness 25 nm. These QC simulations revealed some puzzling interface features. Among others, the activation of any of the three aforementioned grain boundaries could not be correlated with grain boundary energy, or misorientation angle. Additionally, the interface yield stress remains in the same order of magnitude ($\sigma_y = 1 - 5$ GPa) regardless of the mechanism activated during deformation.

The aforementioned simulations were also used to describe the detailed atomic events involved during grain boundary sliding. The deformation mode is similar to a stick slip mechanism. A more detailed description of the mechanism is presented in Chapter 6. From these simulations the following phenomenological law was developed to relate the grain boundary "adhesive" stress to the displacement jump at the interface:

$$\sigma^{adhesive} = \sigma_{crit}\left(1 - \frac{\delta}{\delta_{crit}}\right) \tag{9.8}$$

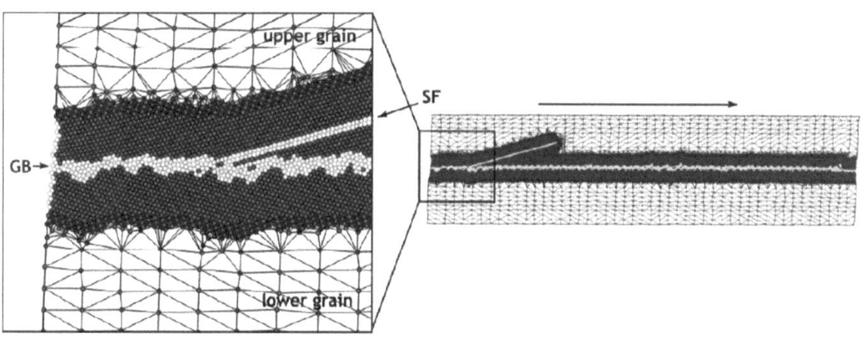

Fig. 9.2 Emission of a partial dislocation from a 25 nm tilt grain boundary in copper

Here, σ_{crit} and δ_{crit} denote a critical stress and displacement. As shown in the example above, atomistic simulations can successfully be used to develop phenomenological laws which are used at a higher scale – usually at the scale of the grain.

9.2 Thermal Activation–Based Modeling

An alternative approach can be used to relate atomistic simulations to continuum based models. Indeed, as shown in chapters 4 and 6, atomistic simulations can be used to predict activation enthalpies associated with deformation modes (e.g., grain boundary sliding, vacancy diffusion, dislocation emission from grain boundaries, etc.). Prior to presenting a model entirely based on thermally activated mechanisms, let us recall some fundamental modeling aspects. Chapter 4 revealed the particular importance of the Boltzmann distribution in mechanics. Among others, the partition functions of the canonical and isobaric-isothermal ensembles were shown to be given by a Boltzmann distribution. In short, when the previously mentioned distribution is assumed to accurately represent the ensemble of acceptable microstates, the probability of activation of a thermally activated event, P, can be written as:

$$P = \exp\left(-\frac{\Delta G}{kT}\right) \tag{9.9}$$

Here, ΔG represents Gibbs enthalpy. For details on the fundamental basis of (9.9) the reader is referred to J.W. Gibbs' thermodynamics treatise [4]. Physically, Gibbs enthalpy represents the amount of energy that must be brought to the system to overcome the energetic barrier – without the support of thermal fluctuations – limiting the activation of the studied mechanism. Clearly, as energy is provided to the system via the application of an external load, Gibbs enthalpy shall decrease. The dependence of ΔG on stress remains a great challenge. To overcome this limitation, phenomenological expressions are:

$$\Delta G = \Delta G_0\left(1 - \left(\frac{\tau}{\tau_c}\right)^p\right)^q \tag{9.10}$$

Here, ΔG_0 and τ_c denote the activation barrier at zero Kelvin and the critical resolved shear stress sufficient to activate the process at zero Kelvin. p and q describe the shape of the energy barrier profile. The following constraint is imposed on p and q $0<p<1$ and $1<q<2$. As shown in Chapter 6, ΔG_0 can be obtained via atomistic simulations; thus allowing to relate atomic scale and continuum scale approaches. This was shown in the particular case of emission of dislocations from grain boundaries.

Most phenomenological models explicitly or implicitly rely on the use of (9.9) and (9.10). As an illustration let us briefly present a model based for the most part on the aforementioned approach [5]. For the sake of simplicity the model is one dimensional. In this model, aiming at predicting the overall response of NC materials and their sensitivity to strain rate and temperature, the following four deformation modes are accounted for: (1) grain boundary diffusion, (2) thermally activated grain boundary sliding, (3) vacancy diffusion within the grain interiors, and (4) transport of dislocation – emitted from grain boundaries – across grain interiors. Chapter 6 presents a discussion on the physical significance of these mechanisms. Molecular dynamics simulations have shown that creep in NC materials is controlled by grain boundary diffusion. Therefore, this first mechanism can be represented by Coble's law as follows:

$$\dot{\varepsilon}_{gbd} = \frac{45\sigma\Omega_a}{kT}\frac{\delta D_{gb}\exp(-Q_{gb}/RT)}{d^3} \tag{9.11}$$

Here, Ω_a,δ,D_{gb},Q_{gb} denote the atomic volume, the grain boundary thickness, the grain boundary diffusivity, and the grain boundary vacancy diffusion activation energy. Similarly, intragranular diffusion is modeled via use of the Nabarro-Herring creep law:

$$\dot{\varepsilon}_{gid} = \frac{10\sigma\Omega_a}{kT}\frac{D_L\exp(-Q_L/RT)}{d^2} \tag{9.12}$$

Here D_L denotes the lattice self-diffusivity. Thermally activated grain boundary sliding is accounted for via use of Conrad and Narayan [6] grain boundary shearing law, which is modified to account for a threshold stress τ_{th}:

$$\dot{\varepsilon}_{gbs} = \frac{6b\nu_d}{d}\sinh\left(\frac{\Omega_a\tau_e}{kT}\right)\exp\left(-\frac{\Delta F}{kT}\right)H(\tau_e - \tau_{th}) \tag{9.13}$$

ΔF, ν_d, τ_e, b denote the grain boundary sliding activation energy, the Debye frequency, the effective stres,s and the Burgers vector. Function H is equal to 1 when the effective stress is larger than the threshold stress and zero otherwise.

The treatment of the effect of grain boundary dislocation emission and penetration is new here. It is based on the idea that the strain rate resulting from the activation of grain boundary dislocation emission can be written as the product of a frequency of emission, ν_e, and of the average strain resulting from slip of the emitted dislocation on a given slip system. Denoting slip systems with α, each dislocation traveling across the grain leads to the following shear on system α:

$$\gamma^\alpha = \frac{b^\alpha \otimes m^\alpha}{d} \tag{9.14}$$

Here, b^{α} and m^{α} denote the dislocation's Burgers vector and the vector normal to the slip plane, respectively. The frequency of emission accounts for the thermally activated nature of the mechanism and is written as follows:

$$\nu_e = \nu_l \exp\left(-\frac{G_e}{kT}\right) \exp\left(\frac{\tau^{\alpha}\Omega_s}{kT}\right) \tag{9.15}$$

Here, G_e and Ω_s denote the activation enthalpy and the activation volume. Note that the product of exponential terms in (9.15) arises from an approximation of Gibbs enthalpy. Combining (9.14) and (9.15) and limiting ourselves to a simple one-dimensional case the strain rate resulting from the activation of grain boundary dislocation emission can be written in the following fashion:

$$\dot{\varepsilon}_{gbe} = \beta_0 \nu_d \exp\left(-\frac{G^e}{kT}\right) \exp\left(\frac{\tau^{\alpha} d}{\mu b}\right) \tag{9.16}$$

μ and β_0 denote the temperature dependent shear modulus and a numerical coefficient in the order of unity. For details on the derivation of (9.16) the reader is referred to [5]. From Equations (9.11), (9.12), (9.13), (9.14), (9.15) and (9.16) the constitutive law of NC materials can be estimated rather simply in a one-dimensional case. With this approximation the materials stress rate is related to its strain rate, $\dot{\varepsilon}$, as follows:

$$\dot{\sigma} = E(\dot{\varepsilon} - \dot{\varepsilon}^p) \tag{9.17}$$

Here, E denotes Young's modulus. $\dot{\varepsilon}^p$denotes the plastic strain rate which is the sum of the averages of the contributions of each mechanism:

$$\dot{\varepsilon}^p = \bar{\dot{\varepsilon}}_{gbe} + \bar{\dot{\varepsilon}}_{gis} + \bar{\dot{\varepsilon}}_{gbd} + \bar{\dot{\varepsilon}}_{gid} \tag{9.18}$$

Here, the symbol bar denotes an average over a distribution of grain sizes. Interestingly, with the relatively simple approach summarized above, estimates of the contributions of each mechanism – at various grain sizes, temperatures, and strain rates – can be obtained. First it is shown, in agreement with experimental data, that intragranular diffusion does not contribute to the deformation.

Figure 9.3 shows the predicted activity of grain boundary emission, grain boundary diffusion, and thermally activated grain boundary sliding during tension at low strain rate for grain sizes ranging from 100 nm to 10 nm. It is found that the contribution of grain boundary emission at the onset of plastic deformation is not dominant compared to that of grain boundary diffusion and grain boundary sliding. Note that the contribution of grain boundary dislocation emission increases during deformation. As expected, as the contribution of these two diffusive mechanisms increases with decreasing grain size such that at grain sizes on the order of ~10 nm, the effect of grain boundary dislocation emission is negligible.

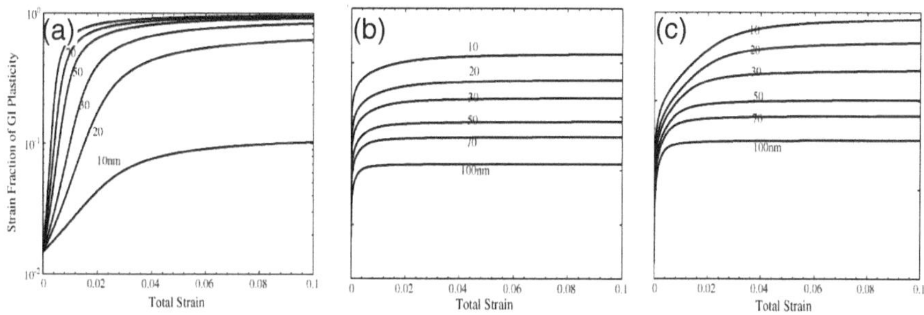

Fig. 9.3 Predicted strain fractions of (**a**) grain boundary dislocation emission, (**b**) grain boundary diffusion, and (**c**) thermally activated grain boundary sliding during monotonic loading of Cu at 3.10E-5/s strain rate

9.3 Higher-Order Finite Elements

The fundamental problem of the treatment of dislocation emission from grain boundaries cannot be entirely addressed with approaches based on continuum micromechanics. Indeed, dislocation emission is a thermally activated mechanism. Therefore, in the loci of higher stresses, the activation energy to be provided by thermal fluctuations to activate an event shall be reduced. In the limit case, dislocation emission could be activated by stress alone. Clearly, knowledge of the local stress fields is required to evaluate the frequency at which dislocations can be emitted from grain boundaries. To this end, finite elements are clearly more suited than continuum micromechanics models.

Although mentioned in the introduction, let us emphasize the role of atomistic simulations in the approach to be presented in this section. MD simulations are used here for the following purposes: (1) identification of the mechanisms likely to contribute to the deformation of the material, (2) estimation of the critical resolved shear stress and activation energies related to a process, and (3) identification of atomistic scale relaxation phenomena following the activation of a mechanism. In the case of (2), it was shown (chapters 5 and 6) that atomistic simulations on bicrystal interfaces can be used to retrieve parameters (e.g., critical resolved shear stress, free activation enthalpy) which can be used as inputs to continuum constitutive relations. Other methods have been developed to perform similar tasks. Among others, the nudged elastic band method – in which, given an initial and final configuration and an initial guess of the path between the two configurations, the minimum energy path can be calculated – is one of the more suitable methods for this end. For a review on the subject the reader is referred to [7]. In the case of (3), it is critical to understand the effect of each mechanism on the local atomic arrangement. For example, the emission of dislocation from (1) a planar grain boundary can seldom lead to presence of a ledge, and (2) a grain boundary with ledge can either consume the ledge or leave it intact. In the latter case, depending on the resulting structure, the grain boundary may be more or less likely to emit additional dislocations.

The finite element method consists of solving the reduced formulation of the system of equations obtained via the application of the principle of virtual work to the case of a system at equilibrium. The unknowns of the system are the displacements and forces which are solved at points referred to as nodes. The latter are defined via meshing of the structure to be studied. In the present case, where it is desired to predict the plastic response of NC materials, the size of elements resulting from an adequate meshing of a NC microstructure is necessarily smaller than the grain size. Moreover, the deformation modes associated with elements representing grain interior regions and grain boundary regions are necessarily different. Disregarding the mechanism of grain boundary sliding, one may consider – in a first approximation – that grain boundaries will deform via emission and absorption of dislocations while grain interiors will deform via glide on primary slip planes of dislocations and could harden via dislocation/dislocation interactions. Assuming a grain size smaller than $\sim$30 nm, statistical storage and dynamics recovery of dislocations do not have to be considered. With the usual finite element formulation, the flux of dislocations from a grain boundary element to a grain interior element cannot be accounted for. However, as shown in work by Arsenlis [8, 9], it is possible to adapt the finite element formulation to account for dislocation fluxes.

As mentioned above, to account for the mechanism of dislocation emission from grain boundaries, the finite element approach needs to associate grain boundary elements from grain interior elements. Therefore, let us consider – for the sake of simplicity – the case of a bicrystal interface as shown in Fig. 9.4. The bicrystal interface is represented by two grains (in blue and yellow) and two half grain boundaries (in green and purple). Each half grain boundary connecting a grain has the same orientation as the neighboring grain.

With this simple representation of bicrystals, constitutive relations based on crystal plasticity approaches can be developed. Let us briefly recall the basis of crystal plasticity.

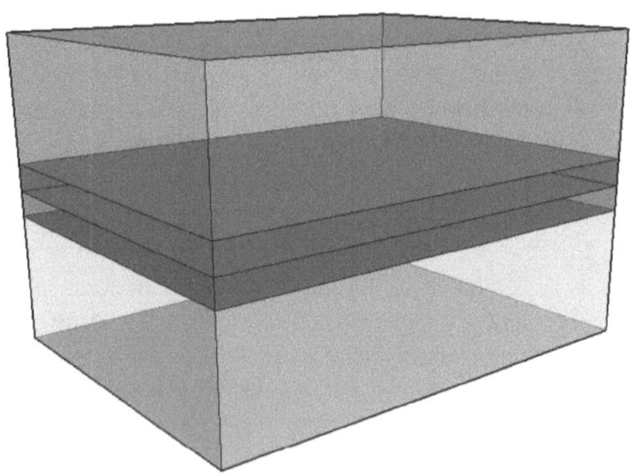

Fig. 9.4 Schematic of a bicrystal interface

9.3.1 Crystal Plasticity

Let x and X denote the position vectors in the current and in the initial configurations. The deformation gradient – corresponding to the spatial derivative of the position vector – is typically written as the product of the elastic, $\mathbf{F}^{\mathbf{e}}$, and plastic deformation, $\mathbf{F}^{\mathbf{P}}$, gradients as follows:

$$\mathbf{F} = \mathbf{F}^{\mathbf{e}}\mathbf{F}^{\mathbf{P}} = \frac{\partial x}{\partial X} \tag{9.19}$$

Geometrically, each slip system denoted α – there are 12 possible slip systems in the FCC system – can be uniquely defined with knowledge of the slip direction $\boldsymbol{m}^{\alpha}$ and normal to the slip plane $\boldsymbol{n}^{\alpha}$. It can be shown that the plastic deformation gradient is related to its rate via the following relation:

$$\dot{\boldsymbol{F}}^{P} = \left\{ \sum_{\alpha} \dot{\gamma}^{\alpha} \boldsymbol{m}^{\alpha} \otimes \boldsymbol{n}^{\alpha} \right\} \boldsymbol{F}^{P} \tag{9.20}$$

The formulation presented in the above will be used to describe both deformations in grain boundary and in grain interiors' elements. This shall lead to overestimates of dislocation activity in grain boundaries. A more rigorous approach would impose latent effects within grain boundaries. The constitutive relation can be expressed as follows:

$$\boldsymbol{T} = \boldsymbol{L} \cdot \boldsymbol{E}^{e} \tag{9.21}$$

T is the second Piola-Kirchhoff tensor and E^{e} is the Green-Lagrange deformation tensor:

$$\boldsymbol{E}^{e} = \frac{1}{2}\left\{ \boldsymbol{F}^{eT}\boldsymbol{F}^{e} - \mathbf{1} \right\} \tag{9.22}$$

Here, the superscript T denotes the transpose of a tensor. The Cauchy stress tensor, $\boldsymbol{\sigma}$, is related to the second Piola-Kirchhoff tensor with:

$$\boldsymbol{T} = \boldsymbol{F}^{e^{-1}} \left\{ (\det \boldsymbol{F}^{e}) \boldsymbol{\sigma} \boldsymbol{F}^{e^{-T}} \right\} \tag{9.23}$$

$\bar{v}^{\alpha}$ the average velocity of dislocations traveling on slip system α and ρ_M the mobile dislocation density, one has:

$$\dot{\gamma}^{\alpha} = \rho_M^{ed} b^{\alpha} \bar{v}_{ed}^{\alpha} + \rho_M^{sc} b^{\alpha} \bar{v}_{sc}^{\alpha} \tag{9.24}$$

The superscripts and subscripts *ed* and *sc* refer to the edge and screw dislocation segments, respectively. Note that, in this approach, within each grain interior

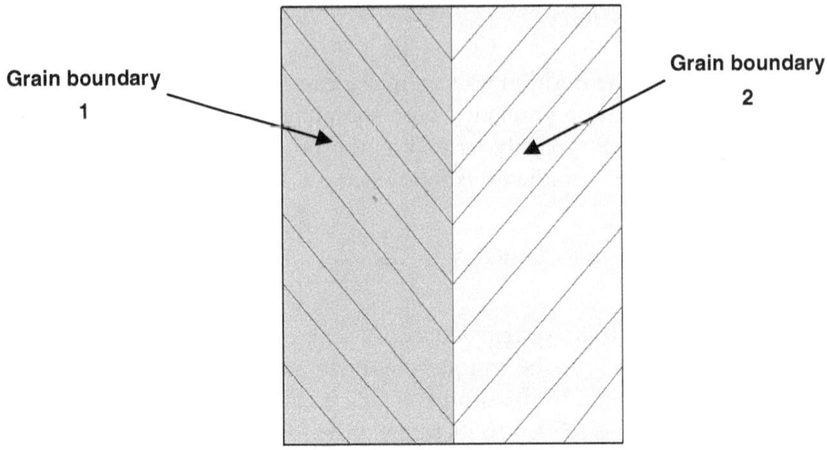

Fig. 9.5 Grain boundary representation

element, mobile dislocations evolve solely via flux. This will necessitate imposing higher-order boundary conditions on each element. The nucleation of mobile dislocations will be generated from grain boundary elements. In order to easily ensure that nucleated dislocation will lie on primary slip planes, grain boundaries can be represented as by two half-unit cells whose orientations coincide with that of the adjacent grains (see Fig. 9.5). Moreover, the resolved shear strain rate in grain boundaries will be given by (9.24).

Within each half-unit cell – corresponding to a single element – the evolution of the dislocation density is driven by the following two processes: (1) nucleation of dislocations and (2) inward and outward dislocation flux corresponding to the dislocation penetration process and to the dislocation emission process, respectively.

With the simple approach presented above, the key problem is that of modeling the dislocation density evolution. In the case of elements representing grain interiors, the dislocation density evolution corresponds to a simplification – source terms must be removed – of that obtained in the case of grain boundary elements. Therefore, the following discussion will be applied solely to the case of grain boundary elements.

Consider a unit cell representing a portion of a grain boundary as shown in Fig. 9.5. On a given slip system dislocations evolve via nucleation and flux. The latter ensures (1) the transmission of dislocations, resulting from the grain boundary dislocation emission and penetration mechanism, and (2) the continuity of the lattice curvature as discussed in work by Arsenlis and Parks [8, 9]. Therefore, dislocation evolution can be written as the sum of a generation term and a flux term:

$$\dot{\rho}_M^\alpha = \dot{\rho}_{nucl}^\alpha + \dot{\rho}_{flux}^\alpha \tag{9.25}$$

The expression above shall account for both edge and screw dislocation characters. Therefore in the case of the dislocation flux, one has:

$$\dot{\rho}^{\alpha}_{flux} = \dot{\rho}^{\alpha}_{flux,ed} + \dot{\rho}^{\alpha}_{flux,sc} \tag{9.26}$$

The dislocation flux corresponds to the integral – over the surface area to be crossed by the moving dislocation – of the product of the dislocation velocity vector by the dislocation density. Denoting $\boldsymbol{n}$ the surface normal, one obtains in the case of edge dislocations:

$$\dot{\rho}^{\alpha}_{flux,ed} = \int_{dS} \rho^{\alpha}_{ed} \bar{v} \boldsymbol{m}^{\alpha} \cdot \boldsymbol{n} dS \tag{9.27}$$

Note that this expression is written in an intermediate configuration. Using the divergence theorem in the reference configuration one obtains:

$$\dot{\rho}^{\alpha}_{flux,ed} = -\frac{\partial \rho^{\alpha}_{ed}}{\partial \boldsymbol{X}} \bar{v} \cdot \boldsymbol{F}^{p^{-1}} \boldsymbol{m}^{\alpha} \tag{9.28}$$

In the case of screw dislocation where t^{α} denotes the screw segment direction – corresponding to the vector normal to the slip direction and to the normal to the slip plane – one has:

$$\dot{\rho}^{\alpha}_{flux,sc} = -\frac{\partial \rho^{\alpha}_{sc}}{\partial \boldsymbol{X}} \bar{v} \cdot \boldsymbol{F}^{p^{-1}} \boldsymbol{t}^{\alpha} \tag{9.29}$$

Similar decomposition as used in the case of flux terms (9.26) shall be used to describe the rate of change in the mobile dislocation density due to nucleation:

$$\dot{\rho}^{\alpha}_{nucl} = \dot{\rho}^{\alpha}_{nucl,ed} + \dot{\rho}^{\alpha}_{nucl,sc} \tag{9.30}$$

As detailed in Chapter 6, dislocation nucleation from grain boundaries is a thermally activated mechanism. Assuming a Boltzmann distribution for the probability of successful emission, the nucleation of both edge and screw dislocations can be described by:

$$\dot{\rho}^{\alpha}_{nucl,ed} = \varpi l_e \exp\left(\frac{-\Delta G_0}{kT}\left(1 - \left(\frac{\tau^{\alpha}}{\tau_{crit}}\right)^{p}\right)^{q}\right) \tag{9.31}$$

and

$$\dot{\rho}^{\alpha}_{nucl,sc} = \varpi l_{sc} \exp\left(\frac{-\Delta G_0}{kT}\left(1 - \left(\frac{\tau^{\alpha}}{\tau_{crit}}\right)^{p}\right)^{q}\right) \tag{9.32}$$

Here ϖ, l_{sc}, l_e denote the frequency of attempts of nucleation of a dislocation, the length of the screw and edge segments necessary for the nucleation to be successful. τ^α, τ_{crit}, ΔG_0,k,T, p, and q describe, from a phenomenological standpoint, an estimate of the probability of successful emission and denote the resolve shear stress on slip system α, the critical shear stress for dislocation nucleation, the free enthalpy of activation, Boltzmann's constant, the temperature, and two parameters describing the shape of the dislocation nucleation resistance diagram, respectively.

9.3.2 Application via the Finite Element Method

In order to use the constitutive framework presented above to simulate the response of NC materials, the finite element method shall be augmented such that dislocations can be accounted for as nodal unknowns. To this end, let us reformulate the dislocation density evolution in a fashion consistent with the finite element method (e.g., reduced formulation of a variational formulation). Therefore, in the case of edge dislocations, equation (25) can be written as follows:

$$0 = \dot{\rho}_{ed}^{\alpha} + \frac{\partial \rho_{ed}^{\alpha}}{\partial \boldsymbol{X}} \bar{v} \cdot \boldsymbol{F}^{p^{-1}} \boldsymbol{m}^{\alpha} - \varpi l_e \exp\left(\frac{-\Delta G_0}{kT}\left(1 - \left(\frac{\tau^\alpha}{\tau_{crit}}\right)^p\right)^q\right) \tag{9.33}$$

Let us multiply the previous dislocation balance equation by a virtual dislocation density, $\tilde{\rho}$, which must respect the real boundary conditions and take the reduced formulation – which mathematically corresponds to a simple integration by parts – of the resulting equation:

$$\begin{aligned} 0 = \int_V \tilde{\rho}\dot{\rho}_{ed}^{\alpha} dV - \int_V \frac{\partial \tilde{\rho}}{\partial \boldsymbol{X}} \rho_{ed}^{\alpha} \bar{v} \cdot \boldsymbol{F}^{p^{-1}} \boldsymbol{m}^{\alpha} dV + \int_S \tilde{\rho}\rho_{ed}^{\alpha} \bar{v} \cdot \boldsymbol{F}^{p^{-1}} \boldsymbol{m}^{\alpha} \boldsymbol{dS} \\ - \int_V \tilde{\rho}\varpi l_e \exp\left(\frac{-\Delta G_0}{kT}\left(1 - \left(\frac{\tau^\alpha}{\tau_{crit}}\right)^p\right)^q\right) dV \end{aligned} \tag{9.34}$$

Introducing the following flux term $\boldsymbol{\Psi} = \rho_{ed}^{\alpha} \bar{v} \cdot \boldsymbol{F}^{p^{-1}} \boldsymbol{m}^{\alpha} \cdot n$, (9.34) can be written as follows:

$$\begin{aligned} 0 = \int_V \tilde{\rho}\dot{\rho}_{ed}^{\alpha} dV - \int_V \frac{\partial \tilde{\rho}}{\partial \boldsymbol{X}} \rho_{ed}^{\alpha} \bar{v} \cdot \boldsymbol{F}^{p^{-1}} \boldsymbol{m}^{\alpha} dV + \int_S \boldsymbol{\Psi} dS \\ - \int_V \tilde{\rho}\varpi l_e \exp\left(\frac{-\Delta G_0}{kT}\left(1 - \left(\frac{\tau^\alpha}{\tau_{crit}}\right)^p\right)^q\right) dV \end{aligned} \tag{9.35}$$

The expression above and its equivalent in terms of screw contributions – which is now written in a form similar to that used in the common finite element method – shall be respected on the 12 slip systems. Clearly, the 24 equations to be solved simultaneously must be added to the equations resulting from the system's equilibrium (e.g., three additional equations). The following three steps shall be followed to implement (9.35) in a finite element framework:

(1) discretization of the dislocation density, (2) global linearization procedure occurring at the element level, and (3) time discretization.

For the sake of clarity, let us consider the case of a 20-node cubic element. In that case, the dislocation density, virtual or real, can be interpolated from the 20 nodal values as follows:

$$\rho = \sum_{i=1}^{i=20} N^i \bar{\rho}^i = \boldsymbol{N}\bar{\rho} \tag{9.36}$$

Where N^i are the second order interpolation functions:

$$\begin{aligned} N^i(\xi,\eta,\zeta) &= \tfrac{1}{8}(1+\xi^i\xi)(1+\eta^i\eta)(1+\zeta^i\zeta)(\xi^i\xi+\eta^i\eta+\zeta^i\zeta-2) && \text{for i} = 1,\ldots,8,\\ N^i(\xi,\eta,\zeta) &= \tfrac{1}{4}(1-\xi^2)(1+\eta^i\eta)(1+\zeta^i\zeta) && \text{for i} = 9,11,17,19\\ N^i(\xi,\eta,\zeta) &= \tfrac{1}{4}(1-\xi^i\xi)(1-\eta^2)(1+\zeta^i\zeta) && \text{for i} = 10,12,18,20\\ N^i(\xi,\eta,\zeta) &= \tfrac{1}{4}(1-\xi^i\xi)(1-\eta^i\eta)(1-\zeta^2) && \text{fori} = 13,14,15,16 \end{aligned} \tag{9.37}$$

Note that in Equation (9.36), the interpolation matrix has dimensions $N = [24 * 480]$ and has the following shape:

$$N = \begin{bmatrix} N_1 & 0 & N_2 & 0 & \ldots & \ldots & \ldots & \ldots & \ldots & 0 \\ 0 & N_1 & 0 & N_2 & \ldots & \ldots & \ldots & \ldots & \ldots & 0 \\ \ldots & \ldots & \ldots & \ldots & \ldots & \ldots & \ldots & \ldots & \ldots & \ldots \\ \ldots & \ldots & \ldots & \ldots & \ldots & \ldots & \ldots & \ldots & \ldots & \ldots \\ 0 & \ldots & \ldots & \ldots & \ldots & N_1 & \ldots & 0 & N_{20} & 0 \\ 0 & \ldots & \ldots & \ldots & \ldots & 0 & \ldots & N_{19} & 0 & N_{20} \end{bmatrix} \tag{9.38}$$

A discretized dislocation vector can be written as follows:

$$\tilde{\rho} = \begin{bmatrix} \rho_e^{1,1} \\ \rho_s^{1,1} \\ \rho_e^{1,2} \\ \rho_s^{1,2} \\ \vdots \\ \rho_e^{1,20} \\ \rho_s^{1,20} \\ \rho_e^{2,1} \\ \rho_s^{2,1} \\ \vdots \\ \rho_e^{12,20} \\ \rho_s^{12,20} \end{bmatrix} \tag{9.39}$$

In the above, the first superscript defines the slip system while the second superscript denotes the node number. Recall that the virtual dislocation density gradient needs to be evaluated (see Equation (9.35)). Using the discretized dislocation density one obtains:

$$\frac{\partial \tilde{\rho}}{\partial \boldsymbol{X}} = \nabla \boldsymbol{N\rho} = \nabla_{(\xi,\eta,\varsigma)} \boldsymbol{N} \left[\sum_{j=1}^{20} \boldsymbol{X}^j \otimes \nabla_{(\xi,\eta,\varsigma)} \boldsymbol{N}^j \right] \boldsymbol{\rho} = \boldsymbol{G\rho}$$

with (9.40)

$$\boldsymbol{G} = \nabla_{(\xi,\eta,\varsigma)} \boldsymbol{N} \cdot \boldsymbol{det} \left[\left[\sum_{j=1}^{20} \boldsymbol{X}^j \otimes \nabla_{(\xi,\eta,\varsigma)} \boldsymbol{N}^j \right] \right]^{-1}$$

$\boldsymbol{G}$ is a [72*480] matrix. Inserting (9.36) and (9.40) into (9.34)and considering all slip systems and both edge and screw components, one obtains the following system:

$$R(\rho) = F^{in} - F^{bd} = 0 \tag{9.41}$$

Where

$$F^{in} = \int_V \left[\boldsymbol{N^T \rho} - \boldsymbol{G^T} \mathbf{H} \cdot \boldsymbol{\rho v} - \boldsymbol{N^T \Phi} \right] dV \tag{9.42}$$

And

$$F^{bd} = -\int_S \boldsymbol{N}^T \boldsymbol{\Psi} dS \tag{9.43}$$

Here, the superscripts *"in"* and *"bd"*, denote the evolution of the dislocation density within the element and due to transport through the boundaries, respectively. In (9.42), $\boldsymbol{H}$ is a [72*24] matrix given by:

$$\boldsymbol{H} = \begin{bmatrix} \boldsymbol{H_1} & 0 & 0 & \dots & 0 & 0 \\ 0 & \boldsymbol{H_2} & 0 & \dots & 0 & 0 \\ & & \cdot & & & \\ & & & \cdot & & \\ & & & & \cdot & \\ 0 & 0 & 0 & \dots & 0 & \boldsymbol{H_{24}} \end{bmatrix} \tag{9.44}$$

Each matrix $\boldsymbol{H_i}$, i $= \boldsymbol{1, 24}$ is [3*1] and its components are given by:

$$\boldsymbol{H_i} = \begin{cases} \boldsymbol{F^{p^{-1}} m^i} \text{ if i is odd} \\ \boldsymbol{F^{p^{-1}} t^{i/2}} \text{ if i is even} \end{cases} \tag{9.45}$$

Finally, in Equation (9.42), $\mathbf{\Phi}$ is a [24*1] matrix whose components are given by:

$$\mathbf{\Phi}_i = \begin{cases} \varpi l_e \exp\left(\frac{-\Delta G_0}{kT}\left(1 - \left(\frac{\tau^i}{\tau_{crit}}\right)^p\right)^q\right) & \text{if i is odd} \\ \varpi l_s \exp\left(\frac{-\Delta G_0}{kT}\left(1 - \left(\frac{\tau^{i/2}}{\tau_{crit}}\right)^p\right)^q\right) & \text{if i is even} \end{cases} \tag{9.46}$$

Finally, one can discretize the dislocation density and the dislocation flux vector $\mathbf{\Psi}$ similarly to equation (9.36). After some algebra, one obtains:

$$R(\bar{\rho}) = F_I \dot{\bar{\boldsymbol{\rho}}} - F_{II} \bar{\boldsymbol{\rho}} - F_{III} + F_{IV} \tag{9.47}$$

With

$$\begin{aligned} F_I &= \int_V \mathbf{N}^T \mathbf{N} dV \\ F_{II} &= \int_V \bar{v} \mathbf{G}^T \mathbf{H} \mathbf{N} dV \\ F_{III} &= \int_V \mathbf{N}^T \Phi dV \\ F_{IV} &= \int_S \mathbf{N}^T \Psi dS \end{aligned} \tag{9.48}$$

The time rate of dislocation density can be neglected if dislocation motion is not hindered by obstacles, which is likely to be the case in NC materials. With the discretized expression of the weak formulation of the dislocation density balance equation, the problem is now that of finding the zero of the vectorial function R. This can be done by using a Newton-Raphson algorithm. At a given time step, the solution $\bar{\rho}^{\,final}$ is obtained as follows:

$$\bar{\rho}^{r+1} = \bar{\rho}^r - \left[\frac{\partial R(\bar{\rho}^r)}{\partial \bar{\rho}^r}\right]^{-1} R(\bar{\rho}^r) \tag{9.49}$$

The tangent stiffness matrix $\frac{\partial R(\bar{\rho}^r)}{\partial \bar{\rho}^r}$ is given by:

$$\frac{\partial R(\bar{\rho}^r)}{\partial \bar{\rho}^r} = F_{II} \tag{9.50}$$

The terms F_{II}, F_{III}, and F_{IV} must be calculated in order to define the tangent stiffness matrix and the residual vector $R(\bar{\rho}^r)$. Finally, the last step consists of defining a time integration procedure. Since the latter is similar to that traditionally used in crystal plasticity–based finite element schemes it will not be reviewed here. The reader is referred to [10] for a detailed presentation of the technique.

As shown with the approach presented above, rigorous treatment of the transfer and creation of dislocations in elements can be accounted for. This allows treatment of the problem of dislocation emission from grain boundaries.

9.4 Micromechanics

As discussed in Section 9.2, grain boundary sliding can be activated in NC materials. Similarly to the case of grain boundary dislocation emission both activation stress and activation enthalpy can be extracted from atomistic scale simulations. The process of sliding was described in Chapter 6.

Conceptually, the challenge is to estimate the effect of this mechanism at the macroscale (see Chapter 3). Clearly, available finite element packages already allow for the treatment of imperfect interfaces via the use of interface elements. On the contrary, Eshelby-Kröner–type micromechanical schemes do not account for the imperfect phase bonding. Moreover, complexity arises from the fact that the materials' response is elastic-viscoplastic. Therefore, micromechanical schemes accounting for all potentially activated mechanisms in NC materials must simultaneously address the subproblems of the treatment of coated inclusions with elastic-viscoplastic behaviors and of the effect of imperfect interfaces on the local strain and stress states of all phases.

The method presented here, based on [11], is articulated as shown in Fig. 9.6. In the first step, a solution for the problem of the viscoplastic response of a composite material – represented as a coated inclusion embedded in homogeneous equivalent medium – with imperfect interface bonding will be derived. Second, the solution will be extended to elastic-viscoplastic behaviors via use of the field translation method introduced by Sabar et al. [12]. The first problem (i.e., viscoplastic response of an heterogeneous medium with imperfect interfaces) will be solved via extension of Qu's work on slightly weakened interfaces [13].

In order to establish a solution of the viscoplastic problem, the procedure shown in Fig. 9.7 is employed. First, jump conditions across the interface between the inclusion and the coating, which is imperfect and allows for sliding, are introduced. Then, the three-phase problem will be solved by consecutively solving two-phase problems. In the first problem, only the inclusion and its coating will be accounted for. Therefore, the coating will play the role of a matrix phase. The bonding between the two phases is imperfect. Mori-Tanaka's scheme is then used to predict the response of this two-phase material, which is now referred to as the "homogenized coated inclusion.. In the second step, the overall viscoplastic response is obtained by using a self-consistent scheme to solve the problem of the homogenized coated inclusion embedded in a matrix phase representing the homogeneous equivalent medium.

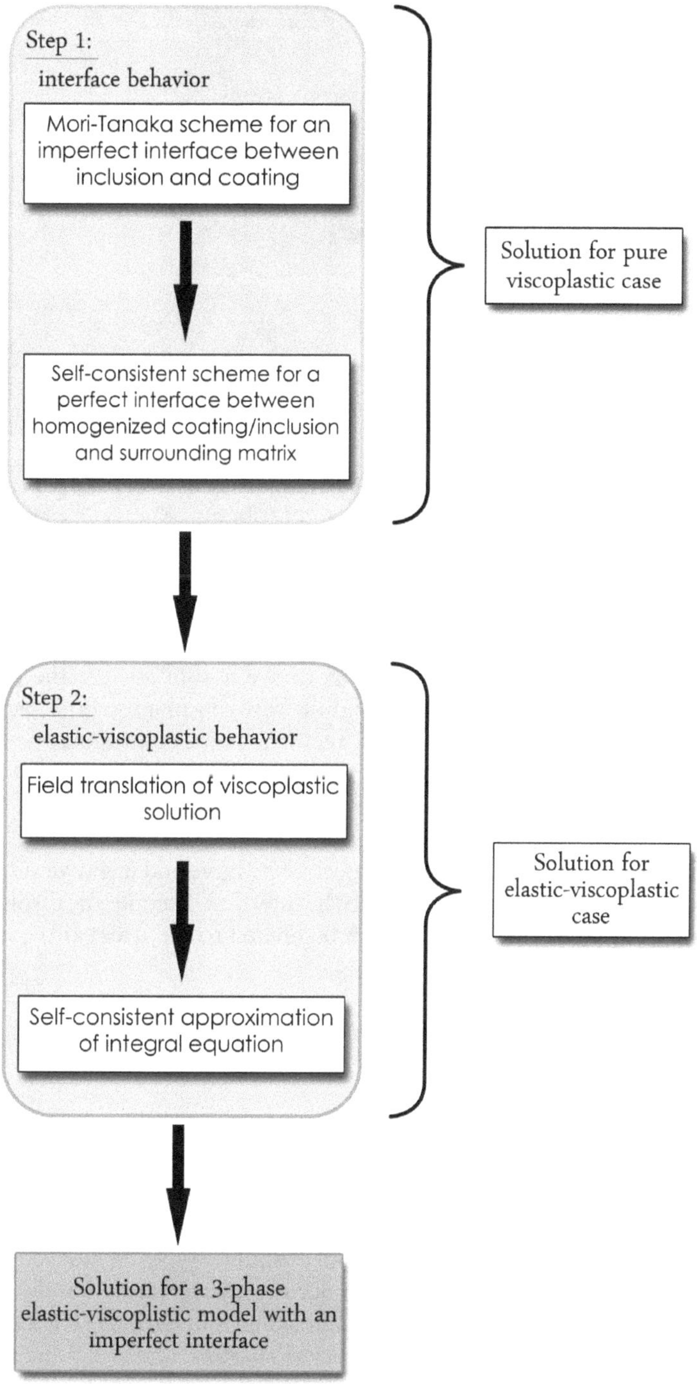

Fig. 9.6 Schematic of the scale transition procedure

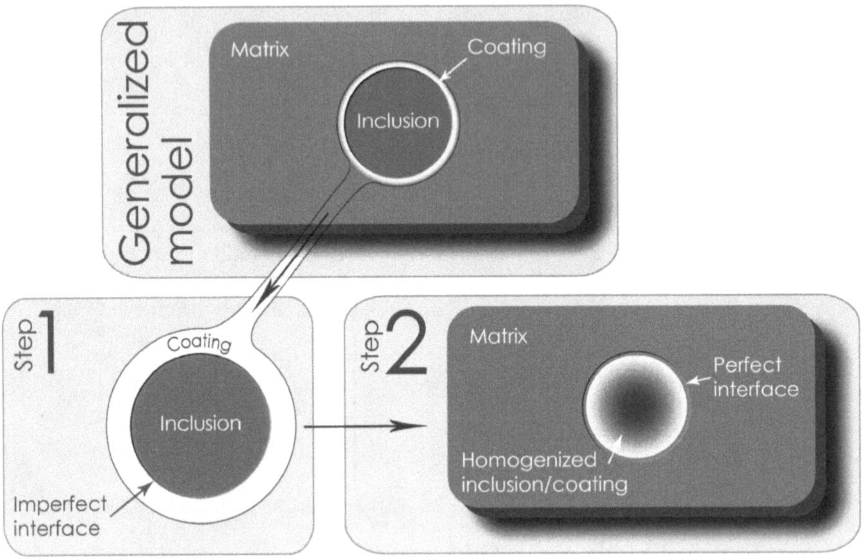

Fig. 9.7 Schematic of the two steps used to solve the three-phase viscoplastic problem

With the considerations above, let us derive a solution for the case of a two-phase material with imperfect bonding between phases. First, across the coating/inclusion interface the traction vector remains continuous, hence:

$$\Delta\sigma_{ij}n_j \equiv [\sigma_{ij}(S^+) - \sigma_{ij}(S^-)]n_j = 0, \tag{9.51}$$

Superscripts + and – denote the respective positive and negative sides of the interface. $\boldsymbol{n}$ denotes the vector normal to the interface. Second, the displacement jump condition across the interface can be related to the stress at the interface with tensor η_{ij} such that:

$$\Delta u_i \equiv u_i(S^+) - u_i(S^-) = \eta_{ij}\sigma_{jk}n_k, \tag{9.52}$$

η_{ij} the interface compliance is given by:

$$\eta_{ij} = \alpha\delta_{ij} + (\beta - \alpha)n_i n_j \tag{9.53}$$

When $\beta = 0$, the relative motion of the coating with respect to the inclusion will not lead to void creation. For the sake of simplicity, let us restrict ourselves to the case where void creation does not occur. α, which describes the interface behavior, can be estimated from quasicontinuum simulations. The latter have shown that grain boundary sliding can occur via stick slip. Adapting the interface constitutive response introduced by Warner and Molinari [14] to the present framework one obtains:

$$\alpha = \frac{\delta_c}{\boldsymbol{\sigma}_c \left(1 - \frac{\sum_i [u_i]}{\delta_c}\right)} \tag{9.54}$$

Here, δ_c and σ_c denote a critical distance and a critical stress, respectively. This equation is essentially a reformulation of (9.8), which was derived from QC simulations. Recalling the topology of this first problem – i.e., the inclusion is embedded in its coating phase – and using consecutively the equilibrium and compatibility conditions one obtains the usual expression of Navier's equation (see Chapter 7 for more details) in the viscoplastic case.

$$b^M_{ijkl}\dot{u}_{k,lj}(x) - \left(b^M_{ijkl} - b_{ijkl}(x)\right)\dot{\varepsilon}^{vp}_{kl,j}(x) = 0 \tag{9.55}$$

x, $\boldsymbol{b}^C$, $\boldsymbol{b}$, $\dot{\boldsymbol{u}}$, and $\dot{\boldsymbol{\varepsilon}}^{vp}$ denote the position, the viscosity tensor within the matrix phase (which corresponds to the coating phase of the three phase problem), the local viscosity tensor, the displacement rate, and the local strain rate, respectively. Furthermore, the displacement engendered by a unit force at point x' must respect the following:

$$b^M_{ijlk}\frac{\partial^2 G^\infty_{km}(x,x')}{\partial x_l \partial x_j} + \delta_{im}\delta(x - x') = 0 \quad \text{with } i,j,k,l,m = 1,2,3 \tag{9.56}$$

Here, $\boldsymbol{G}^\infty$, δ, and $\delta(x - x')$ denote Green's function, Kronecker's symbol, and the Dirac function, respectively. After integration of equation (9.56) on a volume Ω and multiplication of the result by the rate of displacement (in its vector form) one obtains:

$$\begin{aligned} &\int_\Omega \dot{u}_i(x) b^M_{ijkl} \tfrac{\partial^2 G_{km}(x,x')}{\partial x_l \partial x_j} d\Omega(x) \\ &= \int_S \dot{u}_i(x) b^M_{ijkl} \tfrac{\partial G_{km}(x,x')}{\partial x_l} n_j dS(x) - \int_V \dot{u}_{i,j}(x) b^M_{ijkl} \tfrac{\partial G_{km}(x,x')}{\partial x_l} d\Omega(x) \end{aligned} \tag{9.57}$$

Where S denotes the surface surrounding Ω, and n_i denotes the unit outward normal. Alternatively, multiplying Equation (9.55) by Green's function and integrating the resulting expression on Ω leads to:

$$\int_\Omega G^\infty_{im}(x,x') b^M_{ijkl}\dot{u}_{k,lj}(x) d\Omega(x) - \int_\Omega G^\infty_{im}(x,x')\left(b^M_{ijkl} - b^I_{ijkl}\right)\dot{\varepsilon}^{vp}_{kl,j}(x) d\Omega(x) = 0 \tag{9.58}$$

Applying the divergence theorem to the difference of (9.57) and (9.58) one obtains:

$$\int_S b^M_{ijkl}\left(G^\infty_{im}(x,x')\left(\dot{u}_{k,l}(x) - \left(I_{klmn} - \left(b^M_{klpq}\right)^{-1} b_{pqmn}\right)\dot{\varepsilon}^{vp}_{mn}(x)\right)\right.$$
$$\left. -\dot{u}_i(x)\frac{\partial G_{km}(x,x')}{\partial x_l}\right)n_j dS(x) + \int_\Omega \frac{\partial G^\infty_{im}(x,x')}{\partial x_l}\left(b^M_{ijkl} - b_{ijkl}\right)\dot{\varepsilon}^{vp}_{kl}(x)d\Omega(x) \tag{9.59}$$

$$= \begin{cases} \dot{u}_m(x') & x' \in \Omega \\ 0 & x' \notin \Omega \end{cases}$$

Let us apply (9.59) to the inclusion's volume Ω_I. In that case, when r' belongs to the inclusion, the constitutive relation can be used. After some algebra one obtains:

$$\dot{u}_m(x') = \int_{S^-}\left(G^\infty_{im}(x,x')\sigma_{kl} - b^M_{ijkl}\dot{u}_i(x)\frac{\partial G_{km}(x,x')}{\partial x_l}\right)n_j dS(x)$$
$$+\int_{\Omega_I}\frac{\partial G^\infty_{im}(x,x')}{\partial x_l}\left(b^M_{ijkl} - b_{ijkl}\right)\dot{\varepsilon}^{vp}_{kl}(x)d\Omega_I(x) \tag{9.60}$$

Similarly, when x' is exterior to Ω_I, one has:

$$0 = \int_{S^+}\left(G^\infty_{im}(x,x')\sigma_{kl} - b^M_{ijkl}\dot{u}_i(x)\frac{\partial G_{km}(x,x')}{\partial x_l}\right)n_j dS(x) \tag{9.61}$$

Subtracting (9.60) from (9.61) one obtains the following expression of the displacement rate for all x:

$$\dot{u}_m(r) = \int_S\left(b^M_{ijkl}\Delta\dot{u}_i(x')\frac{\partial G_{km}(x,x')}{\partial x_l}\right)n_j dS(x')$$
$$+\int_{\Omega_I}\frac{\partial G^\infty_{im}(x,x')}{\partial x_l}\left(b^M_{ijkl} - b_{ijkl}\right)\dot{\varepsilon}^{vp}_{kl}(x')dV(x') \tag{9.62}$$

Using the compatibility conditions one obtains the expression of the local viscoplastic strain rate tensor. Interestingly, the resulting equation exhibits a dependence on the displacement rate jump:

$$\dot{\varepsilon}^{vp}_{ij}(x) = T_{ijmn}\left(b^M\right)\left(b^M_{mnkl} - b^I_{mnkl}\right)\dot{\varepsilon}^{vpI}_{kl} + \int_S b^M_{mnkl}\Delta\dot{u}_k(x')\Gamma^\infty_{ijmn}(x',x)n_l dS(x'), \tag{9.63}$$

Here $\boldsymbol{T}(\boldsymbol{b}^M)$ denotes the interaction tensor given by $\boldsymbol{T}(\boldsymbol{b}^M) = \int \boldsymbol{\Gamma}^\infty(\boldsymbol{x},\boldsymbol{x}')d\boldsymbol{x}'$. Where $\boldsymbol{\Gamma}^\infty_{ijkl}(\boldsymbol{x},\boldsymbol{y})$ denotes Green's modified operator. The rate of displacement jump can be approximated from (9.52). Neglecting contributions from the derivative of the stress tensor one obtains:

$$\dot{\varepsilon}^{vp}_{ij}(x) = T\left(b^M_{mnpq}\right)\left(b^M_{pqkl} - b^I_{pqkl}\right)\dot{\varepsilon}^{vpI}_{kl} + \int_S b^M_{mnkl}\dot{\eta}_{kp}\sigma_{pq}n_q\Gamma^\infty_{ijmn}(x',x)n_l dS(x'). \tag{9.64}$$

Assuming the stress state along the interface is constant and equal to the stress state in the inclusion and introducing the constitutive law in the inclusion into (9.64) one obtains, after averaging:

$$\dot{\varepsilon}_{ij}^{vpI} = T_{ijpq}(b^M)\left(\left(b_{pqkl}^M - b_{pqkl}^I\right) + b_{pqab}^M R_{abmn} b_{mnkl}^I\right)\dot{\varepsilon}_{kl}^{vpI} + \dot{\varepsilon}_{ij}^{vpM} \tag{9.65}$$

Where $\boldsymbol{R}$ is given by:

$$R_{mnpq} = \frac{1}{4\Omega_I}\int_S (\dot{\eta}_{mp}n_q n_n + \dot{\eta}_{mq}n_p n_n + \dot{\eta}_{np}n_q n_m + \dot{\eta}_{nq}n_p n_m)dS(x') \tag{9.66}$$

Note here that, in the case of a simple expression of the interface compliance, an analytical expression of tensor $\boldsymbol{R}$ can be obtained. The localization can be rewritten in a more usual fashion as follows:

$$\dot{\varepsilon}_{ij}^{vpI} = B_{ijkl}^{vpI}\,\dot{\varepsilon}_{kl}^{vpM} \tag{9.67}$$

The localization, $\boldsymbol{B}^{vpI}$, tensor is given by:

$$B_{ijkl}^{vpI} = \left[I_{ijkl} - T_{ijpq}(b^M)\left(\left(b_{pqkl}^M - b_{pqkl}^I\right) + b_{pqab}^M R_{abmn} b_{mnkl}^I\right)\right]^{-1} \tag{9.68}$$

The viscous compliance matrix of the homogenized inclusion, denoted with superscript HI, is obtained via use of Mori Tanaka's approximation:

$$\boldsymbol{b}^{HI} = (1-f)\boldsymbol{b}^M + f\boldsymbol{b}^I : \boldsymbol{A}^{vpI} \tag{9.69}$$

Here, f denotes the inclusion's volume fraction. With the derivation in the above, the first step of the homogenization scheme is completed. Therefore, one can now proceed to the second step. The latter conceptually consists of embedding the homogenized coated inclusion into a matrix phase with properties and response equal to that of the overall material. This is the essence of the self-consistent approximations. Denoting the macroscopic viscoplastic strain rate with $\dot{\boldsymbol{E}}^{vp}$ one obtains the following localization relation:

$$\dot{\varepsilon}_{ij}^{vpHI} = B_{ijkl}^{vpHI}\dot{E}_{kl}^{vp} \tag{9.70}$$

The expression in the above is obtained by simple application of the self-consistent approximation to the case of a homogenized inclusion embedded in a matrix. The bonding between the two phases is assumed perfect such that the localization tensor can be written as:

$$B_{ijkl}^{vpHI} = \left(I_{ijkl} - T_{ijpq}(b^e)\left(b_{pqkl}^e - b_{pqkl}^{HI}\right)\right)^{-1} \tag{9.71}$$

b^e denotes the effective viscosity tensor. Combining the macrohomogeneity condition and both localization relations one obtains the overall localization relation:

$$\dot{\boldsymbol{\varepsilon}}^{vpI} = \boldsymbol{B}^I : \dot{\boldsymbol{E}}^{vp} \tag{9.72}$$

where the overall viscoplastic localization tensor is given by:

$$\boldsymbol{B}^I = \left[(1-f')\left(B^{vpHI}\right)^{-1} : \left(B^{vpI}\right)^{-1} + f'\left(B^{vpHI}\right)^{-1}\right]^{-1} \tag{9.73}$$

Here, f' denotes the volume fraction of the homogenized inclusion. This provides a complete solution of the viscoplastic problem. Using field translation method of Sabar et al. – the reader is referred to their article for complete derivations [12] – the solution of the elastic-viscoplastic problem can be found; after some algebra one obtains the following elastic- viscoplastic localization law:

$$\dot{\boldsymbol{\varepsilon}}^I = \boldsymbol{A}^I : (\dot{\boldsymbol{E}} - \dot{\boldsymbol{E}}^{vpe}) + \boldsymbol{A}^I : \boldsymbol{B}^I : \dot{\boldsymbol{E}}^{vp} + \boldsymbol{A}^I : \boldsymbol{S}^E : \boldsymbol{S}^e : (\boldsymbol{c}^I : \dot{\boldsymbol{\varepsilon}}^{vpI} - \boldsymbol{C}^e : \boldsymbol{B}^I : \dot{\boldsymbol{E}}^{vp}) \tag{9.74}$$

Here, $\boldsymbol{A}^I$ represents elastic equivalent of the localization tensor $\boldsymbol{B}^I$. Note that another interface condition needs to be introduced to describe the contribution of imperfect interface bonding to the elastic deformation. $\boldsymbol{C}^e$, $\boldsymbol{c}^I$, $\boldsymbol{S}^e$, $\boldsymbol{S}^E$ denote the macroscopic elasticity tensor, the local elasticity tensor in the inclusion phase , the overall compliance tensor, and Eshelby's tensor [15].

Applying the framework above to NC materials, several interesting size effects can be captured. For example, when dislocation glide is accounted for in the grain interior and both grain boundary sliding, via the stick slip approach described above, and grain boundary dislocation emission are accounted for in the constitutive response of the coating phase, which represents grain boundaries, the following prediction of the evolution of yield stress with grain size is obtained.

In Fig. 9.8, K represents a stress heterogeneity factor within grain boundaries. As expected, this simple model predicts that while the breakdown of the

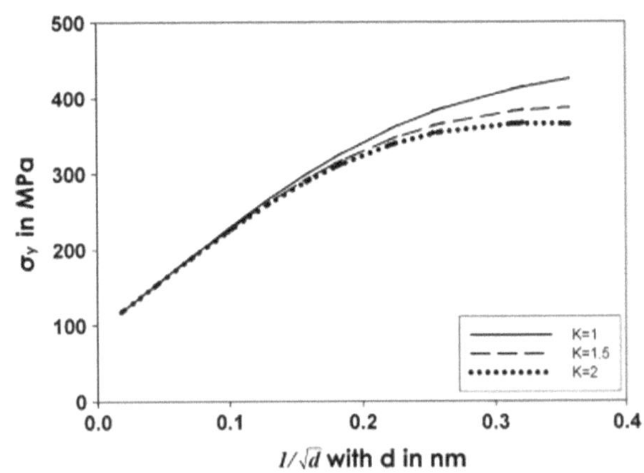

Fig. 9.8 Predicted evolution of yield stress with grain size

Hall-Petch law is not necessarily due to the activation of dislocation emission from grain boundaries – when K = 1, the contribution of grain boundary dislocation emission is negligible and the breakdown in yield stress is due to grain boundary sliding – the yield stress of NC materials shall decrease with increasing dislocation activity arising from grain boundary dislocation emission.

9.5 Summary

This chapter addressed the question of the link between atomistic simulations and the scale of the continuum. While this particular question remains one of the grand challenges of modern mechanics, recent progress in the field is presented.

First, the quasi-continuum method, which allows us to ingeniously reduce the degrees of freedom of a system via the notion of repatoms, was introduced. With this method, large systems can be simulated. Examples of applications to the case of bicrystal interfaces were shown.

Second, the relevance of continuum models based on statistical descriptions of the activity of thermally activated mechanisms was recalled. A recent model allowing estimations of the contributions of each deformation mode was presented.

In Section 9.3, the particular limitation related to the modeling of the activity of grain boundaries as dislocation sources was addressed. To this end, a framework, based on the finite element method, was introduced. The idea behind this framework was to augment the finite element formulation such that dislocation densities are accounted for as nodal unknowns. In turn, this allows us to address the problem of the flow of dislocations with a nonlocal approach.

Finally, a recent micromechanical scheme was presented. This scale transition model is capable of accounting for the effect of weakly bonded interfaces. An example of such an approach was presented with application to the problem of grain boundary sliding as a stick-slip process.

References

1. Sabar, H., M. Buisson, and M. Berveiller, International Journal of Plasticity **7**, (1991)
2. Tadmor, E.B., M. Ortiz, and R. Phillips, Philosophical Magazine. A, Physics of Condensed Matter, Defects and Mechanical Properties **73**, (1996)
3. Miller, R.E. and E.B. Tadmor, Journal of Computer-Aided Materials Design **9**, (2002)
4. Gibbs, W., The scientific papers of william Gibbs, Vol 1.: Thermodynamics, Ox Bow Press, Woodbridge, CT (1993)
5. Wei, Y. and H. Gao, Materials Science and Engineering: A **478**, (2008)
6. Conrad, H. and J. Narayan, Scripta Materialia **42**, (2000)
7. Jonsson, H., G. Mills, and K.W. Jacobsen, *Classical and quantum dynamics in condensed phase simulations*, World Scientific Publishing, New Jersey (1998)

8. Arsenlis, A. and D.M. Parks, Acta Materialia **47**, (1999)
9. Arsenlis, A. and D.M. Parks, Journal of the Mechanics and Physics of Solids **50**, (2002)
10. Meissonnier, F.T., E.P. Busso, and N.P. O'Dowd, International Journal of Plasticity **17**, (2001)
11. Capolungo, L., S. Benkassem, M. Cherkaoui, and J. Qu, Acta Materialia **56**, (2008)
12. Sabar, H., M. Berveiller, V. Favier, and S. Berbenni, International Journal of Solids and Structures **39**, (2002)
13. Qu, J., Mechanics of Materials **14**, (1993)
14. Warner, D.H. and J.F. Molinari, Acta Materialia **54**, (2006)
15. Eshelby, J.D., Proceedings of the Royal Society of London **A241**, (1957)

Subject Index

Springer Series in

MATERIALS SCIENCE

Editors: R. Hull R. M. Osgood, Jr. J. Parisi H. Warlimont

(Continued from page ii)

50 **High-Resolution Imaging and Spectrometry of Materials**
Editors: F. Ernst and M. Rühle

51 **Point Defects in Semiconductors and Insulators**
Determination of Atomic and Electronic Structure from Paramagnetic Hyperfine Interactions By J.-M. Spaeth and H. Overhof

52 **Polymer Films with Embedded Metal Nanoparticles**
By A. Heilmann

53 **Nanocrystalline Ceramics**
Synthesis and Structure By M.Winterer

54 **Electronic Structure and Magnetism of Complex Materials**
Editors: D.J. Singh and D. A. Papaconstantopoulos

55 **Quasicrystals**
An Introduction to Structure, Physical Properties and Applications
Editors: J.-B. Suck, M. Schreiber, and P. Häussler

56 **SiO_2 in Si Microdevices**
By M. Itsumi

57 **Radiation Effects in Advanced Semiconductor Materials and Devices**
By C. Claeys and E. Simoen

58 **Functional Thin Films and Functional Materials**
New Concepts and Technologies
Editor: D. Shi

59 **Dielectric Properties of Porous Media**
By S.O. Gladkov

60 **Organic Photovoltaics**
Concepts and Realization Editors: C. Brabec, V. Dyakonov, J. Parisi and N. Sariciftci

61 **Fatigue in Ferroelectric Ceramics and Related Issues**
By D.C. Lupascu

62 **Epitaxy**
Physical Principles and Technical Implementation By M.A. Herman, W. Richter, and H. Sitter

63 **Fundamentals of Ion-Irradiated Polymers**
By D. Fink

64 **Morphology Control of Materials and Nanoparticles**
Advanced Materials Processing and Characterization
Editors: Y.Waseda and A. Muramatsu

65 **Transport Processes in Ion-Irradiated Polymers**
By D. Fink

66 **Multiphased CeramicMaterials**
Processing and Potential
Editors: W.-H. Tuan and J.-K. Guo

67 **Nondestructive Materials Characterization**
With Applications to Aerospace Materials
Editors: N.G.H. Meyendorf, P.B. Nagy, and S.I. Rokhlin

68 **Diffraction Analysis of the Microstructure of Materials**
Editors: E.J. Mittemeijer and P. Scardi

69 **Chemical–Mechanical Planarization of Semiconductor Materials**
Editor: M.R. Oliver

70 **Applications of the Isotopic Effect in Solids**
By V.G. Plekhanov

71 **Dissipative Phenomena in Condensed Matter**
Some Applications By S. Dattagupta and S. Puri

72 **Predictive Simulation of Semiconductor Processing**
Status and Challenges
Editors: J. Dabrowski and E.R.Weber

73 **SiC Power Materials**
Devices and Applications
Editor: Z.C. Feng

Springer Series in

MATERIALS SCIENCE

74 **Plastic Deformation in Nanocrystalline Materials**
By M.Yu. Gutkin and I.A. Ovid'ko

75 **Wafer Bonding**
Applications and Technology
Editors: M. Alexe and U. Gösele

76 **Spirally Anisotropic Composites**
By G.E. Freger, V.N. Kestelman, and D.G. Freger

77 **Impurities Confined in Quantum Structures**
By P.O. Holtz and Q.X. Zhao

78 **Macromolecular Nanostructured Materials**
Editors: N. Ueyama and A. Harada

79 **Magnetism and Structure in Functional Materials**
Editors: A. Planes, L. Mañosa, and A. Saxena

80 **Micro- and Macro-Properties of Solids**
Thermal, Mechanical and Dielectric Properties
By D.B. Sirdeshmukh, L. Sirdeshmukh, and K.G. Subhadra

81 **Metallopolymer Nanocomposites**
By A.D. Pomogailo and V.N. Kestelman

82 **Plastics for Corrosion Inhibition**
By V.A. Goldade, L.S. Pinchuk, A.V. Makarevich and V.N. Kestelman

83 **Spectroscopic Properties of Rare Earths in Optical Materials**
Editors: G. Liu and B. Jacquier

84 **Hartree–Fock–Slater Method for Materials Science**
The DV–X Alpha Method for Design and Characterization of Materials
Editors: H. Adachi, T. Mukoyama, and J. Kawai

85 **Lifetime Spectroscopy**
A Method of Defect Characterization in Silicon for Photovoltaic Applications
By S. Rein

86 **Wide-Gap Chalcopyrites**
Editors: S. Siebentritt and U. Rau

87 **Micro- and Nanostructured Glasses**
By D. Hülsenberg and A. Harnisch

88 **Introduction to Wave Scattering, Localization and Mesoscopic Phenomena**
By P. Sheng

89 **Magneto-Science**
Magnetic Field Effects on Materials: Fundamentals and Applications
Editors: M. Yamaguchi and Y. Tanimoto

90 **Internal Friction in Metallic Materials**
A Handbook
By M.S. Blanter, I.S. Golovin, H. Neuhäuser, and H.-R. Sinning

91 **Ferroelectric Crystals for Photonic Applications**
Including Nanoscale Fabrication and Characterization Techniques
Editors: P. Ferraro, S. Grilli, and P. De Natale

92 **Solder Joint Technology**
Materials, Properties, and Reliability
By K.-N. Tu

93 **Materials for Tomorrow**
Theory, Experiments and Modelling
Editors: S. Gemming, M. Schreiber and J.-B. Suck

94 **Magnetic Nanostructures**
Editors: B. Aktas, L. Tagirov, and F. Mikailov

95 **Nanocrystals and Their Mesoscopic Organization**
By C.N.R. Rao, P.J. Thomas and G.U. Kulkarni

96 **Gallium Nitride Electronics**
By R. Quay

97 **Multifunctional Barriers for Flexible Structure**
Textile, Leather and Paper
Editors: S. Duquesne, C. Magniez, and G. Camino

98 **Physics of Negative Refraction and Negative Index Materials**
Optical and Electronic Aspects and Diversified Approaches
Editors: C.M. Krowne and Y. Zhang

Kolibri 360 GmbH
Ferdinand-Jühlke-Straße 7
99095 Erfurt, Deutschland
produktsicherheit@kolibri360.de